SMAR

Space and time have long been the province of theoretical physicists and science fiction writers, but space and time are far from theoretical or fanciful—they describe every process that happens in the world. They are the most practical concern of all—they are everywhere and always!

The modern scientific ideas of space and time have been handed down to us from a long history of philosophical ideas, and they have gone through many revisions. Yet many of those ideas have been turned completely upside down by Information Technology, and modern biology. Quantum physics and Einstein's Theory Of Relativity made us rethink them again in the 20th century, and have attached an almost mystical significance to spacetime phenomena—but have we really made too much of their strangeness, and take too narrow a view? Might the much-told weirdnesses of quantum theory and relativity, in fact, have straightforward explanations? Will we meet them again in the growing computing cloud? Evidence amassing in the vast computer systems that power the Internet suggest that this may be the case, as similar phenomena begin to emerge from a far more mundane and accessible source.

Mark Burgess is a theoretical physicist turned computer scientist, who has been one of the pioneers of large scale computer systems. In this book, he offers perhaps the most comprehensive view on space and time yet written, and challenges us to revisit everything we take for granted, using only the most fundamental principle of all: scale.

'...a joy to read...'
–William Louth

'...magnificent; a tour de force of connecting the dots of many disciplines... Marks combination of originality, synthesis and practicality knows no equal.'
–Paul Borrill

'[Mark] flapped [his] wings around 30 years ago and the industry changed.'
–John Willis

'One of the best reads and written by one of the best minds!'
–Glenn O'Donnell (about In Search of Certainty)

'To err is human, to explain is [Mark Burgess]'
–Patrick Debois

'A landmark book in the development of our craft...'
–Adrian Cockcroft (about In Search of Certainty)

'Proud to say that I am a card-carrying member of the [Mark Burgess] fan club. And I think that it wouldn't be too much of a stretch to say that he's the closest thing to Richard Feynman within our industry (and not just because of his IQ).'
–Cameron Haight

'...our whole industry is transforming based on ideas [Mark Burgess] pioneered'
–Michael Nygard

'[Mark Burgess] created the seeds from which the entire DevOps(Sec) and computer immunology movement sprung. The real proof is the number and diversity of tools that evolved from it. Many many people's lives improved from it, and many more can still.'
–Julie Tsai

'The work done by [Mark] on complexity of systems is a cornerstone in design of large scale distributed systems...'
–Jan Wiersma

'Some authors tread well worn paths in comfortable realms. Mark not only blazes new trails, but does so in undiscovered countries.'
–Dan Klein

SMART SPACETIME

How information challenges our ideas
about space, time, and process

Mark Burgess

χtAxis press

First edition published by χtAxis press, 2019.
Corrections to early release edition added, May 2019

Text and figures copyright © Mark Burgess 2018-2019.

Mark Burgess has asserted his right under the Copyright, Design and
Patents Act, 1988, UK, to be identified as the author of this work.

Quotations used within this work are referenced from public sources,
and pose no known commercial infringement.

All rights reserved. No part of this publication may be copied or
reproduced in any form, without prior permission from the author.
Permission is granted to reproduce short attributed quotes for the purpose
of review or citation.

*For ambitious and youthful minds everywhere,
and the many deep challenges they face...*

Contents

1	MYTHOLOGIES OF SPACE AND TIME	5
2	SPACE AND TIME	28
3	ANTI-SPACE: INFORMATION	74
4	GIVE AND TAKE	109
5	TIME AND MOTION	170
6	TURN BACK TIME	247
7	DIAL mv FOR MYSTERY	285
8	SPACES THAT SEE AND THINK	329
9	SMART SPACETIME	424
10	EPILOGUE	494
	References	537
	Index	549

Preface

> Unborrowed from the eye—that time is past,
> And all its aching joys are now no more,
> And all its dizzy raptures. Not for this
> Faint I, nor mourn nor murmur...
> ..For I have learned
> To look on nature, not as in the hour
> Of thoughtless youth; but hearing oftentimes
> The still, sad music of humanity...
>
> —William Wordsworth

What causes change to happen? No question in science is more unanswerable, nor holds us more literally in its thrall than that. Nevertheless, that's the key question to understand about space and time. Since the beginning of the 20th century, theoretical physicists have practically owned the term 'spacetime'. Today, it conjures in our minds popular figures like Albert Einstein, Stephen Hawking, and subjects like black holes and time travel; but, it would be a mistake to dismiss spacetime as being something only of interest to theoretical physicists and science fiction writers. Spacetime is far from being a purely theoretical construct. It is both everywhere and all of the time. What could be more practical than that?

Today, spacetime concepts are finding a resurgence of meaning, in a very different universe. In the mid 1990s, I shifted my attention from a career in theoretical physics of particles and fields to study computer systems—not as a computer scientist, but as a physicist. As a theoretician, working somewhere between physics and computer science, I wanted to address an issue that seemed neglected: to understand the behaviour of computers—not as programmable machines, but as a phenomenon, like any other phenomenon that a physicist might study. This was considered initially to be a silly idea—computers are just machines, they do what we tell them to do—except, now it does not seem so silly. Most people who work with computers know that they do what we intend them to do some of the time, if we're lucky—and, when it comes to massive computer installations, we're not always sure what we intend! There is a need to understand how large numbers of computers in datacentres behave. Their behaviour is quite different from single desktop computers. Why would that be? The obvious answer seemed to be: scale. Physicists know better than any other science how phenomena depend on scale, and how different phenomena occur at different scales. Or do they? Today, physics graduates are embedded in the engineering teams in large scale cloud computing installations, but they don't have all the answers either. I want to reach out to them too with this book.

A core question in this book is: what if space is not like the Euclidean ideal we learn in mathematics, but more like the networks in computer systems? What happens to the perception of space and time that describes interactions as you shrink them? What happens to the ability to measure things as you shrink or expand an observer? Why does Quantum Mechanics use complex numbers? Is it a kind of mystical twist on reality, or simply that we left out some implicit degrees of freedom in spacetime on a small scale (what I call 'give and take' in this book)? These questions are incredibly important, but they are not usually handled properly, because we treat matter and spacetime as separate phenomena, and inconsistently across its branches. We may scale part of the picture without scaling all of it. I've seen how this causes distortions of understanding in the world of computers, and I've come to believe that we may be creating problems for ourselves artificially by clinging on to convenient but inconsistent approximations as if they were truth.

This is a book about physics, it's about computers, artificial intelligence, and many other topics on surface. It's about everything that has to do with information. It draws on examples from every avenue of life, and pulls apart preconceptions that have been programmed into us from childhood. It's a complex book, to be sure. It asks you to set aside what you take for granted about the world, and look at it again with fresh eyes and a simple point of view.

All the authorities that have imprinted a certain idea about space and time can be questioned. I'll re-examine ideas like distance, time, and speed, and ask if we really know what those things are. If they are really so fundamental and universal concepts then can we also see them in computers, or in the growing of a plant? Conversely, can we see phenomena we know from computers in physics? One of the key reference points will be the world of Information Technology (computers). We can learn a lot by comparing the way we describe physics with the way we describe computers—and that throws up a radical view: the concept of *virtualization*, and what it might mean for physics.

In the mid 1990s, I wrote a mathematical text called *Classical Covariant Fields*, in which I lovingly documented the importance of modern symmetry as the basis for classical and quantum physics. The covariant approach is a spacetime approach to physics—championed by Einstein, Minkowski, and others—that puts conventional spacetime symmetry at the heart of the formulation of modern physics. The results have been amazingly successful. I was convinced by the beauty of this approach, but, as the years went by, working with computer systems, I've come to view that view of spacetime physics with increasing suspicion, and I spent 20 years unpicking it. The view of space and time used in physics makes no obvious sense for computers, and is failing to work at the cutting edge of fundamental physics too. That doesn't mean it's wrong—only that the approximations we've attached ourselves to fail to work in all cases. That's to be expected. This book is an attempt to convey some of the perspectives I've developed during my journey, and document the challenges technologists and economists face. For some, this journey may only represent an application of physical ideas to the challenges of information technology and artificial intelligence—a neat analogy—but for those willing to think more deeply, there lie many unanswered questions in these chapters. Might it not just be our destiny to rediscover the fundamental meaning of space and time in these worlds of our own making? By putting space and time on a pedestal, in modern physics, we've come to ignore mundane questions. I hope to change that.

The reading of this book is a kind of spacetime of its own. Time passes each time you pick it up and turn a page, consuming its information. The space lies in the pages and the 'matter' lies in its words. It's rather passive—driven entirely by your own observation and acceptance (call that -). Yet the words came from somewhere—another source process (call it +). The result depends on the successful marriage of (+) and (-). I hope the meaning of these words will become clearer as we progress, but first we have to take a formidable journey—one in which you will question everything you think you know about space and time. We start with an overview of what space and time are, and then take a deep dive into its fundamental questions, always relating these to the world on the small and the large scale. It took me years to understand these matters. It will probably take you several readings and revisitations too—take your time (you'll learn what that means too). I've tried to make it enjoyable as well as informative, and all I can do now is to wish you luck.

I should like to thank a few friends for sharing their time in interesting and helpful discussions over the years, especially: Paul Borrill, Jon Magne Leinaas, Steve Pepper, Finn Ravndal, William Louth. Special thanks to Dan Klein for careful comments. Also, my deep affection to those friends who have been supportive of my maverick activities over the years, encouraging and sustaining them—as well as those who felt unable to condone my disciplinary eccentricity (the meaning of which will be explained herein as a fine spacetime concept).

As a teenager, books like this one inspired me to pursue science and mathematics to the highest levels, as well as to pursue the art of explanation, which served me well as a teacher. In the present day, I miss works like those of Arthur C. Clarke, Isaac Asimov, J.E. Gordon, and P.C.W. Davies (who was my teacher years ago) who were both writers of exceptional talent as well as educators par excellence. Even though I didn't always fully understand them until years later, if at all, they sustained me through an otherwise dry education in the knowledge that understanding was indeed possible with perseverance, and there was a better world at the other end of the educational pipeline. I would like to hope that this book may appeal to a new generation of precocious teenagers, far into their later years, in turn.

Many of the ideas in this book have been told before, by other authors, but never in quite this form. I've tried to give reference to those instances through the extensive endnotes at the end of the book, and I'll ask your forbearance for those references I might have missed. Readers who would like a basic introduction to the idea of Einstein's spacetime may want to consult the excellent and classic references [1, 2, 3] at the end of the book.

–MB, Oslo, 19 February, 2019,
(from sometime in the future)

1

MYTHOLOGIES OF SPACE AND TIME

> It is probably quite generally true that in the history of human thinking the most fruitful developments frequently take place at those points where two different lines of thought meet.
>
> –Werner Heisenberg, Physics and Philosophy

Imagine a field, a parking lot, an empty shop, a shoreline. On the surface, these are just 'empty spaces' with little to characterize them, except perhaps a boundary line that marks their limits—a fence, grass, concrete, sand, or sea. But when we think more carefully, as observers—or agents, who would interact with them—we view these 'spaces' as being qualitatively and even quantitatively different. We assign identities to them, and they mean very different things to us. They play very different roles in the stories of our lives. They are spaces, but not the usual spaces that science refers to.

Around 1687, Isaac Newton (1643-1727) defined a notion of space and time, which has coloured our view for generations. He described space as an empty theatre inside which the events of the world happened, and time as a uniform pull from past to future. Space was deliberately separate from the bodies that inhabited it, and time was deliberately fashioned as a force of nature, doing God's work, like clockwork machinery. His space and time were empty, absolute, and uniform, following the doctrines of the ancient Greeks, and deftly summarized by Euclid. It was filled with perfect bodies that obeyed perfect laws. Newton's view met with plenty of criticism from his peers, but it prevailed ultimately because Newton's mathematics was highly successful in calculating practical outcomes[1]. In a sense, he discovered that science is more about the stories we are able to tell of the world than it is about unknowable fundamental truths. We now know that how Newton envisaged space is not quite right: it can't be considered completely void, even if we dismiss the vast numbers of material bodies floating in it: particles of solid, liquid, gas, and microorganic life particles. At the level of elementary physics, every point in space is filled with invisible processes that we mystify generically under the banner of 'energy'—but perhaps more importantly, processes within space interact with other processes within it, as separate identifiable entities, and recognize one another as different phenomena. They respond to 'function' and 'form', and the 'space' they occupy depends more on their mutual interaction than on what may surround them. What does all that mean?

Although we've been taught to think about the world in terms of places and times, it's processes that dominate the world: everything that happens is a process, or part of a process—and the locations and times affected by these processes have no bearing on what happens unless they are part of the process. Processes don't go on forever like the space and time we imagine: they are localized, which means that they carve up our larger idea of space and time into regions and epochs. Processes occupy identifiable 'contexts'. Those contexts frame the changing events happening within them, and trace out trajectories that could be called spaces in their own right. This becomes important when we try to use space and time to make technologies. So is there really only one kind of space, or many?

We characterize ideas about space in a variety of ways. Some spaces are finite, some unbounded, others are curved, sliced, continuous or discrete. Some may contain uniform fields of grass, sand, gravity, or electromagnetism. However we mix our words, space is never truly empty, which throws into question Newton's view—and challenges us to rethink our lazy acceptance of it. Simply put, in a broader modern view, spaces have *semantics*—characteristics that can be interpreted and assigned a meaning. They have the potential to behave differently, be used differently, and to be assigned purpose from the perspectives of the processes that interact with them. Some of those purposes may even qualify, in our judgement, as being 'smart', as I shall try to explain.

This view of space is certainly different from the traditional role space has in physics, as part of a passive backdrop, perhaps in the night sky—as a theatre for matter to play out its role—but that view is coming under fire from a number of sources, including particle physics, computer science, and biology. As we have become more aware of the importance of information in the world, thinkers pondering its fundamental questions have questioned whether we might have oversimplified the role of space and time by allowing it to dwell in anonymity for too long. In this book, my aim is to raise those many questions and challenge a few of the ones we take for granted.

Since the rise of popular science around the 1980s, it has been the more sensational and speculative frontier topics of space and time that have concerned popular science writers, including black holes, the big bang, hidden dimensions, and so forth. One might easily get the impression from these that space and time are quite well understood: that they are passive bystanders to forces of nature, which are the true mysteries to be solved. I believe that this would be both a mistake and a missed opportunity, for there is ample mystery in the workings of processes at all scales, and it seems increasingly likely that spacetime itself is nothing more than an extension of these processes, represented at a multitude of scales. To understand the role of a modern viewpoint, informed by the scientific and technological advances of the twentieth century, we first need to understand how space and time work, not just in the distant reaches of outer space, but also in the modern world of technology and biology.

HISTORY OF SPACETIME

The scientific view of space and time has undergone many revisions over the centuries, and it still confounds the efforts of our deepest thinkers. Superficially, we all feel we know what space is, and what time is, but it doesn't take too much scratching on the smooth surface to reveal a profound ignorance beneath. The classical notion of space, as an all-enveloping theatre, in which the activities of the world take place, is an orthodoxy that goes back to the earliest times, but its mathematical definitions are—in a sense—just a convenient dismissal that set aside the deeper questions. Through mathematics, we've developed tools to describe and measure space and time, but that's not the same as knowing what they really are. The historical figures, like Pythagoras, who explained the metric properties of space still hold iconic positions in history, and we perpetuate the stories out of pragmatism[2]—but the laws of geometry don't tell us about the nature of space or time.

In the beginning, early ideas about numbers and mathematics were deemed useful to assess the fruits of human spaces: crops, and count sales, and debts. Only later was geometry invented to study space and time, to create maps, to chart the dynamical rhythms of the rivers and seasons, to build structures like pyramids, and to navigate by the fixed stars. It was later still that Descartes developed philosophical and mathematical machinery that we now associate with the Euclidean space and time[3]. But it is surely Euclid whose name is still most commonly remembered and associated with geometry. His chief contribution was to collect the knowledge of Greeks and Egyptians into a number of books. One book *Elements* is best known for its rigour in laying out geometric knowledge systematically as an evolving system of reasoning[4]. He formalized the mathematics of space, and laid the groundwork for defining spacetime.

SCALES AND RENORMALIZATION

Imagine a generalized kind of space—one which captures not only the bland measures of location and state, but which also expresses complex properties and interactions of the processes ongoing within it. It would be a natural extension of our traditional ideas about space, but motivated entirely by the desire to have a common picture that survives any change of scale. If you look at the world maps in an Atlas, you see the major cities of the world as black dots scattered around the terrain. On the world map, a city is a city—a black dot. London, New York, Paris, Shanghai, Tokyo, Venice, Reykjavik are all indistinguishable on this level. The recipe is simple: you mark out a geographical region and call it a 'superpoint', by drawing a ring around a city, could you say anything about what went on in the city? When is a single point just a featureless irreducible point, and when could you magnify it to see more inside? Semantically, we rescale all these very different places to a singular concept—a city, a black dot. Quantitatively, we might express it differently, but with the same idea: a dot with a certain population, a certain fraction of parkland, and so on. Put another way, if we enclose a region of space inside a box, and shrink the size of the box to nothing, would there be space inside the box or not?[5]. If we can't be sure that the region inside an arbitrary ring is empty, what happens when we shrink the size of the ring to zero, using arguments like Newton himself? What would empty space really mean? Things get even more interesting when we apply the

1 Mythologies of Space and Time

argument to cases whose basic form or topology is more complicated, like a cup with a handle.

Renormalization is the term used, in physics and mathematics, for blurring out interior details and rescaling blurred regions into a new kind of unit or 'superpoint'. To normalize something means to standardize it. We can standardize something by different measures: size, shape, function, etc. For instance, take two plates of food at a restaurant and eliminate all of the details by focusing only on the scale of a plate. Anything smaller than a plate gets blurred out. Then all the plates suddenly look the same, ignoring the knowledge that they contain different meals. We have renormalized them, or standardized our notion of plates, and our world simply consists of plates of non-descript food. That's essentially what fast food restaurants do, for efficiency. For some intents and purposes this is a useful approximation: to the servers, orders are simply about delivering plates to seats. The ability to deliver an order does not depend on the name, content, or price of the plate. Renormalization is a way of making irrelevant details disappear—of eliminating information. Different information applies to different processes, so it's efficient to keep only the information relevant to the process at hand.

Renormalization, or changing the scale of focus, is an important idea in science based very much on spacetime ideas, so it has to play an essential role in understanding spacetime in more general terms. Our current views about space and time change drastically as we look at different scales, which could mean that we fail to learn key lessons about what makes spacetime tick. When particle physicists first learned how to compute the quantum interactions between electrons, in the theory of Quantum Electrodynamics (QED), they used a technique called a perturbation series in which they imagined the interaction to be composed of many different processes, with increasing detail. This is a standard computational trick in physics; it's useful precisely because it's usually true that small details don't matter on a larger scale. But that's not always the case. In particle physics, instead of becoming less important, the calculation of detailed processes can become more and more vulnerable to the fine details. Luckily, theoreticians found a way to make sense of the theory by 'renormalizing' their definition of particles to blur out these problems, replacing particles with details by what are called 'dressed particles'. At this sufficient level of detail, the theory made sense and the answer could be verified by experiment.

A similar trick can be used for all kinds of details, whether structural, dynamical, or semantic, and I'll refer to all of these as renormalization. It's reasonable to expect a view of space and time to make sense when we look at different scales, but in our current paradigms that isn't always the case. There are inconsistencies in the way we view processes, because we are often locked into a box of orthodoxy ourselves. The answer to escaping from those boxes may well be to unify semantic (qualitative) properties of spacetime with metric (quantitative) properties, in a way that has previously been frowned upon. Semantics can include things like names or identities (assigned by anyone), properties like colour, mass or charge, and (farther up the scale ladder closer to chemistry and biology) shape and electrical properties. In other words, semantics are rooted in any distinguishing properties that an entity can express. At a human level, semantics may be based on properties like our clothing, our skin colour, our genetic make-ups, our relationships to other things and people, and so on. As we approach biological scales, the number of possible distinguishable properties seems to grow almost without limit.

What makes renormalization or rescaling helpful is that it helps us to separate ideas and concerns—to distinguish 'the wood from the trees'. It allows us to see the world as a number of *modular parts* each with different significance, perhaps different in different circumstances. Just as Newton separated space and time from matter, to enable his highly successful theatrical formulation of physics on a human scale, so we separate many aspects that allow us to focus on one thing at a time. This is clever and helpful, until we reach the limits of that point of view. It can work in both directions: as we blur away detail, we reach a cleaner simpler view of dynamical quantitative measures, unencumbered by detail. Perhaps everything can become a 'particle' moving in an empty spacetime—or cities are just dots on a map. However, as we turn the details back on, what may have appeared empty and featureless then erupts into a tangled ecosystem of significance. We see this in the biology of cells, organs, and organisms, but also in chemistry and information technology too. Not all phenomena can be described separately all the time—that's why it's important to rethink dogmatic viewpoints from time to time.

Distinctions—or, more precisely, semantic distinctions—allow ideas previously lumped together to be clarified, or conversely ideas thought mutually disconnected to be re-unified, in appropriate context. The central issue in these interpretations is a question of *scale*, as defined by limits, borders, and

1 Mythologies of Space and Time

boundaries. Scales alter both our experience of dynamical change, because of how things interact with one another, and of semantics because of the relative capabilities of things.

Staring into Spacetime

It appears that there are more ways to think of space than we first thought. Renormalization allows us to redefine space in different ways, to view it at different scales, or different levels of detail, so if we are consistent about our approach to scaling physical law, we have to take these ideas seriously. What do we mean by space in general? If we stare into it, what do we see, hear, smell, taste? The traditional conception of space is of *empty* space, i.e. of there being an unobstructed path from where we are to somewhere else. Yet, without something material to guide us, how could we even know where we were? Where our line of sight (or other sensory probe) ends, we think of space as ending, or perhaps being occupied by something. But what about when it's dark, or when we cannot reach out? Then we are limited to speculating with the help of a model, based on incomplete evidence.

How about time? This is no better. Time is even more connected to our perception of events than space. We can apparently perceive time without thinking about space at all—though actually that's not entirely true. We might not have to go anywhere to perceive time, but without the ability to discriminate variable properties of space, we could not distinguish change—then from now. Inside our bodies, we would not have a pulse or a brain, and therefore we would also not have anything to measure or experience time with. If there were only featureless empty space, we would perceive no time. There would be nothing to change. We wouldn't even exist to pose the question. Einstein pointed out that observation is key to understanding time, but it didn't stop cosmologists from dissociating spacetime from the process of observation, for pure mathematical convenience, by absorbing the mathematics of relativity wholesale into their equations. Sometimes renormalization for convenience hides the important aspects of one's understanding.

As I am writing this, I perceive the space around me as being anything but empty. I can smell coffee, see sunshine, and taste the aftertaste of a matcha latte. I hear music. I can see a clock on the wall of the coffee house, which tells me that the time is 10:24, and another one on my computer that tells me the time is 10:26. My computer clock is receiving signals to synchronize it with a common source frequently. The clock on the wall is a battery powered clock that receives only irregular manual signals when someone resets the time on it. Of course, the latter is an assumption. It might be that the clock receives regular updates as measured on another clock, or even measured on its own time-face. But if the clock stops, then it doesn't register any time, so it could be asleep for a thousand years before the next tick without being aware. We know that when the wall clock stops, time does not stop for all the people in the coffee shop (at least, I don't think it does—how could I know?). This tells us that people have another clock inside them, which drives their perception of time, and they only use the wall clock as a calibration of their affairs.

At this point, I think it's important to remark that we should not get too bogged down with the space of Newton and Einstein. Think of another kind of space, finite and bounded, modular and functional on many levels: the human body. Inside the boundary of our skins, it is filled with cells and fluids, tissues and organs. It is not a passive space, but a highly active one—filled with processes that mix and interact. It is hard to talk about the human body (or that of any biological entity) without talking about the activities that are ongoing. It is not just space but space and time together: spacetime. At the scale of microbes, the human body is just a convenient habitat, as planet Earth is to humans. A habitat is itself a space to inhabit, relative to the organism that samples it.

What makes the human body a different kind of space is the way we describe its processes—its stories. To Newton and Einstein, who looked to the processes outside themselves, even out into the heavens, the processes they saw were moving carriages, flying projectiles, planets and stars that moved through the air or through the heavens. Their theories of the world were thus based on things moving in a straight line (what physicists call *translation*), or in orbits (*rotation*). They focused on how moving a body from one point to the next didn't really change anything in a featureless void, or that rotating a sphere doesn't change anything either, because of symmetries. These properties are called translational invariance and rotational invariance—they are mathematical symmetries that describe the sameness of spaces. But neither translation nor rotation is typical of what happens inside the spaces of the human body. There are translational movements in the body, to be sure: the movement of blood, for instance. There is translational invariance or sameness too: large swathes of tissue are composed

1 Mythologies of Space and Time 13

of essentially indistinguishable points we call cells. But unlike outer space, the dominion of sameness doesn't last as long—it's more localized and modular. So, if we focus only only theories that can describe large scale boring sameness, we can easily lose the ability to describe more interesting diversity. Physics should not be about artificial distinctions, no matter how convenient they might be for calculation, it should describe everything there is. And, of course, there is much more.

Science tends to favour a certain reality, within the reach of a three dimensional world, at our fingertips—but from arts and mathematics, we are also used to thinking about abstract spaces. We know that there is a world inside the television, and behind the computer screen that has nothing to do with the physical space in which it is represented, or the two dimensional screen it is projected onto. We have also become used to hearing about other dimensions in science fiction, mysticism, or the extra dimensions of space and time relating to String Theory and the like. Dimensions are now second nature to everyone. Even if we don't challenge the meaning of them too much, we are desensitized to the unnatural implications of multiple dimensions. We'll have reason to use this to good effect in the coming chapters.

By force of habit alone, we favour the three dimensions of physical space over any other, treating them as more fundamental and more real; others, we might believe are merely analogy or metaphor[6]. But we can and should go further to talk about space and time as it really matters—as the key *processes* that unfold, rather than the container we believe frames them. We should talk about the political space with distinctions like left and right, liberal and conservative, third ways, middle grounds, etc. These are all semantic distinctions that label and separate phenomena, making them significant in relation to one another, and framing their contexts more specifically than any abstract theatre can. The stage is irrelevant without the play or the audience. In the absence of stimuli our minds are apt to fill a void with their own fantasies. If we can do this, is it any less real than what we can observe? These are questions we should at the very least confront[7]. Semantics are a basic property of the world, most important at a preferred scale, where they best have a chance to stand out. We experience semantics in our human lives in the operation of mechanisms, in interactions with other people, in politics and economics, and so on. As we scale up, higher level agents are desensitized to the meaning of specific interactions, because they get washed out by the bulk aggregation of stuff. As we step back to avoid being overwhelmed, we lose sight of individual meaning. Thus all of humanity's efforts, to a being outside the planet, merely look like the weather, or geology. No one from outer space would look at our planet and care whether one country profits, while another goes bankrupt, whether one country conquers and other or fragments into separatist regions. These are all microscopic concerns, akin to subatomic interactions. Consider whether you really concern yourself with the mechanism of protein recombination in your daily life. You are more interested in the brand of cornflakes you can get from the shop—a purely semantic issue. Until now, science has not been able to describe semantic properties on a par with quantitative dynamical phenomena, but—thanks to information technology—we are now beginning to see how all our separate, modular ideas might just be ripe for a rethink.

TURTLES ALL THE WAY DOWN

We are all confronted by many levels of description about the world. They form a kind of hierarchy of scales, from small to large. A question that has challenged thinkers throughout the ages is: does there exist a smallest scale, an atomic particle of space, which is elementary and indivisible, as we once believed atoms to be? Common sense might suggest that there must be, but so far no evidence to this effect has remained uncontested.

Fundamental questions are theologically loaded questions—surrogates in some sense for the question: who created the universe? If the final answer is God, then who created God? Or if it started in the Big Bang, then what caused the Big Bang, and so on. This hierarchy problem is often deflected by the apocryphal tale of the turtle cosmogony[8]. The story goes that, after a lecture on the solar system, the American philosopher William James was accosted by a lady from the audience, who pronounced: 'Your theory that the sun is at the centre of the solar system, and that the Earth is a ball which orbits around it, is very impressive, Mr. James, but it's wrong. I've got a better theory, namely that we live on the back of a giant turtle.'

James, containing himself with good manners, replied slily: 'But what does this turtle stand on, Madam?"

The woman was having none of it, however. 'You're a clever man indeed, Mr. James,' she replied. 'The first turtle stands on the back of a bigger turtle.'

Taken aback, but enjoying the game, he parried: 'But what does this second turtle stand on?'

'Obviously,' she laughed. 'It's turtles all the way down!'

The story, whether true or not, is a nice parody of cosmogonical origins. It underlines the problem we all face in trying to explain 'why' questions: pushing off cause to one thing after another (I'll come back to this point when we explain explanation in chapter 8). We can always ask one more question and never reach a satisfying answer. This is why reasoning is based on 'axioms' or 'logical primitives' which are assumptions that cannot be explained. They have to be accepted on faith[9]. Requiring an explanation of something, we usually trace our explanations as far back as we can , until there is either some basic fact that we are willing to accept on faith, or we lose interest—and that is what we call an explanation, whether that is religious, scientific, or something else[10].

We can make the same observation about the nature of space, and the reason is clear. We assume that everything has to consist of something, including space, or energy. What could that be? Probably not turtles. If space is finite, then what is outside of space? If we try to define something outside of space and time, we inevitably think of the only thing our brains can conceive of: a bigger space and time, in which the first is embedded. This concept of *embeddings*, like Russian dolls stacked inside one another, is extremely important, and I'll return to it many times.

There seems to be no way to bootstrap[11] the concept of a spacetime without already having a spacetime—or at least some idea of interactions between independent points. This is deeply unsatisfactory. Similarly, you can't explain time without having time, i.e. change that happens. These are semantic primitives, axioms, things we just can't explain. Nevertheless, the concept of embedding one space inside another is one of the most important and powerful ideas we need to explore, because this is the essence of understanding scale.

EMBEDAVERSE

The idea that one spacetime can be embedded within another not only forms the basis for how mathematics models spatial phenomena of differing properties: putting things inside one another is also the way biology has evolved self-contained organisms, and it is how we engineer self-contained technological systems too. Biological organisms are the most obvious example of bounded and localized processes that, when analyzed on the scale of cells, look a lot like the textbook models that introduce the most basic properties of topological spaces. Cells look a lot like what topologists might call 'open balls', used to define mathematical models of Euclidean space, in terms of a kind of foam of close-packed granular regions. These many and competing descriptions of the world, developed over centuries, masquerade as all kinds of different turtles, ranging across all kinds of scales—and most of us can't help but think of them as true rather than as helpful visualizations that are expedient in counting and measuring the world. We routinely talk of Big Bangs, and fabrics of space and time, and superstrings, and all kinds of other mythologies as though they were true—but no one has any basis by which to claim these to be true of the real world. They are stories that are self-consistent: this is helpful, but by no means 'true'. So one of our tasks in this book is to challenge those lazy assumptions. When the number of layers gets to be too much to know intimately, we start to take the stories for granted and believe them implicitly, no matter how far fetched they may be. In a sense, although we have a more sophisticated set of stories, better matched by observation, the modern version of the universe doesn't offer more answers than the turtle cosmogony.

One artificial candidate spacetime, of particular interest, which is manufactured artificially and embedded within our own, is that of computers and the networks that connect them. Computers have locations and clocks that drive processes of change, but on another level they look quite a lot like biology. On a certain scale, networks of computers look like cells connected together into a kind of tissue, which resembles materials that, in turn, resemble homogeneous regions of space formed from clusters of modular cells. In the forthcoming chapters, I'll explore this view of the Internet and its computers as a spacetime, because it is not only fascinating from a philosophical perspective, it is also highly practical. To build a computer system at the scales we need today, we need models of computer behaviour that go beyond the classical models of processors and memory. They are now multiscale phenomena that span wide areas and operate close to the limits for data transmission.

Just as we reached a limit of understanding in physics, around a certain scale of observations, so we are reaching a limit in our understanding of computers as we connect for and more of them together into a larger Internet. I believe that the tools of spacetime physics can be applied to computers and even to biology to offer insights about each of these. Can we translate concepts like position, time, velocity, etc, into the world of computation and biology? Indeed we can, and there seems to be much to learn from doing so. The results are not as frivolous as sceptics might think. Until recently, computers and spacetime didn't seem to resemble one another at all. Today, however, computers are literally filling space in a way that is also seen in biological systems[12]. We talk of the age of 'Cloud Computing', in which there is a universe of computers 'out there somewhere' on the Internet, as if in another universe, forming highly regular patterns like translationally invariant space, connected by tunnels that convey information between them. Data are passed around like quantum particles and fields, and the space it can reach is occupied for processes that result in computation.

This is a fascinating idea, but it isn't just idle speculation. In scaling computer systems to the massive size required by modern services, it is getting harder to understand exactly where data are, and when data arrive at special locations. It is getting hard to know whether all users see the same information, and precisely when they will see it. Some of the quantum uncertainties have analogues in the world of computing. Notions of position and time are highly complex in the modern realm of datacentres and cloud computing. The complications of relativity are far more 'in your face' in computer systems than in our astronomical universe, because computer systems are relatively small, and nearly all motions are very close to the maximum speed of communication (the speed of light). Some of the ideas we take for granted in physics cannot be taken for granted in computers. There is the location of hardware: computers, disks, network equipment, and so on, but then there is also the software that lives in a 'virtual' world, where virtual locations can migrate across physical machines while actually appearing to be in the same place as far as the computation is concerned. There is not one idea of location and time, there are many locations and many versions of time. Without far more sophisticated tools than computer scientists have today, this will become an increasing headache to comprehend.

Never before have we needed to understand space and time better for more practical reasons that today, dealing with the Internet and all of its layers of processes. Computers form a spacetime both in hardware and in software, which can benefit from principles learned in physics. It will be equally fascinating to ask whether what happens in our own universe can sometimes be understood as a version of what happens in the 'dataverse'. Inspiration can go in more than one direction. There are major challenges to understanding computers, even basic issues about time, the synchronization of clocks, and rates of activity. Observing computers is, in some ways, harder than observing the universe on the macroscopic scale; in fact, it is somewhat like trying to observe a nano-scale quantum mechanical system. You can't just point a telescope at it from your bedroom because it's opaque.

A piece of software that starts its life in one physical computer, at one location can migrate to another location for all kinds of reasons. Moreover, computers share resources like CPU, and memory. Software is looking for available slots, like parking spaces, where it can park its programs. What happens when the program starts to grow, and get a lot of data traffic? It needs to expand, so it needs more space. Does it open another branch, or try to move into a larger space? When renovations are needed, the program might be forcibly moved to a new location. The binding between semantic purpose and spacetime location is fluid. All this is mundane stuff in cloud computing, and the method for resource sharing is to create a generic coordinate system of resources, within the computer system, which is what we call the cloud. The coordinates don't refer to our physical universe—it's not about geography. They refer to machine addresses, by IP address, process number, memory location, and so on. The cloud is a bit like a shopping mall, filled with empty units to rent. The units are all similar to begin with, at least until some particular process or business moves in and starts to make specific promises. That differentiates the units, and they become service points: semantically charged.

Space and time express 'functional processes', i.e. processes of interacting parts, where information moves around to predictable effect. Functional processes exist in physics, at the level of 'plug and socket' interactions like positive and negative electrical charges; they also exist in chemistry and biology where certain parts fit into certain holes to make phenomena occur. Obviously they are the very key to all that happens in information technology. We can look at almost any system in this way, in any branch of science. The value in doing so is to see the similarity of issues faced and how to deal with them. From the galactic scales of our universe, to socio-economic and technical systems on Earth, to biology, molecular science, atomic and composite materials, and even fundamental particles and fields: it might not be turtles all the way down, but it could well be information.

How many dimensions?

Mathematics has stringent definitions for what we are allowed to call a space. I want to relax these, in this book, in the interest of simplicity, and in the interest of embracing approximate similarities. Sometimes experts dismiss similarities that cross the boundaries between disciplines, but, in a sense, the crossing of boundaries is exactly what this book is about. If any system of parts has multiple identifiable locations, and ways to move between them, then I'll refer to it as a space. In this spirit, any space has a certain freedom to move around within it. The technical term for this is the number of *degrees of freedom*, or the *dimension* of the space. All school children know that lines and strings are one dimensional, while surfaces and pictures are two dimensional, but the full freedom of open space has three dimensions (or three degrees of freedom): up-down, left-right, and forwards-backwards.

Dimensionality is not only a freedom, it's an important limitation on what behaviours can take place in space and time, but our common understanding of dimension is learnt from books rather than from intuition, so we can easily become confused about how to count dimensions. As we'll see in the coming chapters, the question of dimensions is really not as clear we've been taught. Simply saying the world is three dimensional is inaccurate. Let me give a brief taste of the issue here. Suppose you are driving through an undersea tunnel. Inside there is a straight road. You might notice some local curvature in the tunnel, but you have no real idea about what is outside the tunnel. From the interior, the tunnel is a one dimensional passage. But then you somehow teleport outside the tunnel and can look down upon it from outside, and you see that it twists and winds in 2 dimensions—no wait!—in 3 dimensions, depending on how you look at it. It turns out that our ability to measure *independent* dimensions also depends on memory—which is a startling realization. Two different paths might seem to lead to independent places at first, but an observer needs a certain amount of memory, to recall the path—to know if the two directions eventually meet at the same point, i.e. whether they are truly independent (and over what distance, see figure 1.1).

Dimensionality is all a matter of perspective. The night sky is a safely two dimensional shell around the Earth when you cannot reach it; but if you are floating in space, and you are confronted with drifting away forever, its three dimensions become relevant to you. We should count dimensionality in terms of degrees of freedom to move *within the system* we are constrained by, in order

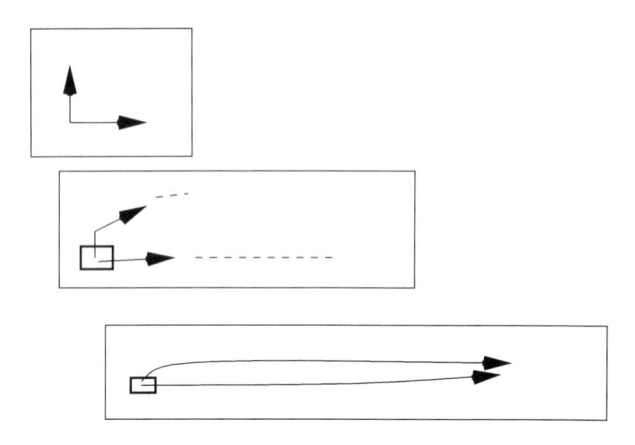

Fig. 1.1. Dimensionality depends on scale. If we magnify, we might see apparently independent directions, along different paths—but, on a larger scale, they may track one another closely and eventually meet. You have to remember the branching points to make sense of dimension—it's a memory process!

to make it an impartial characteristic—but then we have to agree that, from the inside, the tunnel is one dimensional, while from the boat stuck on the surface of the water seems 2 dimensional, and for diverse or ground penetrating radar, it's 3 dimensional—or, is it actually four dimensional?

FROM SPACE TO SPACETIME

Einstein rewrote the textbook on relativity, at the start of the 20th century, when he discovered that there was an implicit limit on the rate at which information could be transmitted in nature, because of the universal constancy of the speed of light in a vacuum. As Einstein's synthesis of special relativity evolved in the minds of the scientific community, it was suggested by Hermann Minkowski (1864-1909) that the roles of space and time did not make independent sense in a relativistic theory. This has led to some profound insights and some profound misunderstandings. The way time unfolds for an observer has to be related to relative processes ongoing in space and vice versa. In other words, space and time are inseparably connected. Due to the assumption of continuity, a continuum model of space and time, like those used by Newton and Einstein, presents time a lot like just another dimension of space—so perhaps time is in fact another dimension, rather than a kind of force in the Newtonian view? This turns out not to be quite right, but it certainly looks that way when formulated without due diligence.

Thus, the notion of spacetime, as a unified entity, was born. The utility of the idea was that it led to a way of writing Einstein's theory, as well as electromagnetism, in a form that made space and time look highly symmetrical, with time playing more or less the same role as any other dimension of space. Einstein embraced this 3+1 dimensional formulation, and it has since become the basis for modern thinking in theoretical physics[13].

In spite of its technical overtones, we all intuitively know what the unified concept of 'spacetime' means. Space and time almost always appear together in our experience. Spacetime is about the union of the two as *processes* that happen everywhere. Perhaps most familiar is motion. Motion is a phenomenon in which we experience different places at different times. Space and time happening together must therefore tell us something about motion. Every time we move, passing certain landmarks, we have a sense of how quickly we are passing these landmarks, based on our internal sense of time. This gives us a sense of speed. But where does this sense of time come from? Each conscious individual has an internal sense of time ultimately generated by our pulse, which drives our biological processes, resulting in brain waves, etc. That interior activity means there is motion within us to compare to the motion outside.

Of course, we could turn the argument around: if we take speed for granted as being constant, then the interval of time between the landmarks tells the time it took. This is how we build clocks: constant motion that points to a set of markers. Interior time (like a heartbeat) is one source of change that complex organisms and machinery can generate from their internal processes, but we could equally change this around a bit and make time an external stimulus, such as passing certain landmarks as we move past them (which is how a sundial works).

Time can come from within or from without. If time doesn't come from within, we can imagine that it comes from an outside source, as if exterior force like a handle that cranks our mechanism to keep us running. This was Newton's view of the clockwork universe. So, suppose we hypothesize that motion takes place at a universal constant speed, and that the passing of each landmark were basically a tick of an exterior clock, then we have a source of dynamics or change, marking out time for us. This is an equivalent point of view. It shows us that how we define time is an interpretation of relative changes—indeed, this is what we mean by relativity. Minkowski spacetime sweeps all of that under a cosmic rug and absorbs the semantics of space and time into an algebraic formalism. Like all beauty make-overs, it's superficially attractive, but conceals underlying imperfections.

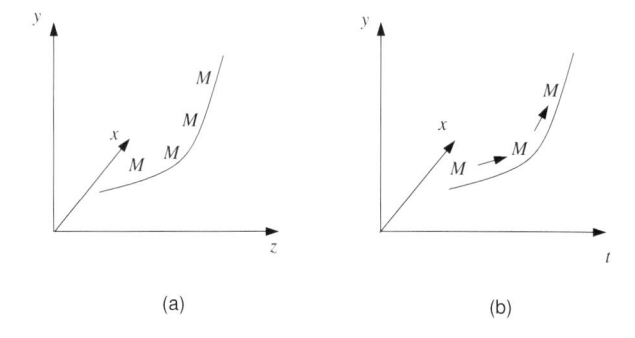

Fig. 1.2. Space and spacetime: (a) A bend in space changes direction, and represents curvature. (b) A bend in spacetime may have no change in direction. It represents an acceleration, and it points along the arrow of time from past to future (each clock step in time t sees an increasing progression in space x).

So where are we? There is a lot going on here. We can say something about space, and we can say something about time, but because either would be impossible to measure without the other, it seems natural to think of the processes of change in terms of a combined picture, which we simply call *spacetime*. Figure 1.2 shows the way we are first introduced to spacetime in school. In the left hand figure (a) we see space: a depiction of a three dimensional region, with traditional coordinate axes x, y, z to guide our visualization. The curve in the image is what we could expect, a stationary object with a certain size that traces out a bending path. When we include time into this picture, in (b), we imagine *spacetime* instead. Because it is now harder to draw more than three dimensions, we drop one of them (z) and replace it by time. So, for every tick of your clock, you advance involuntarily one position to the right—like a film reel. The meaning of the curve is now different. As you bend more towards space, you speed up, because more space passes beneath your feet for each tick

1 Mythologies of Space and Time

of a clock. As you bend more towards time, you slow down because more time passes for each interval of space. Instead of a stationary object, the meaning of a curve that extends over several times is the *trajectory* or 'world line' of a point in the two dimensional space x, y. So, a bend in purely spatial dimensions just represents curvature mixing one direction with another, but a bend in spacetime mixes position and time and represents a change in speed and position: an acceleration.

Time for change

Change happens. No one knows why. Indeed, this might be the deepest mystery of all. The observation of change is what we mean by the ticks of time. Each change contributes to the ticks we experience, but the impetus for those changes remains mysterious, and is only partially related to what we may or may not see. Observing changes involves establishing a connection between a source of change and its observer, each a separate localized process in their own right. Remarkably, both parties need their own separate ticking process to make that connection with the other. There is a lot to unpack in those innocent words, and it will take some time—or, more correctly, it will require a process of some length. Luckily, reading the book will be a quite long spacetime process, so we can get there!

Time and change are almost a tautology, but not quite. As we'll see in chapter 3, this view is not complete: there have to be two kinds of time, relating to our notion of boundaries and renormalized points: *interior time* and *exterior time*. Interior and exterior are relative to a bounded region, like a fortress from which we observe the outside world. The arrival of visitors from beyond the boundary gives us a sense of what is happening outside: exterior time. What happens inside the fortress marks out interior time. This distinction is related to a principle known as *locality* in physics. I'll comment on it throughout the book, because this fortress view is the essence of observation and relativity.

Speaking of locality, there is also *local* time and *non-local* time! Non-local is what physicists say when a phenomenon spans several distinct locations 'at the same time', i.e. when clearly distinguishable locations are involved in the same process. The subject is filled with subtleties and fragile distinctions. The concept of locality has become revered in physics and there are good reasons for this. But its almost hallowed role is curiously misrepresented, because locality is not what physicists call an 'invariant' concept—it is also relative, not to observers of different speeds, but of different scales. If we renormalize space, we may also need to renormalize how we count time.

TIME FOR STORIES—PROCESSES

Our experience of processes revolves around our perception of timelines, which are the paths traces out by a localized entity in spacetime, like the curves in figure 1.2. Journeys, trajectories, or narratives are stories, in which we sew together events into an ordered sequence, which we assign a collective meaning: as an episode. From the interior of one's own castle, an observer sees events relative to its local self—this is what we mean by relativity. Those stories of exterior things relative to ourself are different from stories about exterior things relative to one another, because the ability to observe anything at all depends on interior time. That simple remark is going to play a disproportionate role in explaining why our views about space and time are so muddled.

Storylines, to use a more commonplace term, play a large role in the way we define and organize the semantics of space and time. In Einsteinian relativity, the history of a material body as it moves through spacetime is referred to as a *world line* (see figure 1.3). It represents a path through spacetime, not just through space, by a body forming a kind of journal or history, as viewed by an observer looking on. Different observers might see different paths, or perceive different sequences of events—different stories.

Stories have a strong significance in culture, because they are not only a way to relate journeys, they are also the essence of language. As we'll see in the coming chapters, the tracing of spacetime paths, peppered with semantics at different scales, is what connects spacetime paths to reasoning and to computation. Through the telling of certain stories, spacetime itself can be made to act like a computer!

Scale and detail are once again traded for precision and observability, when watching a storyline from within a bounded location. If we renormalize a story, blurring out the details of the points it visits, then we end up with a simple trajectory—the kind of line through Euclidean space that physics concerns itself with (see figure 1.3). It's the lowest level description, intentionally paying no attention to a larger design. In this view, a classic love story, filled with intrigue and motive would reduce to something like: body moves at 0.2 metres per sec-

1 MYTHOLOGIES OF SPACE AND TIME

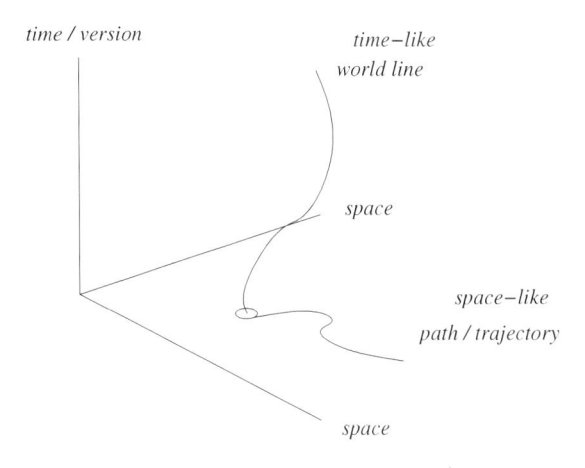

Figure: A 3D coordinate system with axes labelled "time / version" (vertical), "space" (two horizontal axes). A curve labelled "time-like world line" descends from the time axis to a small circle, from which emerges a "space-like path / trajectory".

Fig. 1.3. World lines may be spacelike (stories that tell cached histories) or timelike (histories as they happen).

ond to coordinates (x, y, z) and interacts with body B, spawning a minor body C, and so on. When we blur out semantics, we end up with a quantitative story, or what physicists call a *dynamical* story. Physicists like dynamical descriptions because they are easily quantifiable, but we should not be too eager to ditch semantics in favour of the apparent certainty of numbers. Semantics are important for understanding how information characterizes purpose, cooperation, and compatibility—which is the key to systems and machinery. Think of a story about checking into an airport. It's not all about number of passengers, speed, height, and weight. It's also about intent, having the right credentials, and collective safety.

RELATIVITY SHAKEN AND STIRRED

Newton's idea of spacetime as a purely passive theatrical stage was a view that prevailed until Einstein shook up ideas on space and time by showing that space and time, as they could be defined by measurement, are inextricably linked—at least on the large scale, because measurement is a process. He showed that, because all observations seem to require the passage of light from objects to observer, and that light speed is predicted to be a universal constant by the theory of electromagnetic waves, this must necessarily limit and indeed couple observers' perceptions of what happens. Moreover, since the only measurable reality remote objects have to us is by observing them some kind of light (sound doesn't travel in outer space), we might as well take this as being our definition of reality. Framing this as a Newtonian theatrical performance leads to a view of spacetime that can be squashed, warped, and distended in unexpected ways[14].

Einstein's notion of relativity has garnered a reputation for being weird and counter-intuitive, but it was the result of extraordinarily pragmatic thinking. He did what every good physicist would do: he decoupled what we think we *believe* from what we can actually measure; and he gave primacy to the latter. The relativity of things applies to all areas, so it is an idea that we must also visit in connection with purpose, i.e. the semantics of what goes on in space and time.

In the natural sciences, we have learned to take the view that what distinguishes the semantics of 'spaces', i.e. their qualitative characteristics, is not quantifiable, and therefore irrelevant to natural law. In other words, semantic labels were only thought to matter to those objects living in and around space (space is space, matter is matter, and ne'er the twain shall meet). But, today, this view seems increasingly untenable. We can indeed measure semantics, using a different kind of mathematics than the usual tools of Newton and Einstein. In the language of physics itself, the distinction between space and the matter that fills it (the way we label the meaning of the arena in which we find ourselves) seems like an unnatural legacy of scientific history, made natural only for the benefit of the particular scale of human observation[15]. In physics, what we call matter turns out to be mostly space anyway: microscopy, experiment, and the best theoretical models have revealed this beyond doubt. What we call 'particles', or the granular instances of matter, as imagined by enlightenment philosophers, seem neither to be as localizable, nor as singular of location as once thought. The very concept of an idealized mathematical point leads to problematic infinities when trying to calculate behaviours. All of these disparate facts add up to something of a muddle about what we mean by the distinction between space and matter, empty and filled.

If matter is mainly space, then space is not nothing. Besides, if quantum theory is right, that space itself is far from empty: it is filled with 'vacuum fluctuations'—an oddly ill-defined quantum shaving foam, conjured in the popular physics literature to represent processes that we refer to generically as quantum fields. Take a breath! All these terms, like 'vacuum' and 'energy' are mysterious from an information technology perspective; they seem to have no clear analogue, say in the world of computers. To be sure, they are processes—

1 Mythologies of Space and Time

but are they real?

Einstein showed that matter and energy are not fully independent concepts, which implies that these two may be somewhat interconvertible under the right circumstances; so, a night sky that looks empty, when contrasted against the shining material stars, clearly contains energy and may therefore not to be as featureless as we once thought. The same applies to any volume of space we care to examine. Whatever the truth about fundamental physics, the notion of empty space, which grew out of pre-20th century philosophy, may persist but it has less and less practical utility. Instead of struggling to separate matter and space, with increasingly unnatural reason, it seems more useful to ask of any region: what's going on, and what can be known about it?

Attitudes have changed since our mythologies of matter and void took hold. They would seem to be ripe for a rethink. In the following chapters, I want to show you that the concept of spacetime is far more useful and has far more dimensions (pun intended) than we have conventionally been led to believe. I'm not going to make any radical claims to explain what space and time are (or are not), only suggest that we have deliberately ignored certain questions that have been staring us in the face. Escaping doctrine of any kind is hard, but we've been served with a rare opportunity to return to the core values of scientific and philosophical inquiry—examining with fresh eyes and above all pragmatism.

I want to begin by revisiting some of the most basic ideas, from mathematics and physics, and then by delving into the subject of information, promises, and more selectively into other fields of science. We'll learn to spot common themes, in possibly unfamiliar territory, and then to see the world in a surprising new way. It should become apparent that the fundamental questions about spacetime are not just about counting energies and measuring speeds, but questions also about encoding patterns, memory, and therefore, embedded in this apparently innocent story are also the roots of larger questions about the meaning of information and how intelligence can emerge.

I find it refreshing, exciting, and a rare opportunity, to go back to basics—to try the consistency of long held assumptions about space and time on new phenomena from the modern world. There, one finds that the relative scales can all be rewritten, and that analogues and mysteries abound. These offer profound insights for those willing to dig deep. Only a few years ago this exercise might have been of primarily philosophical interest, but today with the potential impact on technology and economics, it has become of the utmost importance.

2

SPACE AND TIME

'What's the good of Mercator's North Poles and Equators,
Tropics, Zones and Meridian Lines?'
So the Bellman would cry: and the crew would reply,
'They are merely conventional signs!'

–Lewis Carroll, Slaying the Snark

How things are arranged, and how they change, give us our notions of space and time respectively. In this chapter, I want to compare some of the ways we think about space and time, and ask why we have ended up with some rather narrow views in the modern sciences.

To the early philosophers, the idea of space, as an abstract concept, didn't arouse much curiosity. The spaces between things seemed simply empty and were therefore irrelevant compared to the things within or the distance extended between them. In a sense, there was not one enveloping space, but a collection of gaps between significant objects. There were, on the other hand, regions of clearly different purpose: fields, roads, temples, homes, etc. The idea that they could all be part of a larger space called 'the universe' had not yet become a part of the narrative. Even the vast emptiness in the heavens was recognized only as the home of some lesser gods: a twinkling canopy revealed in the night sky, when the mighty sun gave way. Today we still distinguish that 'outer space' as a separate concept, literally elevated above the more mundane varieties of space, and charged with a sense of enormity—of mystery—that inspires some of humankind's biggest thoughts. Spaces may be empty of recognizable features, but they are not necessarily empty of meaning.

Fast forwarding to the modern era, the way we talk about space has undergone something of a transformation. It has a variety of cultural meanings, and has been canonized in mathematics through a menagerie of formal systems that allow science, engineering, and design to treat it almost as a utility. The common theme in all these variants is the apparently featureless continuum discussed by Euclid, Descartes, Newton, Einstein, and others—and yet this view of space is neither realistic nor practical outside of mathematics. Why do we cling to it?

Greek theatre

The Ancient Greeks, from around the time of Thales and Pythagoras (570-495 BC), were the first to document what we would now consider mathematical ideas about space. One of those Greeks was Euclid or Eukleides (ca. 300 BC[16])—sometimes called the most famous of all geometers who didn't discover a single result about the subject[17]. He sought to formalize the notions of geometry into a number of rules that would reduce deep thoughts to mere recipes for reasoning about space. The word *geo-metry* is compounded from the words for *Earth* and *measurement*: a programme to mathematize the understanding of distances for the practical applications of building and civic planning. It was these functional matters that motivated the quantifying of space, it's distances, areas, and volumes.

Euclid's greatest claim to fame was a book, called *Elements*, which painted a revolutionary picture space and its characteristics, in terms of abstract concepts, like points and lines. The book was passed down through the ages, scarcely being improved until the time of Descartes, in the 1600s. Writing a book about empty space, at the time of Greek philosophers, was practically a daredevil act, but its legacy has been profound. Euclid wrote about the relationships between lines and points, as no one before him. Two points, he claimed, are enough to extrapolate a straight line between them, extending forever in either direction. Two lines are enough to define a plane surface, in a similar way. If those lines are parallel, they will never meet, no matter how far they are extended.

Notice how Euclid already presupposed the existence of a continuum of points in the space between any two points or lines identified. This is a significant assumption, which was criticized by Zeno of Sidon through the statement of an apparent paradox[18]. Zeno's paradox suggests that space cannot be infinitely divisible, i.e. that the points of space could not be infinitely small, because it would mean that distance had no meaning. It would be impossible to count distances. Consider a traveller following a journey. After covering half the distance, he would still have half the distance left, then half of the remainder, and half of that... and so on, ad infinitum. The process of walking all of these smaller and smaller intervals would never end, and the traveller would never reach his destination. The resolution of the apparent paradox is straightforward: an infinite sum of smaller and smaller parts can add up to a finite length, when defined as a mathematical limit; however, while the addition he proposed does add up to the correct distance in the end, the cost of counting the lengths doesn't—that takes forever, because it doesn't take half as long to count half the distance. It's a counting issue, but his critique was essentially correct. If you can't count spacetime distance, then what is it?

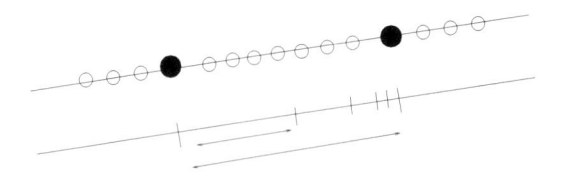

Fig. 2.1. Zeno's paradox versus the continuum limit: in between any two points we can still imagine an infinite number of points. No matter how close together we make the points, there are still more in between to halve the distance to the end. This is not paradoxical mathematically, but it doesn't make sense physically or semantically. Purely quantitative arguments can be misleading.

Zeno's paradox is a fair criticism of the naive use of continuum methods in mathematics, in spite of our ability to make formal mathematical sense of the continuum limit—not to mention the huge simplifications that it brings to mathematics. The idea that space is a continuum doesn't survive unscathed however. It violates common sense, and one of the most important principles in physics: scaling. Suppose we assume, like Euclid, that there is a smallest possible location—a single point—and scale up, by adding more points to aggregate clusters of them into regions. Then we collect together those regions into larger super-regions. Could this go on forever, on smaller or larger scales of our choosing? Certainly, this is perfectly legal in quantitative terms, but not in terms of qualitative meaning.

Think of an atlas map of the Earth, with entire cities represented by mere dots. The Earth is not a continuum of such dots—they are scattered around like a network of train stations, with roads and narrow transport links in between. The different locations no longer 'touch', and they now have different meanings. The idea of interpolating between points no longer makes sense. These meaningful places can't be bisected. Imagine London and New York: we would not think it reasonable to interpolate the existence of a continuous city connecting the two, spreading through the Atlantic ocean, and extending to infinity in both directions. At the smallest scale of physics we distinguish matter from empty space; at the level of cities, most of that matter in between the cities is now the new empty space, and the cities mere points are the new matter. Rescaling a viewpoint in this way is one of the most important techniques in physics. We use the principle to explain behaviours over and over again.

Interpolating a continuum for points and lines, without regard to their role or meaning (i.e. without taking into account their semantics), is what Euclid imagined, and this abstract notion of space prevailed for nearly 2000 years. In a sense, it began a tradition of neglecting qualitative measures in favour of purely quantitative measures. But, when we neglect semantics in favour of purely quantitative arguments, we can be fooled into absurdity.

Roles and maps

The Greeks' predilection for floating above reality and speculating like the Olympians seems symbolic of the association between divinity and perfection that they handed down after them. Perhaps this is what perpetuated the purity of Euclid's abstract thought over the centuries, unencumbered by actual observation, and in spite of reasoning to the contrary. Newton's adoption of a pristine view of space did allow him a convenient theatrical stage onto which any scenario might be poured, and he could benefit directly from Descartes' modifications to geometry, which we know today as the Cartesian coordinates (x, y, z). They formed a latticework, laid out in space, like scaffolding on which to hang any object of study. The simplicity of this device is still compelling, and remains irresistible to many.

Cartesian coordinates were the first of several systems of lines and curves for mapping out the extent of empty space. They were instrumental in developing our modern concept of space. Coordinates are a truly excellent invention for describing locations, especially if you have a helpful grid, such as the virtual grid calculated by a Global Positioning System (GPS or satellite navigation) device; but, before such inventions existed both inanimate bodies and living organisms were able to map out space and time quite unintentionally thanks to their brains.

Newton made formidable advances in formulating the mathematical tools to describe the movement of bodies in space, in part because of the assumptions he made about space and time as an empty but rigid scaffolding, covered in a grid of Descartes' coordinates. Though his work was not without challenge or controversy, in its interpretation, as a method of prediction it could not be surpassed. As we might watch the world, bodies roll down hills, or fall into wells, sink to a bottom, or float to the surface. These are processes that result in motion: they combine a change in space over a change in time. Waves travel on the surface of water, and resound through the air, seeking out the edges of a body of water, or echoing around a canyon. Waves can even magnify images of objects when their paths are interrupted, by casting shadows at a distance from an object, or by passing through regions of unequal resistance that bend their paths to form a lens.

Sensors can detect signals transmitted by these changes and acquire fragments of knowledge about other places in space, with a sense of time determined by the sensors. This is the basic idea behind sonar and radar: you bounce some probe around and see what happens, then add up the results to form a short-lived picture, as an instant of memory. Everything we see is based on this principle too: light from the sun, or some artificial source, reflects off objects or casts its shadows, and from that we are able to build up a picture of what is in its path. The movements of 'bodies' explore space and bring about time; their trajectories tell a kind of story about the region they explore. What Newton's technology allowed was an ingeniously simple way of parameterizing motion, as he understood it, without necessarily understanding how it happened. Eventually, this prescription became so successful at describing motion that people stopped asking whether it was right or wrong.

After the publication of Newton's work, there was increased interest in space as a mathematical entity. A young mathematician Carl Friedrich Gauss (1777-1855) dominated much of the early work, followed later by his less prolific student Georg Bernhard Riemann (1826-1866) who nonetheless got much of the credit for the modern description of curved surfaces and geodesics which led to Einstein's generalization of the theory of relativity[19]. Alas, it strays far outside the scope of this book to tell the fascinating stories of these giants of mathematical history[20]. The theoretical machinery, developed by innovators in mathematics, was one vitally important aspect of the development of space as a concept. Familiar concepts like points, lines, and regions, as abstract entities were refined and rebuilt with the superior calculus of the enlightenment. Today, with the centuries of experience and development in between, we might begin with the notion of a topological space—explained to us in school as the mathematics of doughnuts. Topology is the study of how places are connected— are they solid and contiguous, or are there holes in the middle, like a cup with handle or a doughnut?

Topology introduces the vitally important concept of a place being *next to* another place, which is also called *adjacency*. Adjacency of locations is what we imagine as the freedom to move: from where we are to a place next to us (or adjacent to us). However, we'll see that adjacency is not as obvious and automatic a property as we generally take for granted. Moreover, it isn't really relevant to consider all the places next to us if they are not actually used. A planet might have an entire universe to wander in, but be constrained to move in a simple one dimensional path in orbit. The only idea of 'next to' that matters is the next position of the orbit. Unachievable adjacencies are just speculation. This is a theme that runs through the whole book.

Such was the success of the differential calculus to the divine problems of the movement of planets in the heavens, that for centuries it dominated mathematical approaches to almost every subject related to physics. Parallel to Newton's differential manifesto, were other developments in the classification of phenomena (both natural and human-made), and the role of location played by the 'active agents' within them. Pragmatic natural philosophers wanted to use their knowledge to engineer new machinery: better cannons, catapults, architecture, by applying the laws of 'cause and effect' or *causality* that Newton's mechanics seemed to explain. It spawned the understanding of light, heat, pressure, fluids, and thus came the engineering of earthly devices, from sewage, irrigation, the plumbing of the human body, heat pumps, steam engines, electrical circuitry, and even knowing the weather. Euclidean-Cartesian space and time are surely a natural arena in which to discuss cause and effect: like watching the world inside a computer model, but purely quantitative measures don't tell us how to make meaningful distinctions, or what light or electricity, or gravity are. So those questions were separated from the successes of coordinates through the artifice of the Euclidean space.

FROM QUANTITATIVE TO QUALITATIVE TO RELEVANT

The idea to describe and label spacetime probably began with maps. Mapping terrain, by sketching out landmarks and their relative positions, is as old as prehistory and the earliest cave paintings. Later, civilizations across the world made maps not only of geography, but of civic infrastructure, like town plans, gardens, as well as religious maps of the heavens. Those maps were *functional maps*. They were drawn at least as early as the Egyptians, and new maps were often associated with advances in mathematics and technical skills[21]. Over time, our maps have become filled with all kinds of details that have no relevance merely for getting from A to B, but which illustrate functional aspects of how we might use the properties space offers along the way. Where do we find water? Fruit? Gold? Roman maps were said to be far more concerned with logistics and administrative issues than with dimensions or geography, while the Greek maps used latitude and longitude and astronomical measurements[22]. All space is not equal: some is smarter and bears the fruits of process.

Functional maps introduced the idea of semantics, interpretations located and attributed to regions of space. A mark on a map was not just a position, but a town, a tree, a field, a mine, a well, each distinguished with descriptive names. We name countries, shops, roads, and we mark them with landmark features of particular significance.

Some would say that we only label 'things', not space, but this feels inappropriate as our notion of things is as diffuse as our notion of space. Things come in many forms: is a field a thing? Is air a thing? Is emptiness a thing? The homogeneity, or lack of features within a region is what normally makes us think of it as being 'space'. The contrast of an object standing out against such a background is what makes something an identifiable thing. Of course, it's all very arbitrary, and I want to argue in this book that it is both useful and appropriate to rethink these arbitrary distinctions. As Einstein wrote in the appendix to his popular introduction to relativity[23]:

> The psychological origin of the idea of space, or of the necessity for it, is far from being so obvious as it may appear to be on the basis of our customary habit of thought.

In the biological realm, animals also use a variety of signals to communicate information about space and time. Insects communicate stories using dances, or by singing, and humans articulated language to tell stories about their journeys. We call explanatory stories instructions, *algorithms*, or recipes. They take us from one state to the next—or from one imaginary location to the next. There is thus a deep connection with the concept of different locations (as a realm of possibility) and the concept of the different states of something (such as different colours, different ages, different flavours). For example, suppose you start from this tree, and travel until you see the river, then turn left and keep going to the lake, and so on. Or you start with an empty bowl, break an egg, etc. What is the difference between the place where the tree is and the tree itself? What is the point of an egg, as an active agent in a process, if we only mention its size and shape? These examples add another point: semantics are important, so therefore descriptive language is another topic we must return to in a later chapter. This turns out to be related to spacetime in a fascinating story of its own.

In order to describe relationships between things and processes, we make use of different kinds of map, describing changes in space and time, as relayed to us by accidental travellers, intentional explorers, and proxy signals that probe on our behalf. Which is more real: the stories or the map? They are simply different representations of the same information. But can we say that either one is more like the real thing than the other? That surely depends on what kind of creature you are, and how you sense the world. Stories emphasize dynamics, maps emphasize semantics. The concept of distances and coordinates lies somewhere in between[24].

The Islamic Empire was well known for its detailed maps of the human body, during its reign, but perhaps the most interesting of early maps were maps of the heavens. Clusters of stars, impregnated with meaning, and fancifully embellished, became mythological creatures of the Zodiac. This not only illustrates how we humans tend to fill in details that are not actually present in actuality, based on our preferred beliefs, it also illustrates a theme that recurs throughout this book: the embedding of one representation (a constellation wireframe network) within an embedding space or background space. It's a natural extension of Euclid's idea of interpolating points between measured ones and imagining possibilities from mere hints of evidence.

The story of networks, from constellations, to roads, telegraph, and Internet, can be built up in parallel to the theory of spaces, but traditionally it's more usually treated as a completely separate topic—a cultural partitioning which has spilled over into scientific partitioning too: spaces usually represent physics and networks are used to represent other phenomena in computing, biology, or sociology. Concepts thus split into disconnected cultures. You could even say that they split into regions of an imaginary concept space. We speak of coordinate maps mostly in connection with spaces, while we discuss addresses and the connectivity of networks. The descriptions would work in either case, but they would feel less natural going the other way around.

Today, many of us use spacetime diagrams in the form of assembly instructions, e.g. to understand the construction of flat-packed IKEA furniture. In my youth, electronics magazines were all the rage, though I was never very successful in building working projects. In an electronic circuit diagram, we view a circuit as a very different kind of space formed from important components joined together by less important wires[25]. Similarly, the plumbing in my house that was always leaking when I was a child, was supposed to be a self-contained closed space, bounded by pipes and a complicated topology—but apparently it wasn't as closed as it was supposed to be. It doesn't matter where the circuit components are in the three dimensions around the circuit, it only matters how they are connected in their own world of electric currents. The components are the *locations* of importance because they represent functions, or locations with *semantics* that transform the electrical movements in some very particular way. The wires merely seem to select possible incidental pathways between these transformative events, constructing a flow or story around it.

Circuitry and plumbing tell us something quite deep about processes: that being next to something else is more about the ability to communicate with it. If you don't interact with the surroundings, it is not relevant to the process. You might only be seeing it because you are using a different process to observe the actual process of interest. If we could get inside an electric circuit, or shrink ourselves into the plumbing, we would not observe anything outside them. But it's the active communications that matter. If we place electronic components next to each other, in some embedding space, without wires and the impetus of a battery, nothing at all happens. No new processes take place. The directed communication channels are the *relevant* space for processes—and, if everything in the world is built from processes, then perhaps focusing on ambient embedding spaces, as a separate arena, is not the right thing to do.

EVERYTHING AND NOTHING

According to Aristotle, the function of space was to represent the distances between material things. He did not consider space itself interesting as we do today. However, in inflating empty distances, space also provided the contrast against which to distinguish something from nothing. Within this view lies the first clue that *information* must play a role in what we mean by space and time. To define space, we need something to contrast with the nothingness to define either a boundary or a grid scale. So space can't exist without something? Something, represented symbolically by a 1, and nothing by a 0, form a binary alphabet of states that can be applied to every location. Outer space, viewed as a roof of stars separated by blackness thus becomes a binary map, at this most simplistic level, and the constellations became the first information, or digitally encoded characters.

In this material separation of things, space is treated as being different from the matter and energy, but from circuitry and processes we see a different picture: that it is information communicated from place to place that practically defines the meaning of adjacency, or being next to your neighbour. If you don't interact, you are alone. The view of space essentially as a set of coordinate axes, into which we inject foreign bodies and processes, is convenient for many human scale descriptions, but it isn't really useful or even tenable across a broader range of scenarios, mainly because it doesn't respect an even more fundamental idea about the world, known as *scaling*, whose importance has emerged only in more modern times. To understand why it fails to capture a useful reality, we need to review some of the ideas and put them into a perspective, comparing them to equally important approaches to space and time with other motives and on other scales. This shift from dynamic to semantic is of profound importance to our modern understanding of physics. Let's return to the formal methods of mathematics, and summarize some of its well known formulations.

TIME IS CHANGE

Let's start with time, because—for all its mystery—it's a derivative concept, simpler to describe than space. In fact, some would say that time is only a byproduct of what changes about space; and history is like a sequence of changes to a document stored on your computer. The real temporal mystery is change itself.

Time advances when we detect change, either on the inside of an observer reference frame, like a heartbeat, or on the outside, like a thunderbolt. As Aristotle pointed out, time is a succession of things that happen. If we count changes, we are measuring time. Indeed, this is exactly what a clock does. A clock demonstrates time by tracing out a countable set of distinct states. Whether a clock is time or just measures time is something of a tautology, because the two processes are completely intertwined. On a macroscopic scale, there are so many changes going on that we can't say that a solitary clock is all of time, because there are processes of change happening all around. But, if we isolate a single atom, far away from everything, then the only process going on is what is happening in the atom. If the atom doesn't change, it's fair to say that no time has passed for it.

Much confusion arises from the way we use tiny independent devices as our clocks, instead of using everything we can see as our measure of time. This gives us the illusion of time as passing independently of ourselves. It's an illusion because we are not very good at imagining concepts that involve large differences in scale. We are so used to the idea of being surrounded by constant change that it takes some effort to unlearn what a clock really means. It causes great confusion when discussing relativity, as Einstein realized, and it causes great confusion in computer systems where information has to be synchronized at different locations around the world.

The most fundamental thing we can say about change is that it happens all across space, and any change that affects any observer will be perceived as the passage of time. No one knows why. There is no obvious river of time that drives us all downstream, but there is for some unknown reason something that causes changes to happen, and thus for information to change or be exchanged from place to place. This fundamental process is the real mystery. Everything else about time is a muddle of attempts to capture this simple observation.

Einstein insisted that different times have to imply distinguishably different states—because if you can't tell that something happened, then you can't tell that time has advanced at all—you would be stuck in stasis. In other words, we don't know what changes take place unless we can measure them, and it's meaningless to talk about time without being able to measure it. For Einstein, any changing device was a clock, whether a timepiece, a sundial, or an observer's pulse—and it didn't talk him long to see that this idea totally changed the meaning of what we know about time and motion.

To make a clock, of any kind, we need to be able to count. We take a special set of observable states, perhaps marked out on a clock face, and then we connect them to an isolated and regular process that moves the hands, or advances the digits. We create deliberate intentional change, for no other purpose than to watch the process count. It's a way of calibrating other changes that we can observe 'at the same time'. 'At the same time' is such a casual phrase, yet Einstein effectively tore it to shreds. He pointed out that it's only possible meaning can be: when signals from observable phenomena are aggregated by a single observer at the same location as part of the same process that measures time. That is a subtle idea—and it might take the length of this book to become familiar with it. Essentially: if you have two letters in your mailbox when you check your mail, you can say that they arrived simultaneously. At the 'same time' means 'part of the same observation or transaction'. That's as good as it gets.

As the different positions are counted, by the movement of the hands, we can count, say, from 0 to 12 and then it starts again. But this is not all the change that happens, it is only a tiny fraction of what is going on. The Earth turns and the planet moves around the sun, so there are days and years that we experience from other symptoms, like dark and light, or changing seasons. What a clock measures is not time, but *its own* time. More important to a person is our interior heartbeat that keeps us alive to see changes. Without those changes, we would not see any change at all. This brings us to the first encounter with what I'll call *interior time* versus *exterior time*, mentioned above. Every observer, of any finite size, whether conscious or not, has some kind of boundary between information within itself (like a heartbeat) and information arriving from outside of itself (like a thunderbolt). A single mathematical point, with no internal states, could not possibly experience anything like time[26]. Moreover, because we can't see absolutely everything that happens, and because our ability to know about change is compromised the further away changes happen, our sense of time is dominated by local changes. This is the origin of Einstein's observations about relativity, the effects of which are surprisingly weird should we ever care to confront them in detail.

For tiny localized changes, 'time' as measured by a clock can, in principle, be reversed too, as changes can be undone. This gets harder and harder to achieve, as we include larger areas of space. If everything about the world that was done could be undone precisely, then we could (like Cher) turn back time. But, unlike the universe, which seems to have an ample supply of change at its fingertips, a local agent like ourselves needs to accumulate and spend force and energy to affect change, and we could clearly never be everywhere 'all at once' to undo everything in the right order to reverse time. We see this when working at a computer, for example: when you make a change to a document, you can often select 'undo' from a menu (or CTRL-Z, by conventional key binding). This reverts the state of the document universe to what it was at a previous time. Computer version control systems can do this too. Different versions of a document can be labelled and archived for reference. However, this only reverses time within the document, not on the computer as a whole, or within the entire universe. Because other changes happened on the computer, exterior to the document, the local reversal of time is really an advance of time on a larger scale.

If, by design or by accident, the sum of changes leading to a 'new' version of what we observe actually coincides precisely with an old version, then we could claim that the document has returned to an earlier time, because it would be completely indistinguishable to an impartial observer. In a closed document, this is plausible, but in the wider world it is pretty unlikely that everything would be exactly the same for everyone. As soon as someone making changes to the document starts interacting with it, the states of the two agents get coupled together, and they effectively share a sense of interior time. If several authors write a document at the same time (as one can do in shared writing systems, e.g. like Google Docs) then the reversibility of time is effectively destroyed. Interactions destroy reversibility, because they bring new information into a system that we may not be aware of. This is the essence of 'entropy' though I prefer to avoid that term for now[27]. The entanglement of processes, through interaction, implies that time can never really go backwards from within an open system process. It could only be reset from outside the entire system, in such a way that nothing within the system would ever know anything about it. For this reason, time is usually perceived as having a macroscopic arrow from

2 SPACE AND TIME

past to future, and we can never revisit the same version of the world twice, except by managing and curating localized experience with smoke and mirrors.

SPACE IS VARIABLE

So much for time. Now let's begin the more complicated story of describing an understanding of space. As we understand space there have to be distinct places we can visit, although a single point might be enough for a purely mathematical model. Anything we classify as motion must take us from one location to another. Mathematically we would talk about *points* in space, as Euclid instructed, but it is often necessary to think about regions of finite size, or extended features like towns, paths, and areas with boundaries. One usually thinks of a 'point' as a zero-dimensional dot, of no size, but this idea is too restrictive, since our belief in mathematical points is a bone of contention in physics, and makes no sense at all in other descriptions, like computing or biology. For that reason, I'll try to use the term 'location' to refer to an atomic unit of space, rather than point or position. That scales to any size, depending on our focus.

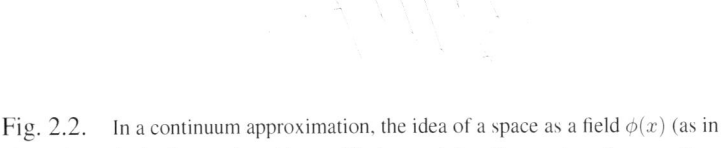

Fig. 2.2. In a continuum approximation, the idea of a space as a field $\phi(x)$ (as in electrodynamics) whose value ϕ is specified at each location on top of a spacetime x, appears as a separation of scalar promises on top of the common denominator of adjacency promises.

In a computer, a location could be a bit, or a memory register, composed of several bits, whose size depends on exactly who judges it. The coordinates of a memory register would then be its 'address' within a memory map. On a higher level, a location on the Internet might be an entire computer, whose coordinate would be its Internet IP address[28]. A mathematician might see a bit as representing a number, which either has no size, or a size of exactly '1'. An engineer might see it as a small region of silicon in two or three dimensions. A programmer might see it as a byte, eight bits wide, and so on. The size of something depends on how we choose to characterize it, and in what embedding space.

The scaling up of a single location is a *region*, but obviously if a location can have any size, depending on our focus, we only have to change our language to see that a region is just a location on a larger scale. Such changes of perspective are what we mean by *scaling*: a change of our units of size or distance, like shifting from millimetres to metres. We talk about London and Paris as being single places, but of course they are filled with smaller places, each with district names, and their own characteristics. Scaling is not only about quantity but also about quality (semantics).

The composition of regions from smaller regions is an important part of explaining space. As for time, the most important characteristic about independent locations is that they are *distinguishable*. This is also somewhat related to whether they run smoothly and seamlessly into one another like the field in figure 2.2, or if they overlap like the patch-like regions in figure 2.3, or are are discrete like the squares on a chessboard or stepping stones crossing a river.

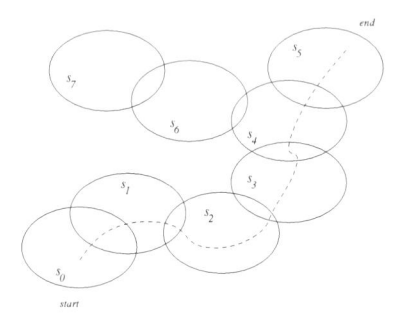

Fig. 2.3. Overlapping patches form a notion of connected spatial regions, and define what we mean by continuity independently of dimension.

Today, anyone who has enjoyed a few classes in mathematics at school, or has watched the news on television, has learned to visualize empty space with the help of coordinate axes (see figure 2.11). This is a common way of representing data, especially in financial news. Coordinate axes are not real things, but they are not pointless either (pun intended): they are imaginary lines that help us to distinguish one place from another by offering a scale against which to calibrate space. We have to anchor the imaginary graduations on the axes to something, even if only to a uni-centric observer looking out into the void[29].

Time is usually measured by a single sequence of instants, like a sequence of version numbers for a document, except that the document is reality. Space describes more than a series of versions of reality, it remembers the details of what changed, so there are more ways to characterize it. We use coordinates to 'span' these possible variations, labelling different locations. the graduations are imaginary measuring sticks, which play the equivalent role that a clock plays, comparing locations rather than different configurations of space. Coordinates are written in a familiar set of 'tuples', or numerical patterns[30], for example:

$$(x, y, z)$$
$$(r, \theta, \phi)$$
```
#RRGGBB
aaa.bbb.ccc.mmm
```

where the variables represent numbers from the coordinate map. The order of the numbers helps us to define the direction of increase or decrease. The first example refers to axis positions on a Euclidean grid, the second a radial spherical basis, the numbers basically equivalent to (height,longitude,latitude). The third is a form of coordinate for colour space used in computer imagery, used to mix red, green, and blue levels into a colour code. The final example is the form of Internet (IPv4) addresses, e.g. 128.239.1.2, familiar to most in the developed world by now. The numbers refer to different networks, or clusters of connected locations joined into a hierarchy. Unlike the other cases, IP addresses are not a complete specification of location in a network, they have to be supplemented by local information. In this sense, they are somewhat similar to Riemannian coordinates, in a curved spacetime. They have only local significance, on top of a global structure.

INTERIOR AND EXTERIOR DIMENSIONS: SCALARS AND VECTORS

In truth, we use the term space to mean different things, some more metaphorical than literal. New age books like to talk of concepts like 'inner space', which is a metaphorical space within our minds. Some also speak of biology as inner space; there is also the biosphere and the ecosystem. These all turn out to be views of acceptable spaces on the inside of a boundary, with a basis in measurable things. However, ideas like new age inner space are 'virtual descriptions' whose meaning has to be understood in terms of the processes that change information in lower level physical representations. This is not just a new age idea—it is normal procedure inside computers. Finally, we all know about outer space, which is the space outside of the Earth's atmosphere.

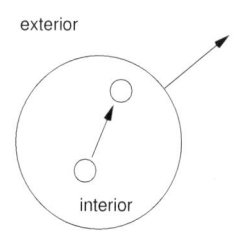

Fig. 2.4. An agent represents a definition of a single location, with a 'boundary' distinguishing between interior and exterior promises. An agent may or may not have interior structure, but if it does, it affords us a way of scaling viewpoints about space: what we perceive to be a single location at a large scale, might be many locations when we zoom in. It might be completely atomic and irreducible, or it might conceal interior structure. Only its exterior promises are visible to other agents.

Mathematics comes (as usual) to the rescue to define some clear properties for space, using variations on the idea of coordinates. Our normal conception of Euclidean space is a version of what we call a vector space. For example, in a vector space going forwards is the opposite of going backwards. But in inner space there is nothing concrete to add up, so its status is not clear.

Vector spaces are a model for exterior space, that builds on the concepts of distance and direction. A vector is like an arrow: it has both magnitude and direction. Forces, flows, and wind patterns are usually represented as vectors, like the patterns one sees on a weather map. We learn about vectors, in school, as representing forces and velocities, and so on. Although we draw vectors at a particular point, vectors are not properties of a single location: they are 'going somewhere'. The fact that we can draw them apparently at a particular point is an illusion of the method by which we define so-called differentials or infinitesimally small distances in calculus[31]. It's a version of what Zeno complained about, with his paradox: definitions by limiting processes. Inner dimensions are different: they are properties that belong to a single location. They are called scalars in mathematics. Scalars have magnitude or value but no direction (or at least not exterior direction, though interiors of points could have directions too).

You might find this slightly inconsistent. We could make an analogy between an interior state that characterizes us (happy, sad, red, green, blue, etc), and we can represent this by a label (just as we write coordinate labels x, y, z. Then we can describe our state as x, y, z, e, c, where e is emotion, and c is colour. This is what we do with the colour space in computing, for instance, which is why it is possible to attribute multiple dimensions to the interior of a location as well as to the larger space of all locations. Remember that space is built up by describing which locations are next to which other locations, or by which regional locations can be nested inside others, like Russian dolls.

The idea of extending the dimensionality at a point was idea used in theoretical physics by Theodore Kaluza, in the early twentieth century, as a way to model electromagnetic fields as a kind of interior space, independent of the normal spacetime dimensions. The idea survives in more modern String Theory—though recently gravitational wave experiments have cast some doubt on the existence of additional spatial dimensions[32]. To picture this in more everyday terms, think of a city like New York's Manhattan that forms a simple grid. The vertical streets are labelled by avenue numbers, and the horizontal cross streets are labelled by street numbers. A building can be located by its coordinates, where the streets cross (5th and 42nd). On top of that, buildings have numbers, and inside the buildings there are numbered floors and rooms may be numbered, like the rooms in a hotel. Those building coordinates are internal to the hotel, and may be repeated in every building. So, relative to the scale of buildings, there are exterior dimensions at street level, and interior dimensions as floors and rooms. The interior rooms are scalar properties of the hotel as far as the rest of the world is concerned. The location of one building relative to another is a vector property.

On the Internet, directions work in a similar way, except that there are is no simple grid plan. Instead, the Internet is more like a forest with branches and multiple overlapping trees. The rule is that most addresses of the form `aaa.bbb.ccc.111` refer to globally unique addresses, a bit like street addresses or postal codes in a country. The numbers don't even have to form a sequence, except for some technical limitations, they might as well be names. But they do have to be unique. That means there has to be a map somewhere else, like with street addresses or postcodes to direct packets to their destination. Like a road network, the Internet uses signposts (called routers or switches) that can be taken one at a time to connect locations on the Internet. In addition, there are some special addresses, like the quite familiar `192.168.1.111` numbers, used by home routers, that can be used to represent an interior network, isolated from the rest of the world (which is why everyone can use the same numbers).

The Internet address has an implicit boundary between `aaa.bbb.ccc` and `111`, which represents a local cluster of computers, called a 'subnet address'. This scaled location, with boundary marked by a router, is a renormalization of the Internet used by routers to avoid having to refer to every computer individually. It saves on the amount of information needed to connect the coordinates.

The Internet scheme is not a vector scheme. The reason why vectors (and coordinate systems) are so important is relativity. Vectors can be added up or subtract depending on their direction, and they still give meaningful information. Names and postcodes can't be added or subtracted, they are absolute not relative. Vectors can reinforce one another or cancel one another out, names and postcode are just names. For example (figure 2.5), a step to the left followed by a step to the right results in no net motion—just as walking down the up escalator, or going backwards along a moving walkway. Similarly, if you run alongside a moving train, you can appear stationary with respect to the train as you wave goodbye.

INTERIOR VECTORS: VECTORIZING SCALARS WITH HIDDEN DIMENSIONS

Scalars do not have the same kind of additive properties as vectors, but even though it's conventional to use vectors only for spaces that extend infinitely, we can establish a basis for labelling any kind of variation with artificial coordinates, which we can represent geometrically too. This is quite intuitive to our human brains, no doubt wired to understand the world in terms of space.

One example of this, which many people know today, is colour. Colour is not a one dimensional property, because our eyes have three different kinds of

2 Space and time 47

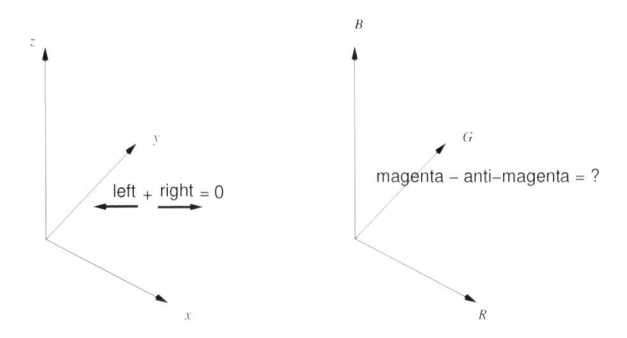

Fig. 2.5. Scalar properties are promises about an interior property of a location agent. Vector properties are non-local properties of a direction between such entities. Sometimes we invent new dimensions of imaginary space to represent interior scalar properties. The difference between a vector such as direction (left) and a scalar like colour (right).

colour receptor, not one. So, while a single object might promise to emit a range of light frequencies, we have three different agents in our eyes that can accept frequencies in different ranges. Our brains then perceive these ranges as red, green, and blue (RGB), forming the colour palette.

Today's computer literate generation also knows that colours can be assigned numbers on a Red-Green-Blue (RGB) colour scale, where higher numbers represent increasing brightness for the colour. This is the basis of colour codes, used in your web browsers and painting software. These numbers are usually given as hexadecimal codes: e.g. `#000000` is black, while `#990000` being a tone of red, `#005500` a tone of green, and `#000099` a tone of blue. So we can draw axes, as if these were really #xxyyzz, but the colours don't satisfy the same symmetries as movement.

We can form these quasi-vectors from colours, but they do not form a vector space in the mathematical sense. Colours can be added: everyone knows that blue and red make purple, for instance. But, painting something blue, then painting something with anti-blue doesn't work. We can speculate about anti-blue but it isn't natural. Painting something blue twice probably doesn't make it twice as blue. So the 'algebra' of colour is not like the algebra of distances.

Another way to represent vectors as a graph is to turn an abstract property into a physical connection to a common location in space (see figure 2.6) that represents that single property, like an anchor point. That interior dimensional

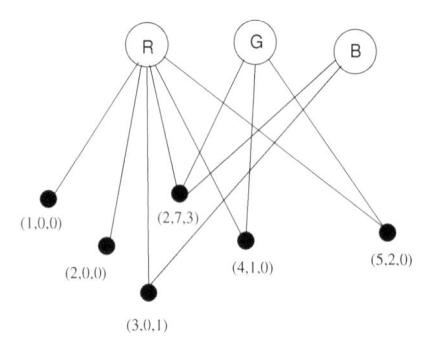

Fig. 2.6. Giving locations the same property is like connecting them to the same source. Why do all electrons have the same mass and charge? Because they are all the same electron, said Wheeler. Scalar properties can be represented as networks with hidden dimensions. Here we show three special hub points that are used as references to build a coordinate map, known as a matroid or basis set leading to three components in each of the coordinate tuples.

location is supposed to represent the source of this property, essentially making all blue nodes members of a blue club. The physicist John Archibald Wheeler famously told Richard Feynman: "Why do all electrons have the same mass and charge? Because they are the same electron!". Many physicists think of this as a joke, yet the interior dimensional construction shows why his reasoning is the only explanation for a calibrating a global property that makes sense in terms of spacetime alone. It gets around the problem of calibration (and so-called global symmetries in physics) by making an actual adjacency, of a different *type* from normal spatial adjacency. There is thus a natural reason to distinguish vector from scalar properties. Scalar properties come from within each point, while vector properties (and generalizations called 'tensor' properties) come from without, or from the links between points. Scalars are interior and local, and vectors are exterior and non-local.

THE SCALE DEPENDENCE OF DIMENSION

To summarize, a space has size, but it also has *directions* that we associate with multiple *dimensions*. Descartes showed us a way to measure direction too, by drawing a rectangular grid of coordinate axes, usually labelled (x, y, z) for the three dimensions[33].

At every point in a space, the different possible directions a traveller could move are sometimes called *degrees of freedom* rather than dimensions, in order to better represent their role. They represent independent alternative pathways in which a hypothetical body could move from where it already is. Of course, there is an infinite number of choices, but not all of them are necessarily independent. I can move to the left or right without going forwards, backwards, or up or down. Similarly, I can jump up and down, without moving left, right, forwards, or backwards, and so on. So these directions are locally independent (we say orthogonal or linearly independent in mathematics).

Whether or not they are truly independent, in the long run, depends on what happens as we keep going in the same direction. Dimensionality is qualitatively different in Euclidean space and in networks. Suppose a traveller arrives at a crossroads; locally, there might seem to be four possible directions to travel from this juncture: forwards, backwards, left, and right. But what if going left and right only take detours through side villages, which are dead ends, or which curve back around to the forward path, slightly farther along? How many dimensions are there then? At the crossroads, the answer is still the same, but looking down from a god's eye view in the clouds would allow us to see that there is really only one forward path and some 'noise' around it.

In a network, we call locations like a crossroads junction a *vertex*, and a road local path an *edge* or *link*. The relationship between networks and continuum space is subtle and not fully understood. Suppose we wanted a concept of travelling North. How can this be understood from an agent perspective? The concept of North-ness is non-local, and uniform over a wide region. In order to imagine continuing in the same direction, we also need to know about the continuity of directionality. Direction too is thus a non-local concept. When we speak of direction, we are implying the concept of something that goes beyond asking which are closest places. In vector spaces like Euclidean space, dimension is a non-local concept—or what mathematical physicists might call a global symmetry. Any agent can promise to bind with a certain number of other neighbours, and even to label its adjacencies to them with the same name (say North, South, etc), but why would the next agent continue this behaviour? How does each agent calibrate these in a standard way? In other words, how do we know that—by following several adjacencies—we don't end up coming around in a loop, back to where we were?

Our notions of direction—and of its more formalized surrogate 'dimension'—

may be *scale dependent*. Because pathways can join up, we can arrive at the same place by multiple routes, and so the routes are constrained by this joining up, limiting their long term freedom. The logical outcome of this is that the more pathways we can follow, at a point the more likely it is that fewer of them are independent dimensions (see figure 2.7). So as we grow a particular network, we might end up with a limiting number of independent macroscopic degrees of freedom that are quite unrelated to the microscopic ones. This kind of over-counting of degrees of freedom is a problem at microscopic scales in physics, where interior 'gauge' symmetries lead to such redundant paths in spacetime processes[34].

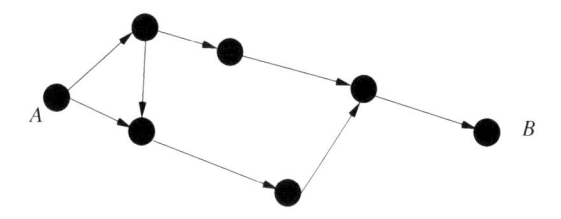

Fig. 2.7. The dimension of a space is the number of independent ways a traveller can move at each point. But as we extend a network, this might shrink as we realize some directions are not independent after all. Starting from A, there are two possible degrees of freedom, or two dimensions, but by the time we extend to the scale of reaching B, we find that the paths end up in the same place, so at this scale the space is only one dimensional, as the degrees of freedom were redundant.

Topology also plays a role. On the surface of a sphere (which is two dimensional), we can travel in a straight line and come back to the same place eventually by going all the way around. The same is true of any direction we take. But none of the other places would be the same. On the other hand, we could travel along parallel lanes on a wide road and many if not all the places would be the same, on some scale.

We see this most prominently in a continuum, with infinite points and infinite directions at every point. A traveller standing at any point can choose 360 degrees of different directions to travel, and even many more directions in between those. But as soon as this traveller moves, we realize that there is an infinite number of paths he or she could have taken to arrive at the same place.

So is the dimension infinite or one dimensional? The answer is complicated and again scale dependent. Over each infinitesimal interval, we can argue that the space was 1 dimensional (space and time move together in a single hop plus tick of the clock). All the other paths would be identical. This kind of step is connected to what we mean by *causality*. But if we take several more steps, we might be able to reach a number of different points along different paths suggesting that there are more than one.

A simple convention was invented around the beginning of the twentieth century, using basis vectors as official signposts. A vector uses the idea of building any movement, in any direction, out of distances relative to 'basis directions'. So a random movement can be considered equivalent to a small movement forwards (or backwards) together with a small movement left or right, and up or down. Since there is an infinite number of ways to get from A to B, why not choose some standard directions as a guide, and say that any movement is equivalent to one or more movements along these special axes (see figure 2.8).

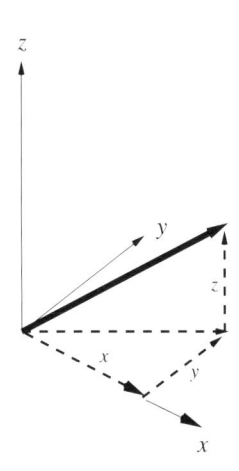

Fig. 2.8. Equivalent movements measured along standard directions make a vector space. The direction and distance of the fat arrow can be broken down into equivalent pieces, relative to a fixed set of coordinate axes. The particular choice of axes is unimportant as long as we can always adopt this strategy.

Using the assumed symmetries between forwards and backwards, left and right and up and down, we can reduce the number of possible axes to a minimum. This is all very simplistic, but also extremely beautiful in its simplicity. The number of standard directions we choose to introduce as vectors is not determined by anything as absolute as we are taught in school[35] Rather we can look for the smallest number of independent directions needed, i.e. in which the motion is non-zero. This gives us a definite answer, but we are always free to add more dimensions that do not play a role, for whatever reason[36].

The invention of vectors, is one of the industrial techniques of modern mathematics that helps us not to think too deeply about space and time, but rather get on with the business at hand. Vectors only work because of the infinite number of paths in a continuum. There is nothing like a vector version of a network in the general case. For instance, if we make any holes in the space, the difference or equivalence of paths may no longer be exactly true anymore. Remarkably, all the effects discussed here are known to play a role in the physics of fields.

The scale dependence of dimension, combined with the role of the fictitious traveller, means that how we define the dimension is a matter of some philosophical ambiguity. In the Cartesian picture, dimension is related to the idea of direction (left-right, forwards-backwards, up-down). In a polar coordinate view, like the geo-coordinates, it is based on height and a kind of spherical grid (latitude and longitude) marked from pole to pole on an imaginary surface of the Earth, all at a constant radius from the imaginary perfect centre.

For example, the Moon orbits the Earth. The Moon is a three dimensional body embedded in a three dimensional space, but the orbit is more or less one dimensional path, that curves around in a two dimensional orbital plane. The embedding space has three degrees of freedom (a galactic bulldozer could push the Moon off in any of three independent directions), but it is actually held in an orbit by gravity so it only experiences two of them, or is it actually only one dimension? Which is the right dimensionality for the system? This is a matter of policy or convention. We can't easily argue which view is correct. In practice, however, the actual motion of the orbit is the important focus, and it's more a question of how we conveniently describe the behaviour than of a correct dimensionality[37].

When we move around in a city, the dimensionality of our behaviours, including interactions between active agents like people and companies, is basically one dimensional. We move around through a series of 'wires', called variously paths, roads, tunnels, and subways, or visit one another point to point, but the way we count their aggregate effect has a different effective dimension[38]. Think of a rollercoaster, diving and twisting through a volume of three dimensional space. Although the embedding requires three dimensions to make sense of the curvature and topology of the space all the dynamics are in one dimension, and all of the adjacency lies in one dimension. Space and time are essentially the same thing in this system (see figure 2.9). It is

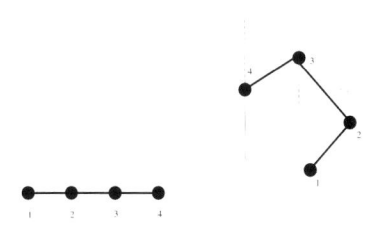

Fig. 2.9. A one dimensional structure can become three dimensional just be embedding it differently. Compare this to the way we make a clock, or write numbers in base 10 in section 6.

instructive to take an extreme case of a simple one dimensional network, and to reinterpret it in a three dimensional basis (see fig. 2.9). By bending the one dimensional space we can make it arbitrarily three dimensional. An entity travelling along it might not notice this embedding. On the other hand, the bends might be associated with forces that the traveller does experience. Think of cars in a tunnel, a helterskelter, or a rollercoaster. Einstein's model of gravity as spacetime curvature becomes something like this idea—in that case, there is a geometrical interpretation in which space effectively curves in time.

So we are left with a confusion in physics: do we define the dimension of a space of possibility (an embedding space)—where a body only *might* go, or do we define dimension as the locus of places that it *does* go to? In the latter case, there is another concept in physics called *phase space* that captures this idea.

INCREDIBLE SHRINKING TIME

In spite of being joined at the hip, space and time are qualitatively different animals. Time is perceived as directional, because it accumulates space as it counts distinguishable configurations (it refers to a region in order to have enough states to form a clock), whereas space is usually not[39] because its interior states are not involved in counting position (they are laid out explicitly), but that can change if something breaks the symmetry, like an external boundary condition.

It is our sense of the involuntariness of time, the imagined comparison to a kind of river current, pulling us along, that leads us to think of time as a force of nature, existing outside of the ordinary world of material things. We could continue to take this point of view (after all, it has worked quite well for centuries), but Einstein showed that it has more than a few problems, and it certainly seems to have led theoretical physics down something of a rabbit hole in an effort to preserve this objective independence. The explicit role of the observer, championed so insightfully by Einstein, has since been suppressed again through layers of tensor formalism, that grant us some reprieve from thinking hard thoughts by industrializing the calculational apparatus. But it was this sharp focus on observation led to some of the most important insights of the 20th century, and it is surely worth questioning whether our quickness to dial it back into a notionally branded 'Fabric of The Spacetime Continuum' is misplaced. I will show, in this book, how simple questions about information lead to some big surprises here.

Time can't be completely outside the world of ordinary things, because without those ordinary material things we cannot observe it at all; moreover, if we try to treat it as immutable and inevitable, we find contradictions in what we observe, at very high speeds, where observers start to race the speed of observation (i.e. the speed of light). In fact, time seems not to be outside things at all, at most scales, but actually inside things. If Einstein was right, then each observer carries its own sense of time with it. This has deep consequences.

As we shrink observers to smaller and smaller sizes, remembering not to fall into Zeno's Paradox, the number of states within a spatial region must get smaller, so the number we can count to using those states must get smaller, until eventually the clock repeats. If you shrink a calendar for the year to smaller and smaller pieces of paper, eventually there is only room for a twelve hour clock. The space to count time is reduced as we shrink the size of a region. A very small observer cannot distinguish time at all. Its clock must either count in cycles, or see change as entirely random pulses.

We still don't understand the source of change, where it comes from, or what causes it, but we can separate its appearance from what we mean by the time we perceive. This is effectively the journey Einstein began with his special theory of relativity, even for continuum spacetime. It is likely that he did not appreciate, at the time, how this might be the only plausible interpretation of time in discrete systems, such as quantum mechanics and present day information technology. But it leaves the question of how time is measured, which brings us naturally to the topic of clocks. Clocks are made of matter or energy, because empty space

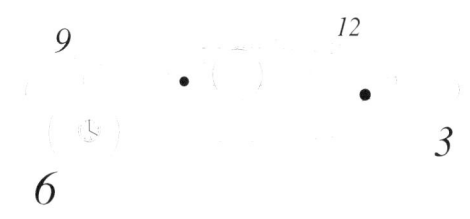

Fig. 2.10. Any changing process behaves as a clock, as long as we can quantify the difference in the state of the process. Interior clocks drive activity in a system, like the heart of the computer's CPU. Exterior clocks are passive markers of activity which originates elsewhere. Which clock tells the 'right' time?

has no apparent features that could be used to mark out time (this might not be true, but at least we can't easily see them). By introducing matter, we have the ability to represent patterns, which in turn allows us to 'write' things down, and remember them. Memory therefore requires material not only so that something interesting can happen, but also to form representations of events, which can record things that happen. Memory is nothing more than the transference of figure 1.2 (b) into (a). By taking an unfolding storyline in time, and making it static for future times, we can remember sequences of events as a structure in space. This is how all memory works, whether in cave paintings, brains, books, or computers.

Is time finite or a never ending story? If we can see that every detail around us is in precisely the same state as it was at an earlier time, is it possible for it to tell the difference between now and then? Can we say that time is independent of what happens within it, or is it actually a function of state? This seems like an odd question to ask, because we can't actually imagine this situation from a human perspective. We carry our own internal clocks with us, which contain many years of ticking power, so we never run out of internal time (as long as we are alive, and perhaps even after that, from someone else's perspective). So even if we could take away nearly all the features that could be used to characterize a space, we would still imagine the passage of time in a system we were part of, because of what's inside us. But not all systems contain humans on the inside, so we need to identify, for every system of interest, whatever source of impetus there is that drives time, as an observable clock, from within or without. What states are there to characterize space? What degrees of freedom do they have to distinguish one moment from the next?

This, after all, is how we build clocks. We create a simplified mechanism that has a limited number of states, and this is translated into twelve numbers; the clock face shows only these numbers from 1 to 12. When another change occurs, and we go from 12 back to 1, everything about the clock is indistinguishable from every other occasion it reached the number 1. Of course, in a daily situation, we now have other *exterior* states to change that we add to the clock time, so that each time we reach 1, something else has changed and we can tell the difference on the outside. The clock cannot. So the single clock is not our entire picture of time? Something about this interior and exterior business seems important. In physics, it is called the *locality* issue.

This pragmatic and functional view of time might seem a bit artificial and picky, but it is very much the version of time that is relevant to information technology and computation. It is also the version of time that must be true for the smallest isolated microscopic processes. In the technological realm, we deal with systems that are deliberately simple, with few degrees of freedom, so that they are easy to predict and control. But with fewer possible states, there are fewer possible things we might be able to distinguish, and so the meaning of time within these limited systems might be limited too. We are effectively saying that time depends on how we choose our horizon. What we can detect of change, i.e. where we look, limits the extent of time that can happen. This is a strange idea, and in fact similar to the observation that Einstein made. Isn't time a universal phenomenon? No, it isn't. Certainly, it is ubiquitous, but that doesn't make it identical everywhere[40].

In any meaningful sense of what time is, we would observe it differently in different places. The point is that, although time is ubiquitous wherever there is change, no observer in any system is causally connected to everywhere change is taking place[41]. So if we cut off a region of space from outside influence, its notion of time becomes independent. The idea that every location is essentially an independent source of causality is called *locality*. Another way of expressing this is to say that every point or place in space makes independent promises[42].

The reason we can get away without asking tough questions about time at the scale of the day to day, is that there are so many states available within ourselves and our mundane measuring instruments that we cannot perceive any meaningful limitation on the resolution of time. On the other hand, when we try to observe very small systems we experience 'quantum uncertainty' for our trouble. Could this be a result of having an insufficient number of states to generate a smooth version of spacetime? I'll return to that in chapter 4. The luxury of having sufficient states to measure with, remains even when we build technological systems on a totally different scale to the quantum world—on a scale comparable to ourselves. It isn't absolute size that matters, but the ratio of information arriving to memory available. Technology today probes faster and faster processes, over smaller and smaller regions, as we stretch the powers of minimization and nano-technology. The result is far greater uncertainty on a much larger scale.

By now, it should be clear that there is no universal way of representing a system of locations and their relative adjacencies. Let's briefly look at some of the common models of space and see what they have to teach us.

The Euclidean lattice and continuum space

Descartes' important innovation of *coordinates* led to a model of what are called *vector spaces*, or spaces in which we can make consistent wireframe descriptions of regions and have them point in a particular direction. This practice has become the default for describing physics, where we need to talk about forces and flows. In Euclidean space[43], points are arranged in a regular lattice, or a kind of regular crystalline structure in some number dimensions, typically 3 labelled using rectangular coordinates (x, y, z). In other words, every point in a Euclidean space has exactly six possible neighbouring points located by adding or subtracting to x, y, z respectively. What characterizes Euclidean space is its presumed *symmetries*. These are called translation invariance (or

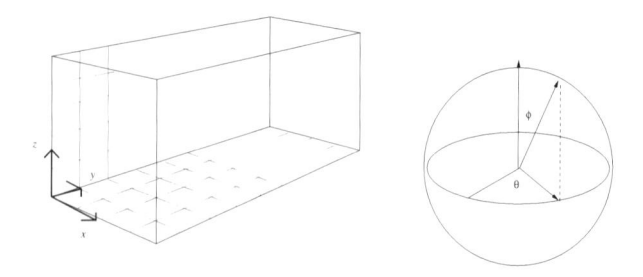

Fig. 2.11. Regular lattice scaling - the dimensionality is constant at all scales due to the group structure. We have a choice of parameterizations: either rectangular coordinates, emphasizing symmetry under motion (left hand side), or polar coordinates, emphasizes symmetry under rotation (right hand side).

homogeneity), meaning that you can't tell the difference, in empty space, if you move your measuring stick by any amount; and rotation invariance (isotropy), meaning that if you rotate by any angle, you can't tell the difference either. When expressed in terms of rectangular coordinates (x, y, z), we emphasize translation invariance, meaning that space looks exactly the same if we move by any amount in the x, y, z directions. Rectangular coordinates make this property manifest, but it remains true of course if we change coordinates, just not obvious. When expressed in terms of polar coordinates (r, θ, ϕ), or latitude and longitude, we emphasize rotational invariance. This is a more centrist or Ptolemaic view (see the right hand side of figure 2.11). The generalization of Euclidean spaces to curved surfaces and volumes is called Riemannian space, after the mathematician Riemann.

LITTLE BOXES, CIRCUITRY, AND 'BIGRAPHS'

The coordinate view of space is useful when embedding map imagery, but it is not particularly helpful when describing *processes*. If we look to technologies or processes in nature, a different representation is natural. Imagine irrigating fields, driving a mill by water wheel or wind sails, a thermodynamic pump, an electrical circuit—all these are examples of functional circuitry represented as spaces whose locations embody functional properties, connected together in a causal workflow, and possibly embedded in a larger context. In an electric circuit, it is the components that are the locations and the wires represent adjacencies. In a computer network, computers are the locations and network cables represent adjacencies. In software, computer programs are the locations, and messages are the adjacencies.

In a functional world, the fact that components are realized and arranged in a larger space in irrelevant, like the bending of the line in figure 2.9.

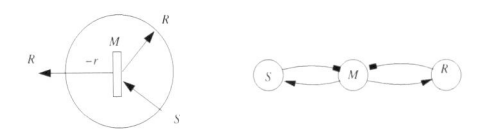

Fig. 2.12. Electrical circuits have significant locations. The symbol for a transistor represents its interior promises, which are drawn using promise theoretic notation along side.

In quantum theory, Richard Feynman used fragmentary circuit diagrams to represent interactions between particles in quantum physics for his famous 'Feynman diagram', which we later developed into an entire diagrammatic language for (exterior time) change and (interior time) equilibrium at different scales[44]. The time-generating (action) parts of circuits are so-called Directed Acyclic Graphs or 'DAG' graphs, common in computer processing. It's not surprising then that computer Scientist Robin Milner emphasized these functional aspects of spacetime in his model of 'bigraphs' in the 1980s and 90s[45]. He described a language for spacetime that mimics the way interactions actually work, but with all the exterior assumptions of embedding in a continuum stripped away. He imagined functional circuitry, analogous to electrical circuitry, in which the locations were 'agents' with certain mathematical capabilities.

In a bigraph (see figure 2.13), active agents are simply 'boxes' in which agency or intent sits. These are connected by communication links that play the role of network cables, but they could be any channel, like molecules coming into contact in solution. This view proved useful to Milner for describing many formal aspects of computational processes, and the usefulness of this view begs the question: how much do we really need the three dimensional embedding space of the real world? The main role played by the solution, or the embedding space, is to be able to reorder the boxes with respect to one another, making them adjacent or non-adjacent in changing configurations.

Milner's model emphasizes the role of relaying of information from location

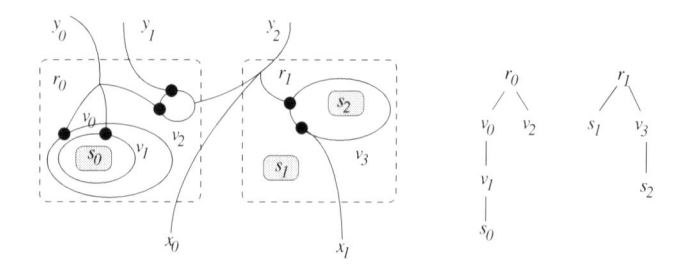

Fig. 2.13. A fully featured bigraph, as described by Milner, mapping inner to outer faces, like a transformer.

to location. Information can't go just anywhere. It is constrained by the existence of wires. Nothing can happen outside an agent, so if we want to send a message to a distant place, we have to send it through all the intermediate agents, just like in a Euclidean space. The obvious difference is that the graphs have wires in between the locations because of the way we draw it, just as a latticework has gaps. But this is only a convention (or a limitation of our imagination in drawing the space), rather than a necessary feature. The question of what is 'outside the space' (the embedding volume) is then left to the philosophers.

One of the interesting aspects of Milner's characterization of computer systems was that it was not based on exterior vector properties like translation as its primitives, but rather on *modularity*, or the semantics of a region's interior properties. This is natural to emphasize interaction[46].

This notion of things being inside other things is not something representable by the classical symmetries of translation and rotation, familiar to physicists. Yet it plays a central role in things like Gauss law. In his work, Milner had made that the fundamental way of mapping spacetime. But concentric boundaries turn out to have a crucial importance to scaling, because boundaries are where information enters and leaves a system. The scaling of time, or how we define a clock, is not independent of the scaling of the way we scale space by coarse graining. Again, while we scarcely notice these effects in the human world, they become amplified in sensitivity in the scaling of computer systems as they get faster and father apart.

The topologies of computer networks are minimal and natural: localized placed connected by straight lines. The topology of Euclidean space, on the other hand, is not minimal or natural for a local picture. It is grossly overfull

Fig. 2.14. Quantifying discrete semantic properties is no harder than quantifying numerical ones, but we need to be careful about the semantics of averages. What is the average of blue and green?

of locations, and has enormous redundancy. What is the point of all that empty space? So in functional models, we strip it all away.

GENERALIZED GRAPHS OR NETWORKS

Milner's bigraphs are a special case of what we know simply as graphs in mathematics, and more commonly as *networks* elsewhere in science[47]. In an ordinary graph (see figures 2.16 and 2.17), there is no assumption of functionality at the locations (called nodes). They are simply places. We can then attach whatever semantics we like to the nodes and the adjacencies or links between them. Links can be made for physical connections, like roads, or for abstract relationships, like family trees and organizational charts.

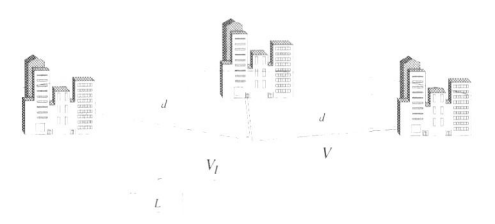

Fig. 2.15. A network of locations connected by links is a called a graph. Networks or graphs are everywhere, embedded within our towns and cities, delivering utilities as well as in biological organisms for transporting blood.

Whereas the Euclidean model of space emphasizes its uniformity under translations and rotations, a graph model emphasizes no particular symmetry. We can surely make a very regular network, like crystal lattice, and imagine it as an approximation to Euclidean space (or vice versa), but this requires us to place a discipline of restrictions on the allowed connections. The true strength of graphs lies in their generality, in having few assumptions—but perhaps, more importantly, in being discrete structures, with a built in minimum scale.

In a graph, locations are discrete places called *nodes* or *vertices*, and adjacent nodes are linked together by the wiring between them, which is made up of edges or links. Unlike a Euclidean space, a link can be directed, meaning that certain directions can be like one way streets.

In a Euclidean space of dimension n, there are exactly twice the number of neighbouring places one could go to at every point (ahead and behind in each direction). But in a graph, from a given node in a graph, the number of adjacent neighbours is arbitrary at every point, as if the spacetime dimension itself is different at every point. Instead of 'dimension', one usually speaks about the degree of the node, which is analogous to direction for a graph. Every new location one arrives at may have a completely new set of directions, but only a finite number of them[48].

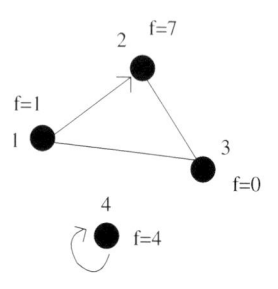

Fig. 2.16. A graph drawn in a more pure notation, where only nodes and links are shown, sometimes with labels. Note that links can be directed or undirected.

Whereas a Euclidean space scales directions due to its homogeneity, and infinite resolution or continuity, a graph does not scale at all in general—it is a completely amorphous structure. This is the reality of most networks in chemistry, material science, biology, and information technology. In chemistry, for example, different atomic elements may act as nodes, forming molecules, and

2 SPACE AND TIME

crystals. Different atomic elements have different preferred chemical valences (node degrees or local dimensionality) which describe their ability to bond with adjacent atoms.

A graph is characterized by locations, directions, and hops to adjacent neighbours, just as in a computer network. The hops are analogous to directions in a Euclidean space, but with some subtlety. For the scaling of regions, imagine

Fig. 2.17. Scaling in a graph. Nodes in the graph can be elementary, or be formed from 'irreducible' clusters nodes connected together. The concepts of connectivity and continuity are not distinct for a network. At the scale of links the arrangement of points seems 1 dimensional. An observer inside a node experiences the degrees of freedom as the number of ways it can choose to leave, meaning that the dimension of the space is potentially different at each point. The average dimension of spacetime can change under a scale transformation. If we put arrows on the links, spacetime becomes like a one way street! This is not possible with the overlap picture, so graphs are more general.

replacing the small clusters by a single node (see figure 2.17), each of which node has an interior star network, now we have a star network of star networks, but each supernode is a scaling of a small graph, which draws a virtual boundary around the smaller graph and treats it as a single entity.

Because of their great generality, graphs can have any connective topology, and can be used to represent almost anything [49]. Their discreteness means that it's easy to discuss their topological properties as patterns. Some common patterns, from the world of information technology, are shown in figure 2.18. The classification in figure 2.18 was originally used to discuss the fragility of a communications networks to broken links. Graphs like 2.18(a) are fragile in the sense that a problem in the central node would destroy the connectivity of the network for all other nodes. Figure 2.18(b) is a hierarchical network of

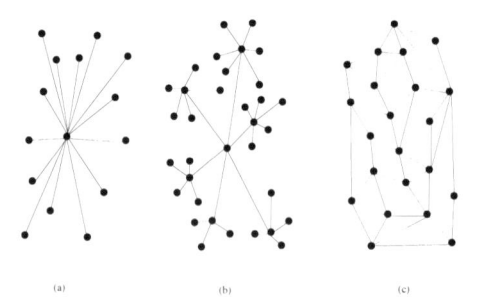

Fig. 2.18. A few simple graphs or network topologies: (a) a star network, (b) a hierarchical graph, or graph of graphs (c) a mesh network.

networks, which is also fragile for the same reasons as (a), but which introduces the idea of local hubs that help to scale. The hubs act like routing points: think of major airports and regional airports, or postal distribution points. It is efficient to route traffic on a network to a regional centre and to distribute locally from there. The Internet was based on a so-called mesh network, as in figure 2.18(c). In a mesh, there are no privileged central nodes that make the network fragile. This makes them harder to navigate and label, but more resilient with multiple pathways to the same place.

As already mentioned, graphs can be both unidirectional, like one way streets with arrows in a single direction, or undirected with arrows going in both directions. An important class of graphs is 'acyclic graphs', which contain no loops or cycles so that travellers can't get back to where they started from, because these represent processes. So-called Directed Acyclic Graphs (also shortened to DAGs) are trees that branch out and sometimes come back together, but never go back on themselves. These describe the typical kind of process we find in data processing and manufacturing. Conveyor belt production lines tend to follow these processes (see figure 2.19.)

Directed graphs can make time and space indistinguishable. Each hop is inseparably a tick of a clock too. Moreover, directed graphs can be unstable if they have loops. A traveller might wander into a region of space from which they never find their way out, forever stuck going around a roundabout, as if time itself becomes a loop for them, This happens in computer networks, both in hardware and software, relatively often, trapping information in so-called 'network black holes'[50].

Process graphs have properties that Euclidean spaces don't have. They can represent more or less everything that Euclidean space can (in the limit of tiny hops), and they can represent dynamics too, explicitly within the same model[51]. A particular class of process graphs is that of so-called Finite State Machines (FSM). Their topologies and connections give them the ability to compute certain algorithms, like a hardwired circuit diagram. In other words, graph spaces can act as symbolic computers.

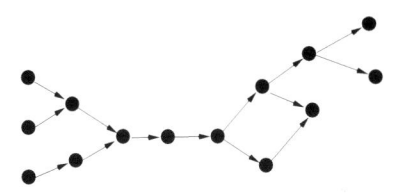

Fig. 2.19. A directed acyclic graph is a typical representation of a workflow, a business process, or a manufacturing or logistics pipeline, where the inputs are on the left and the outputs are on the right.

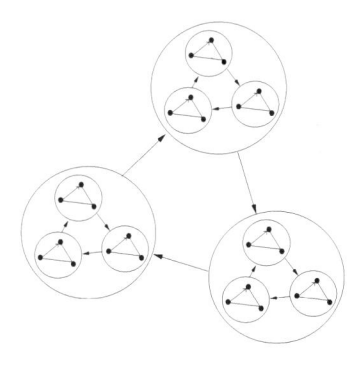

Fig. 2.20. A graph of graphs, forming a hierarchy. This example is also self-similar at all scales. Many cities and social structures form this kind of pattern, of spaces within spaces.

The notion of distance in graphs can be quite complicated because of they have directed one-way streets, and also because of *anisotropy* (difference by direction), and *inhomogeneity* (difference by location)[52]. Graphs can be a model for the way databases work, indeed all information storage. How close are two products in a database, and according to what criteria? Recommendation engines on websites have to make determinations like this. Imaginary spaces of products or animals or diseases are often used to try to compare how similar things are, by replacing similarity with a notion of distance. In such a world, space is not like good old metric Euclidean space.

Underlying the theory of graphs is a wealth of mathematics, and links to topics like Group Theory and Category Theory[53], which lie far outside the scope of this book. Category Theory also overlaps a lot with the next step in the ladder of spatial fundamentalism: Promise Theory.

PROMISE THEORETIC GRAPH

Graphs are important because they remove several implicit assumptions about symmetry and continuity from Euclidean space, and force us to be explicit about them. But there is one final assumption that neither directed nor undirected graphs undo; that's the assumption of complementarity or *unitarity*—i.e. the assumption that a path suggested by a one location must be accepted by the destination it is aimed at[54]. There are no dangling links in graph theory. This assumption is so deeply ingrained in physics that any physicists reading will probably be scratching their heads already. Why should anyone want to remove a symmetry that is everywhere taken for granted in physics? A simple answer is that the assumption doesn't scale. It is not true in biology, or in information technology, and in fact it isn't even universally true in physics either. Promise Theory is a version of graph theory in which basic assumption is the autonomy or independence of spacetime locations: an extreme form of locality. It forces us to document every assumption explicitly, and show how they scale.

The assumption that a point, which is next to me, has to consider me to also be next to it, is uncontested in normal ideas about space, but it needn't be assumed. In fact, it would be wrong to do so. It is obviously true in the human realm that if I offer you food, you do not have to accept it. However, it is less obvious but nonetheless true that similar situations also occur at the lower levels of spatial reasoning. For example, if I am next to you, do you have to be next to me? Could I always reach you, if you can reach me? We built one way streets for traffic, so there is already a possibility of bias between certain directions. In biology there are semi-permeable membranes that allow certain places to be reached irreversibly. Even at the level of empty spacetime, the event horizon of a black hole will admit transitions on one direction, but not in the other. In a Euclidean continuum model, such properties have to be represented unnaturally by discontinuities and singularities that may e hard to deal with mathematically. In a graphical theory, this is a simple matter to accommodate.

Promise theory makes a distinction between something offered and something accepted, and requires both 'promises' to be kept in order to connect even in one direction. If A promises to be next to B, then B has to promise to accept this offer to complete a link from A to B. The same is true in reverse, so both A and B have the last word on connectivity of any kind. This is locality to the max. It acknowledges what we observed earlier, namely that connection is ultimately about communication. You can speak to me and I can refuse to listen, but you might listen to me. There is no guarantee of reciprocity. In physics, reciprocity is normally assumed because of the belief in principles like energy conservation, but such assumptions are automatically tied to very special curated environments. To model the world more generally, we need a formal way of switching such assumptions on and off.

The basic features of promise theory are extremely general, and will serve as a very useful anchor point that helps us to see many different phenomena through a single lens. The nomenclature of promise theory was built around information systems, so we have:

- Agents are locations in space.

- Agents can make promises about their own properties and behaviour, but not about others'. So every location makes and keeps its own promises (but agents can form bindings and cooperate).

- A promise to offer something is labelled by (+), and a promise to accept something is labelled by a (-). A binding requires two agents to make promises: one (+) and one (-).

    This duality or complementarity of $\pm$ turns out to be of widespread importance throughout spacetime processes. For the remainder of the book, I shall therefore use these bracketed plus and minus symbols, within the text, to draw attention to cases where an agent makes available

(+) or selectively accepts (-) some kind of information or transactional behaviour.

Promise theory has an in-built observer-centric viewpoint, which makes it very useful. It treats each agent as a simple opaque boundary. Consider figure 2.21, which shows how agents (circles) make promises (labelled lines), and can promise to bind together into *superagents* (the large circles). The arrows in

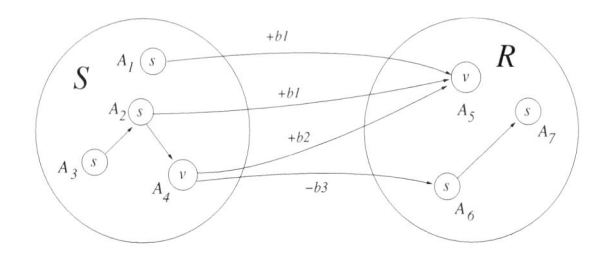

Fig. 2.21. The transformation of promises under a coarse-graining transformation. How promises appear to emanate from the super-agent surface.

figure 2.21 can be pulled out of the diagram and represented as single 'promises', just as vectors can be written separately or be drawn as arrows in a representation of Euclidean space. In the diagram, there are also two boundaries

$$A_1 \xrightarrow{+b_1} A_5$$
$$A_2 \xrightarrow{+b_1} A_5$$
$$A_4 \xrightarrow{+b_2} A_5$$
$$A_4 \xrightarrow{-b_3} A_6$$

and interior promises for the left hand superagent

$$A_3 \xrightarrow{+b_4} A_2$$
$$A_2 \xrightarrow{+b_5} A_4$$

interior promises in the right superagent

$$\pi_7 : A_6 \xrightarrow{+b_6} A_7$$

You can see that, because none of these promises are balanced with (+) and (-) promises, they are all ineffective over distance. They are scalar properties, not vector properties.

You may or may not like the use of the word 'promise', but this is the jargon that was adopted in an information setting. As long as you don't get too picky, it fits reasonably well with phenomena from the microscopic, all the way up to the scale at which anthropomorphism is justified, which is not a bad accomplishment[55].

WHY MAKE IT HARDER?

Why would we deliberately expose ourselves to obstacles in the effort to describe spacetime—obstacles that are normally gratefully discarded? The answer is simple: precisely to better understand the conditions under which discarding them is warranted. Continuity and homogeneity cost quite a lot to promise, and this is the point. In a theoretical model, it might be convenient to ignore unwanted irregularities, in an approximate way, but it is unlikely that we would find such perfection in all cases. It is both unlikely and costly to arrange and maintain.

This proves to be relevant from the quantum fluctuations that disturb the purity of the vacuum to the faults and failures that bring systemic uncertainty to information technology, or even biological organisms. The properties of space should not depend on which model we use. They may or may not include ideas like symmetry, isotropy, homogeneity, and so on. In practice, albeit in an extreme mathematical point of view, it is the extent to which these properties are expressed or broken that tells us everything we need to know about the world. The traditional way to discuss such issues, say in physics, is to start with symmetries, and introduce competing forces that break them. In a spacetime view, we simply tidy away all of that into the properties of spacetime itself.

Distinguishing matter from void does not solve the problem of what we mean by distance. If space is really nothing at all, then given some set of anchor-points to measure against, why are some distances bigger than others? Why doesn't the emptiness simply collapse into no size at all? Mathematics can't really answer this question, except to maintain and account for the distinction or separate identities of points, meaning that space is not nothing: it is defined to be a set of locations. Some believe that the answers to the deeper questions about space and time lie in quantum physics. Indeed, it would be a surprise if quantum theory had nothing to say about the subject of spacetime, but we don't have to reach that far to find important and far more practical aspects of space and time in today's information technology[56]. Information technology offers a rather different perspective on space, that may or may not have anything to do with fundamental physics, but it is certainly one that forces us to confront the most elementary of principles again at an engineering level.

A universal prototype space

Given the importance of engineering systems that span large areas, and involve high levels of interactions, with speeds that are pushing physical limits, there is a sense of urgency to understand how information systems scale. Many of those involved in engineering the largest computer systems on the planet have backgrounds in physics, because they are able to use those tools to look at computers in an analogous way, and say something about how their physical and virtual systems will scale. These techniques have been enormously important in understand material systems, fluid and solid[57].

The language of Promise Theory turns out to be quite useful for probing these issues, as it allows us to bridge natural and imaginary worlds, trace causation, and define scaling precisely by a formal coarse graining procedure. Instead of thinking about the world in terms of a continuum of points and space, with boundaries, it describes a world of agents that are explicitly modular, connected together by communications channels. The notion of a boundary is built into the scaling of an agent by composition. We can imagine a collection of people to be a village, and a collection of villages to be a county, or a collection of computers to be subnet, and a collection of subnets to be an organization, and so on. At the same time, we can imagine that the lowest level structure of even physical spacetime is made up from a network of agents, representing spatial points, that promise certain properties, including being next to one another. We then have to reproduce the standard picture at large scale so that it agrees with experience.

Promise theory makes clearer assumptions about interactions and cooperation that either physics or computer science does. In the latter, signals and influences are assumed to always act with probabilistic reliability, and one may pick and choose when to treat outcomes as inevitable. Even knowing the non-deterministic nature of the microcosm, one does not always question that a force will be noticed, or that a signal will be received. Both physics and computer science tend to be lazy in assuming that what is intended must in fact be accepted. Promise Theory adds back a reminder not to look away from these basics issues[58]. Such a universal spacetime cannot be a continuum, since it has to describe all the cases in section 2.

Semantics certainly seem to be more important than our usual scientific narrative gives credence. Whether we use numbers or names is not the point really the point. These are both valid languages. Using a set of ideas now known as 'Promise Theory' it has been argued that all these formulations are essentially linguistic in nature. Indeed, our perception of processes in general has a linguistic element, because it is through communication that we understand processes, time, and therefore space too. This is quite likely part of the reason why information viewpoints on science are playing an increasing role.

In Promise Theory, there are 'agents' in place of locations, points, or particles; and these agents make promises in order to advertise their capabilities, just as electrons 'promise' a certain electric charge, a certain rest mass, and so on. By using the term 'promise' we can scale this idea all the way up to the level of biology, cities and planets. Agents can also assess one another's' promises, if they have the capacity for such an analysis, however simply[59].

For the remainder of the book, I will refer to the Promise Theory view of spacetime as the general *lingua franca* of spacetime. That choice is based only on the fact that it is the only description that is both elementary enough and doesn't leave anything out. From time to time, we might have reason to compare the view with a more traditional continuum model. By adding essentially a few extra labels and rules onto a plain ordinary network, we win something very useful: the ability to depict both dynamics and semantics together, as properties of the basic infrastructure of the particular universe we happen to be talking about. It can be applied at any scale, and we'll see how this leads naturally to functional notions such as we find in biology and other science.

Physicists and mathematicians have favoured a passive view of space and time, from the traditions of the Enlightenment. But, what if there is another kind of spacetime, one where we can identify distinct locations, and interpret a timeline of events, where things can move around and form patterns just by having space change its spots? That spacetime would be more like a network— and could therefore be described as a promise network.

Spacetime concepts are useful far beyond Newton's bare theatre. There are as many 'semantic spacetimes' as one cares to imagine. The utility of thinking of processes in this way depends a bit on what you want to do—for now, the illustration of a route to unification across multiple scales of space and time seems sufficient.

SPACE	STATES OF SPACE	MOTION
CITIES	WORKPLACES	TRADE
COMPUTERS	DATA	PROCESSING
FINANCIAL SYSTEM	ACCOUNTS	TRANSACTIONS
SUPERMARKET SHELVES	GOODS	PURCHASE/ REFILL
FARM	PLANTS	GROWTH/HARVEST
RAINFOREST	FLORA/FAUNA	POLLINATION
		PREDATION

I'm going to focus a fair bit on computers in this book, mainly because computers are quite familiar now, as a basic tool, and they are involved in nearly all the engineering challenges of the modern world. Nearly everyone has some kind of contact with computers, but—more importantly—there is a lot to learn about spacetime in the realm of computers. Computer scientists are having to confront ideas they never imagined relevant: ideas that first arose in the physics of spacetime. I'll discuss biology too. In biology as well as computer science, a passive spacetime view is quite unhelpful, simply because it is not too useful to describe the status quo aside from simply mapping them out. The phenomena we are interested in involve the collaboration of different points in a fundamental way. Biology and IT are both dominated by dynamical *processes*, or storylines. Semantics enter at many levels. The key topological property is still adjacency, or the state of being next to another, but space is so much more than that.

One of the goals of this book, then, is to compare and contrast the artificial world of computers against the so-called real world of the physical universe. It turns out to be an eye-opening exercise to try to see how they compare, and what it might mean for both. Semantic spaces need not only be based on computers. Information was the province of the human world of cities long before computers were ever imagined. Cities are filled with spaces that are earmarked for certain purposes and certain things and people. Abstraction is the cornerstone of mathematics and physics, and it builds on qualitative or semantic ideas as well as quantitative measures. Nevertheless, there has been an argument against treating the variabilities of humanoria as a true science, because of the apparent difficulty of quantifying them. Much of this problem seems to be in the minds of scientists rather than being an insurmountable problem, a failure of imagination to redefine 'quantitative' due to the intransigent industrialization of science.

Rather than keeping the ideas about space and time locked in separate boxes, as academic disciplines are apt to do, I've chosen to try to bring them together. This is an increasingly important thing to do. Information technology challenges us to understand an entirely new selection of phenomena at a new combination of scales. While there is much to be learned from analogies to material science and biology, only the discipline of proper modelling and methodology will answer the key questions about the limits of the new world we are building on deeply embedded rich information.

3
ANTI-SPACE: INFORMATION

> Let us now turn to discussing a space of agents, based upon *locality* and *connectivity*. It's instructive to reflect how *placing* and *linking* run through existing informatics... [and] can be physical or virtual. These metaphors abound in our vocabulary for software.
>
> –Robin Milner, The Space and Motion of Communicating Agents

No matter how big the Internet may seem to be, counted in the unit of computers or locations on the World Wide Web, that number pales into utter insignificance compared to the typical numbers involved in nature, whether we talk about the scales of biology, chemistry, or everyday physics. Nevertheless, it has grown big enough to have become interesting from a scientific perspective. The Internet is no longer something we can claim to have built: it has now taken on a life of its own[60]. The relatively limited scope of computers therefore places computer science in a world of 'small size' and 'few things', as compared to many natural phenomena, and so—perhaps contrary to expectations—it behaves much more like elementary phenomena in our own universe than physics on the level of our human world. There could even be a chance that we might learn something new about elementary processes by understanding computers better, but that remains to be seen.

The component nature of computer systems—the separation of parts, and the lack of smooth continuity between locations and events—must play an important role in understanding how space and time operate within them. Computer science has a practiced history of describing phenomena in these terms. It is a spacetime of separate components that pass messages and change as they go. What, then, if we applied the basic ideas about computer science to the physical universe too?

The laws of physics that we have come to take for granted, like Newton's laws, seem mysteriously unnatural from the perspective of computer science—in fact, when compared to almost any other situation in science. For instance, the idea that matter continues to move in a straight line in the absence of other forces is something that happens only in physics. When you give someone money, or send them an email, it passes from you to them as a single transaction[61]; it does not continue to move from them onto others, ad infinitum, without some new motivating factors. Similarly, in computers, data do not zap around freely at random, continuing on and on to the horizon: they make limited jumps and then stop. Physics certainly makes no immediately obvious sense in the world of information. Why then is physics so different? These are questions we can begin to unravel only by scraping away at the surface of how computers really work—the subject of this chapter.

The Infoverse

Until recently, we thought of computers either as helpful machinery or as a tool for our entertainment; but, to those who work with the systems of the modern Internet, there are more than a few different spacetimes to be found within this technological marvel: worlds where space and time can be controlled, suspended, and recreated at the push of a button. In physics, a common strategy is to isolate the essential parts of a physical system and describe the closed spacetime alone. The universe as a whole becomes a patchwork of semantic phenomena, embedded within a larger story. We can think of the network of computers in the same way.

Computers deal with information, but they do so in terms of physical processes, so we need to be clear about the separation of these worlds. Whether information is real or abstract to you depends on your point of view, but we always need something physical to represent information (to write it down or pass it on). The interpretation of information may only be imagined, but its existence is real, so in a sense information lies not only the distinction between something and nothing, it is also the bridge between reality and imagination.

We can label every observable and predictable process in the universe by a trajectory from some location, extending over time, and in this way we see that there is at least a version of the world in which everything is equivalent to a form of information. We can describe the universe of the big bang in terms of information too. Several authors have written about this, and it remains a popular topic amongst theoretical physicists. But this is not the story I want to tell in this book. Instead, I want to point out some lessons we learn from actually building computers. Computers have spacetime properties of their own. To compare the world of information technology with the world of our more familiar universe offers a new perspective on both.

Let's begin by trying to draw some lines around what the world of computers and Internet means. As an almost limitless canvas for the imagination, its scope is huge. Computers allow us to challenge our ideas about what reality actually is, through the way information and measurement are engineered. To make sense of space and time, in a computer system, we have to look upon them somewhat differently—in a more elementary, and yet more sophisticated way. We can create imaginary worlds in computer games and on the big screen. Computer generated graphics (CGI) has given life to backdrops and characters that could never otherwise be brought to the screen, and we soak up these experiences as if they were every bit as normal as the physical realm.

There are other information-oriented phenomena, with spatiotemporal characteristics, that we could look to as well. Biology, ecology, even cognitive systems all qualify as possibilities. Would those interacting meshes of space-filling stuff also warrant the term 'spacetime'? We know that organisms, ecosystems, and certainly brains are networks of substantially similar parts, like points in a space. They change on many levels. They may appear much more structurally diverse than the more familiar spacetime of our astronomical universe, exhibiting broad and qualitative differences at different scales. They have differentiated regions that engage in complex interactions, with some parts fixed and other parts changing relative to them. All these characteristics smell of spacetime, just of a more colourful kind.

Is it helpful to think of these radically different phenomena as spacetime too? I believe that there are good reasons to see where this idea takes us. For one thing, it can challenge us to reexamine the things we take for granted, without having to delve into more exotic and speculative worlds of Quantum Mechanics, String Theory, or any of the other wellsprings of modern theory. Information

3 ANTI-SPACE: INFORMATION

sciences describe issues that are often swept aside by those exotic theories, but challenge us with new representations. Organic tissues are uniform and homogeneous collections of cellular points. Biology and computer systems both deal routinely in boundaries, discontinuities, multiple coexisting phases, different kinds of motion—and, of course, a far greater range of semantics. This places them at the apex of the complexity ladder.

Suppose we rank the phenomena according to their semantic complexity, most complex first:

1. Biology
2. Computing Software
3. Computing Hardware
4. Geology
5. Particle physics[62]
6. Astronomy

This is not a very scientific attempt at ranking, but it does capture the heuristic levels of intricacy involved in their phenomena (at least according to my rough thumb measure). Notice that physical scale, relative to the embedding in our universe, is quite unrelated to the ranking of complexity. Astronomy occurs over the largest scale, where there is only gravity that we know of to shape interactions. Particle physics happens on the very smallest scale, where size could well limit what complexity might occur. Then, the most semantically complex phenomena occur in the 'Goldilocks zone' of planetary conditions, where complex chemistry and life occurs. Here there are more opportunities for complex interactions, because these are the scales over which matter itself forms coherent structures—but can still lead us to possible spacetime interpretations. There is also a sense in which all of these are information systems.

To avoid getting confused by all the spacetimes we might choose to consider, let's define the scope and players of the world of Information Technology (IT), and call it the 'dataverse'[63]. In fact, this is not one thing, but a class of artificial systems where the same basic ideas of space and time exist, but with slightly different properties. By examining how they are similar and how they are different, we can learn a lot about the nature of space and time in general—perhaps even see how some of our traditional ideas are unnatural and inflexible for application to the modern world.

THE DATAVERSE

If the Internet is a spacetime, then it must have locations and it must have clocks that both generate and measure time. Let's think about where these are. The locations of the dataverse are computers and network devices, including all the mobile devices like smartphones. There is no smooth idea of a never ending road in the dataverse. Getting from place to place is a bit like hopping from stone to stone to cross a river. Information is transmitted computer by computer along wires that don't go in just any old direction. Not every computer is connected to every other, and they are not all connected to the same numbers of neighbours. The world is much more limited for data than for a fish in a three dimensional sea. There might be multiple paths or routes from location to location, but there is not an infinite variety of courses on a dataverse compass.

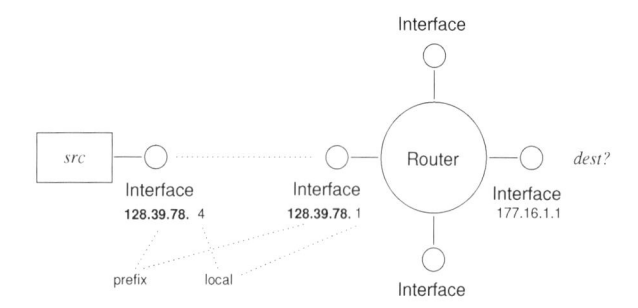

Fig. 3.1. Agents that play the role of router promise to forward communications between endpoint agents. The analogy between endpoint agents and routers in conventional spacetime would be boundary and space. On the Internet, most nodes live on the boundary of Internet spacetime today. The routers are the only locations where motion does not result in a dead end..

Because of this limited wiring, there are several levels at which we can define adjacency in the dataverse. A computer can be said to be 'next to' another computer if there is a cable connecting them, and both sides are listening (see figure 3.1). This is ambiguous though. By chaining computers together, we can create tunnels or virtual network links that make a communications channel between two computer programs that rely on but don't refer directly to a number of intermediate hops in a network. So the combination of links is equivalent to

3 ANTI-SPACE: INFORMATION

a single link on a larger scale, assuming everything works without question.

In science, it is nearly always assumed that the world works flawlessly according to 'laws of nature' all of the time. We know that computers can break down, but we never consider whether points in space can break down, preventing motion from one location to another. We might need to reconsider this, as we learn more about the microcosm.

At each location in the dataverse, i.e. each computer, there is a number of neighbouring directions we can get to that are defined by available network routes. What's even more interesting about this dataverse is that nearest-neighbour relationships are also not all fixed. Smartphones, laptops, and other mobile devices are in a state of changing connectivity all the time. So not only is the dimension of space potentially different at every point, but it is changing from time to time. We simply have no idea whether this is similar to the physical spacetime of our universe at the smallest scales, or whether this is particular to computers, but it is certainly an interesting issue to think about. What might the consequences be?

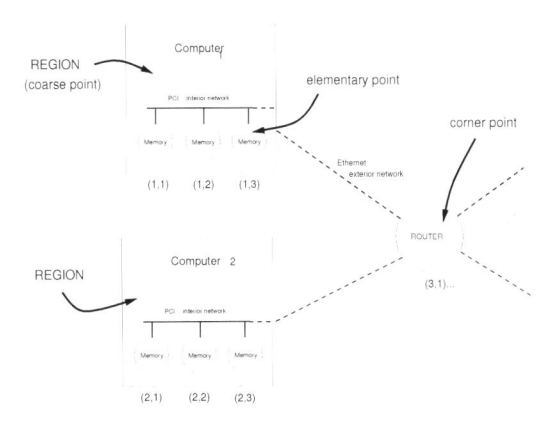

Fig. 3.2. The point locations form a coordinate space (computer,memory), which are connected by different network dimensions. The interior PCI network connects memory, within the same computer. The exterior Ethernet network connects computers and routers (which are also computers). This is not a Euclidean space, because we can't go directly from memory to memory in a different computer, without going through a special point, which is the network interface card (NIC) of the computer..

It will take some readers a significant effort of will to suppress an aversion to this abstract way of thinking: surely computers are not space, they are matter and exist within space! In fact, this is exactly the leap of faith for which I ask your forbearance. We need to make a separation between what we mean by space and the somewhat narrow view we've all been taught. The reward for letting go of familiar prejudices will be an expanded world view, and a powerful new way of thinking about the world. So, please accept my hand, and hold on tight.

A novel characteristic of the dataverse is that there are different kinds of adjacency—different ways for things to be next to one another. This is a richer world than our traditional idea of space so it is already teaching us something that could be new. There are both different flavours of points, and different kinds of paths between them. This sounds a bit like what happens when there are multiple dimensions of space, connecting the same set of points, along different directions, but in this case the dimensions are not symmetrical under rotations in the way we are used to.

In computing terms, these different adjacency types are channels representing different services. Some of these are familiar from the Internet, e.g. HTTP, FTP, RPC, and so on. For all we know, there might be different notions of adjacency in the universe too, beyond the three familiar dimensions, but we simply don't know. For instance, the structure of spacetime with its light signals and its light speed limit is one set of channel types, but there is also a somewhat mysterious channel related to quantum entanglement that no one truly understands[64].

At each dataverse location, changes occur with an internal time generated by the computers' CPU 'clocks'. The rate of a CPU clock (relative to our standard exterior clock time) is the frequency quoted on computer specifications when you buy it, e.g. a 4GHz processor ticks 4 billion times for each second of human time. Changes in software can happen at this rate or slower, and each software program will measure change at its own rate, slower than this.

Because it is a closed spacetime that we have constructed ourselves, the dataverse is obviously embedded within our universe. The location of a computer on planet Earth does occasionally matter. For example, whenever we watch a movie or television broadcast over a digital channel, there is some geo-location service that is able to identify which country the watcher and the sender are located in. This is called geo-identification. Content providers usually restrict the showing of certain content by country due to licensing restrictions. Some governments also use this to track the physical and virtual borders of their countries and citizens. But, as far as the rules of change are concerned, anything going on inside the dataverse knows nothing of the outside world. A software system can speculate about any extra dimensional places, but what happens inside the dataverse can be described independently, as a self-contained system. The planet Earth locations are just data, that could just as easily represent user language settings. In fact, the dataverse is mainly embedded within the two dimensional atmosphere on the surface of the Earth, which itself is embedded within the larger universe. That might sound like a lot of unnecessary nitpicking, but sometimes these details matter to us, as we'll see.

The story doesn't end here. Within the dataverse, and all its computing devices, there are more private universes we can imagine (see figure 3.3). Inside physical machines, operating systems can simulate virtual machines and networks too, using software to close off multiple independent virtual-dataverses, inside the dataverse, inside the Earth surface, inside the universe. It might not be turtles all the way down, but there are plenty of independent arenas in which phenomena find themselves contained. One of the reasons to point this out is to underline the fact that it is the degrees of freedom experienced by particular processes that really defines what its spacetime limits. The idea of one single universe, which contains everything, turns out to be superficial, and there are many cases of phenomena that only exist in surface layers and private networks.

Many readers will have heard about the notion of multiple universes popularized in connection with Quantum Mechanics, which has been a source of particular excitement for science fiction writers, and has been caricatured in many dramas. The universes we are discussing here are not like the 'Many Worlds' discussed by philosophers like Saul Kripke (1940-) as ways of reconciling different points of view, nor are they like the many worlds discussed by theoretical physicists like Hugh Everett (1930-1982) to reconcile the discontinuities of quantum mechanics, and they are not like the multiverses of M theory, and so on. They are not alternative versions of one another, although such concepts do also exist in software version control systems. Embedded spaces are simply regions in which there are processes that are entirely contained within the limits of that region. So, whether that region is embedded or not really doesn't matter, because what happens within it has basically nothing to do with what happens outside it. The reverse might not be the case. If we can look down on a space, like Godlike observers, we can perhaps see into it and make use of it, as we do with communications networks, from the telegraph to the modern Internet.

The dataverse also has its own embedded universes then: software virtuverses, or virtual spacetimes, known more for software than for virtue! There is a hierarchy of embeddings, but for the most part these universes are independent of one another. Once we are inside one of them, the phenomena within are self-consistent.

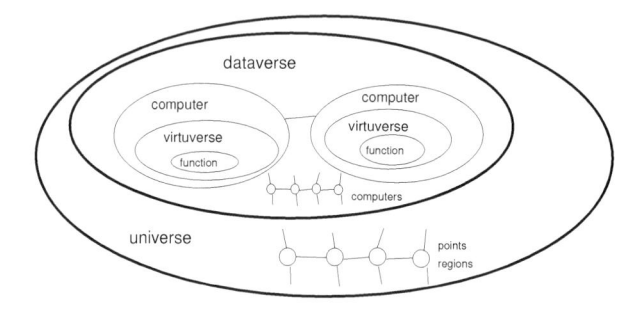

Fig. 3.3. The dataverse has point locations which are memory bits. Just as points scale into regions, so bits scale into words and registers and strings and all the other aggregations of bits that we recognize. Just as the material of the universe is space, so the material of the dataverse is memory..

Time passes in the dataverse in the same way that it passes in our universe: changes occur. In the universe, no one knows why changes occur, but we can sometimes measure them, if they are local. In the dataverse, time is driven by pulses that usually come from a pulse generator within the CPUs of active agents. Later, I'll explain why it must also be true in an information theoretical sense that universe time is generated from within each of its points, independently, too. Just as we don't necessarily know why changes occur in the universe, we don't necessarily know why changes occur in computers either. This is remarkable, because computers are machines that we built. Surely, we ought to know how they work?

In fact, we cannot predict when changes will occur for all kinds of reasons to do with the way processes interact, evolve, and measure time independently. Sometimes change arrives from exterior sources by the motion of information along network channels. The same is true in the universe, except that we have no idea what pulse generator drives time from within a local region of space. The fact that we don't know where it comes from is part of the reason we abstract it as a kind of inevitable force, a so-called 'river of time'.

Time passes and things move around the dataverse and its virtual subparts, in both hardware and software. Signals and messages are sent from computer to computer, from hard disk to hard disk, and processor to processor. Similarly, inside the virtuverse, data are sent through 'pipes' and function calls from variable to variable. All containers and locations are virtual abstractions here, and the meaning of distance is like other networks: the number of hops we have to make from the sender to the receiver (see figure 3.4).

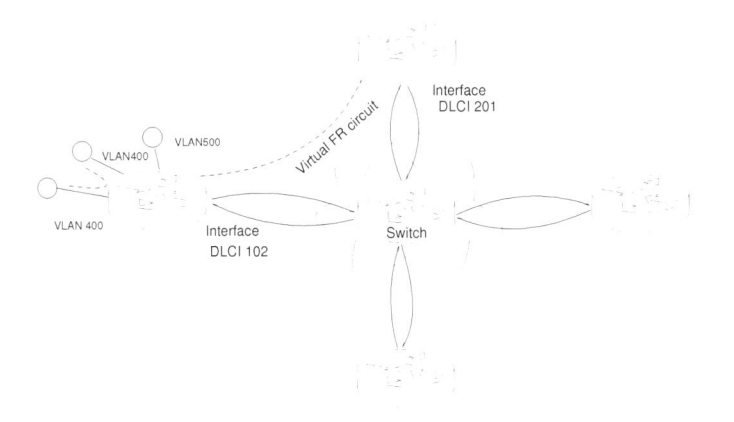

Fig. 3.4. Trajectories that correspond to motion of data from location to location pass through intermediate points, just as they do in a Euclidean space. The difference is that the journeys don't appear smooth and continuous; they are discrete hops.

The structures in the software virtuverse are frequently designed to represent or mimic processes that happen in the real world, because they are made with the purpose of helping us to track them. This means that similar (but not necessarily identical) spacetime structures are found in both, whether they are natural or not (usually they are not). A few of these correspondences are shown in the table below.

UNIVERSE	DATAVERSE
POINTS	REGISTERS / DATA WORDS
SPACE	MEMORY UNCLAIMED
PARTICLES/FIELDS	STRUCTURED MEMORY
DIMENSIONS	TYPES, INDEPENDENT SEMANTICS
LOCATION	ADDRESS
COORDINATE PATCH	ADDRESS SPACE
BOUNDED REGION	URI, NAMESPACE
	CONNECTED REGION
LOCAL TIME	PROCESS CHANGE
MATERIAL MOTION	DATA TRANSACTION
CONSERVATION	MUTUAL EXCLUSION

Scaling in the dataverse

Let's make a lightning review of some of the history of computers, to learn more about how their processes work. The origin of the machinery goes back to the Jacquard looms, and self-playing pianos, of the industrial revolution. They were the beginning of programmable automation. Weaving-patterns and music were input on punched rolls of paper, as you may have seen in old Western movies, each hole representing a note. Modern music software still uses the idea of a 'piano-roll' as a timeline, to control the playing of electronic instruments. As mathematicians took over the description of computing, Alan Turing (1912-1954) imagined computing devices as a mechanism that could read and edit a single linear tape—an idealization of the piano-roll.

The first modern computers began as singular 'monolithic' machines, part mechanical and part electrical. The analytical engine—Charles Babbage (1791-1871) and Ada Lovelace's (1815-1852) mechanical computer—was one of the first if we discount the simple abacus. Vannevar Bush (1890-1974), and later his protege Claude Shannon (1916-2001), pioneered analogue computation, using physical processes to mimic calculations. Electrical analogue computers, meant to mimic differential computations using the theory of electrical circuits, allowed computations to be made as electrical experiments. The first digital computers used vacuum tubes mounted on even larger frames, whose 'main frame' was used to mount or hang the makeshift components in some kind of ordered circuitry. The early technology made these computers large, cumbersome, and quite unreproducible, built from custom manufactured components.

Each component or 'agent' became effectively a separate location in a network, embedded in a room, and connected together by wires. Electrical signals travelled along the wires and the sum effect throughout the total 'superagent' was to promise to transform inputs into outputs. Locations signalled other locations, and the states of the component agents changed, marking out time. At first, time was driven by the inputting of data itself: from the piano-roll of instructions, and then by the changes predicted from computational rules. Each value pushed through the circuit was the tick of a clock. Then later, independent clocks would sample repeatedly, wasting ticks of time in between active processing.

After the invention of transistors, computers got a lot smaller, and could be assembled on much smaller circuit boards. Computers' interior network spacetimes shrank, according to the measures of their embedding space, but did not change fundamentally. They were still like molecules formed from disparate atomic components. It was the invention of chips, or very large scale integration that allowed new techniques to print transistors onto silicon wafers, or chips, that made computer spacetime much more regular and homogeneous. Suddenly, computers looked a lot like coordinate systems instead of wireframe models.

Towards the end of the 20th century, network connections began to extend beyond a single computer, to connect several computers together. At this stage, there were clearly distinguishable entities called 'computer' and connections between them. They could dial each other up to 'join each other' in information space and share some data, then they would move apart again, like independent particles in a gas (figure 3.1). Finally, as the Internet and computers became commoditized, we have built a fixed solid state network[65] in which computers are permanently hooked up to one another, and data are continually flowing from place to place in large and rich spacetime of physical and virtual places.

Today, those ultra-miniaturized chips are stacked on their motherboards into racks, on multiple stories, filling buildings with regular arrays of computing locations that can represent regular arrays of ones and zeroes. The story of computing is one about the scaling of patterns, but it is also a story about steadily increasing regularity. Gradually, we are filling every inch of human space with computational processes.

COMPUTATIONAL CHEMISTRY

So how 'big' are the points of the dataverse? If that sounds like a strange question, compare it to the corresponding question in chemistry. Where are the locations in a chemical process? How big are they?

In a chemical reaction, the active locations are atoms and molecules, and mixtures that span a range of sizes. At different sizes or scales, there are different phenomena. It is not as simple as labelling molecules with a position (x, y, z). In fact those coordinates mean rather little in the context of chemistry (see figure 3.5).

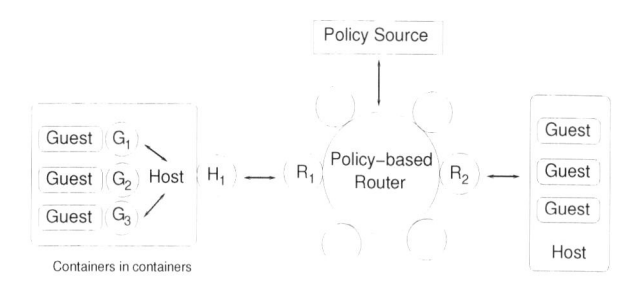

Fig. 3.5. The interior structure of the points at the edges of the network is somewhat analogous to the interior structure of atoms. Computers have a different kind of network on their interiors (such as the PCI bus). At the scale of a computer 'atom' all that interior structure is interior to the principal spatial location.

Due to the nature of process, it's not natural to describe individual molecules as we would describe spacetime, using coordinates (x, y, z). We are more interested in what kind of relationships and encounters they have with other points, i.e. what kind of connections or bonds they form in order to join up spacetime locally as a cluster of points—and for how long.

At every scale, new *types* interactions become possible (scales that can make new kinds of promises), and a new phase of space can emerge. Each individual entity that can make such a promise, or link to a neighbour by matching something can become a new effective agent or entity, a new location at a larger scale. A molecule is like one of Milner's bigraphs, not like a featureless point in Euclidean space. It has interfaces that are more elaborate and depend on a matching context. It has interior properties, and it can link up with neighbours according to certain matching rules. In other words, it can form a kind of chemical bond, and mount a new turtle on the pile.

Computers have their own kind of chemistry. In computer science it's is called *Type Theory*. Instead of atoms and molecules, that make certain promises about electron valencies and affinities, we have software components that make promises to bind and form new molecular promises, usually labelled as different kinds of data, or *data types*. In a similar way, the scales over which these promises apply vary quite a lot. Sometimes they apply to a single region of memory inside a computer, sometimes to a whole computer, and sometimes across a network of computers. Let's take a brief look at what a computer spacetime looks like, both from the perspective of extra-dimensional beings living in the interstitial embedding space[66], and from that of interior beings (i.e. software agents).

EXPANSE

The aspect of space, which is most familiar to us, is the sense of expanse that comes from a freedom to choose our location. We might understand that space is made technically from individual places that act independently, and even that it can be broken down into atomic 'points', but we are less used to thinking of space as regions of isolated machinery than with the vistas of wide open choice.

In material physics, there is a stage in between these two extremes, where individual atoms arrange themselves in an orderly way to become materials. Crystalline materials, like metals or carbon, can form regular arrays of points that are quite uniform, but which are not a smooth theoretical continuum. In the same way, the growth of the Internet has led to clusters of a few computers in a network growing into huge regular crystals of computational infrastructure in datacentres (see figure 3.6). Datacentres look quite like crystalline solids: regular racks of basically indistinguishable computers stacked on top of one another. As a schematic diagram, figure 3.6 shows a tiny region of a datacentre network pattern: the circular locations at the bottom are computers, and their connective links are shown extending into more dimensions through intermediate points in the layers above. You could easily lose your way in the embedding space of the datacentre, and it is basically impossible to know what is going on within the apparently endless three dimensionless grid of computers and wiring.

Arranging computers in regular arrays is more efficient than linking them up in ad hoc clusters, not least because it is easier to assign coordinates to them. The so-called 'translational invariance' familiar in Euclidean space, which expresses the irrelevance of choosing one location from the next, applies on this scale to the dataverse too. We begin to see the real reason for telling the story of

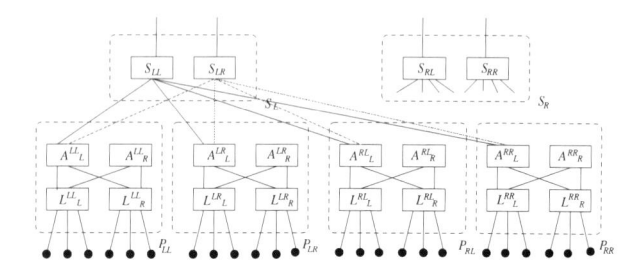

Fig. 3.6. A so-called 2x2 Clos network is an example of the kind of redundant pathway structure now being used to connect computers in more than one dimension in larger mission-critical datacentres, for resilience and efficiency. The natural embedding dimension may be seen by reorganizing the geometry (see figure 3.7).

Descartes' version of Euclidean space, with its rectangular coordinates. We see it in every building, every supermarket, in every warehouse, and now in every datacentre: coordinates make it easy to locate a specific place, when all points appear indistinguishable. Superficially, the rectangular symmetry of the arrangement (when we deliberately renormalize away their interior properties or contents) makes it possible to come up with a standardized addressing scheme based on a regular array of numbers, but it is precisely the lack of symmetry at the locations that makes this useful. If all the points were the same, we wouldn't ever need to find anything. It is the *breaking* of these spatial symmetries that leads to all the important functional characteristic of spacetime: to information—the antithesis of empty space.

In fact, the rectangular geometry of Euclidean space is not natural for computers, because clusters fan out in a radial structure (see the bottom rows of figure 3.6). This unnatural geometric representation leads to wires literally having to be crossed in order to connect systems together on a large scale, with two dimensional multipath redundancy. However, by understanding spacetime geometry more properly, we could easily solve this problem in the future, reducing costs enormously[67]. If we pay more attention to the natural symmetries of a space, then we can use the radial symmetry to reorganize space like the geometry in figure 3.7.

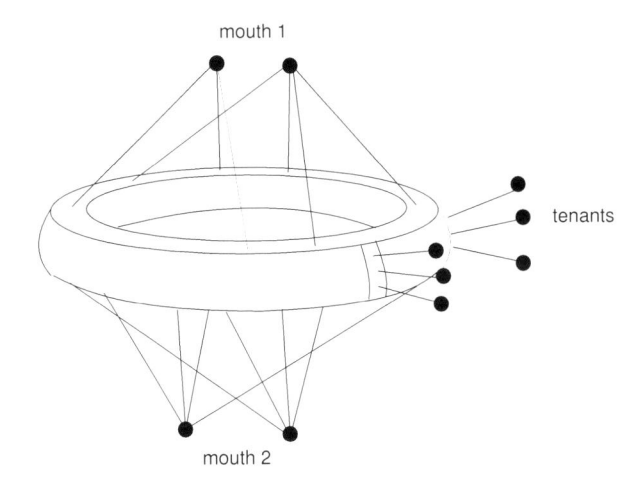

Fig. 3.7. The Clos network looks messy and counter-geometrical in a square geometry, yet is used in datacentres. With a little spacetime insight it can be unfolded into a radial geometry, as part of a three dimensional embedding space. More practically, this understanding enables line-of-sight connections to be implemented in place of folded cables, by direct fiber-free laser optics.

Seeing through a network, darkly

The way the dataverse is connected is not much like the way Euclidean space is connected. For a start, there are no distinct locations in Euclidean space, and there are no wires in between them. This is because Euclidean space is not embedded inside another space. It fills all of what we think of as space already. This is only a detail though. We could imagine Euclidean space as being a three dimensional lattice of cubes stacked closely together, like a pile of bricks. The wires between them could be very short (vanishingly short).

Wires allow us to distort what 'being next to' something else means, according to our prejudices about Euclidean space—but it is useful not to muddle these concepts. Seeing explicit channels of communication can bring a clarity to processes, as we'll see. On a physical level, two computers are only adjacent if there is a wire between them (see figure 3.1), but—on a software level—they are only next to one another if they are communicating, using the Internet message protocol. This helps us to appreciate the limitations of direction with an explicit reminder that something has to be next to a location in order for a direction to exist.

In the dataverse, you can't normally just keep going in the same direction as you can in Euclidean space, because every new place is a randomly oriented crossroads of pointers to new places whose direction has no calibrated meaning. On the scale of individual points, computer networks are not much like Euclidean space, but it remains to be seen if they can look more like Euclidean space on much larger scales. Another kind of space, which looks like something in between a computer network and a closely packed Euclidean pile of bricks is a tissue of cells. If there are enough microscopic capillaries side by side, they can approximate the flow of a vein or artery; if there are enough side roads snaking around residential housing running in parallel, they can approximate a motorway over a large enough region.

The normal geometry of a computer network is a hierarchy of starlike patterns, as shown in figure 2.18, with a switching junction (like a crossroads) at the centre, surrounded by satellite locations. But, an alternative is to connect all the locations to their nearest neighbours, like the buildings in Venice or Hong Kong, creating a route that passes through intermediate points rather than specialized junctions. Such a structure then looks like a cellular tissue in biology, or a crystalline material in solid state lattice.

A lattice structure is also implementable within the dataverse too. The Clos network structure in figure 3.6 has more or less this structure. The mathematician John von Neumann (1903-1957) imagined a model of biology, using computers, called a cellular automaton. This has since been investigated and popularized most prolifically by Stephen Wolfram[68] (see figure 3.8). The pattern looks a lot like the networks of electronic components within integrated circuits used in computer CPUs. Connecting a data centre of computers, in the same way, is an idea that has has been considered and championed by my friend Paul Borrill[69] as a rational way of redesigning computer networks with more predictability, and a stronger notion of spacetime stability. These are not mere fanciful tales of fortuitous analogy—they have very real and on-going implications in the

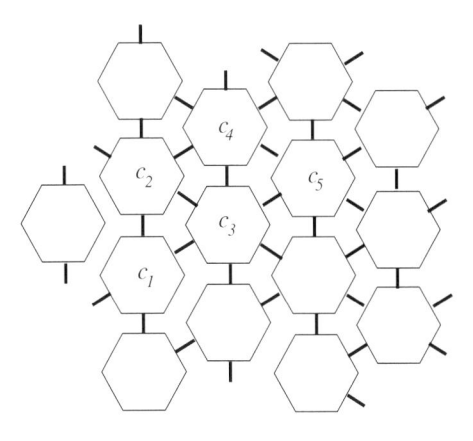

Fig. 3.8.   A cellular latticework space.

shadowy world of information technology.

DATABASES

Let's leave behind the world of physical hardware for a while and think about one of the best known uses for computers in the modern world: the orderly storage of information in an addressable structure, otherwise known as the database. If you are a computer scientist, or perhaps a software developer, then at first glance the interior of almost any space looks like a database: materials, molecules, genes, books, buildings, shops, malls, farms, cities, countries, planets, and so on. Everything is composed of certain regions of space (or memory records), in which certain properties are represented 'promised', and can be thought of as offering information about those locations. The world is essentially a giant archive of information, represented out of the great variety of phenomena all around. The fact that apparently virtual information might be recorded as real things, is no more impressive than the fact that a museum that presents information as actual artifacts, or that a hard-disk that is made of real atoms.

We already remarked that a space, in which the information states can change, is what allows us to create memory in the first place, so this is no accidental similarity. The entire world around us is something that we read from and write to—we use it and we alter it. Even if we restrict attention to IT infrastructure, our networks, PCs, and mobile devices, we are all living users of a massively distributed data system. Space is memory.

A database is a kind of spacetime, then. It has locations, all of which can exist in a number of different states, and there is change. After the previous chapters, you are probably already asking: how many dimensions does it have? Is it continuous or discrete? What is its topology? The Internet itself is a kind of database, consisting of very many locations (computers), each of which has its own purpose, its own services, its own data, and all connected together into a navigable structure from which they can be retrieved. In a database view of space:

- Space is made from tabular records, which play the role of the spatial locations. The different rows in the database tables correspond to different scalar properties and other dimensions of space. If they are empty, they are space. If they contain data, we could call these matter or particles.

- The homogeneity of space, or its underlying *translation invariance*, is called 'database normalization' in database parlance[70]. But this property, like the translation invariance of Euclidean space, is only really true before we put information into the space. Anything we put in breaks the precise symmetry between the points. The templates for the locations may be the same, but the locations are distinguishably different. Indeed, that's the only way space can work—locations have to be distinguishably different.

- Exterior time, for a database, is what happens when an external entity writes to the database. A countable change occurs. This appears to the database as a random exterior process.

- Interior time is what a database does as part of its interior processing, including sampling those exterior events, accepting requests to allowing others to read from it, and accepting and storing the incoming data. It is a countable change on the interior. As usual, interior time limits what the database can sample and absorb from its exterior.

- A search or query to the database is a spacetime process too, something analogous to a measurement probe of the database's interior states. From a spacetime information point of view, this looks a bit like a quantum mechanical measurement. The probe has to be accepted, and it may or may not promise to respond in the way the query hopes.

- Queries may cause copies of recorded data to be repackaged, from one interior representation into an exterior one, and then be transported from location to location before being transferred to another interior space. So a data transfer looks a bit like an interaction between different entities in the natural world, e.g. different quantum fields in quantum theory, different radicals in chemistry, or different protein packages in biology.

- In databases parlance, the structured patterns of things that belong together are called a *schema*. In physics, we might call this a symmetry group, in biology we might call it a cell tissue type[71].

Information starts out as one pattern, written on a region of space, which then moves through space as a second representation, and ends up in another region. This is the story of every process in the universe. For example: we might say that energy starts out as an excited state in an atom, is emitted as a photon, and scatters off a stray electron—or that an antibody starts off as a B-cell which is transported around the bloodstream, binds to antigen on infected cells and is then absorbed by macrophages. We change the words, but the character of the processes is basically the same.

In Information Technology, the practical matter of choosing which of many different flavours of database to use for a problem depends, to a large extent, on what kind of spacetime processes are on-going in the data that it's designed to keep. The way the boundary conditions of space and time are defined, from data catchment area, to whether data arrive as a steady stream of events or in bursts relative to the database service's scheduling, or to the role that time plays in addressing and later retrieval—i.e. do equivalent results overwrite one another and converge on a single answer for all times, or do we keep the full history of versions for each labelled value as a *time series*, like an Electrocardiogram (ECG), or seismic trace? Some databases keep multiple copies of data in different spatial locations, by trying to equalize the data on a number of copies, as a redundant superagent. Such backups are common practice for disaster recovery. Since the process of equilibration takes time, on the interior of a superagent cluster, exterior time (what exterior agents can observe) is halted during these equilibrations[72].

Spacetime adapts at different scales and in different environments. In a sense, it has become 'smart' or functional in that environment. The generality of the concepts is more than enough reason to think quite carefully about what all these common themes mean on a broader canvas.

HOW MANY DIMENSIONS ARE THERE TO INFORMATION?

Those of us who have acquired a passing familiarity with computer hardware over those years in which home computers rose from nothing to ubiquity can be forgiven for tending to leap to a premature conclusion, and thinking of a database as something that exists on the surface of a hard disk (or memory chip, or floppy disc, if you are old enough). In this case, you are thinking about the representation of the database on a physical spacetime, i.e. the embedding of the abstract space of records onto the physical space of the disk surface. Today, the location of data is much harder to pin down 'in the cloud'. Alternatively, you might begin to draw the tables of information and their relationships, in a purely abstract way as a picture, in two or three dimensions (thus embedding it in another arbitrary physical representation).

What dimensionality can we attribute to a database? As with all processes, there is no unique answer. There are both physical and logical dimensions we can attribute. Turing made all computer memory one dimensional, as dots and dashes on a kind of telegraph tape. The interior encoding of data on a disk may be thought of as one, two, or three dimensional, depending on how we look at it. Each track is a serial circular stream, one dimensional. These are arranged in rings, making it two dimensions for some processes.

The encoding in chip memory may also be considered three dimensional: there is the array of chips, each of which contains two dimensional transistor arrays, a bit like a supermarket layout, but there are multiple chips, like different supermarket addresses. The addressing in the computer is only one dimensional, as far as the CPU is concerned—computer addresses start from 0 and count up to the maximum amount of 'simultaneously available' memory in bytes. The addresses coordinatize a 'spacelike hypersurface' in relativity-speak. So in physical space, each memory surface promises an approximately two dimensional available memory (+), but the CPU sees it (-) as only one dimensional slices.

Going even further, during the 1980s, when computer memory was limited by the 8 bit size of the address registers, one could only address 64k bytes of RAM linearly at a time, so manufacturers designed so-called 'sideways memory' or paged memory—again, like splitting up a supermarket chain into separate chains handling different goods. By introducing a second coordinate

3 ANTI-SPACE: INFORMATION

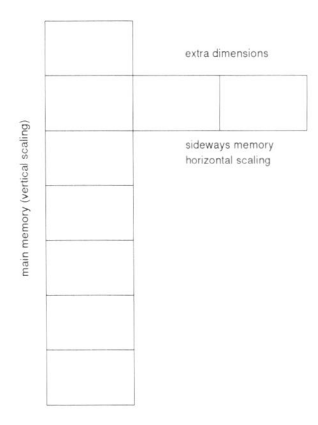

Fig. 3.9. The exception to containment is when containment cannot scale locally (known as vertical scaling), and becomes too restrictive. Then we may scale space 'sideways', introducing a new dimension to space (called horizontal scaling). In the early days of computers this was an important technique as processors could not address more the 64kbytes at a time.

for memory addresses, as parallel data banks that could be switched in and out, almost like tape reels, a second coordinate came into play: the data bank or page number (see figure 3.9):

address → (address, bank/page)

Multi-dimensional addressing in computer memory can be represented by introducing new coordinate tuples, just as in Cartesian coordinates. What is the geometry of space in the dataverse? As in our physical universe, the natural dimension and geometry depend a lot on both interior and exterior properties. These practical technological examples show how dimensionality can be much more complicated than we learned in geometry. Indeed, the success of the Euclidean-Cartesian geometric framework, in a mathematical sense, is very much based on its simplicity and re-usability; but, this is perhaps what many find it hard to swallow in school. I'm willing to bet that everyone who struggled with geometry according to Pythagoras, Euclid, and Descartes has no trouble accepting the much more complicated story of geometry in a supermarket. If you walk a one dimensional path around a two dimensional layout, in multiple stores, without getting lost, then you probably understand everything there is to know[73].

CARTESIAN SQUARE ARRAYS

Descartes' coordinates continue to crop up all over the place in descriptions of spacetime. In information systems from phone books to account ledgers to computer programs, data are frequently arranged in tables, in data processing, because tables have a simple associative structure that makes them easy to read. Rows and columns align with a one to one correspondence. The square geometry of Descartes plays an important role in information technology for scanning through such arrays of data, just as it helps us to navigate the aisles in supermarkets by numbering and indexing. Arrays or matrices are tables of numbers, stacked into related bundles, like the example below:

$$\begin{pmatrix} 1 & 0 & 2 & 4 \\ 3 & 7 & 1 & 4 \\ 1 & 9 & 2 & 4 \\ 8 & 0 & 2 & 4 \end{pmatrix}, \begin{pmatrix} \text{Flight} & 10224 \\ \text{Taxi} & 65 \\ \text{Hotel} & 500 \\ \text{Food} & 77 \end{pmatrix}$$

The second example has the exact structure of an index, which is nothing more than an ordered one-dimensional spatial representation based on numerical page number coordinates.

The analogue of (x, y, z) Cartesian square coordinates is uniquely suitable for traversing tables of data in any situation, and the simple translational symmetry of any grid arrangement is close to Euclidean space, which gives it a kind of universal significance[74]. The main difference is that the coordinate values can't take on arbitrary locations, only discrete integral values. In spite of Harry Potter's famous Hogwarts platform $9\frac{3}{4}$, there is nothing in between aisle 9 and 10 in the supermarket.

The use of coordinates to calculate data, located in a space, whether it is physical, virtual, real or fictitious, leads to something of an interpretational conundrum about the nature of reality. If one invents a space of numbers associated with something the looks like a spacetime, can we say that the spacetime is real, or just a calculational artifact? This is an issue physics faces on a daily basis. When subatomic processes are calculated in quantum field theory, for example, the causal processes embodied in the interactions is wrapped in the invariances of spacetime. This sets up calculations called partition functions, well known from statistical mechanics as computational tricks to aid calculation[75]. But these calculations look superficially geometrical, which led physicist Richard Feynman (1918-1988) to draw them as diagrams.

One could try to claim that, if the indices need to sum these virtual locations, then there must be something there to sum. In a computer representation this would be true, since the mapping is one to one. In spacetime, however, we just don't know the deeper structure, so this is an assumption. It turns out that the assumption is not a terribly good one either, since the limit of summing all these points is usually infinite. Keeping the points discrete and independent cures this ultraviolet divergence problem[76]. It's a strong suggestion that spacetime is really not such an infinite resolution continuum.

NAMESPACES, SCOPE, AND RINGS AROUND THE TARGET

When we need to think about the functional roles played by different regions of spacetime, coordinates take on a different focus. Descartes' coordinates were designed to reflect translation invariance, or the uniformity of space. But functional roles take the opposite view: they are about anchoring properties to an absolute location. In that case an observer-centric view, which defines a region of space and a boundary between interior experience and exterior environment turns out to be the important way of labelling. This process of defining boundaries around specialized regions is called *encapsulation* and each boundary defines a *scope* in computer science parlance. An awful lot of what happens in information processing concerns the management of these scopes or bounded regions. The memory coordinates drawn in figure 3.2 form a simple namespace, sometimes called a Tuple Space, because the coordinates (computer,memory) form doubles, triples, quadruples, and n-tuples in general. We could extend this to separate computers into different named groups.

An obvious example that we all have to deal with in the modern world is the use of 'folders' within filing systems—whether paper folders or computer folders that contain files. Folders (also called directories and catalogues in some computer systems) are examples of containers or encapsulations. Each folder is usually given a name. Objects inside are then effectively tagged with the name of their container, indicating that they belong to a particular category of information, or a particular folder in the space. This is one reason we put things into boxes.

Names are the semantic equivalent of coordinates—or, if you prefer, coordinates are just a naming for a place that uses numbers. There is nothing directly analogous to containment in ordinary Euclidean spacetime, though Einstein's use of Riemannian geometry has something called coordinate patches. Patchiness is not a property that is emphasized in simple, flat, uniform, Euclidean geometry—where lines, planes, and trajectories are assumed to go on forever—but, it's not hard to imagine how it can work. We just need to keep separate coordinates in separate patches or catalogues, like different maps or even different phone books. For example, we could measure latitude and longitude and altitude relative to the Earth's surface, up to a maximum height, and we could do the same on Mars. Completely different places would end up with the same name, but one would refer to namespace Earth, and the other would refer to namespace Mars. As long as labels are unique within their own 'namespace' or coordinate patch, there is no ambiguity.

Nearly every American city borrows its name from a city somewhere in the 'old world' (York, Oslo, Memphis, San Jose, etc). Many states use the same names. In Northern California nearly every town has the same street names: Bryant, Castro, Hamilton, Embarcadero, etc. The only way to know how to locate a particular street is to label it with the name of the town or region it lies within (whose name then has to be unique within its own container). Namespaces thus form nested structures, like Russian dolls. Street names, town names, county names, country names, planet names, and so on. This encapsulation structure is like the section numbering in book volumes, chapters, and their subsections, which we'll come back to in discussing the *linguaverse* in chapter 8.

Each subdivision of a namespace homes in on a specific region. A complete path looks like a filename on a computer, or an address on the World Wide Web:

```
towns:USA/California/PaloAlto/Bryant
```

This *semantic coordinate system* is called a Universal Resource Indicator (URI). When information is processed inside a computer, using modern programming languages, programmers use variables or names that stand for data values. They are structured in this way, and the names of *local variables* are only visible within their containers. the names of *global variables* are visible anywhere as long as you have the URI to reach them. The URI forms a tree structure, with each level of subdivision acting as a branching process, called a spanning tree (see figure 3.10).

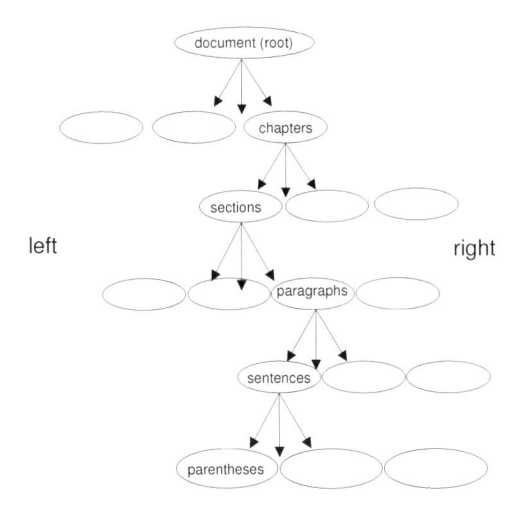

Fig. 3.10. Spanning tree and data hierarchy. Ordering may be preserved by promising a policy of depth-first, left-first recursion.

Imagine a spacetime whose smallest addressable locations were made up of lights, heaters, sensors, and other controllable devices in smart buildings. At the time of writing, smart buildings are those where the building can be monitored and controlled by information technology. The structure of the URI for this namespace tells us what the coordinates would have to look like.

```
device:Country/Building/Utility/Device
```

We could also write this more like a Cartesian coordinate tuple:

```
(Country,Building,Utility,Device)
```

The addressing of smart capabilities can go all the way into the interior dimensions of the devices too, by adding dimensions:

```
Utilities = Heating,Lighting
Device
{
Location = (floor,room,ceiling)
RMS-Voltage = 115,240
}
```

Obviously, the lighting arrangements in the smart buildings are very different, so we do not expect sub-categories to be equivalent. We are used to thinking of spaces as being uniform, but as discussed earlier uniformity is only the 'ground state' or lowest level of description of a location. Namespaces usually diverge in their details, unlike database schemas which are 'normalized' to emphasize the assumption of uniformity or translation invariance.

All concepts begin and end as simple names or labels. Space is nothing but a collection of named entities. How we name them depends on what is most expedient. The fixed schemas of databases often end up being a problem as semantic concerns grow, because nothing static tends to be well adapted to changing contexts in the long run. But, in the past we simply locked processes into fixed state spaces.

Databases are often problematic for scaling IT systems. The semantics of databases pin them to a particular scale. You can add more data, but the scale of a single read or write remains the same. We can change one scale, but not others, meaning that we can only rescale part of an IT system. That means the behaviour of the system has to change as it grows. The same is true of structures in biology, like cells, organs, organisms. You can double the size of a heart or a brain, but it's still made of the same size cells, so a bigger heart is not the same as a small heart. These examples show that there are qualitatively different phenomena of different scales. The structure of a cell only works for one cell—you can't create a giant blood cell to carry more oxygen, because you can't create giant atoms and molecules. Similarly, databases interact with particular applications on a particular scale too, following a particular schema, and these details can't simply be replaced or resized.

What then actually is a scale? It's a region of uniformity, or compatibility, a measure of compatible things, whether by distance, weight, size, colour, etc. There are two different kinds of scale:

- A *dynamic scale* is a representative quantity measure that characterizes the typical measure from a cluster of associated values.

- A *semantic scale* is a representative name or label that characterizes the typical function of a cluster of associated agents.

If a phenomenon doesn't actually depend on a scale[77], it could be universally applicable. But if one of these characterizations plays an explicit role in the promises being made, then the character of the bindings between component parts is likely to change. This is not scale invariance. This is of basic importance in designing machinery of any kind. To do more work, we build a bigger machine to handle the increased workload—but, at the new scale, the machinery might not work in the same way. Imagine trying to wash windows more efficiently by making a cloth ten times bigger, or washing the windows of a skyscraper by making a bigger ladder. The properties of space and its processes are more anchored to particular scales and circumstances than our high school geometry classes care to admit. This is a lesson that follows us all the way back to basic physics.

THE LARGE SCALE STRUCTURE OF THE DATAVERSE

Whatever their detailed properties, we are used to thinking about 'spaces' as large and expansive regions, regular and uniform in its appearance, and where there is plenty of free movement. But a computer network (even one as large as the Internet) feels quite random and claustrophobic compared to the empty regions of the cosmos, or even compared to the vistas of agricultural farmland. Space and time are not only about what happens in a local region, we also have to admit to there being special regions. Shops are known to move goods around so that they can be accessed more easily, or to keep customers guessing by forcing them to walk around the shop hoping to tempt them into new purchases.

In the dataverse, this all starts at the level of information infrastructure, made from computers and networks. In the early days, computers were sparse and cables were long. Today computers are quickly filling the spaces in which they are embedded. Imagine racks of computers as far as your eye can see, like shelves in a huge warehouse. In each computer, stacked within each rack, there are interior processes stacked on top of one another too, on the interior of software agents. This is what the dataverse looks like (see figure 3.11). It has begun to fill space and extend into hidden dimensions too!

The location of computer processes usually only matters in the dataverse, not in the physical embedding space of the real world datacentre (unless they are regulated by local laws). Little parts of a process may be running in different aisles and different levels, and different racks. All of these form the details on the inside of a virtual boundary called a process cluster. The clocks that drive the pulses of time to move the processes are also distributed and they are completely uncoordinated. The average effect of all of these different patches of activity leads to a different notion of time that might be erratic from the perspective of another agent outside the cluster, but communicating with it.

Under certain circumstances, processes might be rescheduled and moved to a new physical and virtual location within the dataverse, leading to a reset of time according to one clock, but not another. During the move, the processes are suspended (they are literally put to sleep by cutting them off from a processor clock, and receive no signals from within or without). There is also an uncertainty in the absolute position of processes too, measured from the exterior of the system—but, on the interior of the dataverse, space and time can continue uninterrupted.

To understand structure and process, and how they relate, conventional notions of space and time have to be rethought. We have to ask: what structure are we looking for? Structure with respect to which promises? A simple concept like adjacency can be physical or virtual. Virtual adjacency need not imply physical adjacency, only long range connectivity. In the world of the Internet it is not uncommon for companies and governments to insist on the use of VPNs or Virtual Private Network tunnels, that connect users directly to their home headquarters, via a virtual link, where information is considered secure. A direct physical link is not possible, but who can really tell the difference when we can't see the cables?

As far as most of us can observe, networking is an atomic and elementary interaction, but we have no idea where or when the processes that we observe on our computer screens actually took place. It begs a simple question: can an observer ever tell if something is fundamental? Is there ever an end to the turtles going down?

SCALING ALL THE WAY DOWN

Observers might not be able to tell what's going on in spacetime at large, but one way to try to understand is to think about how space, time, and spacetime processes scale. Scale is a topic that physicists are trained to think about, but for everyone else, it is one of the hardest ideas to imagine. Think about what imagery you have in your mind for the world of atoms. Now compare this to what you know about the world of cells and biology, to brains, to the Internet, to animals, towns, countries, planets, even galaxies. In each case, you may think about a certain structure that typifies each scale—how locations are

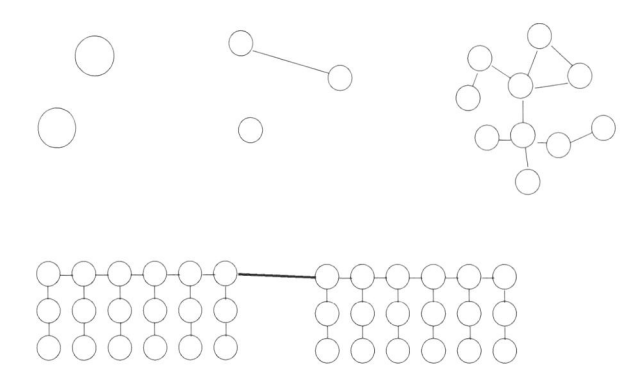

Fig. 3.11. The evolution of computer spacetime, from disconnected computers at point locations, to ad hoc exterior networks forming small molecule-like clusters, to larger permanent networks, and then regular lattices like crystals at the bottom (which is what modern datacentres look like).

connected together. Are they like smooth like liquid or weblike? Are they flat and rectangular like Euclidean geometry, or central and circular like a planetary system? Are the relevant changes and movements gradual and smooth, or lumpy, sudden, and discrete?

As we assemble a library of such ideas, we begin to realize that every phenomenon has very different characteristics at different scales. It would therefore be wrong to compare one scale with another and expect to draw comparisons. Similarly, if we want to compare the world of computers and the Internet to our familiar universe, we should not expect them to line up at all scales. The real challenge is to understand at what scales we might be able to compare them on a fair basis. For instance, figure 3.11 shows some typical geometries in computer networks at different scales. In the beginning computers were separate entities without connections. Later some of them became connected by networks, and some remained outside. This is still the case, on a more ad hoc basis: sometimes we are online and sometimes we are offline. Some computers are mobile, others are more fixed in the embedding space of the real world, but for the dataverse we don't care about the embedding locations, only about the points and connections within the dataverse itself. Either there is a route from A to B or there isn't. Devices literally fall in and out of the dataverse as they are switched on and off.

Can we call the locations of the dataverse 'primitive' or 'elementary', as we think of points in Euclidean space? They have internal structure. From a geometrical perspective this seems like a cheat, but from the perspective of physics this is not important—in fact it is normal. We can still treat the locations as elementary at any particular scale, just by ignoring the inner details. We treat atoms and protons as elementary at different scales, knowing that one is inside the other, and that there are quarks inside protons. Since we never know if we'll reveal a new level of description about spacetime, physics (re)normalizes that is elementary to a given scale of interest. Just as long as we are clear about what scale we are treating as our lowest observable level, we can suppress or renormalize our expectations accordingly[78].

Holes in the dataverse

If locations are not always present—if they can fall in and out of the dataverse, by losing touch—what does this mean for how observers see phenomena? Surely this is not a proper spacetime, like our universe? Well, actually, a better question might rather be: how do we know that this doesn't, in fact, happen in the universe at a small scale? We don't. If it did, several aspects of quantum mechanics might make more sense.

We are not really used to the idea that we suddenly can't reach a certain place from where we are in our familiar universe[79], because our familiar models of Euclidean space are all smooth and connected without holes, like swimming in a smooth ocean that envelops everything snugly. But we *are* familiar with losing connectivity when driving. That involves a different kind of spacetime—a network of roads, with lots of unreachable places, strange topologies, and constrained path combinations. If we take this as a model for any process within this embedding space, it is not hard to see that there may be unreachable places, just as when we are driving through the limited network of roads. Indeed, if we take local interactions into account, then traffic lights or raisable bridges might prevent us from reaching a location according to changes we can't necessarily predict[80].

These mundane examples illustrate a simple point: that we may need to think more flexibly about the world on a scale where dynamical changes interfere directly on the same scale as the phenomena we observe, than we do for larger scale phenomena, where tiny changes don't make an impact on outcomes. At the large scale, details like traffic lights and bridges are not visible, and they average out to something like 'connected but sometimes slower'. As we increase the scale of a system description, statistical renormalizations—or the blurring of these microscopic details—eliminates many of these issues, and brings a kind of stability to our world view.

Time as a series of versions

So far, we've focused on the properties of space, and considered only clocks inside computers (their CPUs) as the generators of an interior time. Each computer has its own sense of time, but we also maintain chronometers on computers, which are kept quite closely aligned with the standard wall-clock time defined by the Greenwich Observatory in London. This is called computer Universal Standard Time (UST), and it's used to put timestamps on files when we make changes. But wall-clock time is not the time relevant to information in the dataverse—it's only an attempt to be friendly to humans on the exterior. It's easily fooled, and often drifts out of synchronization as different computers' chronometers run fast or slow. To understand time in the dataverse in the same way we understand time in the universe, we need to think (once again) of time as that sequence of changes belonging to the processes ongoing within its rule-based machinery.

To its users, the Internet is more than just passive background theatre for random happenings. Processes in the dataverse mean a lot to us. They are intentionally instigated, and their outcomes have functional interpretations that are valuable. We are motivated to preserve these dearly computed outcomes, and we assign a lot of weight to the use of space as memory to make them happen. Moments in dataverse time are slices of computer memory, which therefore have a direct semantic value to a variety of stakeholders—so the progression of time can be quite a complicated issue. We need to decide whether different stakeholders, i.e. the observers in the dataverse, share in the same causal region of space, or if they act in different frames of reference. This is what defines 'relativity' in the dataverse, in a sense that is similar to Einstein's. It's helpful to describe a couple of examples of how dataverse time and universe time interact as users make intentional changes to the data.

Let's take a very general example first: a couple of users, collaborating on documents that are stored in files somewhere in computer memory. This is a very normal situation for anyone using a computer, but the example I'll give applies especially to programmers working on program code.

There are all kinds of spaces in play in this example: from the world of the programmers down to bits and bytes inside the computer, but the relevant one, as far as the intentional outcomes are concerned, is the memory space for the documents, which itself forms a closed space. We can represent it as an interior spacetime, whose clock ticks are generated by every change typed into the files. For instance, we could take the space to be the directory in which the documents are kept. Some filesystems are indeed capable of keeping a history of every single keystroke change to a document, but normally this is much more detail for users to care about, and we register only process clock ticks when we save changes—and then only for changes of sufficient significance. For instance, suppose one of the collaborators makes a correction, and adds a sentence explaining a single problem solved. This change has a clear sense of meaning: it's a semantic change, not just a dynamical change. The users are interested in retaining a log or a history of when and why these changes were made. The timeline has semantics.

There are software systems called Version or Revision Control Systems[81], which can take snapshots of the entire set of documents, as if they were a 'spacelike hypersurface' in Einstein's theory of Relativity. Each snapshot is a complete state of space, and all of its memories, and this defines what a moment of time means in this spacetime. When a user makes an intentional change, they are said to 'commit' their changes to the timeline, by issuing a special command. They open up a 'new world' in the language of quantum theory's Many World's Hypothesis, or Kripke's original many worlds view of logic, which I'll come back to in the next chapter[82].

As long as all users contribute to the same synchronized set of changes through the single change interface—they belong to the same 'frame of reference' in the document spacetime. Their changes then belong to the same interior clock for that frame. The result is a series of versions: 1,2,3,4,5, etc, which represent the proper times in this document spacetime. This is a literal construction of what time is, as a sequence of spacelike hypersurfaces, as in figure 4.3.

Now, it's quite common for programmers to want to work independently, and to stray from a single coordinated region. Say two colleagues decide to work from home. Then both users start making their own changes. They 'fork' the single timeline, by making copies into two different 'branches' of time (say, A and B, essentially like Everett's Many Worlds Interpretation of Quantum

Mechanics). They each make changes at their own rate, so the ticks of the clocks in each branch are quite different. So now there are now parallel worlds, with histories:

```
A: 1,2,3,4,5,A6,A7,...
B: 1,2,3,4,5,B6,B7,B8,B9,...
```

They are completely independent now, like causally disconnected regions of spacetime in General Relativity (known as independent light cones). Should their worlds ever collide again (such as the next day at work), they will be in possession of two independent and inconsistent spacetimes. They have two choices: to maintain two different worlds, or to try to merge them. When two observers come together in Einstein's relativity, their histories also have to merge, which is why space and time often get apparently distorted in the process to achieve consistency. In a discrete space, like a document, where the changes are not just random dynamical ticks, but actual intentional semantic changes, it's important to reach semantic consistency too. The process of merging timelines in version control systems can be done automatically provided they remain independent, and neither made changes to the same region of document space. If they did make different changes to the same region of space, then they have to make a selection to resolve the conflict. The analogue of 'proper time' in Einstein's relativity is a monotonic increment along a process trajectory—it's not necessarily what an exterior observer's clock measures.

There is an analogue here to Darwinian evolution. A species can fork into two independent subspecies, with different environments selecting different changes. For instance, different horses, dogs, or other domesticated animals. Should their worlds ever collide again, and the two remain compatible they could be merged back into a single version, but only by selecting one or the other version of a conflicting gene. Darwinism is thus not only about biology. It's about information, and space and time.

DATA IN MOTION

Because the dataverse is embedded within our own universe, allowing data to move around has literally added a new dimension to our own realm. The data do not move on any time scale that we would identify as motion. They simply tunnel from place to place. Of course, the data do indeed travel at the speed of light in copper or glass as bits encoded on top of electromagnetic waves, but they are effectively carried through private waveguides that prevent them from interacting with the physical world too much. The exception is when we are using a wireless signal on 3G, 4G, 5G, etc, or WiFi.

In IT, timescales have been getting smaller, and speeds have been getting faster, making spacetime notions more subtle and precision more critical. The simple ideas about space and time that have been applied to computers, in computer engineering, are based on their embedding space, not the actual network spacetime they occupy. Defining clear semantics has become harder, and this is starting to cause real challenges in the world's fastest and most critical systems.

So far so good: the IT dataverse has points, dimensions, direction, space, and time. It scales like other spacetimes we know about, and so it qualifies as a spacetime, even though it is discrete and unlike the universe as we believe we know it. However, this is where it starts to get interesting. The relatively tiny number of points available in the dataverse, as compared those we know of the universe means that behaviours that relate to limits and boundaries have an exaggerated importance in the dataverse.

Can we translate concepts like position, time, velocity into the dataverse and computation? In hardware and in software? So far the answer seems to be yes, but with some caveats that we need to explore further. Ultimately spacetime seems to be a set of interfaces and connectors, quite like circuitry. If we take the consistency under scaling seriously, then we are forced to consider this to be a fair representation of space on large scales. An even more fascinating question is: does it work the other way around? Are there cases where we can look at what happens in the universe as a version of what happens in the dataverse—as a kind of circuitry too? That remains to be seen, of course. However it works out, we can and should apply the tools of reasoning to both, as indeed we shall in the coming chapters.

4

Give and Take

> As the creeper that girdles the tree-trunk
> the Law runneth forward and back—
> For the strength of the Pack is the Wolf,
> and the strength of the Wolf is the Pack.
>
> –Rudyard Kipling, The Law of the Jungle[83]

In his poem, from the Second Jungle Book, Kipling notes that, from an exterior perspective, a pack of wolves acts with a collective intent—singular in its nature, even as it depends fully on the behaviours of its individuals. Later in the poem, he makes concession to there being a privileged agent within a pack: a leader, who speaks for the pack in essential matters, calibrating and steering its intent—just as Wheeler supposed that all electrons' promises were calibrated by being images of a single entity. Centralization brings calibration. These matters might seem obvious, but it's rare for physical sciences to speak of the functional aspects of systems. We attribute such phenomena to agents (presumed active) rather than to locations (presumed passive). In this chapter, I want to explain why this is the wrong distinction to make.

The wolves find their order—either spontaneously, from their varied mixture of intentions and behaviours—or perhaps by pledging allegiance to that single source of authority in the leader—even by picking a democratic representative after a vote, if they became subservient or they became statisticians. Such a collapse into order would be called a *phase transition* in physics. It's a collective 'bottom up' process, in which singular parts cooperate to bring about 'order'. In physics, 'order' means a cooperative state with a regularity, on some scale larger than the individual parts. For that reason it's sometimes called Long Range Order[84]. A pack leader may act as a seed for the accretion of the pack to fall in line, with little more than luck to single it out as an initial attractive force. Alternatively, the 'alpha' wolf may apply force, staring down the others into submission, playing the role of an enveloping boundary herding the others by fearful repulsion, as a 'top down' process—two ways to hold the pack together: by give or take.

The scaling of a wolf pack is not just a scaling of quantitative effort, in the way that a larger amount of gas exerts a larger pressure on its surroundings: it is a scaling of *qualitative* behaviour too. That qualitative behaviour is what we sometimes attribute to *intent* or 'agency'[85], which is not to claim that a human or even a brain must be involved. It turns out that intent is not limited to the scale of animals and brains, as we usually think of them. If we only bend the strict meaning of words to see a wider truth, we find that the scaling of intent requires both the cooperation and calibration of several agents' behaviours to assure uniformity of its parts, and it requires that uniformity be received in a dynamically similar way[86]. Neither of these are 'personal' or animate qualities. They can also be identified in inanimate interactions.

When we consider functional aspects (which is what we really mean by intent), a pack of wolves can promise both a wolf-like interaction, at any single location over a larger area (because it is the sum of its interior wolves' promises), and it can also promise new behaviours at its new collective scale that are only possible as a result of several wolves working together, e.g. pulling apart a carcass by cooperating. This confirms what is well known about more complex systems: that, at each new scale, there are new phenomena, not just more of the same. This is what we mean when we say that a system is more than the sum of its parts.

The structure in a wolf pack is quite familiar in the dataverse, where network routing hubs, and centralized services form the seeds to bind and coordinate a cluster of computers. The geometric pattern is not enough to bring uniformity to the behaviour of a cluster of computers; the component agents must also make the same promises and form an ordered state. For example, the hub pattern was shown in figure 2.6, where its role was to define by association to a single point of definition. Connecting to a single source of truth is far cheaper than trying to battle out consensus between every possible pair of agents[87]. Representative hierarchies, based on a single leader, form everywhere in nature when it becomes expedient to treat a collection of things as a single entity.

The lesson of scaling individual agents into clusters of agents is that the changes in behaviours, at larger scales, are not too particular about whether we are talking about wolves, or cells, or atoms, or semantically active locations at any scale. Collective behaviour involves a washing out of detailed information from its interior composition, in favour of larger 'average' exterior trends, depending on what the collective is interacting with. What's promised at one scale may affect an agent on another scale to find different meaning or semantics that depend on the details of the interaction. If you like, what was intended might not be what is experienced. A pack of wolves may appear to be like a single wolf when moving across a plain, or might take on the characteristics of a suffocating noose when meeting prey.

In the previous chapters we've seen space and time through the eyes of mathematics, physics, biology, and Information Technology. These views seem, on the surface, to be quite different, but underneath they deal in a familiar set of themes. Nevertheless, there remains something unsatisfyingly irreconcilable about the underlying formulations that we should like to improve upon. The true nature of the stuff our universe is made from (at the bottommost scale) is and may remain a mystery for decades to come, but the view of the world espoused in Euclidean or even Einsteinian terms fails to answer key questions about its nature, and it doesn't work at all in the realm of information science. Euclidean space seems only well suited to studying certain idealized abstractions that belong to mathematics. Could we have become stuck on the successes of relativity and quantum theory that were able to adapt it for their purpose, afraid of letting a triumphant orthodoxy slip?

As we look at other areas of science and technology, it becomes expedient to attribute more properties to spacetime than is done in the Euclidean Cartesian model—to add richer semantics to spacetime. To see why, we need (perhaps ironically) to get down to very fundamental issues about what we mean by space and time. Let's turn to the most basic question of all—what makes the adjacency of two locations possible, and how this is related on a deep level to how different states of information can fit into what we call a region of space. This is the very definition of what makes a scale, so scaling will play a key role. Simply placing one point next to another does not lead to motion or distance—not without some kind of cooperative linkage. Interaction, or the communicated interplay between what is offered and what is accepted, has to enter the mix. This interaction between give and take, offered and accepted, proposed and selected, breaks with the idea that the source of all behaviour lies in initial conditions of certain entities and a set of uncontested Equations Of Motion, which are handed down by divine right like a kind of driving license. It transplants, instead, *collectively determined interactions* as the principal focus for process behaviours.

This shift from 'physical law' to 'cooperative interaction' conceals an uneasy implication for those who believe in the classical Enlightenment notion of objectivity in science: it means that—no matter the existence of an objective reality—all behaviours observed must, in fact, be coloured by context and subjective to the information of individual observers' circumstances and relationship to the source. What I want to show, in this chapter, that such ideas are not to be feared. They are natural scaled interpretations of the way 'intent' (the matching on functionally specific behaviours) is communicated from place to place.

Nyquist and Shannon

In the 1940s, the initially separate subjects of physics and computation collided, and their fates became forever entwined, thanks to the efforts of a number of scientists including Norbert Wiener (1894-1964), Alan Turing (1912-1954), John von Neumann (1903-1957), Harry Nyquist (1889-1976), and Claude Shannon (1916-2001). In 1948, Claude Shannon wrote his seminal work on *The Mathematical Theory Of Communication*, or what has now come to be known as Information Theory. It changed the course of information science and technology forever[88]. Shannon showed that information could be defined, ultimately in digital terms (as binary digits or 'bits'), by the answers to a set of 'yes/no' questions. It could also be approximated in a language of continuum mathematics, as channel bands (from which we have the colloquial 'bandwidth'), for large scale transmission in electrical engineering[89].

Shannon imagined a simple scenario (see figure 4.1), consisting of a generic sender (the source of a message), a generic receiver (the sink for a message), and a generic *channel* for transmitting the message between them. The model might have been inspired originally by the telegraph (in which case the channel would

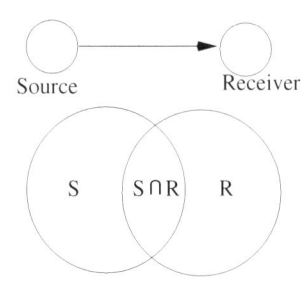

Fig. 4.1. Shannon's communication channel unifies the overlap of sender and recipient interior domains of influence with a simple graphical view of what directed adjacency means in terms of information flow. Adjacency is simply a channel by which information can be shared. It can be defined independently in either direction, by exchanging the roles of sender and receiver. Compare this to figure 2.3, to see how the overlapping $S \cap R$ information defines a mutual and directionless connectivity.

be a wire carrying an electrical potential) or on the wireless radio (in which the channel would be a modulated radio wave)—the point is, it doesn't matter. All information transfer must involve these basic roles, and must follow the same essential pattern. Put another way: if you renormalize away the interior details of communicating agents on any scale, the exterior promises of the parties would be the same. To discuss any particular information transfer problem, just take the Shannon model and fill in the three roles with an appropriate set of agents, and all of the results of Information Theory can be used.

Shannon's contemporary Warren Weaver broke down the problem of communication into three parts, which he referred to as the technical problem, the semantic problem, and the effectiveness problem respectively[90]:

– How accurately can symbols of communication be transmitted?
– How precisely do the transmitted symbols convey the desired meaning?
– How effectively does the received meaning affect conduct in the desired way?

From a spacetime point of view, Shannon's model is particularly useful, because it connects the idea of *influence* with that of *adjacency*, through the concept of a directed graph (discussed in foregoing chapters). It tells us that being next to something is equivalent to being able to communicate with it. Perhaps these two concepts are ultimately equivalent—points are not connected *until* they communicate over some channel, and it is for us to determine what that channel is. It's equivalent to saying that the idea of a pre-connected space, in which processes unfold as a kind of theatre, is not the right picture: rather, space itself is not connected until processes unfold! Each communications channel forms exactly half of a link between points in space, through the influence of a *process*—the directed transmission of information from sender to receiver.

According to this idea, in order for something to change its location, i.e. move from one location to another, there must be a transmission of information. For it to move backwards, a separate channel would have to be set up in the opposite direction, like the two lanes of a road. The generality of the roles, in Shannon's information channel, means that this model scales to any size, and its universal structure makes it a candidate to explain the most basic question in the whole of physics: what does the adjacency of points really mean?

Shannon's communication model describes not only adjacency, but also *causality*, or the link between cause and effect. It defines a direction for information flow, and suggests that the basic spacetime relationship should be directional. There is no assumption or muddle about reversibility in this picture; it's raw and elemental without complication, and thus we can use it to build up any picture of spacetime, just by combining information channels. Some physicists elevate symmetry of forwards and backwards (reversibility) above this choosing of roles, but that view must be incorrect: we can construct reversibility by combining forwards and backwards channels, but we can never break the symmetry of a bidirectional definition of 'next to' once it has been imposed, without simply introducing more 'stuff' that makes it less elementary[91]. This is not really a new assertion. In fact, the same separation of roles was identified in the early days of quantum theory, in a variety of guises from particles and anti-particles to directionality of boundary conditions—a point that I'll come back to below.

As computing and physics have evolved, each along their own paths, this fundamental truth, about the meaning of adjacency, direction, and information, has haunted the fringes of both fields, with deep implications for the nature of every kind of interaction. 'Agents' at every scale can be described in exactly the same way. The only difference between a cell and a computer, or an atom and galaxy, lies in the kinds of promises they make, and the scales over which they are defined[92].

The geometry of a communications channel tends to make us think of a dumbbell structure, like that at the top of figure 4.1, but we should not make the mistake of assuming that the channel of connection between sender and receiver must take the form of some kind of wire or pipe. That would be allowing ourselves to be fooled by embedding space imagery. Shannon's 'link' or channel can also be represented by a Venn diagram, like the overlapping circular regions in the lower half of the figure. Adjacency is simply the overlap of agents' spheres of influence, in whatever sense we care to define. As long as the interior regions of influence of sender S and receiver R overlap, there can be a sharing of information[93]. In other words, the definition of adjacency is connected to the definition of a boundary that separates interior and exterior for each agent. This ties adjacency back to the way we defined observation, and we'll also see that this is crucial to defining what we mean by time. It's ironic too that this simple observation also ends up recentralizing the view of what space is around Ptolemaic central points, rather than straight line geometries.

When two agent interiors overlap, information is shared, and it becomes reasonable to say that agents are next to one another in space. It's a bit like the opposite of cell division in biology. At this level, the only thing that can happen between locations is to share information—but actually, that's all we need at such a low level. The occupying of a position, in traditional language, would be equivalent to an agent's ability to promise some information that distinguishes itself from a neighbour. The movement of something from one place to another would then be like the passage of that information from one agent to another. Instead of musical chairs, by moving occupants, musical chairs by relabelling of the seat numbers from one chair to the next.

What does such a model mean for spacetime, if being next to another agent is the same as having its reach overlap with another agent's region of influence? The first thing to note is that there is nothing in the model that necessitates permanence of the overlapping communications. Agents can overlap at random, like a gas, or be ordered rigidly and permanently, like a solid. So there is also no reason why space itself would not be able to exist in different phases (gas, liquid, solid) just like what we think of as matter. Of course, it's very hard to think of this without imagining an embedding space. Yet the same thing happens all the time in the computer systems of the dataverse, on a virtual level. Messages are emitted by one computer, and may be absorbed by another computer[94] Waves too can pass through such a spacetime medium of communicating agents, in an exactly analogous way to the passage of waves through matter. If matter and fields are just different states of spacetime anyway, and forces between them are just the transmissions of messenger particles (as in the standard model of particle physics), then this is no stretch to imagine at all.

PROMISING COOPERATION

The dual roles of sender and receiver, in the Shannon model, play an important role in spacetime, across a wide range of scales. It therefore seems natural to embed them into the description of spacetime at all levels, however little we may know about the nature of the spacetime that erupted from the big bang. The idea can be applied to the universe, or the dataverse, or any level of description for things that happen. These aspects of Information Theory, as well as others, are built into Promise Theory—which is one reason for using it as a simple reference model for spacetimes. Figure 4.2 shows how two agents need to communicate in order to keep promises. Promise Theory goes further than Shannon's model and makes each node even more elementary, by adding an additional degree of freedom: the autonomy of the sender and receiver agents, which allows them to accept or reject messages sent, as discussed in chapter 2. The outcome of this is that each interaction involves both an 'offer' to be transacted and a mask or filter the expresses the 'acceptance' of a message.

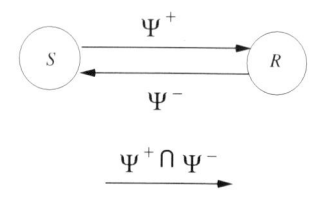

$$S \xrightarrow{\Psi^+} R$$
$$S \xleftarrow{\Psi^-} R$$
$$\xrightarrow{\Psi^+ \cap \Psi^-}$$

Fig. 4.2. The overlap of complementary promises is the projection of one pattern onto another. If the patterns are compatible information can be transferred, else not. This lock and key fit is a simple model for absorption of information on a microscopic level, as well as for the transmission of intent, and access control, on a higher level. What active process is inside the receiver that allows it to sample transmissions from the sender?

Consider figure 4.2: suppose S (which has the role of sender) promises to send a signal or perform a service Ψ^+ to a recipient R (in the role of receiver).

Since the two agents are independent, R is not (and cannot be) compelled to accept the signal, as that would violate locality, autonomy, or whatever we choose to call the causal independence of locations. So, transmission has to become a cooperative process.

It's remarkable just how often we neglect to consider the need for both 'intent to give' and the 'intent to take', assuming instead that one is enough and the counter-promise can be taken as given (by virtue of the cosmic driving license issued by Equations of Motion). This kind of choice to accept or not accept sounds more like the behaviour of a person, not that of a natural phenomenon. Yet that assumption would be quite mistaken. In fact, in struggling to uncover the small scale behaviours of atoms, the elementary physics of quantum mechanics ended up predicting exactly this kind of duality, as we'll see below.

Shannon's model is really a model of give and take, but he didn't make that explicit, because the technology he worked with implicitly arranged for a compliant structure, in which every promise to accept was automatically built in. It's true that that automatic acceptance seems to be built into the 'laws' of 'promises' of physics—the question of why surely conceals important truths about how we choose to model phenomena. The appearance of 'flow' or conservation stuff tends to make us think that 'that which is thrown by one agent must always be caught by the other'. This doesn't apply on our human scale, of course: I can throw you a ball, or call you on the phone, and there is no guarantee that you will catch the ball or pick up the call. Yet this does seem to happen, at least on some range of scales, in physics. To account for it, it's given the name *The Principle of Conservation of Energy*. It's a hallowed principle in physics, even though it's accepted without explanation. What it doesn't require is for only one of the agents to be able to determine when its neighbour must accept that energy. Such behaviour doesn't happen in the dataverse, and it doesn't happen in biology, or socio-economic systems—though sometimes, it's expedient to pretend that pretend that it does, by trying to arrange for the conservation of *information*. All we really can say is that, when mathematics assumes conservation, balancing give and take, the results make helpfully correct predictions on average. They don't say when the agents will be ready.

Despite the simplicity of his vision, Shannon was able to prove remarkable results about a wide range of phenomena, from the simplest idea of information transfer. Many scientists have uncovered deep connections between information theory and other parts of science, making the concept of information one of the most fundamental of all. Shannon's contributions to science cannot be understated. We don't need to delve into the details of his work to appreciate how one of the simplest of theoretical models has unified the descriptions of so many phenomena[95]. The fact that it is embedded in the semantics of Promise Theory, makes Promise Theory a neat and minimalistic encapsulation of the elements of interaction between entities. Having a successful model of communication—i.e. connectivity—is not necessarily the same as understanding it fully, but the implications of the extremely simple model of a *sender* sending messages to a *receiver* have spread throughout nearly every field of science.

THE SOUND THAT FALLS ON DEAF EARS

One of the key results in Information Theory that gets little attention outside the world of HiFi is a theorem by Nyquist about the sampling of information from a stream of events. There are different ways of expressing Nyquist's Theorem mathematically but the essence of the theorem is this: if you want to observe change that happens at a certain rate, or over a certain time, then you'd better be quick! If you are too slow, you'll miss it! Indeed, if you imagine opening and closing your eyes while watching a pendulum swing, you might never see it move at all, unless you keep them open more often than the pendulum visits each location on its trajectory. Nyquist showed that you need to be about twice as fast as the thing you are trying to observe.

To see why this must be true, imagine a scenario you've probably seen in the movies many time before: a thief trying to sneak past a security camera, which is sweeping back and forth to pick up a wider area. Nyquist's theorem tells us that, if a thief can run past the camera in a time t, then the camera should finish its sweep in a time $t/2$ in order to capture to be certain of spotting the thief, i.e. in half the time. Alternatively, imagine a pendulum. If you take a snapshot of its position every hour, you could measure almost any position. If you measure its position every time at the same frequency that pendulum swings, you will measure it in exactly the same position every time, and you will see no change. If you measure it at twice the frequency of the pendulum, you will see it at each end and in the middle of its swing too, and you'll be able to infer the frequency with which it moves (though you still won't know whether it moves smoothly or jumps discontinuously).

Nyquist's theorem can also be expressed in clock time. It says that an observer's interior clock rate (the clock that tells it how often to take a picture) has to be twice as fast as the environment that it can perceive. Or conversely, no agent at any location, can observe any exterior phenomenon reliably faster than half its own internal clock. This clearly sets a limit on how fast something can be observed to move in a continuous manner. If objects could move very fast, they would appear to jump from place to place in a discontinuous manner (see section 5) while the observer sampled[96]. This fact doesn't only apply to human observers, but even to atoms and subatomic agents that expect to receive signals. The ability for an agent to experience time passing depends on its interior activity! We shall refer to this when discussing motion. It turns out to be extremely important in information technology and in biology.

The two halves of figure 4.1 take on a special significance when we think about processes as sequences of change. The essence of locality or autonomy is that independent locations have independent (interior) clocks. The implication of Nyquist's theorem is that no agent can gain knowledge of another's interior clock, except by trying to capture the exterior changes that it sees arriving along a channel of communication between them. A security camera can't capture a thief's intentions; at best it can try to capture the light that reflects off the thief as he moves. If the thief can sneak past the camera faster than the light can be captured, the security camera cannot observe the event. Only processes that persist for about twice the duration of a sample can be observed in sufficient detail to perceive a sense of their temporal evolution. The consequences of not being fast enough can be peculiar and counterintuitive, depending on how they observe signals and make further use of them. The key point here is that sampling (take) is a fundamental part of receiving information given. Nyquist's limit sets a fundamental limit from the overlap between give and take.

Referring back to figure 4.2, Nyquist tells us that the sender must offer information (by a signalling process Ψ^+), and the receiver must similarly accept it (with a sampling process Ψ^-) in order to absorb the signal. The transfer of influence between the two is therefore proportional to the overlap function $\Psi^+ \cap \Psi^-$ (recall the overlapping circles in figure 4.1). The likelihood of a promise (+) being able to propagate its influence to a receiver (-), is about the set of possible outcomes given by Ψ^+ **AND** Ψ^-, which is usually represented by a product of probabilities. But these are not mere static probability profiles—they are sampling processes, meaning that the probability of sampling a signal Ψ^+, which sends information at the rate of a clock within S, must also depend on the sampling rate of R's clock. By appealing to Nyquist's theorem, we see that, unless R's interior clock could sample at twice the rate of S's, it's experience of S's behaviour would be doomed to be somewhat erratic, regardless of how regular S's behaviour may or may not be. Obviously, each point cannot be twice as fast as every other[97], so they will all end up appearing to more or less the same and somewhat erratic, unless for some reason they all promise to change only have as fast as their fastest rate (which includes the transit time for the signal).

Although we have come to suppress Nyquist's considerations for atomic and subatomic processes, the normal laws of scaling suggest that the process of absorption, in the atomic world and below, must also absorb by sampling too. It's simple information accounting. The implications are significant however. What interior processes are ongoing in the agents we refer to as atoms and particles that allow them to accept messages from neighbouring agents? What does that mean for their roles as elementary and indivisible primitives for physics? What does it mean for the distribution and location of clocks that tell interior time—the clocks that enable observation; the clocks that make Einstein's relativity true? These are the questions scaling urges us to ask for processes on all scales from subatomic interactions to monetary transfers in the global economy to gravitational waves in the universe at large.

As the sender and receiver regions of space get smaller, at some point they will not contain enough states to be able to form an interior clock of any significant size on their own, and they will not be able to even distinguish long sequences of change. At that point, there cannot be a viable interior time. A simple bistable pendulum might be enough to drive the sampling of a signal, but it could not distinguish larger intervals. So a human observer could easily use an atom as a pendulum by counting very quickly with the help of some technology, but could the atom itself observe the change in a neighbouring atom during an interaction? There would only be a finite chance of it noticing the change, when the pendula happened to (mis)align. Memory is clearly essential to our notion of time: we have to remember what was to know what is now. As we shrink a region of spacetime, it must lose the ability to observe times and velocities and behave in the kind of way we usually assume that classical physics works.

In the computer networks of the dataverse, computers and routers also need to sample quickly. Forgetting about thieves and pendula now, and thinking instead about the movement of bits of information, around within the lowest levels of computers, it's not hard to arrive at the conclusion that there must be a basic uncertainty in the outcome of a bit transmitted from one location to another. If we give that bit a significance, it could be the difference between an outcome being communicated as true or false. If we base actions on such an outcome, as we often do in software systems, this suddenly seems like a dangerously precarious situation.

Because information is vulnerable to mismatched clock rates, and botched samples, computer systems don't rely on single samples. There are mechanisms for error correction, worked out initially by Shannon himself, to ensure that true and false can be transmitted reliably over an information channel. They work by slowing down the rate of transmission and using multiply sampled confirmations of messages. This is a complex business that extends over many scales in information technology, from the lowest level memory chips, to messages transmitted across the planet.

Seeing may not be believing

The work of Shannon and Nyquist give an unexpected meaning to the subject of scaling data, i.e. observations measured in support of some conclusion. Inference depends on sampling, and sampling depends on independent notions of space and time at independent locations.

When we make measurements, we don't rely on single observations to draw conclusions, because of such underlying uncertainties. The field of statistics was invented in order to try to reach stable conclusions from data by repeating measurements and then looking for a single promise for the whole batch of data. In other words, statistics replaces the wolves by the pack leader, in what we call an *average*. An average is a single representative value that is offered as a summary of the effect of a whole pack of promises. Statistics has developed the definitions of quantitative averages, but we can also imagine semantic averages, in which we give a single interpretation to a collection of individual interpretations, e.g. what's the average of trendy or conservative? How do we summarize the opinion of a country? One way is to hold an election and define the outcome based on counting of certain symbolic categories (left, right, true, false, red, blue, etc). We can't always define the average of names 'fairly', so we use quantitative measures as a way of giving a clear meaning to semantic issues that are a balance between give and take. Whether the numbers are truly representative is a semantic issue in its own right! The promise of an election result (+) has to satisfy the expectations (-) of a population, which is on a large scale, so its notion of fairness has to respect that scale by counting. If the result were only of interest to a machine, one might conceive of other notions of fairness.

I have often referred to statistics as the scaling of the concepts of true and false, because this puts into perspective the relativity of outcomes with respect to scale. Einstein's relativity is also a relativity with respect to mismatched scales, but now the scales are mismatched due to differing rates of change, rather than due to the relative dimension of the promises offered and result accepted. It's easy to mislead ourselves if source and receiver are assumed to work on the same scales.

Time is not only a local concept at each location, it is also a statistical concept in computer systems. That's because our reliance on networks now makes it rare for us to be able to base decisions on only a single source of time. If we mark a document with a timestamp from when it was saved, then transport it into a different timezone, with a different clock, the timestamp is no longer meaningful. As someone who travels internationally quite a lot, it's not uncommon for my calendar to remind me about meetings that I've already had.

Today, thanks to globalization and the international coverage of the Internet, it's far more likely that more than one computer will be involved in every piece of information we rely on. Then we need to ask how the information is synchronized between these sources, and whether we can even define what synchronized means. Which clocks are involved in defining the intent behind information, and its eventual interpretation? In banking, account data may be replicated to several sites around the world, due to the risk of a disaster causing loss of data, and the sampling times of banks may follow the timezone of the country or countries in which they do business. If a customer is travelling between those countries, the updates to their accounts may not follow them in the way they expect.

Time relativity affects databases that want to store observations. When a transaction enters a single master database, it often gets copied to a number of backup databases. But this takes time relative to every agent along its path, because transmission involves the clocks of every agent involved in communicating and sampling it. If all observers of data were limited by the same rules of travel, then this probably wouldn't matter. The problem is that some information might move at different rates compared to others. If all the computers are sampling at the same rate, then—like the pendula—they might not notice changes in their neighbours' data until they happen to wake up and make a sample. Those processes are uncoordinated at different locations. People who travel by plane can be raced by information sent alongside them, electronically; on the other hand, if different parts of the information are sent directly both to the moving plane and across the land, the difference in paths could alter the order in which information is sampled at its destination. The rules of causality seem messed up when processes interact under different conditions!

To compensate for this, we try to base decisions on more than one true or false value. Today, we hear increasingly of 'machine learning' and 'artificial intelligence' methods that are capable of identifying highly intricate patterns in data. The outcomes of decisions based on Information Technology depends on how millions of different decisions come together to form a pack all at the same moment. Due to the size of the 'big data' the computation involves many computers at locations spread across a possibly wide area. The rules of give and take within these 'neural networks' involve possibly thousands of independent channels with independent processing and sampling rates. Data scientists can do little more than trust what comes out of these decision networks, and that their behaviours are repeatable across different computations.

Time, in a neural network, is statistical in nature, just like any interaction of multiple parts. When inputs are fed into a neural network, the data may or may not be considered simultaneous according to the sampling rate of the computer's time. If the size of the computation is such that several computers are needed in parallel to processes the data, then their processes are completely decoupled in terms of their time sources. As processing continues, there may be mixing of the timelines as neural networks blend events from sources in different timelines. As the scale of a computation increases, the amount of time desynchronization may increase, leading to distortions in the causal flow of outcomes. For this reason, artificial training algorithms that calibrate agent responses need to be made insensitive to such changes, but this depends a lot on the details of the algorithm and the problem to be solved. So many details! Imagine a time sensitive activity, such as an autopilot or a high speed self-driving car. The safety of those systems has no choice but to rely strongly on the time-sensitivity of uncontrollable mixtures of give and take. This makes learning processes unstable by nature, and casts significant doubt on the idea of 'truth' as a concept independent of scale and information.

COMMUNICATION MAKE US CLOSER

Our conventional view of spacetime presents locations in a space like an open canvas of pixels, onto which changes can be painted—changes that may then be interpreted as information. But this doesn't include a picture about how the painting happens. Moreover, it doesn't explain the role of scale in explaining coordinated changes (how do we create a wolf pack?). If we zoom into a point, we might find that it is not really elementary and point-like at all, but is in fact a region with further detail revealed, so the concept of information must surely be related to renormalizations of scale. With Shannon's view of information, we realize that our ability to know anything of those things depends, in fact, on the communication of—not one, but—two parties, and on the balance of give and take between them.

How does this help us to understand what it means for two points to be next to each other? Imagine being in a room in the dark, unable to see, you may want to know if there is anyone else nearby. You might speak out: is there anyone there? If you receive a reply, you know there is at least one other point in your spacetime that you can reach. You can't touch them, because you are space itself[98]. But your information can move from place to place. The converse is also possible: there might be locations nearby without our realizing. We can look at someone and not see them, if they do not have the intent to be seen. Two devices can be connected and they may not exchange any information because they do not have that intent. The universe can also play hide and seek, and this can extend all the way down to the lowest levels of spacetime.

The channel is everything. We don't see anything without light; we don't hear anything without sound; we can't touch anything without...what? It's tempting to answer 'without proximity'. But here is the real question: what is proximity? If we zoom into our fingertips with the microscope, we find that touch is mediated by electromagnetic repulsion of electrons between our fingertips and other objects, which in turn is mediated by 'light' electromagnetic radiation. So what do messages have to do with the structure of space? They are the only way to interact with whatever might be next to us. If we can't interact with anything, even at the smallest scale, then there is nothing next to us at all. The idea that talking brings us closer is a cliché that has been used in advertisements for telephone use for decades; but it is literally true in the sense of measuring distances in different spaces. If nothing is listening, then nothing is connected.

We have assumed that space was something other than a process—that it was a timeless and immutable phenomenon in its own right; but the information perspective suggests that it is actually indistinguishable from the emergent phenomena that lead to connectivity is every other branch of science, higher up the hierarchy of scales. We only treat it differently because of the vastness of the scale mismatch between those processes and ourselves. We might not know what the method of communication is that lead to adjacency, but there is a growing suspicion that the answer lies somewhere close to the answer to another mystery, concerning quantum entanglement.

In the dataverse, these claims are nothing new. Being next to another computer, in practice, doesn't involve a cable: it involves a process of communication. If nothing is said, computers are just as isolated with or without cables. This is true of the underlying infrastructure of the dataverse, but it is also true of the virtual 'excitations' or processes that run on top of it. Processes are shunted around all over the fixed physical infrastructure, and are completely unaware of their virtual status. They can claim to be in a cluster if they are talking to one another. Virtual computers are processes that look exactly like the fixed machinery, but have no definite position, and so relativity has to be taken into account. In a sense that's a concrete example of how relativity does not contradict the existence of absolute space[99].

Probably the most important implication of Shannon's work is that the information theoretic view of spacetime suggests that we should separate the concept of A being next to B from the concept of B being next to A. Although, this may sound like a truly strange idea indeed, it's a lot more reasonable than first impressions. We experience an asymmetry in 'next to'-ness, or adjacency, all the time on a macroscopic scale, from simple one-way glass, to turnstile barriers at subway stations. It's simply the nature of interactions. What's more, this idea extends all the way down to the level of quantum mechanical process, where promising to measure particles in a certain location does not guarantee reciprocal the cooperation of particles actually being where they promise to be.

Before probing nature too deeply, let's get more familiar with the idea of this non-reciprocity with some more mundane examples. In systems with functional adaptations there is a clear benefit to being able to go only in one direction: it can bring survival value. In biology, there are semi-permeable membranes that allow movement in a single direction—why raisins absorb water and become fat when cooking. The process is used inside lungs for oxygenating and scrubbing carbon dioxide from blood, as well as directional processes for removing pathogens from circulation by active filtering involving many kinds of cells. There are elementary processes, and processes at scale, where events unfold in a specific direction, a particular order. There are many many examples, both of directional preference and complementary (+) and (-) promises being needed, in the dataverse too. They include obvious topics like security barriers, firewalls to prevent intruders, switches, and information filters[100]. If you mute the microphone on your phone when talking to someone, you create a semi-permeable one-way channel. You accept (-) the voice offered (+) on the phone call, but you don't return the offer (+) even if they promise to listen (-). Or the other party could refuse to accept (-) the call that you impose on them (+) in the first place.

It's interesting how spacetime plays a key role in the effective functioning of security matters like access control. Before so-called virtualization became commonplace in 'cloud computing', there was a tendency for technologists to confuse the locations of their assets in the dataverse with their geographical locations in the embedding space of the physical world. Firewalls were built to be physical gateways—a literal box that acted as a selective interruption on what was sent along the wire, like a passport control or Checkpoint Charlie within an iron curtain around a region of physical space. Today, there is a greater realization that locations in the dataverse do not have to correspond to any particular physical geography, so access controls have to be specific to every individual agent, just as Shannon's model predicts[101].

Today we are inundated with channels of communication relating to all kinds of processes. The promise to bring data is not necessarily matched by a promise to pay attention or receive. If you don't have the right password, or passport, or if you arrive at the wrong place, you'll meet a dead end. Access controls are a necessary part of any spacetime on a semantic level, and thinking about boundaries as walls around regions misses the point about the nature connectivity. If we control the basic processes of spacetime, then we don't need to build walls: we can simply leave out the promises that make points adjacent to one another in the first place. More pragmatically, we can't always tell the difference between an intentional rejection for access and an unintended fault—because intent is itself a figment of channel cooperation.

Barriers are not only for safety, they also have functional roles to play. Processes may route particular messages to particular locations in order to perform a function—say, separating needles from hay. Just as processes can give spacetime locations different properties and interpretations—i.e. locations may keep different promises—so they might be selective in what they do with the information that arrives within their borders. Subroutines in computer programs are structures that contain such partial processes. They are scaled clusters of micro-processes (superagents, or packs of wolves, if you like) with semantically delimited activities. By packaging a process with known semantics into a labelled cell, machinery can be copied (reproduced) and reused in a variety of contexts, just like electronic components in circuitry, or differentiated cells in biology. The tags or labels that identify membership of a particular process can be implemented for each wolf in a pack, or by creating a fence around the pen for all the wolves implicitly, provided they are connected by communication[102]. This is how it is possible for services on the Internet to appear with a single name or point of entry, by typing in a name (its URI), while in fact being distributed all over the planet, and for every MacDonald's or Starbucks to act as a single organization, with a single organizational boundary, distinguished by keeping exactly the same promises no matter where they are embedded around the globe.

Being next to something is the same as being able to keep a promise to pass information (influence) to it. Whether that is a digital bit, a photon, or a punch on the nose is only a question of scale. We naturally think about the motion and adjacency of physical things in terms of our experiences, but this is a limitation of our human ideas of physical reality[103]. If the phenomena that we measure are only 'promises' that move around virtually on top of an underlying set of agents—promises that enable neighbouring clusters of agents to function collectively to create the phenomena of matter and energy (like a pack of wolves), then there is no contradiction between the picture based on Shannon's observations, and the conventional ideas about matter, energy, movement, and the principle of relativity. This may require us to change our views on what is real.

Adjacency is all about the communication of information. Messages are more fundamental than the idea of touch, because messages can go in one direction without having to go in the other, whereas touch is always mutual (because Newton's third law applies on that scale). I might be able to see or hear you when you can't see or hear me. If there are two points in space, one can be next to the other without the other being next to it. Normally, we don't describe space in this way, so it sounds strange, but sometimes we have to find ways to express this. We've all seen one-way glass, one way streets, doors that open one way only, garbage chutes. All of these effectively create spaces in which A can influence B, without B being able to influence A[104].

The reason for singling out promise theory as helpful model is that it accounts for these matters in a basic way, with a simple rule that's compatible with an information theoretic view. That which is promised as an offer by one agent, called a (+) promise, is independent of that which another agent may promise to accept, called a (-) promise. In order for influence to pass from one to the other, in just a single direction, there must be a (+) and a (-) promise made between the two.

BOUNDARIES, INFORMATION

Regions of space can't be expected to go on forever. They may have edges or boundaries where properties change abruptly, or *discontinuously*. These edges play the role of the senders and receivers in Shannon's model of information. That makes edges particularly important: they are where information enters and leaves a region of space.

We draw semantic boundaries artificially, based on changes in how we interpret the local properties of space (see figure 4.3). Boundaries are used to demark political regions (countries), governance regions (counties), resource or functional regions (cities built around ports or mines, etc), and they are used extensively in human organizations to create departments, and offices of special responsibility. Boundaries are used to package physical goods as well as services—that's how we make boxes and containers. In computer science, modules play the same role, and in biology there are cells, organs, organisms, and habitats at various scales.

Not all of these examples have intentional boundaries, as an organism does: some are simply 'sudden changes' (as defined with respect to a particular scale of interaction) in the properties of space, as in the case of habitats. There are simply changes in the promises offered by different locations, connected by different channels of cooperation. On the other hand, this is really the same thing: an intentional change is really nothing more than a change in one region of space or time that is complicit in precipitating a change in another region of space or time, In computer science, this is referred to as a *dependency*. The use of the word 'promise' is more relevant than our initial prejudices suggest. Indeed, when physicists talk about 'locality' to reflect the idea that the source of change is preferentially close by, rather than far away, they are effectively drawing ring-like boundaries, like contour lines of influence, around an observer. This centric view has a critical role to play in understanding physics.

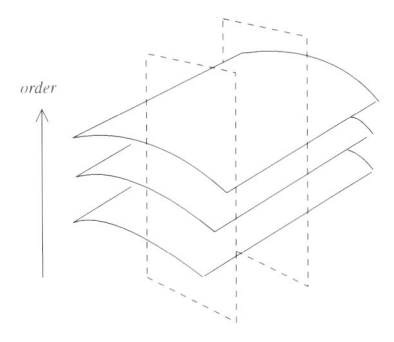

Fig. 4.3. Hyperplane (subspace) decomposition of a space. The hyperplanes are boundaries, and they are also subspaces of lower dimension, embedded within a larger space of higher dimension. Spaces being within other spaces (though we may prefer to call them points, lines, planes, volumes, etc) is an important concept in geometry.

The concept of embedding one region inside another is extremely useful, and is so often used to explain spaces, because it really relates to the freedom to switch on and off certain channels of information to the outside that may or may not be useful in thinking about a system. It's fair game to invent new ones as long as they don't play an important role in outcomes. The difficulty arises when those habits become so ingrained that we start to believe that they are unavoidable. There are many examples where physics muddles the convenience of an embedded problem formulation with the reality of a continuum, and then has to scrabble around to patch up the result.

One of the famous laws of electromagnetism in physics is Gauss's law. It relates the amount of charge in a region to the electric field produced from within it. In other words, what shines out of a boundary is a result of what is on the inside of it. Or the information emerging is a result of the sources within. The exterior promises are a result of the promises on the interior. The strength of the pack is the strength of the wolves.

Not all boundaries are enclosures. Suppose half of space is yellow and the other half is green. Then there will be an interface between these, or a sharp boundary between the two regions, separated by a line, a plane, or a hyperplane depending on the number of dimensions to space. The boundary itself forms a subspace of lower dimension, i.e. one that fits inside the larger space, as in figure 4.3. Examples of boundaries are what we refer to as 'boundary conditions' in mathematics: they are the parts of a process description that we match to reality, as anchor points, when trying to predict what happens in between. Again, it is the boundaries of a region where actual information is input[105].

By definition, symmetries and invariant aspects of space don't change, and therefore contain little or no information. But they are beautiful and we are clever to have seen through the pattern. Boundary conditions on the other hand are the information that starts a process and tells us how it ends. They are not invariant under those same symmetries, by definitions. They come from outside the rules of the beautiful idealized system, as they must. They represent 'exterior forces' to godlike observers. Boundary values are where the real power lies to play with the invariances in order to send messages. Moreover, just as a castle wall or a hundred suits of armour can be boundaries that play similar roles, it would be wrong to assume that a boundary must be a ring of fire around a huddled enclosure. How we define 'next to' depends entirely on what process we are talking about.

Intent with probable cause

Knowing a state, and knowing the story of influences and circumstances that led to it, are quite different things. Causation is the narrative of how change unfolds, but stories unfold on many scales. Once again, ideas about space and time are complicated by scaling. We can speak of causality from event to event, at the level of elementary processes and their intermediate states, or we can extend it to the broader arcs of human history. As is always the case with scaling, the larger the scale of the state we try to make inferences about, the less resolution we can incorporate, or infer. Causation is related to both computation and reasoning at a basic level, as we'll see in the chapters to come.

In computer science, the concept of causality is summarized neatly by the idea of necessary preconditions for an event. Some physicists define causality as a partial ordering of spacetime points along a trajectory—meaning that what comes before, in a time ordering, leads to what comes after it[106]. That sounds a bit like the idea of a precondition, but in fact it's a much narrower definition. The structure is something like the flow of a river, with tributaries contributing from side channels along the way.

Causation relies on chains of evidence that lead up at particular events. As such, it involves the give and take of information, passed along a sequence of agents locations. The result is to combine singular locations into strandlike superagents. These superagents are more or less one dimensional, but if there are tributaries, then they form tubular bounded regions of spacetime, connected by *dependent interactions*, whose width depends on the strength of influences from the environment.

In special relativity, these rivers are called worldlines. They are not really independent lines, because they are all part of a kind of fabric that adds up to what we call spacetime, but we can follow individual threads in the skein to imagine a pictorial view of history. Each link in the chain of interactions is effectively ordered by its prerequisite dependence on the changes it samples from neighbouring agents. As the changes (+) are sampled (-), each subsequent the order of the events creates its own form of time, in which the clock is the strand itself. By being sampled, the neighbours become 'earlier' and the sampler becomes 'later'. If each location had persistent memory to remember the interactions, then an imaginary deity looking down from exterior embedding space could trace these causal threads throughout the cosmos, as in figure 1.3.

Because it creates chains of dependency through spacetime, causation is a non-local phenomenon—it depends on more than a single location. It involves the joint give and take of information channels, with senders and receivers, communicating over an extended path. For that reason, it is impossible to speak about *the* root cause of anything that happens in spacetime. There is a range of scales over which we can discuss the causal evolution. There is the 'proximate cause' (the previous agent or agents in the chain, at some scale) and then there can be particular causal changes that are identified as having 'special significance' (+) relative to the interests (-) of particular parties. All of the previous events that latter events depend on can be considered partial causes for them. This is what we mean by causation. Information is flowing in all directions, even in the simplest situation.

The lack of a proper 'single point of cause' is a semantic headache in information systems, where we are trying to design processes and machinery with simple causal properties, e.g. pull the level to cut the paper, turn the wheel to steer the car, and so on. When a part of a process unfolds in an unexpected way, i.e. a part of it doesn't keep its promise, technologists often look for the 'root cause' of the problem[107]. They imagine that there must be a particular agent at fault, or a particular promise that needs adjusting. The detailed semantics of causation in processes are complex, and therefore unfamiliar to most of us. Moreover, in society, causation is often muddled with the idea of *responsibility* for the outcome. In accident investigations and in law, probable cause is an indication of a malicious intentional breaking of the social contract. The fabric of society, as a kind of spacetime made from interactions between people and things, is held together by promises that are restrictive. The laws and rules we make for social order are the Equations of Motion for processes within that social spacetime, but in a network of social complexity change is non-linear and unexpected outcomes are not only the effect of change at intentional boundaries, they may also result from the feedback self-inflicted by emergent boundaries.

Even at a simple level, causation does not follow the simple model of there being an unambiguous cause. There might be an isolatable *source* for a chain of events, but that's also subject to relative viewpoint. For example, suppose I shoot you (give +) and you are wearing armour (don't take -), then my intended or unintended imposition (+) does not cause you any effect: there is no channel completed that connects sender and receiver. If you remove your vest (restore -) and die, was it me or you who caused the death? Who was responsible?

Similarly, suppose a fire breaks out and a fire alarm sounds (+), and you are wearing ear-plugs (-). The imposition of the alarm doesn't affect you. If you remove the ear-plugs, you hear the alarm. So: did you cause the alert or was it someone else? Who was responsible, who *caused* the alert? Today, the proliferation of portable music sources means that a large number of people stumble around cities with loud music blocking their senses. If they walk into traffic and die because they didn't hear something coming are they responsible? The chain of evidence can be traced back as far as we like. Do we blame the driver of the vehicle for hitting the person, or the person for stepping out, or the earplugs for blocking the sound, or the manufacturer of the earplugs for making them in the first place?

It feels natural for the direction of causation to be from (+) to (-) (sender to receiver), but the order of events that enabled this condition was (-) to (+) (receiver to sender). Promise theory clears up this conundrum, because it makes causation (propagation of influence) a property of a link not its endpoints. Intent to harm (+) usually trumps negligence (-). But if changes are considered random, then negligence (-) may itself be considered harmful intent.

The dynamics of causation might be simply about the order of sequences of give and take, but the semantics are complex, because they pertain to the larger contexts of the spaces in which events transpire. Crossing the road is one such context. Society expects pedestrians and motorists to keep certain promises to avoid such accidents. The law uses *obligations*, or the imposition of an expectation that everyone keeps a certain promise. Of course, we know that we can't really impose (+) any such desirable behaviour on someone. Instead, all we can do is to promise to interpret the lack of a promise to act responsibly as negligence. We often present these matters in rather simplistic ways, but there are complex webs of promise channels involved in how responsibility and blame are assigned.

Looking forward or looking back

In physics, it has been understood since the start of the twentieth century that causation can be described in two different ways—by what is called *advanced* or *retarded* boundary conditions. Sometimes the methods are complementary and sometimes one is more suited than the other to describe a particular phenomenon. They are directly analogous to the question of whether demand (-) arises before supply (+) or vice versa in triggering a process. The interesting observation is that when the information about demand is what that triggers events, it has the appearance of time reaching back from the future to pull events forwards. The causal structure has been both utilized and mysticized in quantum theory, but there is nothing particular to quantum theory in this observation. Again, it is a simple consequence of the information structure of causal channels.

A retarded boundary condition anchors its predictions to information provided in the past, and the consequences propagate forwards into the present according to the promised process. For instance, you start with an initial clean slate, and gradually fill it with writing. An advanced boundary condition fixes information in the future and its effects are computed backwards to the present, like working backwards from the answer: e.g. suppose you know what the outcome will be but you want to know how the journey will look (this is called a desired end state model). For example, you want to put a person on the Moon, but don't yet know how. Information enters on the boundaries of the process. If we have information about the end state, then we can work backwards. If we have information about the starting state, we can try to work forwards. In both cases, we have to assume that there are no other edges where information can enter (such as signal noise). Advanced boundary conditions do not imply reverse time travel—they are just an approach to calculating the chain of promises backwards from the end to predict starting from a hypothesis or final desired outcome.

This method is used a lot in technology. Perhaps we know what we want to build: we formulate a blueprint and then we work backwards: set up assembly, order the parts, make the parts, get the materials for the parts. Think of a restaurant: we take the order, which does not yet exist. The order determines a desired outcome, which is handed back to the kitchen, which figures out a recipe, which gets the ingredients, and then once each dependency is identified, the chain rolls forwards. It's called *recursion*—or a last-in-first-out (LIFO) process stack—in computing. Shall we claim the cause of the food was the customer order, or the presence of the ingredients? This depends a bit on how we connect up the dots in the causal graph (i.e. the DAG discussed in chapter 2).

This intentional method of reverse causation is well known to computer programmers as a way of avoiding unnecessary effort (e.g. when building software using a Makefile), and to logistics industries as Just In Time delivery. Whether the (+) or the (-) promise comes first doesn't matter to the outcome, but it contains information about the conditions at the boundary—i.e. about the bigger picture. The local decision for each agent to give or take has no prerequisite order, because there is no common time on which to measure them until both promises exist to create a causal link between them. In economics the independence of supply (+) and demand (-) means that mismatched intentions have local sales implications. In business, mismatched incentives (-) and rewards (+) fail to start processes. In security, a mismatched lock (-) and key (+), or encryption (+) and decryption (-) prevent the passage of unauthorized influences. Starting with the (+) is what we mean by imposition. Starting with the (-) is how we understand and define intent—it has little to do with free will.

Amongst the algorithms for data processing in a computer, the give and take of information have to be carefully balanced. The old aphorism 'garbage in—garbage out' is not quite right, because if a (-) promise is given to interpret data, without a compatible (+) promise, the computer will still try to make sense of the garbage and print it as non-garbage letters, or (worse) execute it as unintended commands. Hackers exploit such careless mismatches to precisely this effect, in order to coax a computer to interpret information unintended by the designers as commands and pervert the course of the process. This is also the behaviour of a virus in biology: rogue molecules (+) that partially match a receptor (-) can gain entry to a cell and pervert its machinery to reproduce the virus.

There is clearly more to causality than time ordering (the order of prerequisite dependencies). An almost unchallenged assumption that runs through science places the seat of causality at the origin of an arrow pointing in the direction of change, rather than going backwards from its destination. Nothing could be more natural than to begin by thinking that time and causation flowed like a river from downstream from a source[108]. Some writers have become fond of dismissing the existence of causality, based on the observation in statistics that 'correlation does not imply causation'[109], but this assumes that partial orderings and directional relationships cannot occur, which is easily refuted. Others claim the apparent arrow of time is related to the entropy of the universe, but this ignores that fact that microscopic individual processes exist independently of entropy. Causation goes beyond discussions of the 'arrow of time' in the universe at large, to something even more fundamental. Without the intrinsic directionality of Shannon's information channel, the entire notion of an ordered process is impossible, and there would be no spacetime. There is no denying that the partial ordering of information is unidirectional, but whether it goes forwards or backwards according to your particular measure of time depends quite a lot on how you look at it, and which clock you are running on.

How we explain problems, using stories, can be quite complicated. It might involve a chain of dependencies—a full blown story involving many concepts over a long and distributed confluence of histories. If we want to prevent faults, errors and deliberate sabotage, i.e. stop a problem trigger state from being causally connected to any outcomes, or isolate the space of contexts in which the problem can gain traction, or contain the process that unfolds once the problem has occurred (managing symptoms) by responding quickly, then we have to acknowledge that all of these are spacetime related questions. Central to all of these matters is the simple duality of give and take implicit in Shannon's information channels.

It could seem that, with such an abstract and overarching principle at work, based on the connection of events by channels of influence, that the impact of these observations could only be trivial and cosmetic. Yet the implications strike at the heart of our ability to manipulate processes to intended outcomes, everywhere information plays a central role.

Quantum measurement

I couldn't leave a discussion of spacetime and information without mentioning Quantum Theory and its (by now) infamous and peculiar properties. The interpretation of Quantum Theory has dogged physics since it was pieced together from fragmentary phenomenology to understand the behaviour of electrons on the atomic scale. Many physicists would immediately dismiss a discussion of quantum phenomena as something entirely specific to atomic systems, and therefore unlikely to be relevant in a book about multiscale phenomena, but I believe there is a case to be made for comparing the structure and concepts of at least one interpretation to what is now taking place in the dataverse.

There is a tendency to afford Quantum Mechanics an almost mystical and even final significance, but I think one should be quite careful to adopt the premature assumption that Quantum Mechanics is the ultimate fundamental truth about all scales below the atomic level. It is ultimately a set of constraints, built on top of a model of spacetime, just like many others. Most likely, it is only a stepping stone that bridges the scale of atomic phenomena and quantities from classical physics[110]. Although Quantum Theory is highly technical, and we have no hope of doing it justice here, we can get an inkling about some of the ways in which space and time are involved in its predictions and its interpretations, and especially how the concepts of give and take are fundamental to its interpretation. We can then compare these to spacetimes on other scales—especially in the dataverse. Some of its concepts, concerning motion, are best left for the next chapters where we can unravel them step by step, but first we need to introduce the basic essence of Quantum Theory from the perspective of channels of give and take. By presenting it in a light compatible with Promise Theory, it becomes surprisingly easy to compare its causal structure to what happens in modern computer systems. That in itself is an exciting prospect for potentially helping to understand both.

I should begin by pointing out that what popular books tacitly call 'Quantum Theory' is not actually a single unified train of thought, but rather a patchwork of rather different, fragmentary theories. There is Quantum Mechanics, which describes whole regions of space, focusing on single particles in the frame of reference of a stationary observer; then there is multiparticle quantum mechanics used in material physics. Farther up the mountain, there is Quantum Field Theory, which emerged from the desire to incorporate Einstein's Special Relativity into quantum ideas. It ends up describing something rather different from Quantum Mechanics. Initially, relativistic Quantum Mechanics appeared to be an inconsistent theory, because relativity is not about whole regions of space, but rather about different relative descriptions of processes. Later, its realm of applicability was realized: a relative theory is only consistent in the presence of interactions (where relative time enters)—and this spurred on the quantum understanding of interactions between light and matter.

Quantum Mechanics begins with classical ideas of particles, positions and momenta, and then tries to transplant these into the world of the atom, where they seem to be somewhat ill-suited and contradictory. It takes the classical notions of position and momentum of particles and goes 'full Einstein' on them, by saying that these concepts only make sense in terms of what an observer can measure—not necessarily as objective realities for all. This introduces the notion of (+) and (-) roles and a number of information channels between spacetime as a source and the spacetime of observers—almost as if observers should have to stand in line to read information from a quantum system.

Following the discovery of electrons and nuclei, in the early twentieth century, it was assumed that electrons would orbit around nuclei like the planets around the sun; however, classical electromagnetism predicted that they would then radiate away all their energy until they crashed into the nucleus. Moreover, the absorption and emission of light from atoms revealed a flaw in the continuum theory of the electromagnetic field: rather than seeing every possible frequency of light emitted from atoms, only very particular frequencies of light could actually be absorbed and emitted, corresponding to 'allowed energy levels' of electrons in the atoms. To account for these facts, the view of an atom as a continuum system was replaced by a model of an atom as a 'composite superagent', with unexpected interior properties. The discrete 'jumps' of quantum measurements were puzzling. They can't immediately be seen as a consequence of the discreteness of a physical assumption. Quantum Mechanics takes place in the same Euclidean Cartesian embedding space as Newton's laws. Rather, discetness emerges from an intermediate 'overlay' process, whose geometrical constraints lead to discreteness at a higher level.

Formally, Quantum Mechanics begins with an equation for a continuum field called the 'wavefunction', which is the effective 'overlay spacetime' that describes its dynamical and semantic behaviour. It contains the distribution of properties of the quantum system that are compatible with conservation of energy. As we'll see, the crucial thing about this overlay is that it isn't like the underlying embedding space: rather, it looks like a channel for communication from source to receiver, which acts as a proxy for the distribution of energy[111].

Erwin Schrödinger's famous equation determines this wavefunction, usually written $\psi(x,t)$—defined over absolute embedding space x, and in some cases involving change over times t. The absence of time from the original Schrödinger model followed from the equilibrium assumption, used to look at static, single atom energy models. It was later extended to allow time dependent interactions, which introduced a partial relativity between different subsystems, but not in a way that was fully compatible with Einstein's relativity. Non-equilibrium interactions introduced a measure of exterior time through time-dependent potentials. Quantum Mechanics is not a limiting case of the relativistic theory, it's a completely different theory that asks different questions—but we'll stick with it, because it is a very interesting construction.

The wavefunction $\psi(x,t)$ is the central variable of Quantum Mechanics, but it has no counterpart in classical mechanics. It is an extra layer of dynamical content, like a machine overlaid on top of it. A lot is made of the use of *complex numbers*, for representing the wavefunction—as if this confers some mystical depth to the theory. In fact, I see this as a natural expression of the reinstating of 'give and take' degrees of freedom that are smudged out on the classical level. Properties, like velocity or momentum, that can be treated as approximately local on a large scale, fail to make sense in Quantum Mechanics, because they simply aren't local and Quantum Mechanics knows it (see chapter 5). The quantum wavefunction has an analogue in the dataverse, in the layer of software between the physical infrastructure and the virtual processes that run on top of it. I'll return to that too shortly.

Quantum Mechanics bridged Newton's classical concepts of position and momentum, inserting them into a non-local model, by replacing them with linear operators (*matrices*) that sampled the wavefunction's interior degrees of freedom at each spacetime 'point' x. Substituting those operators into the basic machinery for energy conservation led to solutions for the wavefunction that predicted the energy processes, and from there the stable energy levels of the hydrogen atom—all with unprecedented accuracy. Nevertheless, the non-locality of the construction and the results was unexpected and confusing.

Let's think about why. Up to that moment, equations of physics had been 'local' by assumption. That meant that the behaviour of each point-like entity depended only on its own position. But this is really a fiction that results from Newton's limit method of approximating points as having zero size. The accidental discovery, by Max Planck (1858-1947), Einstein and others, that light emissions from atoms had both a discrete and a wavelike characteristic ended that possibility. Spacetime phenomena were discrete, and points were no longer the primary variables of the theory: only measurements were. Those measurements could not be shrunken down to an arbitrarily small locale. In Quantum Mechanics, operators for momentum and energy became non-local to explain the quantized wave characteristics of electrons that, in turn, led to discrete emission frequencies.

The derivative and matrix operators, which replaced classical variables, represent cooperative changes between adjacent locations, i.e. *gradients*, that span multiple points—energy channels that redistribute discrete amounts of energy non-locally. However, the classical property of a particle position is, by definition, local. The continuum limit trick of Newton and Leibniz (known as differential calculus) could initially paper over the non-locality, but—for these two things to be consistent—the notion of spacetime processes, including position itself, have to become non-local[112]. The concept of location (the coordinates referred to in the theory) and concept of position (the result of a sampling of particle position) became dissociated.

Heisenberg's Uncertainty Principle also follows from this non-locality. In fact, it's equivalent to the Nyquist uncertainty in resolving a distribution by sampling when the distribution is composed of modes with a particular range of energies. Heisenberg's principle shows that the smallest observable position, compatible with a classical concept of quantum momentum and energy, must tie the concepts of position and momentum together as if they were wavelike. From the continuum limit perspective, this seems weird; but from the viewpoint of momentum as something only defined over the ends of an information channel between 'points', it's necessary for consistency. A similar phenomenon follows in the dataverse, because of the virtual layers of processes.

The theory tells us that we can't get away with assuming a false level of precision, afforded by continuum spacetime, as long as energy processes are distributed in such a way as to lead to discrete quanta. That continuum has to fragment into cells, from the graph of constrained processes. This makes samples themselves non-local, and thus effectively *probabilistic* across the overstated locations of the system machinery. In a sense, this is an artifact of the assumption that $\psi(x,t)$ covers the whole system, with the boundless detail of a continuum, but the problem solves itself as the solutions of the energy constraints single out the only consistent possibilities. Mathematics has a way of showing you what you did wrong—but we still do it anyway, because the mathematical tools are so well known. In the future, we might find other tools that don't encourage this misrepresentation in the same way. It's not uncommon for authors to say that the electron itself 'is' the wavefunction as well the point measurement, and thus has weird behaviour (+), but as we'll see it makes more sense—from an information perspective—to attribute the weirdness to a measurement *channel* between a coarse grained process with measurable properties and an observer. The distinction is important, for reasons that will become clear in the next chapter.

Non-locality, in the outcomes of quantum measurements, is effectively forced onto the theory by deliberately trying to shoehorn classical concepts into it—in order to bridge a *correspondence* between them. Position and momentum are classical concepts, characterizing processes above the atomic level[113]. Correspondence is fair, because all our ideas and measurements are based on those concepts.

Physicists and philosophers still argue about what this all means. From a scaled information view of spacetime, some of the weirdness becomes less objectionable by thinking more carefully about the natural boundary that describes what local and non-local really refer to, as we'll see in chapters 5 and 7. If we try to measure something classically defined as a 'point property', when the definition of a point is itself under scrutiny, then we will likely muddle interior and exterior degrees of freedom, leading to a channel sampling mismatch. That, in turn, means that we may only obtain a definite value by projecting out a single case with a partial probability. It's a situation unprecedented in physics, but not in engineering.

QUANTUM GIVE AND TAKE

Quantum theory is full of examples of Promise Theory's give and take rule. In fact some of the most important discoveries from Quantum Theory involve the matching of dynamical agents (+) with possible outcomes (-)—or, what Quantum Theory calls the matching of particles with available states. Some of these will have to wait until the next chapter. For now, let's focus on some signature cases, where the role of give and take prove central to the behaviour of processes.

The discovery of antimatter, and the oddly related phenomenon of 'holes' in the conductors that make up all of our electronics, is one such case. Atoms in solid form regular crystals whose wavefunctions overlap and smooth out the discrete lattice of points on the scale of the atoms. In a conductor, the energy levels line up to allow the electrons the roam freely and conduct electricity, instead of being bound to their atomic locations. An effective continuum forms on top of the discrete atomic spacetime—yet another twist in the story of discrete and continuous space. On the whole, conductors are electrically neutral, but locally there are variations in the density of mobile electrons, compared to the fixed background space of the atoms. A possible location or state in which an electron could exist, but is absent, is known as a hole. Interestingly, holes (gaps in the sea of negatively charged electrons) look a lot like positively charged electrons or *quasi-particles*, from a relative perspective. They play a surprisingly important role in describing the behaviour of electronic conductors and semiconductors, even though they are only a 'virtual' process that runs on top of the actual fixed spacetime of the crystal.

In 1928, as physicists tried to reconcile Quantum Mechanics with Einstein's theory of Special Relativity, physicist Paul Dirac (1902-1984) derived an equation in which the channel for sending and receiving energy was compatible with Einstein's generalized energy conservation rules. The generalized solution of his equation turned out to be the proper equation for electrons and other fundamental particles like quarks that can move around relative to one another and 'communicate' by interacting with other particles. Dirac's equation predicted something new: not only did the measurement channels connect spacetime regions, they revealed new properties on the exterior of these regions: electric charge and a new quantity that looked like an interior rotation and was thus called 'spin'. Moreover, the symmetries of the channel between send and receive, revealed that by running it backwards there must be another form of matter: with opposite charge, which came to be known as antimatter. Physicists Feynman and Wheeler pointed out, in turn, that antimatter looked essentially like normal matter travelling backwards in time, with advanced instead of retarded boundary conditions. They formulated a theory of particle mechanics with absorption and emission, the precursor to Feynman's famous diagrams, and saw how the ordering of events could create 'hole'-like antiparticles, which came to be called positrons[114].

Once again, the directional roles, in the interplay of information between emission and absorption, sender and receiver, lock and key, or occupant and state, became a key part of a process description. Just as an electron could fill a hole and make it disappear, releasing some electrical energy, so an electron and an anti-electron (positron) could annihilate one another releasing some spacetime energy in the form of a photon. Alternatively—and surprisingly—the opposite was also allowed: spacetime could open up a spontaneous channel, creating an electron and a positron at the ends of a pair of channels, on the interior of a region of space which could then collapse and vanish—a quantum loop—unless furnished with energy from an exterior photon channel: that would allow the pair to separate forever, and pass beyond to the exterior of the region.

Spontaneous pairs are known as *virtual particles*—though we should rather call them virtual processes, since their 'particle' nature is both irrelevant and inaccurate. It's often said that they borrow the energy to create particles for an amount of time, before paying back the debt to spacetime. This monetary metaphor is an interesting one that I'll discuss in chapter 9. Physics offers no real explanation for what this means; however, the analogy is flawed in the sense that there is no clock exterior to a region of space that can measure this borrowing and repayment. Whether the process acts like a loan that takes measurable exterior time seems unobservable, emphasizing that a naive view of time is only a cover story up for the existence of ordered processes, not all of which belong to our own timelines.

Matching local holes and occupants explains quite a lot of phenomena in material science, and in Quantum Theory. The information about what crosses some boundary to enter a region, and what comes out, is what we call energy

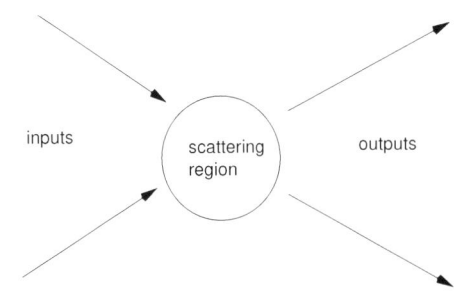

Fig. 4.4. In physics, a scattering problem is a generalized transition problem over a bounded region. It takes the form of a black box, where the scattering takes place. All we know is that something comes in, and something goes out, with different velocities, and some interactions happen in the scattering region. This could equally well describe a computer program that accepts inputs and produces outputs.

scattering (see figure 4.4)[115]. Dirac's equation showed how matching properties of promise roles (+) and (-), in the relative description of the electron field led to new semantics in the way electrons occupied available energy levels, a new degree of freedom called 'spin' (with only values 'up' or 'down'), and the possibility of channel separation of matter and antimatter (electron and hole) provided the absorption of a photon could sustain the channel between them. In each case, the matching of these 'lock and key' combinations led to channels of influence in the theory associated with the communication of energy in matching forms. The framing of the phenomena is not so much an explanation as a pointer to what is concealed. The entire meaning of absorption and emission remains a mystery in Quantum Theory. Where exactly does a photon go when it is absorbed? From a functional perspective, this is incomplete. But there is plenty of room at the bottom of physical space to hide the details.

The examples of Dirac's field are about the coming together of matching (+) and (-) promises at the same spatial location. This is unlike the original Schrödinger formulation of Quantum Mechanics which described all of space at the same time. In a relative theory, opening a spatial separation required the intervention of a third signal, like a photon to sustain it, and introduced relativity, which was not present in the wavefunction formulation.

An interesting side effect of Dirac's equation was the way in which spinor fields took up space.

The depth of a point

There is something quite odd about space, no matter whether we are talking universe, dataverse, bioverse, or something else. For example, if you fill a glass of water, you can't fill it again. Once it is full, the space is occupied. But if you send a disturbance through the space, resulting in waves travelling through the region, there seems to be no limit to how many waves you can send through it; in fact, they mainly pass through one another unaffected. This latter property is called *superposition* of states. States or disturbances can be added and subtracted, leading to the well understood phenomenon of 'interference' of waves. If you run a program on one computer, it's busy so you can't run another program at the same time and place. But, you can relay apparently several messages or files through a computer that don't affect each other[116].

Both water and waves are forms of energy, in the language of modern physics—or forms of information, in a more general sense, yet they behave with very different semantics. Processes and messages, in the language of the dataverse are both aspects of computation, but with different semantics. It's as if space has some kind of tenancy arrangement that only works for matter or processes, but is happy to allow the free passage of waves or messages. But how deep does this right of passage go for waves and messages? Is there a limit to how many are allowed? In other words, in the language of Shannon, what is the channel capacity of a point?

In a computer we can answer these questions quite easily. A computer fills its resource space through memory and processing. At the hardware level, transistorized chips occupy non-overlapping network locations. At the software level, memory is occupied by data exclusively too[117]. Processes can be interleaved by 'time sharing' or multitasking, but each chip can only work on one timeline per chip. In biology, too, cells can't overlap, but they do divide into more cells during growth! Interior cellular processes are performed by molecules and limited by access to the genes: busy or free. Superposition is timesharing.

In elementary physics, the concept of occupancy is stranger. Waves don't interact like a chemical interaction, material things bumping into each other or binding to form a different substance—waves pass right on through one another and keep going. Two waves can occupy the same location, and they flow through one another, adding and interfering at each location, then eventually moving apart. That suggests that—unlike the states that make matter—there must be a *quantitative* representation of the information they impart, not just a tag, which can be added and subtracted like numbers. That suggests that there is something about space—even for the most fundamental turtle—that can sometimes add and subtract quantities like a computer's timesharing, and which other times merely assigns them semantic labels: occupied or vacant.

In solid state physics, the effective spacetime of electrons is a field of quantum states or 'energy bands' generated by the parent atoms in the material. Electrons fill up these bands, but in superconductors the electrons form 'molecular' pairs known as Cooper pairs that behave more like waves and can pass through one another by superposition. If the bands become too full, electrons can break out of the solid spacetime altogether, at which point the model breaks down. Then, the quantum fields of electrodynamics and the Standard Model, ostensibly fill every point in our universe. What could possibly be the mechanism for occupancy and filling for matter and energy in the world of quantum fields?

At the start of this book, I discussed the idea that space and matter need not be two different things—that matter might simply be a kind of state that space finds itself in, like a program running on a particular computer. But the two things seem to have different semantics. If a discrete region of space is occupied, then we can't occupy it with something else, without displacing the original occupant—unless we could interleave both on a much larger scale so that we couldn't tell the difference. In the dataverse, this illusion is performed by design. We can imagine a computer being busy with one program so that nothing else can find room to enter it. Simple yes/no semantic *type* labels fit with the idea that you can fill the space once—just turn the lock on the stall to occupied—-but such labels can't explain how excitations like waves could pass through a point, unless the location were not simply tagged, but were used to record a quantitative property. The way to do that is to mimic a continuum of states with a lot of discrete ones. How big does a bucket of glass beads need to be to look like a fluid?

Occupancy is a semantic property—the answer to a yes or no question. Waves, on the other hand, span many points like rolling hills—and, at each point, there is an amplitude, which is the quantitative magnitude of the wave disturbance as it passes through different locations. An amplitude is like a real number in a computer program, like a concentration of chemical agent in a cell. Its magnitude can change—and change is the same as saying that time is implicated. If locations—the agents of spacetime—are thought of as memory, like a simple computer, then there would be a limit on the memory you could fit into each point. That means there would be a limit on the largest value that could be represented at a point: a maximum amount of energy that we could feed into it. At some point, waves could not be superposed. Does that mean that every kind of spacetime has to break apart when we try to fill it too much?

These two kinds of behaviour can easily be observed, at any scale. However, it's interesting that Quantum Mechanics predicts them in the form of two kinds of spacetime phenomena called *bosons* and *fermions*[118], which emerge from different spacetime considerations, such as distinguishability of spatial configurations during statistical equilibrium. The solutions to Dirac's equation for relativistic electrons, with their spins, turn out to be fermions. Electromagnetic photons turn out to be bosons. Bosons can be stacked up like waves, as many as we like, whereas fermions can only be placed one per unique state of spacetime.

An analogy I like to use is that bosons are like voices of a song. They can fill space and travel around, overlapping without trouble. Fermions, on the other hand, are like the choir: they take up space, so that only one choir member can be in each position on the choir's bench space at a time. Fermions are what make up matter sources and receivers, as far as we know, and they are the endpoints for all signals. Signals, on the other hand, including waves, are bosons. The image is not perfect however, because fermions can be paired up to act as bosons (as in Cooper pairs of superconductivity), and spin waves can travel like the waves crowds make in sports stadiums.

The idea that superposition is like addition and subtraction can be mimicked with matter too. You can build one building on top of another, e.g. a roof extension. The spires on churches were sometimes built as afterthoughts (often without due consideration of gravity). But this requires space to extend into. So when matter operates additively, it's 'state space' consists of the dimensions of ordinary spacetime, because the unique occupancy of each location means that increased magnitude has to be encoded by taking up extra space. What makes waves different from matter is that they implicate other degrees of freedom at each spacetime point. Waves occupy different amounts of interior states, as well as exterior locations.

If we think about waves that pass through matter on our human scale, such as the disturbances of water or material that lead to the surf and to sound transmission, then there we also observe limits on the magnitude, determined by interior details. Water waves occupy the two dimensional area of ocean, and their amplitudes extend vertically into the third vertical dimension. They reach a maximum height determined by the depth of the water available to add or subtract at each location. Sound waves, whose amplitudes are movements in the positions of atoms, straining against their bonds, or as bulk pressure variations. They reach a maximum magnitude when the bonds break apart and shatter the material, or the pressure variations fail catastrophically, as in a sonic boom.

The amplitudes that characterize electromagnetic waves, like radio, light, and X-rays, do not involve three dimensional distance, but rather electrical and magnetic field strength (a kind of energy). The electromagnetic field could be thought of as a kind of interior dimension at each spacetime location, like the depth of water, but its size is then a question. How many electromagnetic waves can be superposed on top of one another before no more is possible? What is the maximum size of the electromagnetic field at a point, and why does it have one? It was once thought that electromagnetic waves also needed a material medium to travel, known as the luminiferous aether, but this was replaced by the notion of a field acting as its own energy-buffer medium. Now even gravitational waves have been detected, which—according to Einstein's General Relativity theory—use the structure of spacetime itself as a medium for buffering wave energy.

In each of these cases, one can imagine that, as the energy density of a wave process reaches a certain threshold, other avenues of possibility open up for the process—that perhaps break the usual assumptions of a spacetime, or are simply amputated or 'clipped'. In quantum theory, an energetic photon wave can change into particle-antiparticle pairs, for instance. Tunnelling of particles between locations that are not classically adjacent happens too, meaning that new spacetime channels open up. Moreover, these effects become more and more probable the higher the energy density. So there may be limits to the amount of superposition that spatial agents can tolerate.

It appears that the connectivity of space and time may not be constant: rather it may depend more subtly on the limits of interior accounting at each location. The significance of this is open to interpretation, but it's not unreasonable to claim that this is another indication that energy, communication, and topology (adjacency) might all be aspects of the same thing.

Although bosonic and fermionic characteristics are conventionally thought of as being somehow exclusively quantum statistical phenomena, there are plenty of analogues in the dataverse, especially in connection with the scaling and virtualization of processes. Recall that in the 'computing cloud' the absolute spacetime of computer hardware runs software processes that can be relocated and form a separate virtual spacetime, where relativity is important. When a computer is occupied by a process, no other process can be accepted, because the processor serializes execution. In other words, its interior time generator is a serial computation—indeed, serial change is what we mean by time, though sometimes we might aggregate parallel points into a single scaled region. Memory is also exclusive. We can't record multiple memories in the same location at the same time without destroying one or the other. So, computer processes and memory occupancy behave like fermions: only one process can occupy the same state at the same time. If interacting processes were to be exchanged, then the roles of sender (+) and receiver (-) would be reversed.

On the other hand, messages passed between processes, travelling along exterior network links can sometimes overlap. This needs qualification though. Just as a single photon is absorbed or emitted at a time, so individual packets of message travel serially. However, as they travel along waveguides and signal cables, they can be interleaved and sometimes even travel along parallel channels (using a variety of technologies from multiple frequencies of light carrier to duplicated wiring). In other words, there is a physical reason why messages can be superposed, but not processes—it has to do with the way spacetime resources are shared. Messages play a different role. They are ephemeral states that don't need persistent memory. Processes, on the other hand, have interior states that require persistent memory.

The virtualization of processes creates new uncertainty. Virtual connections are a mirage if they are not mirrored by an underlying physical layer. Recall the network arrangements in figure 2.18. In a virtual representation of spacetime, any pair of end nodes can imagine a direct link to any other as long as they are part of the connected networks. However, if a number of messages have to share the same common underlying infrastructure in order to get from source to receiver, then there may be contention for that resource. This is true in computer information networks because, while messages might appear like bosons on a virtual level, they might still physically have to share physical resources, leading to contention between competing processes. Similarly, when mail order companies make sales promises that include delivery, it appears that any point in the country can be connected to any other, and a delivery estimate could be made based on a definition of the distance from the source to the destination. But if there are roadworks on a key road link, or there is sudden congestion due to an unfortunate confluence of occupants on the transport link, the transport time will be uncertain for invisible reasons. The observer will be left with a probabilistic view of the transaction taking place, just as in a quantum measurement.

Virtual layers know nothing about underlying physical transport processes. They see only the results of their own sampling, according to their own clocks. In other words, a message link might appear to support superposition on one level (a virtual level), but may actually involve a link via a form of memory that can only be occupied by a single kind of state at a time at the lower embedding level. Such matters, belonging to processes and their effects, are easy to see in the dataverse, where they cause real issues for information and financial systems, but we can only speculate about the bottommost layers of the universe. Delays tend to have cumulative or 'knock on' effects, so they propagate as systemic faults. In physics, there is no known notion of physical dependency except at the matter level. Defects and dislocations in materials do affect bulk material properties like strength or conductivity in this way.

The discreteness of a configuration space plays into these matters in a decisive way. Finite resources can be shared between multiple processes provided resources are discrete. In computer systems, both tasks and memory are shared by a method known as context switching or paging. The active resources involved in a process (the CPU or generator of interior time) can be used for one process and then flipped to another, as long as they are kept in different scopes or different 'worlds' that don't overlap so that they can maintain their integrity. If the resources do not have a fixed size, then there is no way to swap them back and forth. We couldn't suddenly put a larger process back into a smaller slot. Think of parking spaces. Parking spaces have fixed sizes and locations are laid out like a regular lattice so that they can be swapped and easily found. By constraining space, we make the process of interleaving time easier. In computers, proximity is important to being able to swap occupants and perform timesharing. Memory is handled a bit like valet parking. When some information arrives at a location, it might occupy some active memory (which is why the process is busy). However, the information can be parked somewhere freeing up the resource for something else. If the information is used again, it can be brought back by the valet. If not it might be moved out of the immediate cache to an overflow region. Process timescales lead to a gradient of memory spaces, like drop-off, short, and long stay parking at airports. Space and time are fundamentally joined through the mechanics of processes, but the link only makes sense when there is a fundamental, fixed scale for space (the size of a parking space), and a fundamental unit of time, which is the change that switches between states.

SUPERPOSING SPACE BY ENTANGLEMENT

A second example of give and take, in Quantum Theory, concerns the nature of the wavefunction, and a kind of communications channel known as quantum *entanglement*. This is where physicists usually bring up the issue of quantum non-locality, even though the entire theory is, in fact, non-local on a basic level. Entanglement is a property in which certain quantities like spin or momentum, 'promised' by regions of space, are accounted for over an extended region of space, and which are inseparably bound together by the process that created them. For example, when particle and anti-particle separate from a spontaneous creation event, spin is conserved—meaning that if a new particle with spin is created, an anti-particle with the opposite spin must also be created. This is not a problem as long as they begin from the same location, but once they move apart the theory starts to seem incomplete.

What happens between the 'particles' themselves is one issue, but what an observer can see about each is more interesting still, because is mixes together the channels of measurement between particle and detector with the actual channels steering the quantum interaction. If these were virtual, how much could we know about what happens underneath? When an observer measures the spin of one, he or she may then immediately know that spin of the other, even though it might be far away. When two agents become entangled, they cannot be separated without breaking the promise they keep together. It's as if two separate agents remain a single distended superagent rather than two separate particles, i.e. the two apparent spatial points are, in fact, a single point[119]. In information language, two agents are entangled when their state is co-dependent. Einstein initially objected to this interpretation, because it appears at first as though this could be used to send information faster than light, though this has since been shown not to happen—the subject is too complex to discuss here.

The concept of mutual co-dependence is not particular to quantum physics either. My friend Paul Borrill of Earth Computing uses information technology to simulate this concept of entanglement by mimicking the properties of interior and exterior spacetime using technology[120], with the aim of accounting for exact consistency of information transmitted between databases at different ends of a channel. Consistency of changes across spacetime regions is a property that finds increasing importance in the dataverse, as the world becomes more globalized. We distribute information, and expect processes to be spread around the globe far more than before. The scope of information accounting is now larger. The point is that what goes into an interaction region must balance what comes out, like an accounting ledger. If we treat the electron as the sender, by measuring it, and the positron as the receiver, then the act of measuring the electron's spin would seem to open a channel to the positron, determining its spin. This property causes problems for the measurement model of quantum mechanics. According to the standard interpretation, the wavefunction determines the possible constraints on the dynamical variables in the quantum theory, but it doesn't define a unique solution, because there are no boundary conditions to anchor the outcome to any known information. That happens when a measurement is made.

According to the canonical interpretation of Quantum Mechanics, the act of measurement causes the wavefunction to 'collapse' from being a cloud of general possibility into a very specific configuration—everywhere across space and all at the same time. The problem with this view is that it seems to violates common sense (and Einstein's theory of relativity, which predicts that no energy can travel faster than the speed of light). Why should a single measurement have the retro-causal effect of resetting the state of the entire universe? When particles separate, but are also constrained by their promises, there is no way, within the framework of known spacetime, to reconcile an instantaneous 'collapse' of all agents. However, if 'collapse' is assumed to determine the value of one particle, then it must also determine the opposite spin for the other too.

The confusion lies in the non-locality of the wavefunction and the assumption that it behaves as an almost classical causal force; but the wavefunction is only a statistical summary of energy distribution. It has no causal significance. The state information of both electron and positron resides in the globally defined wavefunction—whose time to settle into a new equilibrium is not accounted for. The entanglement promise contains a prerequisite *dependency* (which we've already associated with the definition of exterior causal time). Quantum Mechanics must be incomplete because it has no method by which to reconcile the local process of selection (-), by the antiparticle, with the local process of selection by a third party observer (of both particle and antiparticle), when both are mutual dependent on an instantaneous global function. There is a triangle of promise channels, all of which are supposed to be instantaneously connected to the global wavefunction, but this is instantaneous by construction.

The two values on the entangled link are joined by the establishment of a measurement channel. Electron and positron are co-dependent, so there are information channels from electron to positron and vice versa. The electron promises and samples the positron, and vice versa. How exactly would a 'collapse of the wavefunction' occur, and in what order? The co-dependence of end-points, in the entangled pair of states, is a symmetrization over the direction of the ordering of dependency. There is insufficient information to perform a causal collapse. If we try to match the constraints between all three agents in the problem, there is no solution: both ends (with respect to the states and the states and observer) cannot happen before one another on the same clock. It's what is called a deadlock in computer science, or a static equilibrium in physics. Just as Einstein's resolution of Special Relativity involved acknowledging the locality of space and time, as perceived through a constant speed of light, deadlock can only be reconciled if time itself runs differently for the different observers, i.e. on the interior of the electron, the positron, and the observer. This sounds problematic, from the perspective of our common wall-clock time, but as long as none of the agents are aware of one another's times it's not a problem. Any signals that pass through ordinary spacetime channels will still be limited by the normal rules of Einstein's relativity, so nothing can break those rules.

Experiments show that the state changes referred to by collapse of the wavefunction—if we insist on calling it that—do *appear* to be transmitted from one end of the channel to the other is instantaneously non-local, but only as far as the observer can tell. This suggests, quite strongly, that the type of channel between the states is not like the channel from the states to the observer. It lies in a lower layer, under a virtual process. There must be another form of spacetime adjacency that links co-dependent clusters of agents—one that is not like normal spacetime adjacency. Physicists may struggle to bring themselves to say the words—that there must be other channels to the fullness of the universe than are represented in the spacetime processes we know, but no better explanation has been offered to date.

The geometry of entanglement is bound up in the geometry of non-relativistic quantum mechanics, and what might seem like a careless oversight: there are two local parties in a process, making the give and take of input and output processes different in general. The issue of wavefunction collapse is not a local phenomenon at all, if you take into account the channel separations involved between the entangled agents and the observer, but we assume it is, so the result violates the normal meaning of the coordinates for x and t.

So far, I assumed the existence of Shannon-like information channels between the states at different spacetime locations. This is not a conventional interpretation of the theory, but it is a consistent one. Let's complete the foray into the quantum world by explaining why this point of view is applicable.

Transactional interpretation

No one would deny that the physics of Quantum Mechanics feels strange compared to mundane classical Newtonian physics. The unfamiliarity has sustained an on-going philosophical discussion about its 'interpretation'. Key figures like Niels Bohr, John Archibald Wheeler and Richard Feynman played important roles early on in this discussion. The interpretation I've been leaning on in this book, is a synthesis based on lessons learned from Promise Theory and Shannon's notion of communications, but it turned out to not be an original one[121]. In the 1980s John G. Cramer (1934-) pursued an interpretation of Quantum Mechanics, based on information transactions, known as the *Transactional Interpretation*[122]. It has been developed and embellished by a number of authors, and was rooted in the original idea of Wheeler and Feynman, known as 'absorber theory'. This was the seed for Feynman's famous diagrammatic approach to quantum processes. It made absorption look, in essence, like time reversed emission.

Suppose we ask, what really is the mysterious wavefunction? Most physicists would probably choose to answer in technical terms, quoting Hilbert spaces and formal properties of mathematics, but this is fatalistic and unhelpful in understanding how we can learn from the experience of quantum theory, and compare it to other scenarios. The wavefunction is clearly a distribution, based on the conservation of energy. The Schrödinger equation that defines it has characteristics similar to diffusion equations, rather than the more expected wave equations known for light. That means it depends more on the pathways of adjacency, intrinsic to spacetime, than a wave process carrying its own non-local motive force with it does. Its interior information seems to get projected out of it by applying filters in the form of dynamical scales (represented by 'quantum operators' for each observable) that extract measures from of the distribution, compatible with conserved energy. It's a bit like an analogue computer that works out the economics of energy transfers across a closed space, and whose only source of time is the exterior clock of the observer inducing disturbances. As a cross disciplinary theoretician, I am always excited by parallels and new ways to think about difficult problems. It turns out that there are plenty.

From a promise theoretic perspective, what we would immediately look for is an interpretation of processes from a give and take (channel) perspective. From the fundamental idea of information, one would expect to identify how Nyquist sampling takes place, with its give (+) and take (-) counterparts. Such roles don't immediately jump out of the theory of Quantum Mechanics, but that doesn't mean they aren't there. Quantum Mechanics began by focusing on the properties of single particle systems at fixed locations, which appear superficially local, and where it is easy to overlook the non-local aspects of communication involved in observation of microscopic systems. When one looks at scattering experiments, there is a definite and more obvious separation of source and receiver—but that is a distraction from the most important channel of all: that between source and the third party observer, as well as between receiver and the same observer[123]. Unlike a classical theory, we have no godlike ability to see into a system from outside its spacetime without also being party to its limitations from the inside.

The wavefunction is imagined as a field that lies atop a space that stretches across an entire system, like a mask, or a factory for energy processes. It isn't a physical field, like the electromagnetic field which represents a new physical property, but rather an information field that describes spacetime measurements and energy accounting. It was built around classical concepts of position and momentum, but the link between those quantities in the classical and quantum theories is tenuous and is never fully explained—rather, it is brushed aside using dimensional arguments that essentially represent scaling. It covers the whole of space, according to the strict interpretation of the theory; but that may be an oversight in formulating the theory. Originally, the span of space in a quantum system was expected to cover only a single hydrogen atom, reaching as far as some surrounding asymptotic region. However, the theory doesn't set any such bounds, so it seems that it could represent the entire universe, as some cosmologists have speculated[124]. As an equilibrium system, one can view space along different cross sections than Einstein did: as flat slices of the universe over which energy is constant, and time is irrelevant.

The wavefunction that evolves in Schrödinger's Equations of Motion describes not a particle position, nor even a probability for finding a particle, but any fully fledged information channel, guided by the conservation of energy, whose input and output spans the whole of space for the physical system concerned. The wavefunction is the solution to a macroscopic constraint on energy conservation, so it answers the question: how can information be distributed such that it maintains the integrity of energy conservation? Measurable observables are represented (actually approximated) by coarse-grained phenomena that mimic classical Newtonian quantities. The sender 'agent' is effectively the entire non-local wavefunction itself, over all of space, and it's usually denoted $\psi(x,t)$. The receiver agent or output is the same space, just at a later 'time' (usually defined by the act of measurement[125]), but is oriented to receive, relative to the start measurement. It's called the adjoint wavefunction, and is denoted $\psi^*(x,t)$ or $\psi^\dagger(x,t)$. Viewed as entities, these may not resemble a promise theoretic system. Viewed as processes they do.

A slice of spacetime (see figure 4.3) plays the role of a single agent, in Promise Theory parlance: a source that promises a particular distribution channel for energy. The source emits information about the state of the system[126]. If there is a kind of emission of information then local promise duality tells us that we should be looking for a corresponding acceptance (absorption) distribution. In Quantum Mechanics the formalism for this acceptance is built into a 'Hilbert space' normalization, by the existence of a wavefunction $\psi(x,t)$ and its adjoint $\psi^\dagger(x,t)$[127]. As long as energy is conserved, these two are bound together as a loss-free channel—that which is offered is fully accepted at each location, but only with a statistical probability. The process $\psi^\dagger\psi$ is the 'square' of an amplitude because it is the probability of independent events: emission *AND* absorption[128].

Quantum mechanics is an algebraic formulation of subatomic physics, in which the algebra's principal goal is to account for the distribution of energy; it was set up to mimic the kind of algebra that counts similar measurable quantities in Newtonian physics. The interpretation of underlying processes was a secondary consideration. The fact that it worked so well, without a proper understanding of the underlying processes, may either be considered a bonus or an unfortunate handicap, because physicists have been arguing about its meaning ever since. Some physicists, like Niels Bohr, argued that we can't know what is really going on—but that is an argument about the theory, not about the physics. Theory has muddled this distinction ever since. As I'll discuss, in subsequent chapters, claiming that quantum mechanics is necessarily opaque is not an acceptable form of human reasoning, so obviously it feels strange and non-intuitive. The subsequent elevation of Quantum Mechanics to the status of fundamental truth then basically told physicists to back off from trying to find more answers.

In Cramer's Transactional Interpretation, ψ is called the offer wave, and $\psi^\dagger$ is called the acceptance wave. Whether processes start and end at the same location, or at different locations doesn't matter (our interest depends on what question we are asking), causality is always ensured by the combined cooperation of both sender and receiver. This is a promise to give (+) and a promise to take (-), and this feature alone might be the most significant feature of Quantum Mechanics. The pair of a sender wavefunction ψ and its 'adjoint' or complementary receiver $\psi^\dagger$ is a causal communications channel, like a process queue in a computer network, promising to propagate the distribution of the states in the wavefunction in a single direction. We can call that direction time (by convention), but it is not driven globally; it evolves by changes or fluctuations that redistribute energy locally. No one knows the source of these fluctuations.

The mathematical notation used by Paul Dirac, for quantum processes, is particularly interesting, because it takes a form a lot like an information processing queue in computer science. For historical reasons, channel processes are written from right to left as if they were 'queues' of events waiting to be processed by the spacetime machinery. So a passive channel, in which nothing happens, is written:

$$\psi^*\psi \quad \text{or} \quad \langle\psi| \leftarrow |\psi\rangle$$

meaning

$$\langle\text{receive}| \leftarrow |\text{send}\rangle$$

or

$$\langle\text{output}| \leftarrow |\text{input}\rangle$$

just like a scattering problem. In the language of figure 4.2, we could write this:

$$\psi^{(-)}\psi^{(+)}$$

Because the wavefunction covers the entire region of space for the system being described, there is no interior or exterior to begin with. The channel is therefore a view taken from an interior space to the same interior space. In scattering problems, inputs and outputs are from one observer's interior to a different observer's interior, via an exterior channel, written:

$$\psi_2^* \psi_1 \quad \text{or} \quad \langle \psi_2 | \psi_1 \rangle$$

It is the absence of a clearly defined observer boundary that prevents the formalism from describing the information relativity as a clearly formulated channel. This might be because the focus on translational symmetries for energy distribution is at odds with the centric geometry of observer measurement, which breaks those symmetries.

All these elements leave us with a mixed bag. The overlap function between one system and another expresses the mutual information in a Shannon-like information channel (see figure 4.1). The Equations of Motion, on the other hand, describe how energy conservation constraints lead to the emergent distribution of the wavefunction. This, in turn, implies a certain likelihood for the outcomes of causal processes. But, what does any of this have to do with classical ideas like particles, positions, and velocities? Those concepts, insofar as they exist, have to be matched to operations to extract the information from the causal channel. In order to measure an outcome of the equilibrium energy distribution, from sender to receiver, one introduces 'operator channel filters' that metaphorically 'shine a light' on the quantities we'd like to observe. Now things become awkward, because the quantities we expect to find are those emergent classical quantities, like position and momentum—but scaling arguments tell us that we are unlikely to find something precisely like those quantities at a very small scale, as I'll explain in chapter 7. Algebraically, we can simply try to engineer this by seeking the closest possible fit for a particular physical observable quantity, by inserting a best-guess filter into the process queue that projects out the state of the wavefunction relevant to that measurement (the guess is constrained by dimensional analysis, which basically tells us how to count consistently[129]). Max Born (1882-1970) proposed the interpretation of the channel

$$\psi(x,t)^* \hat{M} \psi(x,t)$$

as the *probability* of measuring an event M at different locations x, for some matrix M. The structure here is a lot like an information filter, such as you might apply to a Google search. In other words: don't tell me all the websites in the dataverse, just tell me those to do with the search operator $\hat{M}$.

The mathematical properties of matrices are closely related to *graphs*, in the sense of chapter 2. Graphs represent directed processes, or undirected filters. Self consistent solutions of the filter measurements are called the *eigenvectors* of the matrix. The channel output represents the flows in a graph, for which the channel information is conserved[130][131]. In the transactional interpretation, the answer is the same, except that it embodies the probabilities of both source 'AND' receiver in causal collaboration.

Even though a particle might not exist, per se, we can give meaning to 'the position of energy' by introducing a mathematical matrix $\hat{x}$, which we queue up in the channel to extract a localized property:

$$\psi^* \hat{x} \psi \quad \text{or} \quad \langle \psi | \hat{x} | \psi \rangle$$

Similarly, to measure a momentum-like property:

$$\psi^* \hat{p} \psi \quad \text{or} \quad \langle \psi | \hat{p} | \psi \rangle$$

Many of the unusual aspects of quantum mechanics arise from the fact the order of operations matters, i.e. the operators for position and momentum do not *commute*, i.e.:

$$\langle \psi | \hat{p} \hat{x} | \psi \rangle \neq \langle \psi | \hat{x} \hat{p} | \psi \rangle$$

This also fits. Just as applying information filters in different orders leads to different results, in a search, so the order of operator filters in Quantum Mechanics matters too. A simple analogy might help to understand this. Suppose you have a document as your metaphorical wavefunction, containing the text

```
The previous state of a quantum system
```

and you perform two filter operations, by searching and replacing:

```
R₁:   state → promise
R₂:   pro → pre
```

The result of applying 1 then 2 is $\psi^\dagger \hat{R}_2 \hat{R}_1 \psi$:

```
The previous premise of a quantum system.
```

The result of applying 2 then 1 is is $\psi^\dagger \hat{R}_1 \hat{R}_2 \psi$::

> The previous promise of a quantum system.

The difference between these versions is $\hat{R}_2\hat{R}_1 - \hat{R}_1\hat{R}_2$ is called the *commutator* of the operations, and it occurs everywhere in Quantum Mechanical predictions.

VIRTUAL INFORMATION PHYSICS?

The virtual aspect of quantum theory suggests ways to relate it to very practical problems in the dataverse. It used to be that, in the world of computers and data processing, we knew exactly where jobs and processes were running within a computer system. We would assign a job to a specific computer, at a definite location. Today that's no longer the case. Jobs are submitted to a kind of robot, which allocates the task to virtual computers running on real computers. During the course of execution, the virtual computers might be moved from one physical location to another, while maintaining the same virtual adjacency in the 'cloud of computers'. Just as there is a veil of obscurity (officially called a cloud of probability!) in Quantum Mechanics, so there is a probabilistic nature to the cloud of computers (the overlap in names is fortuitous). In both cases, there is no compelling reason to suppose that a comprehensible underlying process cannot exist; rather, there is a practical limitation about the access to process details, caused by observer limitations. It could all be related to the limitation on relative sampling rates due to the scales of observers' interior clocks.

Let's look at day to day example from cloud computing. A common case is a kind of process called Map-Reduce, in which a large scale computing task is scheduled in the cloud, and requires resources from a large number of computing locations in parallel. There is a 'mapping' process, which fans out jobs to fill available space, allocating task fragments to different computers (+). The distribution of jobs over space is analogous to a kind of wavefunction for the cloud, but not exactly analogous[132]. Then, there is a complementary 'reduce' process of collecting the results (-) to a localized output channel that measures at one specific location—not across the entire space. The structure is something similar to the overlay of a function across space, like the 'wavefunction to wavefunction' information channel—a kind of scattering process. The outputs from the process fragments are then integrated in the eyes of a single observer, by a process-observer channel. Certain jobs might happen in different orders, depending on the number of transactions required to get their data from multiple sources to receiver. If they depend on one another, the order of delivery is constrained by those relationships—but also, each of them has an independent interior clock and the local sampling rate, which allows data to be transacted, may vary for several reasons. So different routes through the cloud will take different times due to relativity of clock rates. The queue of tasks, some parallel (simultaneous, each on its own disconnected clock) and some serial (sequential in local observer time) form an implicit job queue, distributed across the cloud, like operations acting on the 'virtualization wavefunction' infrastructure:

$$T_1 T_2 T_{3,4,5,6} |\text{cloud}\rangle$$

When jobs come together, they are aggregated by a single agent that collects data from exterior processes, temporarily integrating the many clocks according to its own interior time—by sampling. The precise time it takes for the job to finish according to any observer is quite uncertain. To get an idea of what this looks like, think of sending out your Christmas shopping list to different online sellers around the world, and waiting for the full list to return. As the goods come back, they will not arrive in the same order that you requested them because there are too many independent processes in between to mess with that. Precise knowledge of where your goods or money are, at any given moment, is inaccessible. This is a feature of a small scale network—it would never happen in Euclidean space, by Newtonian rules, where you would be able to observe the motions along their entire paths. In a restricted network, the information still exists, but you can't reach it, let alone assemble it until all the goods arrive with you. The arrival of the result is simply the first occasion on which any information about the process arrives at the observer location. If you get impatient and try to find out where packages are, you might call different postal agents by some faster channels (telephone or Internet, say). Then their uncoordinated information might claim that a package is actually in two places at the same time, or nowhere at all (in transit). This is not unlike the experience we have in quantum theory. It's characteristic of a process network.

While computers are processing their tasks, they are occupied, and cannot be occupied by other tasks. However, the tasks can be interleaved by 'time sharing', in which case the processes are suspended in time, as if in entirely parallel spacetimes. They can also be moved from one location to another, suddenly jumping to a new physical location, without affecting the overall scheduling of the 'wavefunction' for the process. Just as the properties of position and momentum are virtualized by the attempt to reconcile local energy accounting of the wavefunction with non-local definitions compatible with waves in Quantum Mechanics, so there is a similar effect of loss of predictability in the dataverse: we can't know precisely where a job will be located or the likelihood that it will move again over an interval of time, because the larger patterns of information flow are based on a holistic non-local model. The expectations for position and migration are both dependent on underlying information about the accounting of resources (analogous to the energy counting in Schrödinger's equation).

Today we see computer installations into the millions of computers, where statistical phenomena can begin to rival the unpredictability of small quantum systems. The state of each independent job proceeds not according to the exterior clock for the entire cloud, but according to the interior clocks of each local process. Only when jobs interact, by depending on one another, do the clocks overlap. The scale of the overlapping clocks varies depending on job size, but it could only approach 10^6. Clearly, nothing compares to the scale disparity between quantum phenomena and whatever the discrete structure of space and time might be[133]. So whatever strange effects of scale we identify in the dataverse, we can expect almost infinitely more scope for the effects of scale as compared to the very bottom. The scale mismatch between spacetime and atoms is unimaginably vast compared to the scale of a computer compared to the world's computer infrastructure—and yet, still we see parallels just from the causal separation of channels and the effects of independent give and take.

Uncertainty is a principle

Suppose an observer, without inside knowledge of the causal processes, tries to identify where a job is running. He or she (most likely it) could try to sample every computer in turn to see if it were there. Given that it might be moving around, there is a certain chance of hitting it, or not. If the job actually occupies multiple locations—like a superagent—then its collective position would also be fluctuating. If you think the job might be travelling in approximately a straight line, then the uncertainty reduces to a finite distance interval along that line, as in the Heisenberg case (which is built on Newtonian correspondence). In the dataverse, the concept of motion in a straight line has no immediate meaning (see chapter 7). Since the observation process and the allocation processes are independent, the probability of measuring the job would be a combined probability of both allocator AND observer sampling the same location.

The quantum 'probability cloud' and information technology's 'computational cloud' are both causal channels that propagate processes step by step, counting time as they go. They both have layers of obfuscation between absolute location and measured outcome. The parallels are many. There is a finite speed of communication in the infrastructure, but there are certain correlated pairs (such as public-private key pairs used in encryption) that can lead to instantaneous reasoning, a bit like entanglement. The dataverse forms a set of fixed locations, an absolute spacetime where processes are executed at locations determined by another process that also runs on same computers. The so-called virtual machines are the locations of processes relative to the absolute space of the cloud infrastructure.

Can we calculate how long a process will take in the dataverse[134]? Can we be certain that changes will be made in the correct order? How does this affect the certainty of the outcome? The processes executed in the distributed computer system, or 'computing cloud' form Directed Graphs, as mentioned in chapter 3. The understanding of smart spacetimes in computer systems is nowhere close to being on a par with the physics of our universe. The techniques are poorly developed, mainly because there is a basic cultural difference between physics and engineering. Physicists want to understand; engineers want to force an outcome. Alas, today it's clear that we have reached a scale at which that kind of certainty of outcome is simply impossible to engineer (as discussed in my book *In Search of Certainty*).

Let's summarize some key points. The meaning of classical quantities like position of an agent or particle in Quantum Mechanics is not the x that appears in $\psi(x,t)$ (the latter is a dummy variable). There are two completely different definitions of space in play: one is an absolute space x, and the other is a set of observable locations occupied by dynamical activity, represented by the filter operators like $\hat{x}$. The wavefunction covers all possible x, and the channels are summed over all the spacetime points in the channel, just like the for-loop index variable examples in the previous chapter. The important distinction between this and the concept of a position, in the Newtonian sense, is that $\hat{x}$ is sandwiched in a non-local distribution of energy throughout space, so it represents a smearing of process energy, not a sharp localizable point-like action in the Newtonian sense of a particle. In Quantum Mechanics, signals are communicated between all positions and the observer, and accumulation of energy at each x tells you the relative likelihood of measuring a property M

at that location. The non-commutation of filtering operations is not particular to Quantum Mechanics, as the example illustrates. It's a pervasive issue in computer systems and in business change management, especially when trying to patch software for errors. The order of application may be critical there is overlap between the affected spatial regions. Clearly, this kind of causal dependency will have a big impact on any system in which properties are distributed.

This is all neat and tidy, but of course it's nothing like Newtonian mechanics. In Newton's view, things move within space and time, and their positions change according to certain rules that count the movement of energy. Quantum mechanics has a set of rules for the distribution of energy, without an explanation of locality or position—just recipes for estimating scales associated with them. It turns out to be enough information for most purposes. Quantum theory is remarkably successful at predicting the possible transitions and distributions of matter and energy, in atomic systems. Treating the iconic wavefunction ψ and its conjugate $\psi^\dagger$ as offer and acceptance promises at the ends of a causality channel, might be just semantics, but that is indeed what theories are for. We should not be embarrassed by that. The role of semantics is growing in science, as it turns out to be of fundamental significance to nearly everything else that happens, in chemistry, biology, polar fields, directed communications, information technology, filtration, restriction of access, privacy. The list could go on and on.

PTOLEMY RETURNS: MANY WORLDS AND THE DATAVERSE

A common theme, in the give and take between processes, is that each agent in a spacetime sees the rest of its universe through a bubble between its own interior and its exterior—no matter on what scale that boundary happens to be relative to the processes under discussion. This is the only meaning we can attribute to locality. The interplay between scale and observability is complex and frequently non-intuitive. In quantum theory, this issue has been ongoing for a century. In computer science, the state of understanding concerning measurements and their interpretation lags far behind the corresponding state of affairs in the natural sciences. To compare the situations in physics with the dataverse and beyond, we need to think more carefully about how motion is defined and observed at different scales.

The so-called *Many Worlds Interpretation* of Quantum Mechanics, also discussed in connection with timelines in chapter 3, claims to resolve the issue some people have with the collapse of the wavefunction brought about by the 'act of measurement'[135]. The Many Worlds solution of placing everything in a separate namespace is very much like an IT model of scope. By squirrelling away details into its own private space, we separate concerns. This isolation or mutual immutability has become a methodology for modular separation in computing[136]. The branching of multiple information channels, containing different local samples, is analogous the the branching of 'worlds' used as an interpretation of Quantum Mechanics. It's as if every agent adopts its own centrist view of the universe, placing itself at the centre—like Ptolemy. Each separate module acts as the analogue of a virtual process in quantum theory. What makes this feasible for IT, preventing an explosion of realities, is that its function calls are ephemeral, whereas Many Worlds timelines go on forever. Computing state collapses are local, and thus are analogous to what we call quantum loop corrections, rather than ongoing branching universes.

This forking or branching of state-spaces is the origin of modularity in information systems. It has huge implications for the meaning of innovation, like genetic mutation and selection. Innovation happens when worlds collide and information gets mixed. If new processes build from this, they may be selected by other processes, forming channels that propagate—or not. Either way, the possibility to maintain a kind of information integrity in order to propagate in a primordial soup of agents, depends on there being some isolation for the gestation and execution of the process.

In computer programming the passage of interior time also follows different threads, with separate clocks. Each 'time' the programming logic enters a new function, or module, with its private scope, the time inside the contained execution behaves like a single agent with its own interior time. As long as there is no causal connection between the process within the container and the process outside the interior details cannot affect the outside. The outside may perceive the complete execution (the keeping of all the interior promises) as a single tick if its execution is suspended to wait for the result. Or, it might execute 'asynchronously' meaning that the interior and exterior processes can continue side by side in separate timelines which will later rejoin one another as a sampling of the interior result occurs.

Clearly, information generated (+) in 'different worlds' has different interpretations or meanings to observers (-). Without an expense of effort, timelines that have diverged cannot be brought back together consistently. A different interpretation (-) leads to a semantic separation that might be physical or virtual—in the computing sense of the terms. On the other hand, if we look at space and time as information processes, then there is no difference between the two. Cryptography is an obvious example of how there is a real semantic significance to protecting against the accidental overlap of worlds. Privacy is becoming harder to arrange in the era of ubiquitous information. Encryption (+) and decryption (-) is sometimes an entanglement. Both Shannon and Turing worked on this problem in the early days of communications and information.

In recent years, the call to fragment system spaces into modules has reached a peak around the concept of 'microservices' in IT system design. Although

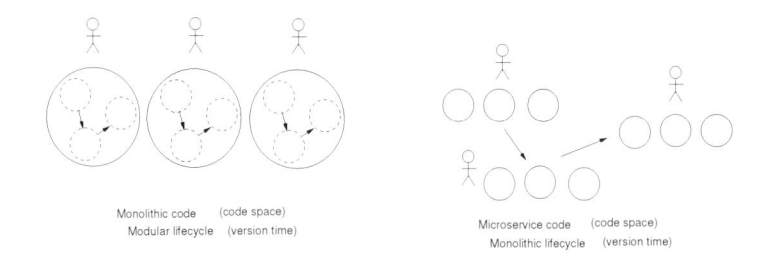

Fig. 4.5. A shift from monolithic code to microservices involves a corresponding shift from separation of lifecycle concerns to integrated lifecycles. If the dashed lines represent modular agents of code, and the full lines represent the independent agents deployed, i.e. the boundary between interior and exterior, the left hand side shows so-called monolithic design with wires hidden, and the right hand side shows a microservice design with wires exposed. So the IT manifesto to shift from monolithic design to microservices is just a rotation from spacelike integration to timelike integration, in spacetime adjacency language.

there is no consensus on what a microservice really is, the general idea is that a microservice represents a maximal separation of programming or 'code' concerns (see figure 4.5). Microservices came to fame at Netflix , when there was a shift from a style of system in which most software code promises were integrated behind a single monolithic boundary, but the lifecycle of the process (deployment and maintenance) was coupled only loosely through human promises (running and maintaining the service for customers), to a model in which all the lifecycle promises (maintaining, deploying, running) for the process were tightly integrated and functional promises were divided and spread into maximally separated modules. In other words, microservices advocated a shift from functional renormalization to lifecycle renormalization (a spacetime rotation from a single lifecycle or version clock, to multiple clocks). Its boundary placement went from spacelike integration or 'monolithic' to timelike integration or 'microservices' (micro referring to spatial extent, in the sense that software code takes up space).

Modularity clearly plays a grossly underestimated role in spacetime processes. It would seem to be ripe for a thorough reinvestigation.

Birds of a feather

Although they arise from quite different considerations, the similarities between Quantum Theory and the more generic Promise Theory (a set of skeleton principles applicable to any localizable processes) are quite striking. They boil down to a preoccupation with locality—a centrist observer viewpoint—albeit for different reasons. The importance of interior and exterior information, relative to arbitrary surfaces is what leads to this focus on locality. It plays a key role in how each region of spacetime perceives spacetime.

In the interaction picture, the description of spacetime that we attach so much importance to looks very much like a pointer table in a computer[137]. The coordinates we refer to in the equations of quantum mechanics do not correspond to the locations of matter, as in Newton's theatre—they are only a convenient set of labels for a sea of underlying energy transactions, like virtual pointers that provide linkage and addressability to the information in neighbouring agents—much more like a computer system. The concept of an agent, as an active location where virtual processes may be situated, seems ever more relevant than a simple coordinates representing a passive point of presence. This may sound like heresy in physics, where we have come to believe in the elementary nature of things, but that elementary nature seems to contradict the semantics of a field and its interactions. Could it not be that it's only a figment of the story of an independent spacetime theatre.

With this version of spacetime, built more like a computer network on promises, the emphasis shifts away from what space and time are, to how promises can be labelled. This is not a new problem, per se, we still have it even in the physics of ordinary phenomena such as statistical physics (where space and time variables are not needed). Distinguishability or indistinguishability have been treated as abstract issues, but they might be more fundamental than spacetime. With promises, the focus shifts to labelling.

Promise Theory begins with a completely irregular discrete collection of generalized 'agents'—a scale-free model where things happen. Each agent is initially independent, but can interact with other agents by making promises about its own behaviours that may or may not be kept. A close analogy is the way chemistry begins with a discrete space of independent atoms, and then explores their interactions into larger structures. We then add a few formal axioms:

- Every promise theoretic agent is 'autonomous'. Autonomy is a semantically laden word that means something very similar to what physicists mean by locality: it means that whatever properties an agent has and whatever it ends up doing, that behaviour comes from within itself, not from any other agent. Put another way, no agent can make a promise about any agent other than itself. Already this can be applied from the smallest scale through the human scale to the largest scales.

- Promises of any kind imply a virtual network for communication and transmission of 'influence'. We needn't know how it works.

- No transmission of influence can be promised without both give and take—i.e. a promise from one agent to offer something (+), and a promise by the receiver to accept (-) the first promise.

Promise theory captures the main features of space and time for many purposes. At the same time it challenges us to understand how to reconstruct the familiar concepts of our very scale-dependent experience: from subatomic particles, through material science, to cellular biology, sociology, and all the way up to astronomy.

For whom the clock ticks

The theme of this chapter lies in the balance between what is offered by a sender and what is accepted by a receiver—what, in the language of communication would be called the potential for distortion. What does it tell us about how time passes for different observers?

From the basic give and take of Promise Theory, we would expect a basic principle to hold: namely that the advancement of any change has to be a cooperative (interaction) property between a sender and a receiver. A promise to signal change in one place must be accepted and observed in another place. Then there is the question of how causation depends on this change. Is an agent's sense of time driven by sampling from a clock within, or driven by events from without (like a plant growing in response to raindrops). This is why scaling is so important to interactions. Simple processes, like clocks, (those with few degrees of freedom, or few working parts) tend to be characterized by a single predictable scale. More complex processes (like biological things) have any active scales, from overlapping sub-processes. Observers can see these differences in the give and take of their interactions with them. When you press a button, a light comes on 'immediately', but when you taste a cake or a glass of wine there may be an immediate sensation, say fruity, that later gives way to a coffee or chocolate aftertaste. The promises contained in the cake (+) are received (-) by different receptors, each with their own chemical processes and inherent timescales. Coffee is slower than fruit. The collective process of tasting is a superagent of sensory circuits. Here the variation is in the receiver (-) end of the observation. In music, we can see cases where the source timescales contain the variation instead: when a musician hits a drum, or blows a trumpet, a sound is formed very quickly; but, when the string section plays with its bow, or gong is struck, a sound forms over several seconds—much slower—which makes playing in an orchestra quite challenging, because all the instruments have their own intrinsic timescales, which may or may not be compatible with the music. The observer groups the sounds into a timeline according to his or her own interior listening process.

A consequence of Nyquist's sampling, and the ability to detect change, there is a natural suggestion that time has to come from within each spacetime point. There is an *interior time* that has to be the source of all response—locality could not be consistently driven by events that arrive from distant locations. Exterior time consists of interruptions, but unless interior time is capable of sampling exterior events, they could not even be sampled. This suggests that, in order for locations in spacetime to interact at all, time has to be come independently from within each agent. On the other hand, a location in space can't distinguish different times without an interior cumulative memory, or immediate access to a persistently observable exterior state, in which case it would be natural to renormalize those exterior states to be a part of the agent itself.

The idea that spacetime properties originate on the interior of spatial agents is compelling on a number of levels, and on all scales—as if every spacetime location were a computer with its own CPU and memory.

The watched pot never boils. Why not? Because, as we wait for something to happen we sample faster and faster, afraid we are going to miss a detail. Thus the interior clock runs faster that exterior time, and it all seems to take forever. When arrivals are unpredictable, we tend to sample information channels more often. For example, if you don't know exactly when you will receive a message, you tend to check your messages more often. Social media companies use this to try to draw more of our attention to platforms where they sell advertising[138]. Conversely, the arrival of a signal after a long absence is anomalous, but eventually if we don't expect messages we will sample less frequently, lose transmission fidelity, and perhaps even miss signals, by Nyquist's law. When we know that something might arrive, but we are not quite certain when, we may oversample.

Telephone users, before the Internet saw a slower world. Time only ticked when a new telephone directory arrived in the post box, so in the world of that information, a new printout of the telephone directory is a tick of its clock. There are no intermediate states visible to an observer. Whatever mysterious processes took place to cause those changes, they are not available to the observer. Processes communicate and time advances when they establish links by message passing. Even wall clocks have to be synchronized by this method.

As the scale shrinks down to the smallest possible scale, the capacity to measure by interior distinction—to sample and recall—must presumably shrink too. In the next chapter, I want to think about what effect this must have on our ability to describe meaningful processes at the most basic level.

5
TIME AND MOTION

Parsifal: Ich schreite kaum, doch wähn' ich mich schon weit.
(I scarcely move, yet already it seems I have travelled far.)
Gurnemanz: Du sieh'st, mein Sohn, zum Raum wird hier die Zeit.
(You see, my son, here time becomes space.)
– R. Wagner, Parsifal

So far we've been discussing the structure of space and time in a number of different scenarios, but something obvious has been missing: *motion*, i.e. what happens when changes in spatial position and time occur together. Motion, as we normally think about it, involves some kind of material body that sweeps out a sharp trajectory, taking it from one location to another, through a smooth succession of steps. The generalization of the motion concept involves many more kinds of process, in which information initially expressed at one location gets reconfigured to occupy a succession of locations, one at a time—like renumbering the musical chairs to move people around virtually instead of physically. All turtles can be employed to effect the outcome! Motion phenomena are not about space or about time independently, they are about *process*—a mixture of the two. We've mentioned motion in a number of contexts, but we've been skirting around why it's a tricky issue. Like many of the things we have grown up taking for granted, simple ideas about motion begin to fall apart when we deconstruct the very fundamentals of what it means. It turns out that motion lies at the very heart of the nature of space and time and a closer examination of what seems to be a simple phenomenon reveals some deep surprises.

5 TIME AND MOTION

DEFINING MOTION

What we know about motion has been almost completely rewritten in the 20th century. From uniform motion in a straight line by perfect bodies, to stochastic Brownian motion, to curved spacetime, to Heisenberg uncertainty clouds, to packetized transmission of data along channels. The list goes on. If we extend the idea of change in space and time to other kinds of spacetime, then motion includes processes that unfold in nature, biological systems, and—of course—the dataverse. Philosophers have talked about the motion of matter for millennia, and science has gradually found ways to describe the common cases. The way we think about motion—and *mobility*—in the world of information systems is quite new, and it is very different from our experience of motion in the physical world too.

What is motion? It's about how something changes its location over an elapsed time. But, this conceals many deep questions that we normally take completely for granted. We already need to ask: what is a thing, what kind of locations, and whose time are we talking about? Our normal assumption would be to assume a contiguous sequence of adjacent locations, that we can travel to ourselves, and a normal wall clock that proceeds at a constant rate relative to our own bodily processes, but these are matters we shouldn't take for granted. Does it matter if we are trying to measure our own motion, or the motion of something else that we observe? How can we detect the motion of either? What can it be measured against?

In mathematics these questions are dealt with quite easily: you draw an imaginary grid of coordinates, with a certain scale, and you imagine or define a uniform time by choosing a clock (see figure 5.1). As the location changes on the grid, the clock changes too, and by dividing the change in grid position by the change in the clock position, you define the speed of the motion over that interval. Then you have motion in a Newtonian sense. Something moves from point to point, in a certain time, and that is motion. You just assume that you know where all the points are, and that you can tell when the moving object has reached a certain point, and that time is represented by the clock. You *trust* the integrity of the measurements completely, even though you might not be able to verify them! The problem is that this simple view glosses over all kinds of issues that we might be able to disregard in our daily lives, but which we can't ignore when making observations in the modern world of science and technology—and trust is at the heart of that acceptance. In a flight simulator,

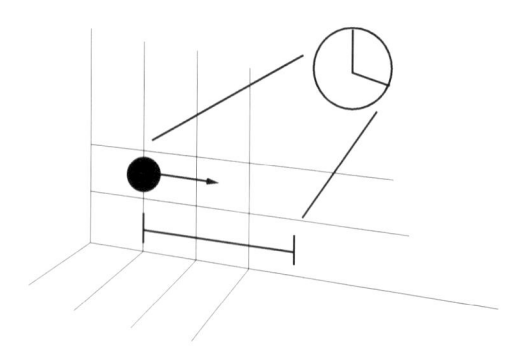

Fig. 5.1. Our 'take it for granted' view of motion. We assume the existence of a natural measurement grid and clock—but where do we find these in empty space, or a spaghetti of cables?

pilots experience an abject lie: they are fed information to make them believe they are moving, when in fact they are not. How can anyone really know?

Science and technology now probes the world beyond cases where we can implement this trusted model of distance and time—to situations where there is no natural background scale on which to draw a coordinate grid, and no way to get inside the system to really make an objective measurement. When we measure the distant stars, we can't visit them to check if the distribution of points at all locations is homogeneous in between. We end up having to trust a theoretical model of how we believe space and time behave, because we can't actually measure it directly.

RACING CHANGE WITH CHANGE

Using the dataverse as an example is a helpful reality check, because it is a lot more accessible than tiny scales or distant places where physics lets us down—at least, in principle—and yet its structure may be more like the physics of small things than anything else. What if we wanted to watch the analogue of motion on the Internet: could we do it? What would move, and relative to what? We can't simply get inside the computer with a ruler and a clock to measure it, just as we can't visit distant galaxies to establish coordinates—so we have to rely on probes.

One way to follow a packet of data moving through the Internet would be to ask the locations it passes through to report back to us on the positions of the data as it goes. The locations are computers of different kinds, and could easily signal us about the signals they receive and transmit, as a kind of diagnostic trace. But sending signals about signals is not sustainable. If all moving phenomena sent signals from every location to every observer, we could certainly detect motion, by seeing the sequence of locations visited, but how would those locations know to whom they should send those signals? And— once received—the observer would know when the signals arrived, but not when they were sent. In the dataverse, these locations are all computers and switches that are easily capable of sending confirmation receipts to some location. Each time a packet of information arrived at the next computer, to be relayed to its destination, the relevant computer could send a signal to every observer on the Internet to say where it was—a bit like the ambient light that allows us to see objects around us. But this would be madness, of course. Whereas there are plenty of vacant pathways to send ambient light around in the physical world, the world of the Internet has only a few cables to route its messages through. The Internet would immediately be flooded with signals tracking other signals—the receipt signals would have to use the same medium as the signals they were reporting on. Every signal would have to send multiple signals to report the positions at every point—including all of the signals to report on signals that report on signals, and so on. The result would be an explosion of information, completely impossible to interpret[139].

Now, compare this to the way we watch and experience our familiar surroundings. We are used to being spoiled by a view of rich information, served to us by a very fortunate coincidence of scales. We happen to live in a very slow moving world, which is quite empty of interesting points. The only points that are detectable are those occupied by matter. The signals we use to observe material objects (light and sound) are abundant and travel comparatively quickly along a practically infinite number of possible pathways. That grants us a unique privilege: we can watch things moving from location to location, as signals fly back and forth. It's only when objects approach the speed of those observations (of light or sound) that we get into trouble. You can't watch a light ray move, because you'd have to race light with light to see it—like asking a carrier pigeon to report to you about the movements of another carrier pigeon far away. Unless one was a lot faster than the other, it couldn't work. Similarly, we can't observe a spacetime location directly unless it signals us explicitly, like a beacon. There are processes in nature by which this can happen, e.g. with particle-antiparticle pair creation, but the events are so rare and sparse that they are not a useful probe of spacetime structure.

To speak of motion, or even to observe it, we need the basic elements that we know from Galileo and Newton. We need to be able to mark out distances, or markers in space, and we need to have an independent sense of time, like a storyline that orders the events. Classically, motion has been considered uniform if, for every tick of some clock, a body moves the same amount according to a measuring stick, across our line of sight. We have to trust the clock's process to be uniform, and that we are in a position to see change. If something is coming straight at you, you might not be able to tell the difference in position at all. If you are standing in a line and someone calls out, you can only tell who it was if they are standing directly adjacent to you, in front or behind. If the message is relayed by a middle man, you have no idea where it came from. Information from far away is always relayed by a middle man, so we live on trust.

If we don't have a clock or a measuring stick, it's hard to say anything about movement at all. We have to rely on inference, informed guesswork, or extrapolation—did the sound get louder, did the light get brighter? Without processes of change that allow time to be measured, by us, we would never see any change, because observation itself requires a change in our instruments (or our minds), and therefore the advancement of time. Our minds and every other physical process are part of a process that we call time too.

It's all about information. If a change happens in one place, how does that change get observed somewhere else, by copying the information or moving it somewhere? So, what matters to our perception is whether the processes at source and receiver are correlated.

IF A TREE FALLS....

Suppose we want to track a moving body in order to know its position or its velocity. How can we do this? The usual way would be to watch it. In order to be able to see it, we have to receive a signal from it. Observability of processes is the root of the answer, but what does that entail?

If we throw a ball, we can watch it move, following a trajectory until it lands, and we intuitively believe that we know what happened—what that motion means (see figure 5.2). We see the ball move because the ambient light, emitted from the ball continuously (as far as we can tell), reaches our eyes much faster

5 TIME AND MOTION 175

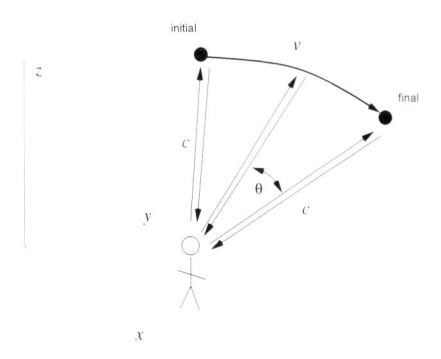

Fig. 5.2. A conventional macroscopic continuum view of a trajectory. The observer can effectively sample all angles θ (unless there is a line of sight obstruction), by sending a probe signal which returns so fast as to appear instantaneous ($c \gg v$) to give the impression of a continuous motion with velocity v. For a continuum space, dimensions of spacetime are shared concepts. Based on the rules of trigonometry, the observer can calculate a value for v based on θ and its clock time.

than any process associated with the signal motion or our internal physiological processes that make up interior time in our brains. In short, light can travel there and back so quickly that we can potentially sample far more information than we could ever hope to process. This saturation of our ability to sample leads us to conclude that there is continuity, an infinite well of information to tap, and that we see a body moving at a constant speed.

This perception, of a coherent body moving smoothly along a continuous trajectory, builds atop a conspiracy of scales that will take some time to unpick—however, by the end of this book, I hope you will find this mundane experience as mind-boggling and fascinating as I do. The conspiracy of scales allows observers to believe that observations are instantaneous, i.e. that the path length from the observer to all the points of the trajectory (see figure 5.2) is irrelevant to the measurement of the motion. This is a key assumption, because it is one of the first to fail when trying to define motion in other situations. The scaling also convinces us that motion has properties of its own, i.e. that speed or velocity is something that is carried with the body along its trajectory, rather than being a function of spacetime between locations. These are natural choices, given our everyday experiences, but they are not the only choices we could make.

Now let's compare this classical idea about trajectory with an experience on a rather different set of scales—a situation where spacetime is far from being

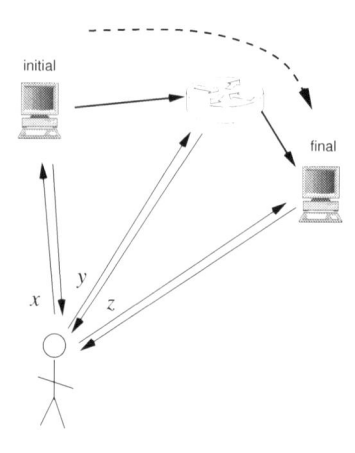

Fig. 5.3. A microscopic discrete view of a trajectory in the dataverse. The observer can only even stand a chance to sample those points for which there is a route between the observer and the point through which the trajectory passes. The observer has no concept of continuous 'angle' only orthogonal dimensions. In network spacetime, there are literally only three points in the figure. A message might arrive at a variety of answers for a velocity along different legs of the journey.

continuous, and information is in short supply. In the dataverse, you might type some search phrase into your computer (figure 5.3), and packets of information are sent out along a very limited number of pathways, guided by one of the few actual cables that connect cities around the world. You can imagine trying to watch the request move from your browser into the network, enter each switch and router along the way, ending with the server, and then come back, for each new page. But you can't see any of this. All you can see it that the result arrives and there is a sudden transition from one page to another on your screen. What happens in between is not visible; we see only discontinuous changes as events, like blips on a Geiger counter, or quantum detector transitions. This is a totally different experience and exercise, but why is it harder to see data in motion in the dataverse, rather than balls in motion in our universe?

As it happens, there are tools to provide engineers some visibility into the intermediate locations on the Internet. Most users will never use these. The application `traceroute`, found on most computers, asks each computer nicely to return a confirmation of the path taken by data packets between two points to the origin—but this is a voluntary matter. If the computers in between don't promise to accept (-) and respond (+) such requests, then we can't observe them. This differs from spacetime because the empty spaces of the universe have no processes that accept queries and promise to reply to any particular origin, hence they are invisible[140].

In the physical universe, there are two ways we track moving bodies: in daylight, by observing ambient reflected light (where we use the surplus of preexisting signals), or in the dark (where we need to provide some illumination of our own—think of detecting vessels in air traffic control, or ships at sea by radar and sonar). We can use a little promise theory to reason about this. A vessel can promise to send a signal, like a beacon or transponder to others nearby indicating its position (see figure 5.2).

- What if the vessel does not promise to signal its location (+)?

- What if others do not promise to look out for the signal (-)?

An alternative would be to try to provide ambient daylight, like a radar does, by bouncing light off the vessel. This is effectively trying to intentionally impose a signal, or induce the vessel to transmit a signal involuntarily:

- We could try to bounce a signal off the body.

- What if the body absorbs the signal or is transparent to it, so that it doesn't reflect back to the sender?

We use these methods to probe our bodies with medical scans using X-rays and sound waves that pass partially through media, and we use seismic waves to infer geological structures in the Earth. These methods work only because the spaces the signals pass through are not completely transparent—they do partially accept (-) the signals and return a measurable change. If all methods to observe space fail, then the position of spatial structures is non-observable—avoiding detection like a stealth aircraft. Without verification, observers have to theorize, perhaps based on partial evidence and inference based on the assumed 'laws of physics'.

Because computer networks are closed, we can only use agents within them to observe other agents within them, and only using signals that can be sent within them to try to follow where the data are. For example, consider figure 5.3. The observer is trying to watch data move from an initial to a final destination, through an intermediate router device. However, there may or may not be a path directly from the initial location to the observer, so it might not even be possible to observe the initial state at all. Alternatively, a signal might have to first travel the same way as the data (to the router), and then through the router to the observer (assuming there is even a path from the router to the observer). By this time the data would presumably have travelled at the same speed to the router and beyond. Actually, that would be based on speculation, because we don't know if motion is continuous and preserved when we aren't looking. By default, we tend to assume that all things happen like the continuous tracks on a vinyl gramophone record, rather than as discrete jumps on a digital CD track, but this is only our human-scale bias at work. Without the ability to see (and see very quickly) the data at each of the intermediate points along the path, the observer simply cannot tell what is going on—and has to trust in an assumed story or model, based on occasional glimpses of the journey when it happens to be observable.

Suppose, however, a best case scenario in which there is indeed a path from every point along the path (from initial state to router and to final state) to the observer. What then? In this case, we still don't really know how quickly the signal can travel to the observer, compared to how quickly the data can travel along its trajectory. To measure the speed of the motion, the observer has to be able to sample the signals and compare how long it took to make the transition from one location to the next, and the only clock that matters is its own. It does so according to its sampling rate, which is driven by its own interior time. The observer has no information about how long observers at other locations believe the transition took. The question means nothing. Similarly, the observer has no way to measure how long the signals took to reach it from the points along the trajectory. At best it can try to form a theory about how fast the signals travel and how far away the points are and make a calculated assumption—such as we do with the distant stars in the cosmos.

Indeed, this is what happens in physics, because—with the assumptions about spacetime as a Euclidean theatre—position and speed are all easily calculated according to Newton's prescriptions. But we can't actually check whether this is correct or not—we trust the results, and sometimes the models turn out to be wrong. All the observer sees is the arrival of the signal; it has no independent way to measure its journey too (because it is already trying to use the signal to measure something else). We can only assume a constant speed of signalling according to the observer's clock. It isn't too hard to imagine that this experimental setup could conceal all kinds of realities that are simply not observable. We might find it hard to believe for our own spacetime, but this is all directly verifiable in a computer network.

Figures 5.2 and 5.3 illustrate how important a high density of paths is to our intuitions about spacetime, and how very fast signalling makes it possible to obtain an apparent continuum of movement (which might be entirely illusory). It is also this high density of points that allows us to make use of magnifying lenses in cameras, telescopes, and binoculars[141]. Conversely, a paucity of paths or a limited signalling rate (of about the same speed as the motion we are trying to measure) leads to a disjointed and bizarre experience of motions—such as the experiences of quantum measurements, or packet arrivals on the Internet.

Motion and velocity on a limited network

Although we take our day to day experience of motion for granted, there are more than a few different ways we can imagine motion. When we try to extend ideas that we take for granted about motion, in a Newtonian sense, to places like the dataverse or even the day to day world of the econoverse, we find that we can't trust them. We need to think about what happens point by point, and be quite clear about the limits of observability.

Consider the figure 5.4(a). When we observe something on a human scale, we take as given that all the bodies involved are operating on the same clock— the same clock that we the observer sees. We can make that assumption because, our own capacity to perceive information is probably the slowest of all the clocks, so whatever we perceive is more or less instantaneously observed by light travelling from the moving body to us. When a body moves past by moving from point to point, it appears to move continuously because there are so many points that light can travel through, light seems to travel in straight lines. As the body moves from left to right, the angle light rays arrive changes continuously too. With an infinite variety of angles and paths the motion seems continuous, and our presumptions about the constancy of the number of points, and therefore the constancy of the speed of the signal along significant segments of the path, combined with the homogeneity of the space allows us to interpolate a smooth motion. The quasi-local concept of a velocity, which is a property of each location, is born!

But suppose the signals are not travelling through a continuum, rather through

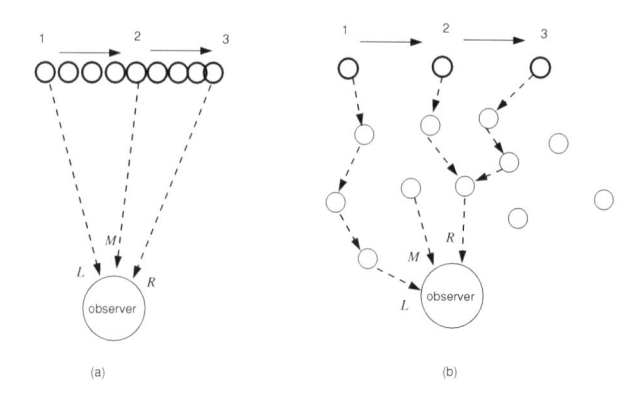

Fig. 5.4. Uniform motion in a straight line? What does this mean? In (a) the density of points is so high that the lines look straight and we can't see any limitation in position. In (b), we see the discrete version, more like the situation with an information network, in which there is a limited number of points available for relaying messages. In what sense can we talk about a straight line in such a space?

a network of discrete jumps, like a computer network, or perhaps spacetime on a tiny scale, as in figure 5.4(b). Where then is the time that matters? Without all this infinite resolution and the assumption of very fast signals, we need to rethink what happens. Either all the points have their own clock, or all the points *are* their own clock. No agent perceives anything until a signal arrives. There is a lot of difficult thinking here, so let's take it slowly. There are many possible combinations of possible outcome, that all fade away in the continuum limit. But in a scenario like a computer network, we really have to think about these issues. An observer has to try to build an objective model to provide a single point of calibration.

There are two kinds of change going on in figure 5.4: signals reporting the state of the locations 1,2, and 3 to the independent observer, and the movement of some observable property from 1 to 2 to 3. It is only when the signal reaches the observer that the observer knows the state of each of the locations 1, 2, and 3, by seeing different signals appear in order along its similarly ordered receptors left L, middle M, and right R.

- Actual locations: 1,2,3.

- Observed proxy locations: L, M, R.

Let's assume that the signals are sent regularly and are sampled at a faster rate than the motion. Implicit in this assumption is that the observer has an internal clock of its own, ticking away measuring the changes and possibly doing other things. In this case, the observer stands a chance to be able to see the motion from location to location, because it is sampling fast enough, a bit like the continuum case. But, now it gets weird. Because spacetime only has few points, there is not an infinite variety of paths the signals could take. In fact the observer can only distinguish two different directions (L and R), by accepting signals relayed to it from two neighbouring points, not the middle position. So it cannot detect all the motion, because its spatial resolution is not great enough. It's middle receptor M is not connected to signalling paths: there is a hole in space, so it must see a sudden jump. Similarly, the same thing would apply if any of the signals failed to be relayed by the intermediate points and thus failed to arrive. The state might be migrated at 1, 2, and 3, but the observer would not be able to see the changes. When the scales are all similar and there are limited options, it changes the basic mechanics of the relativity of motion. This is the situation in the dataverse, for all programmers and operational engineers trying to follow what is going in the computing cloud.

Figure 5.4 illustrates a difference between online 'services' and 'data pipeline' movements, in IT systems. In a service interaction, when a user or source agent queries a service, the result comes back to the same source. In a pipeline motion, a source agent forwards data to another pipeline element, the output goes to a different agent location. A service says 'return to sender', a pipeline says 'move along'. What the observer sees is only calibrated to a single clock in the service interaction case. Indeed, everything seen in the system depends on the exact software versions in operation for the trajectory of the data. If one were to send the same data through the combination of locations in the pipeline after software updates, there is no guarantee of seeing the same result. The interior clocks of change bring immediate uncertainty to non-local processes. Observing third party motion, based on voluntary or ambient signalling, is fundamentally impossible—it can only be inferred from theory.

We can consider the extreme case, in which the observer has no independent clock of its own, i.e. it can only wake up and act when it perceives a change internally or through its two neighbours, then the signals become a part of its perception of time. This is what happens, for instance, when a news service reports the news to us, or when a computer diagnoses its state with 'log messages'. In those cases, a distributed reporter or sensor reports its information to a central location, which may redistribute it according to a different clock. We might hear about the events of a political crisis or famine months after they actually happened, because the news logging service is our only channel of observation, and it takes control of the motion of that information, according to its own clock. Half the information could be missing, and you could never know about it. This is a series flaw in the design of a lot of computer systems, and yet we continue to build systems based on the assumption of perfect connectivity, perfect reliability, and one common timeline for all.

Before leaving these basics, consider the related issue of direction in figure 5.4. Unless the freedoms to move are uniform and homogeneous at every point in space (this is what Euclidean space means by vector dimensionality), we cannot define a velocity *vector*. In a network or graph, there is no such homogeneity over a region of decent size. We can define a point velocity, but there is no comparable notion of direction at each point at all (see figure 5.5). The observer can only measure the angles between L and M, and M and R,

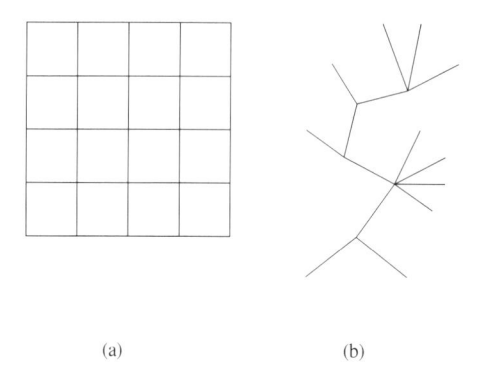

(a)   (b)

Fig. 5.5. Homogeneous lattice of Euclidean space give vector dimensionality (a), compared to an irregular graph with no consistent dimensionality (b).

so it has to rely on a theoretical model for what is happening in terms of those measurables. The quaint idea of a uniform Euclidean space is nowhere to be found. What one calls the x-axis or the y-axis at one point, might be L and R, or the red, yellow, and pink axes at the next point, followed by cow, dog, sheep,

5 TIME AND MOTION

bison, directions, and so on. Even if the next agent labels the axes according to the same convention, we cannot assume that their interpretations are the same (they could be interchanged or twisted).

When data pass through the Internet, through routers and switches, these issues become essential. We can only wonder what happens in our physical spacetime far below the level of particles and fields we discuss today. How do the points of spacetime know that they need to maintain these conventions in a uniform way, with long range order? We only take it for granted in mathematics and physics. Of course, it might be one of those unanswerable questions. On the other hand, if we are serious about understanding the nature of space and time, filled or empty, these are not questions we can dismiss lightly.

THREE PROCESSES THAT CREATE MOTION

The kind of motion we've assumed so far in this chapter is quite close to our familiar idea of matter moving through space, but as we pointed out in chapter 2, this kind of distinction between matter and space is not altogether natural, and is at the very least not the only possibility. It's hard to define the idea of motion just in terms of space and time—we need properties that we can interact with, i.e. that can be measured. Beyond this, it's also not natural to talk about space and time as an arena independent of events and properties that we could measure. In the physics of elementary quantum fields that describes elementary particle interactions (which it does with great accuracy), the references to spacetime coordinates in its mathematics play only a bystander role in the calculations—as dummy variables for labelling processes. This is one of the criticisms sometimes levelled at the 'Feynman diagrams', used by particle physicists, as a picture of reality. Certainly we need some parameters to label the mathematical processes, but to identify those generators with familiar space and time could very well be a mistaken assumption that only leads us down a blind alley[142]. The concepts of space and time could be removed entirely, for much of it, leaving only the idea of interaction as an elementary concept.

If we go back to the most elementary representation of a discrete space, from chapter 2, based on promise graphs, then we can rethink motion too. In spacetime where adjacencies are not rigidly fixed, there are three ways for motion to be defined[143].

1. The first kind of motion (figure 5.6) is to label one of the points that we

want to consider as moving, and then to swap the order of the points, using what is called a bubble sort algorithm in IT. This is exactly the kind of motion one finds inside the software data structures of computers all the time. A 'precedence relationship' means an ordering of points. The points are joined into a structure known as a 'linked list' to form a queue of agents acting in the role of points. Agents effectively move through the queue by swapping places.

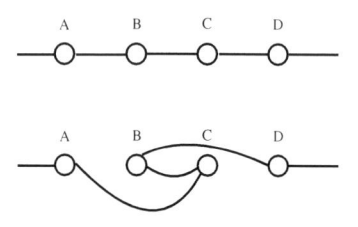

Fig. 5.6. Motion of the first kind: displacement, as in a fluid aether. Individual points change their order, as in a bubble sort, by swapping places with their neighbours.

2. The second kind of motion (figure 5.7) is to imagine two different kinds of agents—fixed agents and mobile agents, one of which is attached to the other, a bit like an electron attached to a nucleus. This is the model used by cell phone towers and mobile phones, for instance, as well in electron configurations of chemistry and covalent bonding.

3. The third kind of motion (figure 5.8) does not involve motion in the traditional sense—only a transaction of information about the exact promises each fixed agent position should make. Nothing moves as such, but an observer probing the local promises would see a change that looks like motion. This is virtual motion, such as takes place in the dataverse for software systems[144].

Cases 2 and 3 are similar but there is a subtle semantic different between the two, which could seem *ad hoc*—much like the arbitrary distinction we make between matter and space. In 2, matter and space are treated as explicitly different agents; in 3, they are different states of the same substrate.

5 TIME AND MOTION

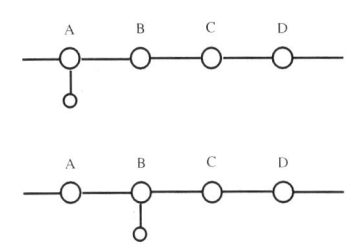

Fig. 5.7. Motion of the second kind: handover or musical chairs, as in a solid crystal which hosts covalent guest agents. The moving agent binds to a fixed solid-state background to yield a locally absolute motion, without prejudice about large scale structure.

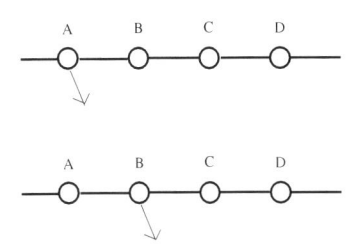

Fig. 5.8. Motion of the third kind: message propagation, or relabelling. Motion involves the transfer of a state (scalar propagation) from host to host, along a solid-state crystalline spatial background.

THE LAST TURTLE?

By studying these examples, we start to understand something frustrating and confounding: there are phenomena that localized observers are not well adapted to comprehend, because our sensory model is designed to deal with a particular scale and a particular range of speeds. Not only do fundamental processes happen too quickly for us to see, but the very ability of any observer (no matter how big or small) to sample the world and detect a change is also the very thing that the observer experiences as time, the very changes that make it tick. We have reached the end of the scaling of our simple picture of reality, because perception requires a minimum scale. On the scale of the very small, our usual story simply doesn't work.

This is effectively a restatement of the observations that Einstein made in his theory of special relativity, but emphasized in a different way. He talked about clocks, as all relativists have done since, but this analogy only works if your mind's notion of time has already been expanded to understand that a clock is a lot more than Big Ben or a Rolex. A clock is an interior (local) process. In his special theory of relativity, Einstein discussed two principles:

1. Absolute uniform motion cannot be detected by any observer.

2. The speed of light (signals) is independent of the motion of the source.

The latter principle is a typical characteristic property of waves when they propagate relative to a medium of some kind. In the case of light, the medium is the electromagnetic field, or local virtual displacement currents. The first principle of relativity is true for any virtual process that has no direct access to the underlying turtle of its physical infrastructure. So, for example, it applies immediately in the dataverse, but only on the virtual layer: computer programs are unable to know their physical location, even though they have an absolute physical location in their embedding space. Could it be that our measurable phenomena are only virtual phenomena on top of an absolute substrate that we are prevented from observing?

When theorists find ever 'deeper' theories of matter and energy, they assume infinite resolution of observability and representation, or they tend to assume that space and time are infinitely divisible and have an independent existence: turtles all the way down. If this were not the case, it would not even make sense to use the quantities x and t in their equations. This seems to be a conspicuous blind spot in theoretical physics.

Unless the observer has interior processes that are faster than what is going on outside, it will never be able to sample anything (near or far) fast enough to observe it. Now add to that the problem of seeing: that in order to see anything, we have to receive a signal from the thing, which has to travel across space. This involves a sequence of changes across space, which itself acts as a clock. How fast do those changes happen relative to the interior changes we (or any observer, big or small) can perceive?

We are trapped in circularity. Time is only a narrative about a series of samples received in a certain order at a single place (the observer). We may choose to attribute additional semantics to this observer, as physicists have sometimes tried to do in quantum mechanics, but this is a kind of bottom line.

5 Time and Motion

Phantoms of continuity

Sudden, rapid jumps between locations that we assume are smoothly connected have been considered disconcerting to physicists, especially at the quantum level, leading to all kinds of explanations of phantom probabilistic motion. As a frequent global traveller, I suspect my friends and family experience the same consternation about my own movements. I send a message from Oslo, then I am unobservable for a number of hours—perhaps while they are sleeping—only to send a later message from China or America. Since they were sleeping or inactive, there were no relevant ticks of their interior time relating to sampling of our text conversations, so they were unaware of the number of ticks passing them by in my own processes. The result is a kind of probabilistic behaviour in which I am observed at random locations with different probabilities at consecutive waking ticks, as they experience their exterior reality. This could be attributed to some mysterious quantum effect, but actually it's a phantom of mismatched scales in the sampling rates between observer and observed. No wavefunction collapsed, no universes branched off—someone's perception of time was just paused during sleep.

We get around the problem of gaps in perception by 'cheating': by trying to fill in the gaps by sewing together disparate experience from a number of observers over a non-local region, which amounts to agreeing to measure time according to standard clocks, instead of using interior processes. This opens the door to a wide variety of confusions. Information technology builds plausible approximations to reliable clocks, in the sense of having a process that ticks away in a stream of events, and we label these events with the regular designations on a clock face. So all computers have a time and date function, which is calibrated when you install the computer and also gets reset by standard time services called NTP (Network Time Protocol). When you change timezones, by flying from country to country, that calibration between interior ticks and exterior time gets muddled, and can cause problems. On the whole, we've learnt to fudge our way pasts these minor issues though—they affect relatively few people. But we've also started to reach the limits of the simplest naive approaches we've used so far. Our clock time is based on the rotation of the Earth relative to the sun, but that is not a uniform process either, and the units by which we count hours and days are not really compatible with the units for atomic processes or for years—hence the need for leap seconds, and leap years on our global clocks. Don't even get me started about Daylight Savings Time.

Another phantom of observation relates to limitations in the boundaries and surfaces over which we sample information. Our mental model of the world, based on our experiences of three dimensions, compensates for the fact that we only see a two dimensional projection from each eye. Our brains do some fancy interpolation to convert the angle swept out by the motion (figures 5.2 and 5.3) into an effective location, based on our assumption of distance. We know that these perceptions are non-unique and easy to fool, as every photograph of someone holding up the leaning tower of Pisa, or holding the moon in the cup of their hands, illustrates. Moreover, we assume that our observations are instantaneous—that the things we observer are happening in real time, with no latent time-lag, because that is more or less true for things on the Earth. But we already know that what we see in the heavens may have happened billions of years ago, what we see in the dataverse happened *somewhere* and *somewhen* that is hard to explain, and that we have trouble swatting flies because our brains are slow—by the time we see that in one place, they are already somewhere else. Our mental model of space and time are based as much on a model of assumed scale as they are on observation. There is no type of signal that can locally inform us about the absolute locations of objects and the times at which they occupied those locations.

As humans, we have come to think this is the normal state of affairs, because we've evolved to perceive precisely this kind of behaviour in our environment. But it extends even to the instruments we use to expand our sensory capabilities. We can create better sampling machinery, oversampling then try to slow down films in order to see the granularity of image samples in a film, frame by frame—or we can take time-lapse photographs of slow processes and speed them up to look continuous. In each case, we play with relative clock rates in order to alter our model of spacetime, but in each case our assumption is that some version of spacetime ought to be a continuum that we can observe on the scale of human cognition.

The effects of mismatched (+) and (-) in signalling and sampling are the same set of effects at work in Einstein's special theory of relativity, just dressed up a bit differently. Einstein didn't really discover that time runs slower for some than for others—he showed that we will observe that—if we could observe the interior time of a faster object—it would appear to run slower than what we would expect for similar processes in our own interior frame of reference. Observers, in their own frames of reference (unable to directly touch those moving bodies) are unaffected though—as long as the motions and configurations of the particles are part of the clock the measures local time, we can only ever measure one rate of time. This is the case inside a computer network, and in every process where we use signals to observe other phenomena. The awkwardness comes from a simple reality check: we can never interact with other moving objects without experiencing the mismatch, because our only link to them is via an interaction whose interior clock is fixed to run at a constant rate relative to all locations.

The distortion of perception, like a hall of mirrors, reaches a hiatus in connection with gravitational acceleration—in which the consistency of these conclusions turns out to imply some actual intervention in the energy accounting of how processes proceed at different locations within a spacetime distortion. Whatever the source of interior time may be, it seems to be affected by gravity and other forces associated with acceleration. The ubiquity of the concept of force allows it to remain unexplained in physics—hiding in plain sight, and making its inexplicable nature something of a tautology. The sense that this is all very unsatisfactory is what makes it seem strange.

OBSERVABILITY VERSUS PREDICTABILITY

Whether in outer space or the Internet, relativity messes with our senses. Let's return to our understanding of motion. How do we know that a body is at a particular spacetime point, unless we can observe it? We might be able to glimpse the position of a body at occasional locations and obtain partial information about it, as part of a coordinated grid of locations that have previously been measured and calibrated, like a chess board. From there, we could try to estimate its direction and assume that it visits all the locations in between at some imaginary time, but that is the kind of speculation that physics generally frowns upon.

Today, on the Internet, people say 'send a photo, or it didn't happen'. We are starting to realize that reliable reports are needed when we can't see everything from a privileged position. But, of course, we interpret those photos according to our own norms and expectations too. What is promised by the photo (+) has to overlap with what we derive from it (-). There is basically no situation in which observers can avoid using a mixture of both observation and inference based on some assumed model.

It is physically impossible to see everything that happens, because seeing requires processing by a brain or a computer, by a process that may have a longer duration than the motion we are trying to observe. So at best, we have a maximum sampling rate, limited by the machinery of observation. We should forget even about the idea of there being a continuum. Shannon's picture of information channels makes the idea nonsensical.

Space and time may only appear as a continuum because that is the illusion our brains make for us. When we try to swat the fly, we miss most of the time, because the fly can change its motion at about the same speed as neurons can conduct our vision to our brains. So we are always lagging behind. By the time we see the fly, it has already moved on. Was it every really at any points in between? Yes, we can take high speed camera film and slow it down to observe that space has greater resolution than we can observe directly. But how many times can we do that before there is no more resolution to observe?

Our brains are quite good at predicting the position of moving objects as long as their paths don't change at random, as they do for insects. We rely on the continuity of motion to make inferences that shore up our perceptions of location. We can also use the assumption of continuity to diagnose our own failings. If we know how a process will continue, and we miss something, we might either infer that we fell asleep or that there was a sudden fundamental discontinuity in the motion. In the same way, continuity plays an insidious role in physics. It is assumed that experience about the past will be a good guide to what we are going to experience in the future. That very sentence implies concepts that span a multitude of different assumptions about scale and aggregate meaning. Defining continuity of reason seems like a hopeless task, but what else could we do? It's practically an axiom of science—so, when it begins to break down, everything feels up in the air. However, we can try to be more open minded. Science has tripped itself up over the centuries by holding onto dogma after doctrine, convinced that it was doing the best possible job. Yet, a surprising number of arguments about observation also make hidden assumptions about biased human faculties. The 'observers' of science are expected to think about the world as our brains do, operating at a certain speed, and experiencing certain preferred scales, to be able to see the effects at work. It seems like bad science to assume that the observer in a physical phenomenon must be a brain (a structure many orders of magnitude larger or smaller than the phenomena where most of these observational challenges take place). No wonder observations appear odd if we try to mix phenomena on quite incompatible scales.

MUDDLING OBSERVATION WITH CONTINUOUS INFERENCE

To illustrate some of the problems that arise from muddling the assumption of continuum space with discrete space—not to mention muddling qualitative and quantitative measures—consider, for a moment, a habit that students are often chastised for, when dealing with data from observations: interpolating and extrapolating samples in an unwarranted manner.

Suppose someone measures some local value at two or three sample points in an experiment, plotting these as in figure 5.9. There is an almost irresistible urge to draw a line through the points. It helps to make the individual samples more visible, but it may also tempt us into unwarranted assumptions. By succumbing to this lure, it is easy to believe that all points on the line are equally valid and can be used meaningfully. I deliberately chose an example to make a nonsense of the idea; other cases might be less obvious. The mistake lies in the tacit assumption that everything we want to measure is smooth and linear.

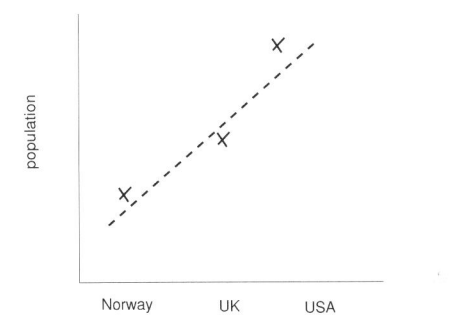

Fig. 5.9. Abusing interpolation when it doesn't make sense. There is no meaning to the spaces between the country names.

Take a look at figure 5.9. The measured samples are population figures for the countries. Three samples are taken for three countries. Countries are discrete entities (as labels based on semantics they don't form a continuum). Thanks to our presentation (-) of the supplied data (+), they happen to lie on an approximate geographical line, which happens to be embedded in the continuum of the Earth's surface. The sample populations do not represent any single point on the Earth's surface or in Euclidean space: they represent huge regions, with completely different shapes and sizes. Summarizing a country's population by a single number at a point is analogous to the Newtonian trick of attributing a mass of a body to a single effective point (the 'centre of mass'), which is the average position of all the interior masses belonging to a body.

The country locations are just names representing the regions, each of which may involve multiple discontinuous regions (islands, and disjointed territories, and so on). The straight line drawn seems to suggest the inference that there are relevant locations in between these named locations, e.g. halfway between Norway and the UK. The straight line connects the points on both axes: position versus population, so it seems to imply that there might be several million people residing half way between Norway and the UK. But where is this location? Is it in the middle of the North Sea (where there may be a few people on boats and oil rigs)? Or does this imaginary place represent an abstract location, such as people who are half British and half Norwegian? In what space are we interpreting the distances between these samples points? On what basis do we merit continuity?

The mistake is a common error of semantics. When we misparameterize observations, which are really discrete, as if they were continuous, we can be misled. For example, if we use GPS coordinates instead of names for countries, we might be tempted to use mathematical properties of the real numbers that are not appropriate in the context. Numbers are nothing more than a systematic scheme of names that happen to fall into a certain order. We can postulate the existence of real (decimal) numbers, ratios, and fractions, but they only are only defined in a relative sense in nature. The world is discrete. Interpolation is what the brain also does for us, when it can't resolve tiny details—so we are well used to seeing this illusion made before our very eyes. We sample something at a few points and our brains immediately start their natural tendency to smooth out the picture into a complete and unbroken perception. We want to draw a curve through the points and believe in the curve.

The limitation extends beyond human faculties to any measuring device that translates measurements into something we observe. Human perception is a slow process, so we see motion as continuous because our biological processes are slow compared to the speeds of light and sound, which are the primary ways we perceive the world. Light signals may well have to travel through empty spacetime to reach our eyes, but then signals interact across a sampling boundary that defines its own interior and exterior separation—with its own clock. On their interior, our eyes start chemical processes, which in turn have to travel far more slowly through a variety of sensory tissue and electrochemical channels called nerves, to a brain, which uses yet another bounded technology to make sense of the data. Because the brain has evolved to process different aspects of imagery through specially adapted machinery, it separates colour, motion, and shape through different sets of cells[145]. Colour information is quick to compute, then motion, and finally shape is more complex—all taking different amounts of time. In other words, the brain's separate awarenesses of colour, motion, location, etc, all lag significantly behind what is going on in at the actual locations. A fast moving ball might have moved several metres before our brains compute the imagery.

Our conscious experience is cleverly adapted to take into account this blurry out-of-focus arrival and to predict the movement's true position. As long as objects continue to move in simple predictable lines, this works quite well. However, small erratic bodies like insects that move quickly and change course unpredictably can stay ahead of our prediction capacity, because their speed of change (+) outstrips our ability to sense (-) and compute—making flies difficult to swat. Clearly, the sampling rate of our sensory processes is not just about the arrival of light, but about the speed of propagation of a data through a multi-stage data processing pipeline, completely analogous to the artificial data processing pipelines we build in the dataverse.

Humans are confronted with a composite illusion, caused by data processes of the brain[146]. Now that we are employing data processing for purposes that span human and non-human timescales, this issue is starting to become a critical one for our daily lives. Computers that drive autopilots or self-driving vehicles accept data through sensors along a succession of boundaries, each with their own clocks, and the sum of those mismatched processes adds latent time (or *latency*) to the perception of the final decision making parts.

Again, when watching a ball or an insect fly through the air (figure 5.2), even assuming we have line of sight between the observer and the ball, we see the ball pass through an angle of our visual field. The fine graining of the available paths for light is far greater than the resolution of our eyes, and our brains fill in with the illusion continuity, not knowing any better. This allows us to build a systematic prejudice: assuming the rules of geometry, and the speed of light, we can (in principle) measure the distance of the ball and the angle it moves through, and we can measure the time taken. In this way, we can calculate the speed it must have moved at. Eventually, we don't need to verify these assumptions—we take it for granted that all movement follows this pattern—until something else changes and the assumptions fall apart.

We don't need to rely on our evolutionary limitations. All this can be done with machinery faster and simpler than the human eye-to-brain connection too, so we can recount experiments in 'slow motion replay', and verify our model over a wide range of scales. But in each case, we still try to translate the measurements into a model of the world based on inferred continuity. This is how we develop inferential knowledge of spacetime on a human level. This is also what allows us to fall foul of optical illusions.

What happens inside the computer's dataverse? As we already noted, the spacetime of the dataverse has far fewer points, far fewer locations, and processes are almost as fast as the signals that move between them. It is closer in a scaling sense to a quantum system, and selects an entirely different set of scaling ratios. An observer, inside the computer network (figure 5.3), would not have such an easy time of seeing data move from an initial to final location—because there is no ambient signalling to rely on, because there are only certain points at which the data could be observed, and because the spacetime of a virtual computer network only connects the observer to the initial and final points, while a physical one may have only one routing station in the middle. There is simply no way to observe data at any other intermediate location, and no reason to attach any meaning beyond that—even if we chose to invent an imaginary continuum of locations in between.

Figures 5.2 and 5.3 compare the basic problems we face in building up an intuition of motions in the universe and in the dataverse. Figure 5.4 explains why the two can't match. Some of the differences we perceive are the result of the velocity of motion v approaching the same order of magnitude as the velocity of observation carrier c, expressed by the dimensionless ratio v/c. It also depends on the extent to which position and time can be measured independently by the observer, and across how many locations. In the two figures, the observer is far away from the trajectories themselves, so he or she has to infer the distance travelled somehow, or find out separately.

To build up the technology for predicting motion, it takes quite a lot of information that we take for granted—not all of which may be available in all situations. That means we cannot transfer our intuition about figure 5.2 to the scenario in figure 5.3. The deeper implication shouldn't be read as computers

AN ABSTRACT BELIEF IN MOTION

If it isn't already obvious, then let's state it plainly: what we mean by motion, in different situations and on different scales, is qualitatively different. In the dataverse, we see a series of hops or transitions that may seem to spring out of nowhere. If we try to imagine the dataverse as our universe, data might seem to be almost anywhere with a certain probability, something like what we have come to believe about Quantum Mechanics. In the universe, we imagine a smooth sliding of bodies, with such continuity that we define the effective changes using differentials—quantities like speed, velocity, momentum, and so on, by analogy with Newtonian systems. Speed and momentum are treated as *effective vector fields* that assign values belonging to several places and times—on 'non-local' measurements at multiple points—to a single 'local' coordinate point! This is a travesty of relabelling! In doing this, classical physics creates a schism between observation and causality that occasionally comes back to bite us. I'll return to this issue in chapter 7.

We can tell the classical story of motion only because we assume plentiful information and a high spacetime density of paths. The idea of motion in quantum theory, for instance, has nothing like this to rest on. Although quantum theory also talks about positions and momenta (velocities), they are defined only implicitly—not based on the same kind of observation at all. We are unable to observe motion, in the sense of figure 5.2. Instead, we can observe something at the ingress of a region and something at the egress of a region, and we can only infer what happens in between. This is how *scattering* problems are couched in quantum mechanics (see figure 5.10). Measurements are made as transactions, much like money transfers from input to output. We have certain information at the spacetime boundaries of the region (ingress and egress), an assumption of energy conservation within the region, and a model of what could happen to the energy, how it could be transformed and redirected along different channels. These events are called interactions, and they are represented as 'vector fields' as if they actually happen at different mathematical points in space and time (just like centre of mass or population of a country).

When we compare our mundane view of motion to a computer system, we

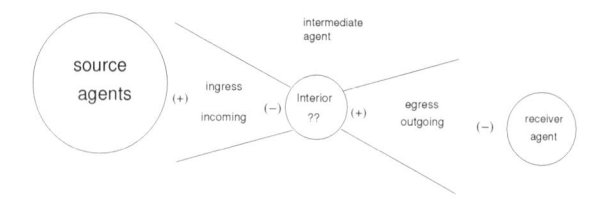

Fig. 5.10. With a scattering interpretation, we have incomplete information, and we can only see what goes in and what comes out, but not what happens in between.

start to appreciate the depth of our assumptions. A pretty common way of processing data in organizations today is to feed data into an input, pass it through a number of computer programs, and see what comes out at the other end. These are called 'data pipelines', because the process is more or less linear like data flowing through pipes, or current flowing from component to component in an electric circuit. Data move from the input to the output only because the computers are programmed explicitly to pass data from place to place: the previous computer in the chain makes a transaction available to the next, the receiver accepts the available data, matching (+) and (-), at each stage (as in figure 5.10).

Unless you have interior observability, you have no idea where data are at any moment, from a perspective in the exterior. If we look only at the ingress and the egress, we can sample a kind of flow rate of processing or data momentum according to a clock on the exterior, independent of all of the locations along the pipeline. We have no idea of what sense of time is ongoing within the pipeline, whether data go only in one direction, or whether there are conversations going back and forth between points that may or may not exist on the interior. All that is speculation, but we can *define*, by assumption, the existence of a momentum at every imagined ordered sequence of points along the way.

If we took a *statistical* or *probabilistic* interpretation of what goes in and what comes out, as we do in experimental science—based on a model of what we imagine spacetime looks like on the interior—then we would build up a picture in terms of fragments of evidence repeated over many experiments, based on what happens to data going in and coming out. The intermediate agents or regions in the chain of evidence are unknowable, so we have to trust our knowledge or our guesses about what transpires inside. As soon as we shift to this model, we give up on knowing the details of what might be happening on the inside (which is pragmatic if a little disappointing), and replace the entire data pipeline with an effective state transition model—just as we forget about random atomic motions in all directions when water comes out of a tap. We allow the large scale boundary conditions to dictate a model of what is happening on a smaller scale. Again, the narratives of scale play a key role in our modelling.

The concept of motion in the dataverse makes most sense at the process level, not so much at the level of physical computers (except where there is wireless mobility). This is significant because we might only be able to fully define motion as a virtual process on top of an underlying fixed set of embedding locations. Could it be that motion is not an elementary process at all? Whatever corresponds to the matter and energy that move around our universe must be virtual rather than fundamental and elementary, otherwise it's turtles all the way down. By inference, we have no way of knowing whether the peculiar physics we know from Quantum Mechanics and upwards corresponds to virtual processes on top of a fixed substrate, or whether they are truly elementary. As far as I can tell, there is no way to exclude that possibility.

These examples focus on elementary scenarios, like computing and elementary particle physics, because their simplicity makes them easy to understand, but there are plenty of other examples. Biological motion happens on many levels and scales—so many, in fact, that it is sometimes difficult to think of motion in elementary terms at all. There are both virtual and physical processes running side by side, with Newtonian and non-Newtonian behaviours. Immunological responses and gene selections are virtual. Chemical bindings and flow processes are physical. This is typical of a complex system, where scales are intricately mixed. Sometimes we have a trajectory model of motion, as in the spreading of a drug throughout a patient. Other times we have only an interaction model: what goes into and comes out of a kidney or a liver. It was surely the linearity of classical mechanics that allowed Newton and others to separate space and time from matters of scale in the first place, leading to our received view of spacetime as a theatre. Biology shares some aspects in common with limited information systems, because of its modular or cellular construction. Modules are important structures for reinterpreting information: this leads to specific functions. The counterpoint is that modules reduce observability, by deliberately decoupling spacetime agents. Boundary conditions describe the constraints on a process outcome at the edges.

Biology is complex because it operates on a multitude of scales. Economics too has a similar challenge. For instance, when talking about the economy, economists replace all the microscopic exchanges, sales, bankruptcies, successes and failures with a simple measure of GDP (Gross Domestic Product), and thereby turn our view of the economy from a turbulent rollercoaster of millions of interacting agents into a fantasy about smooth sailing at approximate equilibrium by a single superagent. Little wonder that the economy experiences large discontinuous fluctuations that can't be explained by the models. Any theory that bases causation on transitional outcomes at the edge of a region will be an incomplete description—a statistical estimate of an imagined ensemble of subprocesses. It might well be time to compare these issues of partial information across all scales, from top to bottom, rather than inventing new explanations across the board—but that itself is a symptom of modularity of the spacetime process of science. Models blind us to unnecessary details on purpose.

BACK TO THE FUTURE: RECURSIVE EVALUATION

In descriptions of causality, it is generally acknowledged that causes precede effects. This fits our ordinary macroscopic continuum picture of how the narrative of the world proceeds. The trouble is, it is simplistic. We think of causality as a deterministic sequence in which the past advances stepwise towards a future, but according to whose clock? There are plenty of cases where information about an inevitable future outcome effectively determines what happens leading up to it. We discussed this in chapter 4, using the term advanced and retarded boundary information, and I want to return to it here too.

Advanced and retarded boundary descriptions of cause and effect play a large role in the explanation of process motion at the semantic level. Let's start with a mundane example. Suppose I order a new phone online. The manufacturer looks to see if they have any in stock. If they don't they need to build some. Industrial logistics calls this 'Just In Time' delivery—essentially propagation of a process by an advanced boundary condition. However, in order to build a phone, they need a load of components, so they look to see if they have the parts to assemble. If not, they need to order them, then someone else needs to check their stores, and possibly make a new batch of components to send on. It's as if someone in the future went back in time to make sure that the correct sequence of events actually took place. Process direction determines what is forwards and backwards, breaking the symmetry by its semantics. Causality can flow forwards and backwards in the sense of *desired end states*.

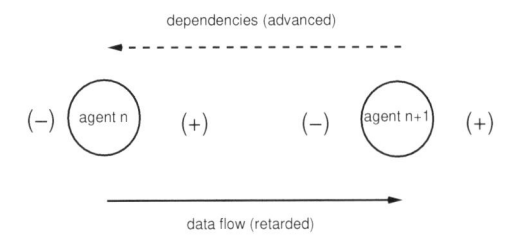

Fig. 5.11. A chain of agents making conditional promises that break the symmetry of 'reversibility'. Direction of causality is multifaceted. Reliance on prerequisites reaches backwards from the end to the beginning, while the flow of change (data) goes from start to finish. Each agent in a chain promises to accept (-) something from a predecessor and to promise forward (+). When the (+) promise is completely conditional on the (-) promise, the chain becomes a deterministic state machine, and one has 'forced push' semantics.

When I order a phone, the request effectively makes an acceptance promise (-) to use any matching offer of a phone from a manufacturer (+) it can get. It signals 'backwards' along the axis of delivery, transportation, and manufacturing (see figure 5.11). In a transactional quantum interpretation, you might say that the absorption wave precedes the offer wave, guiding it into port. If there is a longer chain of dependencies (factory, component suppliers, raw materials, etc), the manufacturer can only match the request with a new promise to supply phones conditionally, because each promise depends on prior promises in the chain. In other words, there is nothing deterministic going on in the dynamics—only in the chain of semantics or intentions.

A manufacturer can promise phones based on availability, and availability can only be promised dependent on supplies of components, and so on—reaching backwards all the way to the raw materials, in a complex chain of logistics. We might prefer to see the manufacturing of a telephone as a purely forwards causal process—a deterministic one, but this would be misleading. The actual assembly process might be a chain that goes forwards in time from the raw materials until delivery, but the *determinism* of the process goes backwards, 'recursively' from an imagined time in the future when we hope the phone will arrive. The key point to emphasize is that time, in the sense of a prerequisite dependency, is not clearly oriented until we have information to propagate causally. Causality and time are not the same thing.

In my role as a technologist, I've been one of the advocates for understanding computational processes, in the dataverse, from advanced endpoint boundary states, rather than by their retarded starting states[147]. The 'Just in Time' viewpoint is a bit like baking a cake from a recipe or from a cookbook. If you follow a recipe blindly, without knowing anything about the desired end state, you follow retarded boundary information: you start with a complete set of starting conditions and you follow the causal steps, adding ingredients and baking until something emerges. Without knowing what the result should look like, you can only hope that the result is indeed what was advertised. To believe that, you need some 'channel isolation' from causal errors and mixing of signals—what if someone interrupts you, or there is a power cut while timing the oven? A cookbook, on the other hand, offers some advanced warning usually with a picture of what you are trying to make. It encourages you to emulate the final result, with advanced boundary information. This is a form of backwards in time causation, that makes you adjust your process trajectory along the way towards the final result. This is what Claude Shannon referred to as inline *error correction*. This identification of symbolic information with boundary value problems is rarely given much attention, but this is now changing thanks to the computing cloud.

Advanced causation is also used more implicitly in cybernetic *feedback*. Feedback is a common method of regulation in dynamical systems. The current output state of the system reaches back as if it were a (partially) desired state in order to stabilize itself by comparing the actual outcome of past changes with new changes arriving, and mixing them together[148]. When you select a destination in a GPS satellite navigation system, you have defined a location as the desired end state of your journey process. The car compares its input location with a route computed from a desired end state. If the car begins to deviate from the direction of the destination, the GPS recalculates a course correction.

The method of 'Just In Time' delivery isn't particular to logistics chains in industry, or even cookbooks. It is also extremely important in computer science, where it goes by the name of *recursion*. Computer programs are written ac-

5 TIME AND MOTION

Imperative

```
Let X = 1
   ↓
Add 2 to X
   ↓
Let Y = 3*3
   ↓
Add Y to X
   ↓
Print X
```

Recursive

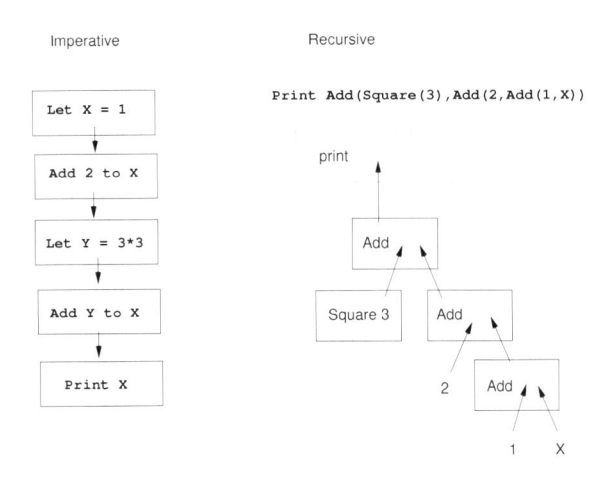

```
Print Add(Square(3),Add(2,Add(1,X)))
```

print ↑
Add
Square 3 , Add
2 , Add
1 , X

Fig. 5.12. An imperative (retarded) program versus a recursive (advanced) program. On the left, time flows simply from top to bottom; the start is fixed and the end is derived. On the right time starts at the top, then jumps to the bottom, and returns step by step from bottom to top; the end is fixed and the start can wander.

cording to both advanced and retarded models of causation. Retarded causation is known as *imperative programming*. It's the normal kind of programming we learn in school: assembler or machine instructions one used it exclusively, as did the original BASIC programming language. Programmers make a list of commands: do X, do Y, and so on, forming a 'flow chart' (see figure 5.12). Imperative programming is well known through popular flow charts also used in business processes. Its limitations were quickly realized however. Today, every language from machine code, assembly language, BASIC, Pascal, or Python, use a mixture of the imperative model and an advanced recursive model.

Imperative (retarded) programming is sequential, deterministic programming: instructions proceed step by step and lead to an orderly queue of transformation, just like any other data processing pipeline. Functional programming (advanced) is *recursive*, meaning it works by stating the desired result—anticipating it—then reaching backwards to pull the result out of its hat (see figure 5.13). This is called recursion. For example, we might say "print result", where result has not yet been evaluated. The computer says—ok, hold that thought, while I go and figure out what result is. Encountering a name without a value, the computer looks for a causal rule to make 'result'. It makes a note of what it was in the

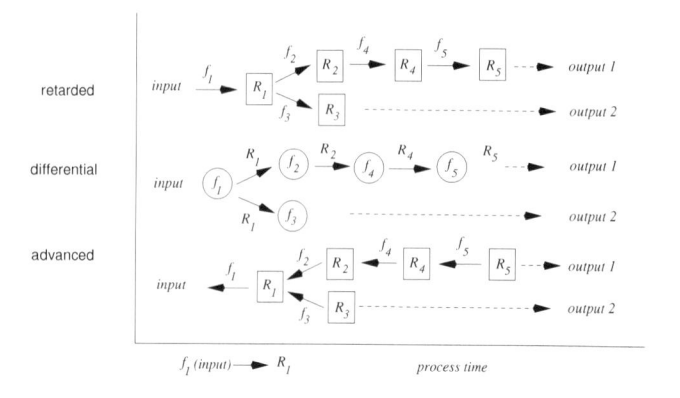

Fig. 5.13. Three descriptions of a process graph: the retarded view that starts with boundary information in the past and expands into a set of future results, the differential view, which considers only the transformations from start to finish, and the advanced view which starts with the desired result and relates it to its causal history.

process of doing, and goes off to compute the result. Along the way, it might encounter other things that haven't yet been computed, and do the same again. It keeps a pile of references about what it was doing before being interrupted, called a *stack*[149], which implicates *memory* in causal boundary conditions, so it can record how to resume once it can substitute a number for the promise of a 'result'. Each time an implicit result is referred to, the computer puts its current process on hold and sends out for a finished result (as if sending out for pizza). This recursive method is the basis of *functional programming*, which allows causal processes to express semantic relationships more easily than the imperative builder approach. It works by cake photos rather than by cake ingredients.

The importance of advanced boundary information, i.e. recursion, also turns out to be of great importance in *linguistics*, which is another form of advanced and retarded data processing. Flexible causality is one of the basic mechanisms that organisms use to scale processes, by recycling a finite number of 'words' or capabilities. We'll return to this in chapter 8.

In biology, emergent phenomena are processes that happen in complex systems, where many retarded causal factors (+) come together to make something new that happens to fit or find acceptance (-) in another advanced context—what we call an evolutionary niche. For instance, the evolution of a frog cannot be said to be predestined, but it has some advanced characteristics. Within the dynamical space of possibilities for all the parts that come together to make a frog, there emerged an 'attractor', or a set of previous causal circumstances making that outcome favoured over others. It has to fit into the larger environment. The realization of the attractor appears to be in the future, but its really a phantom of forward directed processes, like a goal or intention. It's important to understand that the fact that it is inferred does not make it unreal. Possible outcomes are well defined states of a system. This is another way of thinking about Darwin's basic insight about evolution as a 'blind watchmaker'[150]. We can't necessarily say that the 'intent' behind the future state was decided in the past. Some of it lies in the confluence of independent parts from an entirety of independent timelines. Some 'attractor' acceptance waves seem to lead inevitably to definite outcomes, through the action of deterministic or biased processes—but the conditions leading to those outcomes were set up in the past (i.e. as a prerequisite dependency). The flow of past to future influence is unidirectional only as long as we account for it in a deliberately unidirectional way. But, in doing so, we risk glossing over complex interactions that we use every day in our understanding of phenomena.

THE ARROW OF TIME VERSUS THE ARROW OF DIRECTION

Unlike multi-directional processes, our human idea of time seems to move in a single relentless direction—the so-called Arrow Of Time. I say 'seems' because the whole idea of what is important about time is intertwined with a variety of agendas and perspectives. If it is true that time only goes forwards, why should it be? Space seems to have multiple directions, why should time be different?

Advanced and retarded information tell us that causation and time are different things. Causation implicates space in time's arrow through process, leading us to confuse the direction of time with the direction of motion in space. After all, we use the motion of clock hands to make clocks that tell us the time, so it's not that surprising that we get confused. Boundaries play a role here too. Boundary information governs how changes propagate through space, from start to finish, and the main direction of interest to an observer is from its definition of exterior to interior and vice versa. These are arbitrary partitionings of non-local regions into something we renormalize and relabel 'local'. The discussion of causation is really about where boundary information comes from. Another way of putting this is to ask where 'intent' is defined: what selects a directional arrow in the first place? Where is the fixed information that anchors our understanding about an outcome, on some scale, to some fixed state? What determines the direction or *arrow of time* at different scales?

In physics, there is a clumsy discussion around the 'Arrow Of Time', which goes to the heart of these issues, but gets somewhat muddled along the way. The observation begins with the Second Law of Thermodynamics, also known as 'the law of increasing entropy'. This law was originally imagined to explain the behaviour of steam engines, but seems to be generally applicable in physics. It says that the total amount of 'entropy'—to be explained—in a system always seems to increase over time. Just as climbing a gradient points a direction in space, so a gradient of 'entropy' (a degree of mixing) can be associated with a direction for time. It selects no particular spatial direction on an average level, but acts as a large scale arrow for evolution. An ordered succession of macrostates (superagents' aggregate states over a wide area) is also a kind of motion, at scale, from which we can build a clock. It doesn't go in any direction, but it is statistical change in all possible directions. Curious that physics should use that as the model of Newtonian time. The law of entropy was initially mysterious until entropy was understood in microscopic terms, as a statistical concept. And, like all statistical questions, we can't make sense of it without confronting scale.

The entropy law is simple enough: on a microscopic scale, systems tend to go from what's usually called an 'ordered' state to a 'disordered' state— simplistically, it's more likely that things get messed up than tidied up. Many physicists think that this average change is more significant than the microscopic changes of elementary components, which seem easier to reverse or undo, but I contend that that is more of a fictional narrative about which symmetries are interesting than a reality. The macroscopic expression of 'mixing' is that a mathematical quantity—called the entropy—always increases over long times[151]. Or put another way, the orderly placement of matter and energy in space tends to decay by spreading out from a region of something to a region of nothing—from a state of minimal descriptive information (something and nothing, i.e. 1 and 0 is the smallest possible description of a state), to a state of maximal descriptive information (a big spread out mess of somethings and nothings that requires a lot of information to describe)[152].

An information picture can be applied to everything: different types of matter and energy get mixed up, so that the average appearance of all locations in space ends up looking the same, at a certain resolution, both dynamically and semantically. But if you look closely, all of the microscopic spacetime behaviours are still there. Only bulk 'superagent' regions shift from appearing distinct to indistinct—but that is a story of selective give and take: directional mixing (+) and selective choices about what we care to accept as distinguishable (-). By bringing scale into the picture, the history of change to a system—*non-locally* across a region of space—looks like a series of ordered epochs that can be associated with a partial ordering of non-local states—and this evolutionary motion is what we are supposed to associate with time. This sequentialism is not motion in the sense of physical spacetime, but rather is a virtual descriptive space of states called phase space[153].

In a large spacetime—large, that is, relative to the changes that contribute to entropy—the unilateral increase of entropy is easily explained by probabilities: order and distinguishability are states of structured organization. A pile of bricks is more unique and distinguishable than a random collection of bricks spread all over space (by the very definition of random). Since there are many more places for bricks to migrate to from the pile, there are far more ways in which a pile of bricks can spread out in space by toppling and crashing down than there are configurations in which all the bricks are at one location. As long as we assume that microscopic processes are random (non-oriented in space and without an advanced attractor), then this argument is true. There are millions of possible outcomes that are *not* a pile, but there is only one way of having the pile at a particular location. So, any small 'random' perturbation of the pile of bricks is more likely to destroy that unique state than preserve it (or assemble it from bricks lying around). A perturbation of all the bricks in the same direction, by the same amount, will not lead to a condition of greater entropy. The entire story revolves around what we choose to distinguish and what we choose to call indistinguishable. It is built as a grand tautology on how we express average symmetries of space and time. Thus it is about distinctions, which is exactly what we mean by information.

For emphasis, let's examine the story in more detail. The probabilistic argument for increasing entropy and time is the only non-spacetime-directional explanation for why processes go more in one time-direction than another. You make spatial direction go away by making some arbitrary choices about scale and semantic distinguishability that are helpful from a thermodynamic perspective, but are misleading from a microscopic perspective. Space and time are connected through processes, so any direction of time has to mean a direction in space too on a fundamental level. But—it is argued—we have to exclude directional arguments because the laws of physics are 'reversible' and the time that appears in Newton's laws can go forwards or backwards.

This argument is so pervasive that I want to proffer a quick example to put it into perspective. Imagine such a thing as a law of shopping. We postulate that shopping is driven by a boundary condition called a shopping list, and the price of the shopping is proportional to the length of the list. This is generally true on any shopping expedition. You don't need to say what is on the list to make this observation, but you wouldn't claim that physics denies that particular choices exist. A shopper is free to take or leave any product in the shop, but once a shopping list breaks that symmetry by deciding, there is a process direction supplied. A shopper is not doomed to pick and return items to and from the shelves in a meaningless display of randomness, guided only by the forces of probability. Prior information bounds the process by which you interact with the shoppingverse. You only leave the details out of the general law because you don't know what they are yet.

The reasoning about reversibility denying other explanations for the arrow of time is wrong. It's like saying that the laws of physics forbid you from driving to the city because the law does not specify in which direction you have to drive, on the same road. Or, equally absurd, the laws of physics don't care whether you drive on the left or the right side of the road, so you can't drive on one or the other. It's a facile observation about partial or underconstrained solutions to constraints. Microscopically, what selects one direction or the other is the presence of a boundary condition: information about the 'intent', i.e. what, in neutral language, we call *symmetry breaking*. If we are honest, we still have no explanation for what breaks that symmetry on the microscopic scale, but there is no reason to use this as an incorrect explanation.

The Equations of Motion that are associated with physical law are not the complete laws of physics. They describe a realm of constrained possibility. They can be made specific by completing them, in a spacetime sense, with boundary conditions. The association of 'law' with a partial description began with the Natural Philosophy of Newton's time, in which the desire to see The Divine obsessed philosophers with the idea of there being a wider beauty in the form of general 'laws' of nature, rather than a litany of individual cases. The apparent reversibility of the laws expressed by Newton is a figment of the symmetry that he built into their formulation, in order to cleanly separate concerns. Little did he know that those concerns are not generally separable in the wider world of physics—but that's another story[154].

There are plenty of problems with the hallowed status of reversibility in physics. While it is true that irreversible systems may lead to an increase of entropy on average, this would not be possible without the existence of a microscopic arrow for time too. It ignores important directional phenomena: a movement of a clock in one direction counts as a clock tick and thus as a new time. Without this there could be no physics at all. It is the act of receiving or sampling an observation that breaks the symmetry and advances an observer's time. Recall the earlier comment about forces: the phenomenon of force as a prerequisite for certain changes is space-directional, yet force also goes unexplained along with time. Vectors, from source (+) to receiver (-), like force, are defined in terms of spatial channel direction rather than statistical direction, implying change that propagates, selecting a microscopic arrow of time. The idea that some directional changes could be undone is no more probable than a reversal of entropy—it's just a question of where unexplained information comes from (ordered or disordered).

The ability to talk about reversibility already assumes that there is a difference between forwards and backwards—which implies that there are already two arrows for motion, and the argument is about which of them gets selected. Symmetry breaking is the term used to describe the latter. The time that appears in equations of motion is microscopic time, and this is not the same as the arrow of time that thermodynamics and entropy are talking about (macroscopic time). In other words, this is really about scale. What should be apparent from these remarks is that disorder, or mixing entropy, is inseparable from ideas of scale in space and time, and can't be used as an explanation for it[155]. Entropy can be defined when a distinction between states becomes unimportant, i.e. a boundary groups agents into a superagent region, fixes a scale, and replaces detail with common information on the boundary. If you return to exactly the same configuration or fixed state, then you have returned to the same time, by definition of what observability and information mean.

If space were tiny, so that there could only be one possible location at which a pile of bricks could exist, then the argument about disorder falls away. Everything would already be as spread out as it could be. In a very small spacetime, there cannot be any change at all—not enough degrees of freedom. The implication that time is related to spreading out is wrong. Time is related to change on any scale, including the perturbations that lead to the spreading out of stuff. The large scale sequence of coarse changes has a kind of partial ordering—not in any direction—but in an imaginary orientation that can be associated with large scale time. The problem with the notion of entropy as the measure of time is that, in virtue of being non-local, entropy is mainly exterior to every observer. However, Nyquist's sampling rule suggests that the time that leads to propagation of information should be based on interior properties. Are there two kinds of time? That is unnecessary. Microscopic time is sufficient to explain macroscopic time, but the reverse would violate the rules of scaling.

Having taken some trouble to make these distinctions, it is worth noting that our human perception of time on a macroscopic level is guided largely by the exterior entropy that forms the backdrop to our experience. The states we observe around us are incorporated into our interior sampling clocks, and we use them to tell the time in an informal way. They have no effect on the rate of interior sampling, so entropy has no effect on time as a degree of freedom to change. However, we use large scale thermodynamic state to gauge the *storyline* of our cognitive processes, so the arrow of time does is deeply interwoven with our observation entropy until we descend to the level of microscopic isolation.

In computer science, biology, and other phenomena, we can also talk about direction and entropy. In computer science, the equivalent of equations of motion would be algorithms. In biology, there are processes with broken symmetries: antigens and receptors—patterns and counterpatterns. How does your body know when you caught a cold, and when to start its immune response to the viral infection? The presence of the virus is distributed on a large scale, but the details of the virus are highly directional and localized on a small scale. How does your body know which way to pump blood around your arteries? At the bottom of the phenomenon of change is an on-going language of give and take, as discussed in the previous chapter, from which all oriented information stems. Matching promises are needed on all scales in order to provide a channel for adjacency and functional significance. Spacetime of all kinds has remarkable lock-and-key relationships, in which some property is effectively 'offered' (+) and may therefore come to be exploited (-) by others. A version of this story is the essence of Darwinian environmental niches, of security access controls, and of conservation of resources, to mention just a few examples.

As spacetime processes propagate from a starting state, they implicate more and more locations that are available to them, regardless of whether they mix indistinguishably with other processes or are superposed independently. This makes long term change harder to reverse, with a burden of memory that can't be maintained. Similarly, the finite range of influences due to latent delays in communication means that the farthest reaches of a process may pass over a horizon of reachability insulating exterior time from local time. Memory plays a key role in the long term passage of time. The so-called Arrow Of Time frequently leads to a basic muddling of two very different concepts: spacetime orientation and thermodynamic orientation—i.e., directional irreversibility versus irreversible relaxation of a distribution. The former is a local phenomenon, rooted in the elementary properties of spacetime. It is encoded in the difference between sender (+) and receiver (-). The latter is a non-local statistical concept, which can't be extended down to the level of microscopic behaviours, because non-local eventually turns into local. I am not offering any new explanation for time here—what causes change in the first place is the greatest mystery of all.

PREDICTING SPEED OR VELOCITY

After that brief digression on direction, causality, and time, let's get back to motion. Since scale is really the hero and the villain of this book, let's examine its role more carefully. The basic scaling relation between distance, speed, and time that we all learn in school tells us that, starting from a fixed origin, the distance travelled by something, at constant speed, equals the speed multiplied by the time it has travelled:

$$d = vt$$

where d is the distance, v is the speed or velocity, and t is the time measured relative to the start and end of a measurement. There are many assumptions in this relation that warrant further examination, such as our ability to measure distance and time in the first place, but we'll return to that shortly. For now I want to point out that the equation implies that these three measures are not independent. We have three related quantities, so only two of them can be independent, and the third will be determined by the other two. How do we choose which to measure and which to infer from the formula?

Our choice naturally depends on whether we have other information to go on. If we can measure one of the variables independently, e.g. we have access to a measuring scale for the distance, or a stopwatch for the time. In practice, measuring time has been considered easier because classically we don't have to go anywhere, but this assumes that we can observe events quickly and faithfully without delays due to the experiment.

For example, suppose a race master calls out 'on your marks, get set, go!' to a runner, and someone starts a stopwatch. The runner runs, and after passing the finish line, the linesman calls the moment at which the runner passes and the watch is stopped. If we assume that the distance is fixed, and the time was measured accurately, we use this to measure the speed. This is a normal set of priorities, but one could turn it around and say that if we know the runner always runs at exactly this speed at all times (with or without porridge for breakfast) then we could use him or her as a standard measure. By measuring the time to run between two points, we could measure its distance.

If any of these assumptions is uncertain, it throws up a whole avalanche of questions. What if the positions moved—how could we know? What if the signal that told us when the motion started and stopped was delayed? How long did it take the sound of the shout to reach the stopwatch holder? We might be able to compensate for the delay if we know how fast the sound travels, and it that itself is constant. All of this seems unproblematic when we have a god's eye view of a system, or the illusion of one resulting from a fortuitous scale advantage, such as that discussed earlier. A problem arises, however, when are stuck inside a spacetime with limited degrees of freedom, such as in the dataverse—or simply over inaccessibly long distances.

In a sense, a measurement's duration is already built into the distance-speed-time relation as an assumption. It is assumed (from our prejudices about time) that we have a ticking clock of some kind by which to measure Δt. To show that this is relative to the start and end of the experiment, we usually write the Greek 'delta' Δ symbol which indicates an change interval:

$$\Delta d = v \Delta t.$$

The definitions are not independent, so we can't separate time and space unless we can fix something—like assuming the velocity to be constant over this interval. In physics, there is a fortuitous reason to believe that the speed of light in a vacuum is constant, because it depends only on properties that are believed to be universal about the electromagnetic field, but we don't always have that anchor: we could also turn it around and say that bodies can move at different velocities, provided we can agree to fix something else, such as the time interval Δt, or the spatial interval of adjacent points. In the case of light, the refusal to do that due to what we know about the electromagnetic field leads to the curious time effects of special and general relativity. This led directly to our present day concept of spacetime.

WAVING THE MESSENGER THROUGH

Speed is an interplay between time and space: a disturbance in position propagates from place to place, marking out time as it goes. For instance, the hands on a clock ticking are a reference system that uses motion to count time. Is that reference system's ticking all we have to measure time by, or is there another source of change we can use to calibrate the motion? This turns out to be a vitally important question in information science, and it rears its head in unlikely places, both in real and virtual spacetimes of the dataverse.

Let's look at this more closely. Speed is defined by the ratio of the distance moved to the increase in time over the interval:

$$\text{speed/velocity} = \frac{\text{distance}}{\text{time}} = \frac{\Delta d}{\Delta t}$$

There is nothing in this simple formula that suggests speed has to take on any special value. But, that presumes that we can measure distance and time independently—which we've already discounted. In elementary systems with limited options, we are often forced to make our clock from the same processes as the thing we are trying to count? Without a local boundary, the interior and exterior clocks are the same, and the clock is driven by the process it is measuring, then it is not independent. The clock ticks only when a message arrives, so how shall we interpret the result? This is what happens in news reporting, factory worker clock-ins, and message logging, and it sometimes leads to confusion in computer systems.

A constant speed of transit has an analogue in any discrete spacetime without scale variations. In a computer network (the dataverse), distance is measured in hops or links between computers. The length of the wires in the embedding space is not relevant as such. Similarly, the clock that captures and samples signals may be driven by a computer's internal clock, or by the arrival of packets themselves (by interrupt). Thus the speed of a every message in a computer is

$$\text{network speed} = \frac{1 \text{ hop}}{1 \text{ tick}} = 1.$$

The speed is always constant, according to both sender and observer, who share the tick caused by the transmission event. We simply cannot measure any other speed. So, unless an observer has an independent clock, based on an entirely different process, this sampling of data transfers is limited to a constant velocity over any distance. We know that this is false from the viewpoint of the embedding space, but it shows that our ability to define space and time at the ends of the link, from point to point, is more complicated than a simple Galilean viewpoint would credit.

For Einstein, a clue about the problems of measuring in space and time spacetime came from looking at the equations for wave propagation. Waves obey an equation for the displacement of some medium, usually material, but waves also exist in the electromagnetic field. Waves are special in a sense because they are agents that travel with their own clock. The motion of the wave is expressed in the medium of space through which they travel, but the wave itself ticks and tocks as it goes, borrowing local space to provide the material memory that represents the ticks, and creating its own interior time, represented by the crests and troughs of the wave. Because the waves interact with the observers who see or feel them, they become part of the space of the and so all see the same speed, because the changes happen according to the interior clocks of each location. This can change from location to location, leading to the refraction of waves. The speed of waves, as observed by different observers, can be inferred from theory and, in each case, it is shown to be a property of the bulk material through which the waves pass[156]. In the case of the electromagnetic waves (which include light and radio waves) the only definable speed turns out to be a universal constant, intrinsic to electromagnetism.

Wave velocity does not depend on the individual properties of single locations[157], it is a non-local phenomenon that depends on the properties of bulk regions. Waves are 'collective excitations'. Although waves may appear to be fully local in our averaged field theoretical formulations, this is a figment of the description, which while beautiful, conceals the non-local nature of collective behaviour. When Maxwell showed that light waves travelled at a constant speed relative to the bulk field, Lorentz, Einstein, and others realized this had very specific implications for the *process of observation*. We are reliant on light to perform nearly all experiments over a significant range of distances. If we use light, the fastest and most reliable signal to observe the motion of bodies travelling in the universe, then the constancy of its speed, no matter what it is relative to, means that our notions of space and time have to adjust to different relative circumstances instead. If speed is fixed, we have to accommodate relativity by adjusting its dependents: space and time.

To be sure, the electromagnetic field is not spacetime, in the direct sense usually meant by physicists, but as long as it's our main channel for observing phenomena, we have to use it as a proxy. Indeed, we have built that into the theory of relativity that defines our ideas about what space and time are[158].

We always prefer to have some simple and independent source of ticks, supplied at regular intervals on which to base the rate. But how would we know the ticks happened at regular intervals? Naturally, we would have to measure them with another clock (and thus it's clocks all the way down, ending in pure belief). At some level, then, we simply have to trust in a process that we can use to generate clock ticks in a regular way. The search for the most trustworthy and most accurate clock is the search for the largest turtle. The concept of speed in terms of space and time intervals is thus fundamentally flawed. It only works if our independent measurement technique is so fast, and so pervasive, that we effectively have instantaneous total knowledge of what is happening. If we have *incomplete information*, we have to be a lot more careful.

LATENT SPEED LIMITS—THUNDER AND LIGHTNING

The scaling of motion has played, and continues to play, a major role in shaping the worlds we inhabit. If you can't get your intentions and observations to and from the point of action fast enough, the whole nature of interaction is changed. In biological and sociological realms, there are explicitly functional imperatives needed to maintain functional behaviours—channels of (+) and (-) influence that form composite and collective behaviours. If the functioning of any part of a system is compromised, relative to its environment, a chain of activity will fail to behave as expected by other parts of the system. Without stable promises and expectations, processes cannot maintain their role in a larger system and it will fail.

In a functional world, timing is important. A signal that arrives too late to be accepted is not counted. Businesses speak of Service Level Objectives and response times. Everyone will be familiar with the distortions of perception that result when there is not one but two signalling channels: say sound and light. We see the lightning strike and then count the seconds to the thunder roll. This same effect is a problem in audio-visual entertainment systems, where we chain together devices: a computer source sends data over your home WiFi to the television which outputs sound to a digital decoder, which is passed to a pre-amplifier and a power amplifier. By the time the pictures and sound reach your eyes and ears, the lips and the voices are no longer synchronized. It's as if the sounds get weighed down by a kind of 'mass' in passing through the additional circuitry. Indeed, this is basically true, as I'll explain in chapter 7. The same phenomenon results in rainbows, when light travels along different paths with different velocity for different frequencies, resulting in selective refraction or bending of the trajectories in media light rainy skies or glass prisms. Instead of single mixed white light, we see the different frequencies splayed out, trading time for space, in a failure of synchronicity—a rainbow.

For a unified perception it's critical to keep the of multi-staged signalling through queues of intermediate agent devices short. When response times get too long, dependent services queue up and lead to congestion—just like traffic gridlock. The more limited the spatial degrees of freedom for a process, the more susceptible it will be to congestion and delay. If multiple pathways are all equally usable (i.e. they are functionally indistinguishable) then processes can proceed uninhibited in parallel, but may be subject to distortions, like rainbows. When the paths are not identical, they may lead to differences. Choosing one path or another, is what we call 'decision making'. Distortions of perception and decisions are closely related, as we'll see in chapter 8.

In the dataverse, delays are quite important, because the small number of points involves effectively amplifies the importance of each leg of a trajectory. Obtaining access to information resources can be a tricky issue, because the waiting time to move data from one location to another may vary enormously depending on network spacetime topology—and different parts of a message might be splayed out through the network, like a prism. The term 'latency' (from the Latin term for 'hidden on the interior') is used for unexpected delays in the arrival of expected events. Just as in microscopic physics, there is no guarantee of delivery on time—only various probabilities of getting a round-trip response within a range of intervals, as measured by a receiver, because all the points in between are independent and their behaviours are not completely predictable. The observer who is trying to get access to a resource has no way of distinguishing between delays caused by slow agents or because signals travelled along longer or shorter paths between them. In the dataverse, dealing with this uncertainty is a day to day issue that plays a huge role in the functional reliability of computer systems.

Response times for computer services are notoriously unreliable due to the small number of paths they have to rely on. There is a branch of computing known as 'real-time systems' that claims to make computer responses eliminate delays for mission critical scenarios[159]. This is perhaps the most pernicious misnomer in IT history. Of course delays are unhelpful to technologists who try to use phenomena as part of a reliable process. If your autopilot is flying into a mountain, and you request a course change, you don't want to hear 'processing, please wait' or 'busy, come back later'. In practice, there is always a delay, but it needs to be kept within promisable limits. The art of managing spacetime is to anticipate the magnitude of delays and allow for them. Locality (short distance to travel) and avoidance of motion and its latencies go hand in hand, especially in processes where synchronization of events over a region of space is important.

The semantics of the message might be a response to a flight emergency, the time it takes for a banking transaction to clear, or the time it takes to get access to a vaccine, racing the spread of a virus. At the fundamental level, 'information' is delivered to regional observers in an uncontrolled way. The narrative of constant velocity is only helpful for the purpose of trying to make average (large scale, coarse grained) predictions, but on a small scale it's flawed and inaccurate. We have little control over what events we receive and when, as measured on the clock of an arbitrary observer—but each observer, with enough interior resources, can exercise some control over how we process them.

Lenses, prisms, and load balancers

Trying to homogenize service latency is the goal of 'high availability' computing architectural design. The idea is to scale up functional regions of spacetime, by spreading information processing out amongst as many clocks as possible—so that there is a guaranteed Shannon channel capacity between user and service, meaning sufficient parallel redundancy in agents that can respond. Greater size brings coarse grained stability to the response rate.

For example, service companies, like Amazon mail order, and video streaming providers, like Netflix, have to deal with process routing when making delivery promises to customers. They spread incoming orders and requests over a larger number of 'servers' in order to handle the load. The servers might be computers or manual workers, depending on the job. There are several ways to do this. Traditionally, services define a single point of service, or a 'point of presence' (POP) in modern jargon. Everyone would send their requests to the same 'single point of entry'—just one phone line or computer, which was the bottleneck. This forms a queue of requests. That order processor would then try to hand off the requests to different computers at the backend, in a chain of dependent promises: a salesperson taking orders for the factory on a single phone line, or a hardware device that fans out requests to many different computers in a datacentre. Another approach is to use a service directory, often called a 'Yellow Pages' lookup[160]. Service providers register with a directory and keep them informed about their workload, and when a user wants the service, the directory hands them the address of an available provider. Providing services for lookup and duplicating data distribution is a major Internet function today, known as Content Delivery Networking (CDN).

In the globalized economy of the dataverse, as our appetite for data grows alongside its availability, we are learning to play with scale, just as explorers of the natural world played with optical scale using microscopes and telescopes, since the time of Galileo. A lens is a technology for bending light so as to either focus a large area into a small area, or vice versa (see figure 5.14). Lenses transform our observation of space, bending signal paths so as to map points to points at different densities. They perform information resampling process.

If you are trying to capture a large number of points in a small area, e.g. to fit into a camera, then the density of samples has to increase, or information has to be discarded—you might hit a limit of resolution. This is what happens when digital photographs become 'pixellated'. The opposite process is to take a strong dense signal and spread it out to make it less intense, such as when you project a movie on a big screen. A blinding source of light is spread out to reduce it to normal brightness, over a wider area to make it available to a larger audience. This is also what companies try to do when sharing their orders amongst factory workers. Each worker has a certain capacity that can't be exceeded before dropping requests.

The modern equivalent of this process, in the dataverse, is called load sharing or *load balancing*. There are many examples of its use in the computing cloud, to spread the load of incoming demand across a battery of suppliers, from Internet companies like Google to Netflix and beyond. Nearly all companies using the

Internet need to use some form of load sharing to cope with customer levels today, because the computers that perform the services are small, commoditized devices that were designed for 'personal computing' not for big data processing. When data are spread out over a wider region, expressed as a cluster of discrete locations, it's called 'sharding' of the data. The shards of a database can be deliberately located so as to reduce the latency for transporting data back and forth, as long as there is a natural spacetime criterion for separating the data.

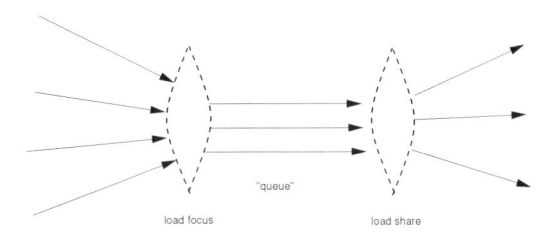

Fig. 5.14. Focusing signals from space, trading space and time by queueing and load sharing.

The movement of data in computer networks now takes place on such a scale that it saturates the channel capacity of spacetime locally for many jobs where there is a significant amount of 'lensing' or feeding of data from a large source into a local region. We speak of an information 'superhighway'—and, like a regular highway, it need multiple lanes to accommodate the traffic. The inability to focus traffic through a wide enough opening means losing sight of some of the bigger picture. For a business this means lost customers. The inevitable distortions of those lenses lead to misperceptions in just the same way as optical distortions.

Computers are connected together by networks, but they are also built as networks at all scales. Every chip and motherboard of a computer is a network of electrical components serving different functions. These functions work together to create the virtual processes that we perceive as computation. There are natural boundaries that lead to interior and exterior regions, over which data interactions experience relative issues of give and take. What makes computation tricky is that data can be channelled along different parts of these networks unevenly, and at different rates on all those scales. Unlike light passing through an empty but homogeneous space, with plenty of spare channel capacity, signals travelling along wires in the dataverse are guided and constrained by them. Different wires conduct signals faster than others, limited by the maximum speed of light in a material medium (which is about two thirds of the speed of light in a vacuum). A pipeline model of data flow is a crude approximation to this state of affairs, but it helps to visualize the processes.

The challenge we face now in processing large amounts of data, or large numbers of customers, in the dataverse, concerns how to avoid transporting data slowly over large distances, i.e. through too many intermediate hops of the large scale connective network. Although the speed of exterior networks is now much faster than the speeds of the interior networks, computers still work better when data are kept as far as possible on the interior networks (the PCI bus), rather than spilling out into the exterior networks (e.g. Ethernet or even wireless). Nearest neighbour interactions are acceptable though, as they don't contend over the same paths. So the more you can do without involving intermediate agents along the network path, the quicker and simpler the processes. However, modern jobs can't easily fit inside a single computer[161]. There is not enough interior state capacity (memory) or computing capacity (interior time) in a typical computer. Data sizes are much larger than computer program sizes, so it makes sense to try to move software to data rather than the other way around.

Recall that cloud virtualization leads to a different experience of adjacencies than in the physical underlay (see figure 5.15). When data are needed by a process, localized somewhere, they have to be fed in from a storage network. The shorter the distance data have to travel, the quicker the system can change to carry out its function, relative to whatever clocks it is interacting with. The problem is that you never really know what is 'local' in a cloud computing network. That network might share the same spacetime paths as other communications, or it might be part of a 'parallel universe' shadow network, using its own cables (often fibre channel). You can try to copy the data from their resting locations by going through ordinary spacetime channels that involve a lot of intermediate agents with interior clock issues, or you could try to rearrange spacetime to connect the data as a direct neighbour of the computational process. This can be done physically, actually bypassing the limitations of one spacetime (perhaps analogous to the hidden channels of quantum entanglement?), and it could be done virtually (see figure 5.15), in which case the problem is not avoided—it's just rendered unobservable.

5 TIME AND MOTION

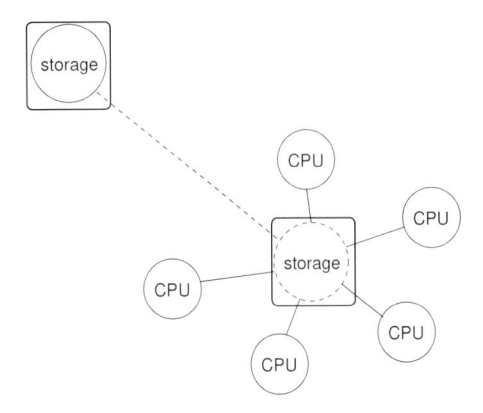

Fig. 5.15. The illusion of virtual adjacency or presence. Sharing resources in a computing cloud. What exactly is the meaning of local? Are computers real or virtual. Are networks real or virtual? When resources are virtual they may move physically at random, while appearing to remain adjacent at the semantic scale of the processes ongoing.

Cloud providers like Amazon Web Services, Microsoft Azure, and Google handle these matters differently, making the experience of users unpredictable—with practical consequences. The computational clouds represent semantically different kinds of spacetime, with different promises and limitations. Running multi-cloud software services presents businesses with interesting challenges that confront the phantoms of relativistic distortions. Attempting to average away the differences between clouds is the goal of further layers of network virtualization called 'service meshes'. When in doubt, technology piles on a few extra turtles.

So, returning to the load balancing story, we can now see that a naive flow-based load balancer is the semantic equivalent of a magnifying lens. When you send a concentrated beam of light into a movie projector, the lens spreads the light out more thinly over a wider area making the image appear bigger, but allowing all the individual visual receptors on your eye to deal with a smaller amount of information each. This is how we avoid overloading a region of spacetime with too much information.

Load balancing and lensing are spacetime processes that transform a concentrated longitudinal (timelike) progress into lateral (spacelike) volume. The buffering of arrivals into a queue is the opposite: it focuses discrete lumps of information from possibly multiple spatial sources into a smaller volume, turning them into an ordered train of inputs to be handled over a wider range of times. If a load balancer is a lens, then queue is a valve or transducer that trades a volume space for an increased duration of time. Like the quote of Parsifal at the start of the chapter, we trade time for space and space for time through process management. A queue has to be located somewhere, to avoid dropping the information. This requires memory. Space itself is the memory capacity for physical processes. Messages or customers are not allowed to pass through one another—and they can't be superposed by going through parallel channels if there is only one checkout. Sometimes systems burst their confinement and information is indeed lost. For example, the Internet TCP/IP can drop packets of data that it cannot handle. Pipes burst. Perhaps even the universe ruptures when a black hole gets too massive—no one can know for sure, or come back to tell the tale.

Shortest path

When processes proceed along their own path trajectories, in parallel, their relative time and relative order have no meaning unless the streams become re-aggregated later. When confluent process events arrive at a single point of aggregation—say like customers merging into a single checkout—they may keep the same order relative to their arrival at the single point, but when the channels are entirely parallel and disconnected, it makes no sense to ask which process may be faster or slower. That determination can only be decided by arriving at a point, calibrated by the same clock. At an airport, for example, there are parallel lanes for processing travellers at nearly every stage. If customers have to pass through multiple checkpoints, ordered serially, say check-in, security, and passport control, then the final receiver of the customers—the aircraft boarding gate that aggregates and serializes the parallel channels again—can't expect any particular order in the arrival of the passengers. Who knows how many clock ticks each passenger was subjected to while passing through parallel channels?

Anyone who has visited a supermarket and tried to get into the faster queue knows that time does not necessarily pass at the same rate in every aisle, once the lines are separated. The best of both worlds—well known in queueing theory—is to form a single queue with multiple servers, rather than several parallel queues. This is the load balancing problem. With a single queue, arrivals are processes First Come First Served (FCFS), and as quickly as any of the servers can handle the issue. This only works if all the servers promise the same service though: if only certain servers can accept certain arrivals, the potential for time speedup model breaks down. Distinguishability is your enemy if speed is your friend. Entropy has its positive value too sometimes. Any information process is potentially more efficient if it can disregard the semantics of information.

When discrete information from several sources has to pass through the eye of the needle, by waiting for dispatch in a single queue's bottleneck, it has no alternative: it must take longer than if each source had its own channel. Whatever process is involved in motion, it cannot escape the additive nature of a queue, as long as the information packets cannot pass through one another, or be superposed with impunity. Serialization of distinct information means that the clock at the queue's 'checkout counter' has to tick more times than it would if there were many more parallel checkout locations available to share out the time. By opening up the process into parallel spacetime channels (the many worlds theory of checking out), in which time can run in parallel on multiple clocks, processes can behave like a discrete superposition of states. This is the challenge faced in modern computing clouds. By creating a spacetime of many many computers connected together with a powerful shared network, one creates the illusion of almost unlimited resources. Cloud providers made this initially quite cheap to rent, but the actual costs are hard to assess. When a cloud infrastructure is utilized fully, i.e. when no computers are idle with 'wasted capacity' then there might be significant contention (interference) between parallel processes. The virtual layers on top of the physical resources can see this interference: it only affects their interior and exterior time perceptions. So, while there is an economic efficiency of scale from sharing common management resources, the effect on computing processes might in fact be suboptimal. Time is a serial process, and serial processes are only efficient when they are sparsely utilized[162].

Airports are quite interesting places. Some of the most efficient and carefully designed functional processes on our planet are found in airports. On a recent visit to Zurich, I noticed that the terminal building was designed with parallel pathways for different semantics. For passengers who wanted to go quickly to their boarding gates, there was a fast moving walkway running under the main concourse. For passengers who wanted to dawdle, shop, and eat, there was a slow moving walkway. Different semantics, different interactions, different channels, different speeds.

In physics, Fermat's principle, or the principle of least action, postulates that physical processes try to minimize their path length in spacetime, or equivalently the transaction cost of the trajectory. This is an interesting way of formulating the patterns we observe in physical processes, because it seems to act as a simple unifying principle. It says that given two equivalent paths whose only difference is length, a physical process would tend to take the shorter, cheaper path. If we add semantic distinctions, however, we weight the cost computation of such choices with more complex criteria—so there may be no simple analogue of this beyond elementary physics. Computer networks route along quite complex policy determined paths. Higher level systems in biology and IT have many criteria relative to different functional imperatives. The criteria to try to find cheapest or best solutions to constraints may suffer conflicts of interest and end up in a local minimum that is suboptimal with respect to any one condition[163]. In any spacetime, the challenge of optimization is to play with spacetime transformations and relative rates, defined by interior process clocks, and to seek solutions that speed up costly processes.

On a human level, time is costly because our processes are pinned at timescales that are limited by the need to feed, sleep, and learn. We measure all processes against the clocks that keep ourselves alive: those biological processes are complex and not easy to change. Simpler systems, like machines of our own making have different costs and are therefore sometimes preferable, as they can be fed at different rates. A technique that's growing in importance in IT is 'Machine Learning'—a part of what we call Artificial Intelligence (AI) today. The idea is that, instead of having to experience the world at normal speed, on a human timescale, one can instead collect data over long times and with many observers, and store them up front in a big reservoir (so-called 'big data'). Then you can play back the data to a machine faster than a human can experience time, thus training the machine using this compressed experience. A wide volume of experience is funnelled into a single 'learning network', which absorbs the experience in a certain way in order to play games with spacetime. This is how facial and voice recognition work. The challenge in scaling this as a machine process is: how do you collect the experiences of a lifetime—or several lifetimes—and feed them into a small machine in a matter of minutes rather than years? It's a bit like asking, how can you take a photograph of a lifetime? The answer is that you can't necessarily do it. It depends on the relative rates, the proper use of lensing and frame-by-frame capture.

Dopplergang—playing with time distortions

Modern computers use a system of timesharing, so that multiple contending processes, can run alongside without getting held up by one another. By slicing up tasks into fixed batches the time they need to occupy a shared resource can be interleaved, exactly like time sharing an apartment between multiple occupants. Individual batches belonging to different processes can move through the queue in turn, assuming they can't pass through one another. It happens so quickly (on a timescale of milliseconds) that everything normally appears continuous to users. Time is absolutely discrete in the dataverse, but it still seems continuous to human observers, because our brains are unable to sample any process as fast as the switching between tasks. We humans even timeshare our lives throughout the working week: Monday to Friday we're Working Nine To Five, while in the evenings and at the weekend we schedule hobbies and family time. Each week is a multichannel semantic experience.

Any incoming signal process that's unlimited by semantic constraints can be 'magnified' or 'reduced' in space, just as images can be transformed by lenses. This only requires there to be a free destination available for the serial or parallel trajectories. Just as guests can find lodging at a hotel as long as there are plenty of parallel rooms, we can spread them out without colliding in space or time, by interleaving slots of space or time.

In the computing cloud, the most pressing issue is often how to get data from storage to CPU across a network, without suffering contention over timesharing slots. Sometimes the density of signals is so high that the transmission capacity of the information channel becomes saturated, and processes that need to use the same channels begin to *contend* for the usage. To avoid this, we build private channels, in the form of shadow networks dedicated to data transmission, with ultra high speed switches. Even then, sharding of destinations is the key to parallelizing channels to avoid that contention—spreading out the signal more thinly over the available pathways. As the pressure to solve the bottlenecks increases with the growing amount of information sent over the Internet, datacentres are exploring extreme form of lensing into discrete 'microshards'[164], to optimize further.

This common pattern of divide and conquer crops up again and again, but on a technological level, it is expensive. If you need ten computers instead of one, obviously it costs more. Even if you have infinite money, you might not have enough space—computers take up a certain finite amount of room, and although we are always trying to reduce their size with Very Large Scale Integration (VLSI) in chip design, we are approaching the limits of Moore's Law.

If a serial queue can't be shared out, because of particular semantic constraints (perhaps your airport flight only has a single service counter), then the process has no choice but to cram more events into the single stream, to be sampled one by one by each location along the path—effectively slowing down the perceived timing of events. This phenomenon is familiar to us on a human level, waiting in line for a service, but it is a basic spacetime process that we know elsewhere too. The *Doppler effect* is the archetypal example of this. Everyone who has heard an ambulance or emergency siren passing at speed, hears the difference in pitch of the waves when the moving source is approaching and receding. The same effect happens with light, and is observed by astronomers as 'red shift' or 'blue shift'. Sound waves travel a lot more slowly than light waves, so the effect is exaggerated and we can experience it readily on Earth[165]. Recall also the general relativistic effect of signals passing close to a gravitational source that constrains the spacetime paths, thus causing signals that count time to bunch up and appear slowed down in the gravitational field. As far as the relative observers can tell, sounds or light travel at the same speed, but their spatial features become compressed by relative motion. The spatial features of a wave are its wave peaks and troughs. These arrive more quickly, relative to the clock of the receiver, due to the relative motion, and are thus raised in pitch. In astronomy, this gravitational lensing tends to make light appear to come from a halo around heavy bodies, as it gets bent and focused closer to the gravitational centre. In the dataverse, data coming from load sharing lenses seems to be focused at the load sharing queue nodes that are bottlenecks for the traffic.

The slowing of time is not unlike the time distortions caused by a tape recorder or analogue film reel running at the wrong speed—though in our digital age, perhaps not all readers are familiar with these technologies! Imagine slowing down an 'analogue' tape recording, if you can. As the tape runs slower, all the pitches are reduced, because the tape contains a direct analogue representation of the waveforms. When one slows the tape, virtual time is slowed relative to the receiver's physical time rate. The wave peaks and troughs arrive farther apart. When sped up, the opposite happens and the pitch rises. The speed of sound does not change, but the speed of sampling does. We can imagine slowing the tape until the highest frequency is audible to someone with poor hearing. The clock driving the output is the message medium itself. The song will then play at a slower rate. Today we can reduce pitch digitally without changing the virtual time rate relative to the receiver's clock, with some approximation, because the clock driving the playing of the sounds is the CPU clock, not the tape itself, but rather an independent time-keeper. The sampled waveforms can be transformed, in finite intervals of spacetime, with some frequency correction that may or may not be noticeable to listeners.

In order for queueing effects, like the ones described above, to happen, sound signals have to maintain their integrity. Information can't just disappear, be reordered, or redefined. Think of a supermarket checkout again. By forcing information arrivals (i.e. customers) to flow through a narrow specialized region of space, we effectively require the clock sampling rate of the region to tick more times to handle the flow of information at a faster rate. But if it can only tick at a maximum rate, then the process will take longer to process a longer queue. What if each customer carried another piggyback? Could the queue be processed faster? That would depend on whether the checkout could distinguish them as one or two customers or not. If they had twice as much shopping, then there would be a quantitative giveaway. If they had no shopping, then the process could be sampled faster—but then we have to ask: would piggybacked customers be forever doomed to be treated as a single person, or could they separate again?

The example sounds absurd for humans, but phenomena do exist at all levels from elementary particle physics, to biology, and Information Technology, where the type and packaging of information can be transformed leading to compressions of space and time. Just imagine combining two half-full shipping containers into a single superagent container, which are then sampled by a single crane and passed on to a single truck for delivery. The separation of the contents requires the package to interact with another process that resamples the superagent into two separate agents, with distinguishable contents.

The speed of customers moving through the checkout, or of containers moving from ship to shore depends on what kind of process clock does the sampling, and counts the time it takes to process them. There are two possibilities that corresponding to the concepts that I refer to as interior and exterior time.

In a moving flow of information, a downstream sampling agent (a checkout server or an offloading crane) might not have its own (interior) clock, in which case its sense of time is driven by the (exterior) arrival of an event. The arriving customers or shipping containers drives the sampling of downstream locations a bit like a water wheel or a windmill is driven by the flow it samples. The arrival of each customer, packet, or signal passed through the spacetime checkpoint, signifies a tick of the sampler's clock, and the information moves one step forwards for each tick. Thus the velocity of the signal is always constant:

$$\text{velocity} = \frac{\text{one item}}{\text{one tick}} = 1.$$

There is no way to define or measure any other value for the velocity as long as time is part of the queue, and the agents of the queue have no separate clock. If no customers or containers arrive for three weeks according to the checkout crane operator's Rolex, time is wasted, but a coin-operated Rumpelstiltskin at the checkout or offloading crane has literally been dormant until the arrival of a customers or container. A human, with an interior heartbeat, will get bored waiting, but a machine that only wakes up on an arrival doesn't.

A checkout crane operator, with an interior clock that ticks faster than the arriving customers, may have spare ticks of time to sip at a cup of coffee or read a book in between seeing customers move through the checkout. With interior time, the sampler is partially decoupled from the queue process, and timeshares its interior while interacting with the outside world, and can count more ticks than customers. This means the sampler will observe a speed that is less than an agent with no interior time, who sleeps in between.

$$\text{velocity} = \frac{\text{one item}}{\text{several ticks}} \leq 1.$$

The effects in the rate of time for two observers are the side effects of the interaction between give and take. Sender(s) and receiver(s) may experience time differently. The experience may be relative to an absolute spatial theatre, as in Newtonian mechanics, or it might be relative to an absolute speed of communication as in Einsteinian mechanics or the Rumpelstiltskin checkout crane operator. When dynamical processes rather than static locations set the standards for change, our ability to observe both remote space and remote time could be distorted by communication.

Queueing is a spacetime phenomenon of great importance in the functional world. The time rates may not be important, but the limits on spacetime process must be. In principle, if packets arrive too fast, along multiple routes, and then

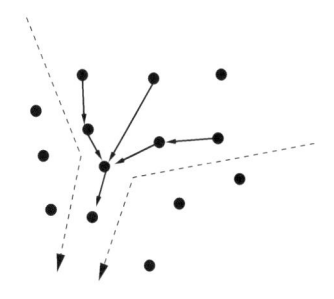

Fig. 5.16. Queueing funnels may lead to signal congestion and delivery delays.

get funnelled into a single queue with its own rate of time (see figure 5.16), the processing will be too slow, and the packets will be delayed. Jobs will bunch up and the queue will grow. Whether there is or is not an absolute spacetime, on some turtle's back, doesn't matter—what counts is the processes that can observe one another. When the sampling process and the signalling process become entwined with the process of observation, distortions of give and take may have strange effects. In a continuum spacetime, at a large scale of quasi-continuum spacetime, the result is essentially Einstein's theory of Relativity. The sampling rate gets compressed or stretched out, similar to the Doppler effect. The observation of remote events travelling at a relative speed is also compressed—at least as far as we know up to the limit of observer sampling rate. As far as I know no one has investigated the effect of scale on temporal sampling rate in relativistic changes and found any loss of observational sampling ability manifested as a change of probability for absorption, but that might be an interesting experiment in quantum relativity.

QUEUEING VERSUS SUPERPOSITION?

In chapter 4, we mentioned two kinds of process that are distinguished in elementary physics: bosons and fermions. Bosons are messengers that can occupy the same state or location at the same time (like the voices of a song). Light and sound waves are bosons. Fermions, on the other hand, cannot occupy the same state or location—they experience Pauli Exclusion, and therefore take up space (like the singers in the choir). They can't pass through one another, except by tunnelling, so they have to queue up for their supper. It's fascinating to speculate on what spacetime processes might be at work to enable superposition or Pauli exclusion in bosons and fermions respectively. The question is clearly related to the question of why messages need to queue up in spacetime, rather than passing through each other, or disappearing altogether.

Serial and parallel channels have these properties in the sense that queueing is fermionic and superposition is bosonic, for discrete information channels. How does a spacetime process know whether signals should queue up or be superposed on one another? To pass through the same point, there must be multiple lanes for overtaking, like lanes on the highway, or an arterial bypass operation. Superposition is a form of parallelism. If two cars tried to pass with only a single lane, a physical transformation of the cars would have to take place, or a renegotiation of trajectories (a collision with recoil). These sound like mundane matters, in our Newtonian world view, but we'll see in chapter 7 that they are far from obvious phenomena. Indeed, there is a lot more to motion than the simple issues of defining it covered in this chapter.

There is nothing in information theory that corresponds to continuum superposition of waves other than parallelism of multiple channels side by side (or internal memory for computation). Routing through parallel channels enables discrete superposition, but the concept of continuum superposition is something of a unique phenomenon in the physics of waves. Could this mean that quantum superposition is only evidence of a large scale effective phenomenon of some much deeper discrete scale? We have no idea whether spacetime has a limit to the number of its highway lanes. There is another possibility, of course, that every plumber knows: if a channel can't take the pressure queued up, then leakage and burst pipes can change spacetime radically. Computers break if information density exceeds their maximum processing capacity. Animals break if their heart rate exceeds its maximum capacity. Is there a limit on the absolute amount of energy passing through a spacetime trajectory, forcing some critical threshold transformation? Is this what happens in particle pair creation? I don't know the answers to these questions, but it seems clear to me that credible answers should be found.

Local distortions of spacetime

Spacetime distortions are not the sole preserve of physics. In biology, our bodies have adapted to play games with the sampling of space and time too. Confronted by a sudden change in the rate of events in their environments, animals are forced to respond quickly. The response is a faster heart rate that delivers oxygen supply through arteries of a fixed size. Since space is fixed, we have to decrease the time between oxygen supply events to scale the process queue. The layout of the biological transport network is limited by spacetime too: how the arterial network fills the embedding space of the animal tissues around it has been shown to follow a predictable law based on the sharing of blood amongst the locations where tissue fills the embedding space[166]. This is a form of load balancing. Increasing the flow rate can only work if we breathe more deeply to magnify the surface area of the lungs (a kind of magnifying lens for oxygen intake). When the external stimulus has passed, the heart rate begins to fall since there is no exterior time stimulus (+) to drive the sampling (-) clock of the organism. It means that time is decoupled on the interior and exterior of an observational causal boundary: an animal can only adapt its metabolic rate because it is a separable bounded system with its own interior time that is capable of sampling the exterior fast enough according to Nyquist's law. Obviously the complexity of a biological organism permits these scaling feats, but more resource constrained processes, like an offloading crane, may not be able to alter their interior time rates in the same way. When that happens, their experiences of exterior reality must become compressed.

While we marvel at the effects of relativity in astronomy and physics, such distortions are less acceptable to us in the world of critical services, for life or business. Even a stream of video arriving on your TV has to arrive at a certain rate in order to provide the promised experience, because we can't easily rescale human time perception (though it does change when our heart beats faster during a horror movie). You can't watch a movie at half speed, or twice speed except in special circumstances. However, you might be willing to accept a less frequent or delayed delivery of a newspaper subscription because of a large number of customers. The same principle applies to other processes that make delivery promises. Mail orders make certain service level promises—such as guaranteed delivery within a few days. If your order takes too long, its usefulness expires with the context in which you ordered it—say you ordered perishable goods. Global trade relies on this ability to manage Just In Time delivery, whether across the physical world in container ships, or through the dataverse in Internet packets.

Process distortions wreaked by trusting intermediate agents, whose embedded processes participate in information delivery, can play havoc with arrival times—and therefore with signalling too. Motion itself is a sequence of send and receive events. Remarkably, we have no unified description of observer phenomena, not even across physics let alone across science and technology in general. In the science fiction series *Star Trek* it became something of a cliché over the years to blame any new phenomenon the starship Enterprise on a 'localized distortion of the spacetime continuum'. However, all kidding aside, spacetime inflicts such distortions (the less spectacular queues, load balancers, and lenses) all the time to corral the information passing through it.

Motion, or the consumption of information in trains and parallel lanes, through distorting lenses and compression schemes, also leads to scattering of information along different paths. If a sender tries to break up data into batches and send the parts independently, they may travel different distances with different speeds, and arrive in a different order at some final destination. In his formulation of Quantum Mechanics, physicist Richard Feynman actually proposed this to be the correct way to envisage the processes of motion in quantum theory[167]. As paths separate into independent causal chains, they become differently clocked regions—effectively different causal universes—and the order of relative events might come out differently after travel. This is a normal state of affairs on the Internet; indeed, it's a central design issue in the dataverse with the Internet protocol known as TCP/IP. The effect has been observed in quantum mechanics too[168], but I don't want to dwell on that, because there are practical cases that have a profound importance to how we transmit information—such as in the global financial system. In TCP/IP, every packet of data sent along a channel is numbered with a sequence number. If you try to send a long document, it can't all be sent in one go. Instead, it's broken up into chunks that are numbered, sent individually, and reconstructed at the other end—just as if you had to pack your belongings into boxes during a move, then unpack everything to reconstruct it at the other end.

The history of breaking long streams of data into short batches, for efficient processing through channel timesharing, goes back to engineer or computer scientist Leonard Kleinrock, protegé of Claude Shannon, and his work on queues that led to our modern packet switching networks and the TCP/IP[169] protocol. The idea is simple: if you break up work into chunks, smart infrastructure can compute and route the optimal delivery of a result for you—effectively creating a load balancer at every point. If faults occur on the way, you can take an alternative path (assuming one exists). This is the timelike analogue of dividing a space up into parking slots or cubicles for predictable sharing.

The Internet was designed to look like figure 2.18(c), so there should be plenty of redundant routes in Internet spacetime. The problem is that, when you split a contiguous message into separate packages, the packets may end up travelling along different paths and at different rates. Imagine posting a long letter in three installments, each in a separate envelope. The receiver has to wait for all the packets, by queueing them with some buffer memory. In fact, in the dataverse, packets may even be lost and need to be retransmitted. The result is a dialogue that effectively mixes advanced and retarded boundary conditions on the interior of a transmission, an on-going basis.

The preservation of message order can't be taken for granted in any situation, though it's more likely to be predictable in systems where very little is going on. To preserve data during transmission, each spacetime region must buffer the transmission of data with enough memory to hold disordered packets in waiting until slower packets catch up. Simple transport is easy enough. But functional systems go far beyond this in combining dependent information in particular sequences. In the dataverse, and in chemistry and biology, it's the mixing of combinations of agents, each making different promises, that allows one to build something new (think of genetic mixing). If the parts don't come together in the right order, the 'logic' of the assembly will not fit together.

This is all very interesting, but what does it mean for the scaling of our international payment system? If payments do not arrive in the same order at different locations, different parties will disagree about who has money in their account, how much interest they are owed and whether they are overdrawn or not. If music, movies, or documents are corrupted in transit they will make no sense, and might lead to missed opportunities.

The 'computing cloud' that forms the backbone of a growing amount of computational activity across the planet is not a worldwide uniform environment as its name suggests. It consists of regions of closely packed computers in datacentres—from thousands to millions at a time—connected by trunk wiring, and linked up to many satellite clusters of computers spread across the planet. There is no simple way to predict the rate at which a particular location's computer CPU clock will be ticking, sampling and processing events to propagate information. Computer processes have to spend a lot of effort on 'locality management', trying to keep data and computational processes close together, even as virtual turtle layers beneath them reconfigure the ground on which they rest. Distances and times are in frequent flux, making prediction quite difficult.

Computer Science has to deal with issues of this kind every day, on all scales—and our increasing reliance on information technology makes a reliable solution of utmost importance. The problem we now face is that, for the first time, the speed and scale of the global Internet is reaching a level on which the effects of relativity, discussed in this chapter, are beginning to shows cracks in the methods we've been able to rely on thus far. Just as Newtonian physics breaks down as an approximation on some set of scales, so data science too is beginning to bend under the weight of increasing scale.

Logical clocks: the Treaty of Common Time over space

The conundrum stated above leaves one final puzzle I want to address, concerning the relativity of information in the dataverse—a problem that information technology now strives to overcome with impunity, in spite of all the odds being stacked against it. It concerns a simple issue: how to scale the transport of information, so as to make a collection of independent copies change as a single virtual copy over a region of larger size. At first blush, it seems to be a simple matter of copying data from a source to a number of receivers that hold copies. The problems begin when we try to ask: at what time will observers see the different copies updated?

Writing data once, then reading it many times is easy. This is what we do with static media like CDs and movies. We can print as many copies of the information as we like and create a collection of independent agents that contain the information, which acts as a large scale pervasive database that is completely consistent. Once printed, the data are frozen and immutable, so anyone can observe the CD or movie as many times as they like and they will receive exactly the same information (+), even when several observers watch in parallel. Of course, some might perceive it (-) differently, but that's a separate issue. However, the story falls apart if the data are not immutable. When a bank tries to publish its ledgers to its many users, who may be located anywhere in the world, the payments and receipts are changing all the time. There is no end to money transactions, no tranquil period of rest during which the bank could print account information on a CD. The records are being updated without rest, just as the payments are. Moreover, the data might be handed to different payers and payees by different routes that reorder payments or even delay them.

5 Time and Motion

Different versions of updating may be seen by the same observer at different locations at the same time. How could that observer choose which version was 'right'? In fact, can we even claim that any version is right?

The in-depth solution to this subject is complex and the details are beyond the scope of this book, but the main outline is nonetheless interesting because it touches on the physical processes involved in information replication, and touch on basic thermodynamics. Clearly, we have to define a right answer to the problem, even if it is arbitrary—it's less important what answer we choose, as long as it applies to all observers equally.

The basic solution to the problem of consistency involves the idea of locking information, by denying access to change it. A bank, for instance, collects transactions into batches and clears the incoming and outgoing values on different timescales. This is why payments may take several days to clear, even though the information travels in a split second. Once the state of its ledger is updated, it becomes an immutable part of the past. There is an arrow of time in the bank ledger associated with memory of the transaction log.

Computer systems do something similar, by simulating concepts of mandatory access control (MAC) is and discretionary access control (DAC) to the limit access to a region of memory space concerned. This is not a phenomenon we can associate with a description of Euclidean space, or any other model of space in physics—with the partial exception of the quantum wavefunction with its transactional matching wavefunctions. All observers have unfettered access to space and time, but not in the dataverse. Because the pathways of the dataverse are more limited, it's relatively easy to cut off an interior location from exterior signals at the boundary between exterior and interior, by forming superagent clusters with a 'monitor' process, or 'kernel' executor.

The purpose of locking is to effectively rewire spacetime so that the same data can be effectively adjacent to several locations that are far apart. It's an issue that is, in some ways, particular to the human realm—it doesn't happen naturally, because the aim is to maintain the illusion of determinism in data transactions for technical and social convenience—which is not an issue anything 'cares about' elsewhere. The catch with the locking approach is that it relies on all the data being at the same location. It's quite hard to imagine locking access to multiple locations in space for a common interval of time, when the common standard of time depends on distance and speed of communication along potentially unequal paths.

As explained earlier, the basic approach to replicating any kind of service would be to build a magnifying lens, with a load sharing configuration to spread data changes to multiple memory locations—but this is not enough. We also need to spread the locking mechanism and coordinate it to ensure that all copies of the data agree for all observers. Actually, we can't ensure that every observer sees the same data at the same time—that's not even a meaningful statement. But we can try to ensure that the ordered story of events is the same for all observers over time, regardless of when they actually read it, or where they have their bookmarks.

This is the *distributed consensus* problem in computer science. It's important for disaster recovery as well as load sharing, because when you lose data locally you need to know which copy is the 'right one' to restore from. Copying data while maintaining equilibrium is the key to sharing data. For example, banking data, legal documents, and even software versions. One way is to print CDs, but today we use Internet copying directly. Two models used for amplifying data to many destinations, called 'centralized server copying' and 'peer to peer copying'. In a centralized server copy, there is a single source of information which forms a directly adjacent channel to every point it wants to copy data to. This might be a physical channel or a virtual channel—and, given the number of layers of process involved in computing, can we even say what is physical or virtual anymore? In this direct approach, we effectively rewire spacetime. With or without load sharing, there is a logically central source and we establish a directly virtual adjacency called a socket-pair, possibly using a 'Virtual Private Network', or a 'tunnel'. This is a one-to-many copy, fanning out data in a starburst pattern. Data are exchanged at full channel speed, but the source agent has to shoulder the load of sending to every location, unless it can scale its efforts with a lens. The alternative approach is one of diffusion through an existing network. This is called 'peer to peer' copying. The data start from a single source, and pass it on to a number of nearest neighbours. Those neighbours then pass it on to their neighbours, who pass it on to their neighbours, and so on. This is a form of 'viral' or 'epidemic' spreading—so called because this is exactly how transmission of disease works in the bioverse. Peer to peer networks were made famous by their involvement in media piracy, but have since become used for all kinds of technical solutions including Bitcoin and Skype. An epidemic process spreads like ripples in a wavefront, and therefore travels at its own speed, slower than channel speed but with less burden on the source agent.

In both cases, the transmission of data takes a finite amount of time, on any observer's clock. In a peer epidemic process, the time for dissemination can be quite long, because there are many transition events during which data pass through the checkout queues of intermediate agents. With a direct channel approach, only two clocks are directly involved but their relative rates of sampling are unknowable. This can be a big problem in being able to predict who knows what across a wide area network.

In 1978, computer scientist Leslie Lamport realized that the problems of sampling and signalling at finite rates, discussed by Einstein for astronomy, would also apply to computer systems that were distributed across many locations and joined by a network. Each computer, stuck within its own skin at a location determined by the cables that connected it to others (before the days of virtualization), could experience signals as 'events' from 'out of the blue' each time it sampled the cables, but non-deterministically. So each computer would see only partial information about its own locale, and end up with a potentially different version of history, just as each of our life experiences sum to a different set of events. This would not be a problem as long as third parties did not expect the different computers to agree about events.

If the world's banks had different information about our bank accounts in different regions of the world this could be catastrophic for the bank and for its customers. The order of payments is quite important. We want to know information about money transfers from bank to bank—that they happen quickly, and in the right order, to avoid a situation where a credit and a debit came in the wrong order, leading to an overdraft, or even instantaneous bankruptcy!

The equilibrium of information over a region of adjacent locations is also what leads to the scaling of energy states in physics, explaining thermodynamic behaviour, and the steam engines and refrigerators that make the wheels of the world go around. In the dataverse, the corresponding problem for information keeps the financial wheels turning. If a third party computer asks a question of two remote computers, at the same time according to its clock, it can get different answers from both. At the very least, we need to understand what it could mean to get different answers, if the two databases promised to be synchronized at all times (by everyone's clocks).

The concept of equilibrium in physics is related to the concept of entropy, mentioned earlier. It applies to a large-scale indistinguishable bulk phenomena, not to small regions of spacetime with specifically distinguishable states.

Two reservoirs of heat (say water tanks) are said to be in equilibrium if their temperatures T_1 and T are the same at T_2, simultaneously. The temperature of a reservoir is a sugeragent bulk promise that no individual molecule of water can promise alone. It has to do with the processes between the molecules on the large scale. If we mix two reservoirs, they reach a common temperature by mixing, and it must be true that T_1 is the same as T_2 at the same moment—but only after a certain reorganization of the processes called a 'relaxation time'. The idea that this takes a finite amount of time is well known from the physics of heat, for instance, where different bodies are said to agree about the temperature if they are in equilibrium with themselves and also with one another.

Equilibrium is a coarse granular spacetime property. If a region of space is in equilibrium with itself, it has a certain temperature. If the same region is in equilibrium with another region or 'body' then they both have the same temperature. Similarly, adding a third body or region, all three must be in equilibrium to all have the same temperature. Put simply, if $T = T_1$ and $T = T_2$, then $T_1 = T_2$. This makes perfect sense if T, T_1, and T_2 are numbers, but it is more complicated if they are statistical concepts like temperature, especially when the statistical concepts are also based on spacetime. The temperature of a body is related to how fast all its interior processes are, which is about space and time. Suddenly what we can take for granted, while standing still in a local region, may be more complicated when different bodies are moving with respect to one another.

Equilibrium is a dynamical process, of the kind we expect in spacetime, not just a static property, and being able to have two changing processes agree about any property of both is a non-trivial issue. Indeed, in Einstein's special relativity this becomes quite complicated. The idea is easier to understand for databases. The analogue of temperature for two databases is a state in which their data are exactly the same and change in step, i.e. the processes of updating each other are in equilibrium. The problem is that this nice stability in physics is a result of huge numbers of identical particles involved in bulk process—which is not the case for computer data. No one wants the amount of money in everyone's bank accounts to reach an equilibrium state evenly distributing the amount of money amongst every account. If that ever happened, the economic processes would probably stop, just like a steam engine would stop without a heat source. Processes need inequalities in order to sustain behavioural change—hence the thermodynamic notion of an Arrow of Time.

General equilibrium is not good enough for a bank—where detailed microprocesses need to be exactly in step, down to the lowest level details. Instead, we want all databases to be identical in every detail at all 'times', and according to everyone's clocks. This means that databases have to send messages to each other to exchange information. If we add more than two databases, the same thing applies, but now over more than one path, with possibly different distances and different speeds of message propagation, subject to relativistic distortions. The properties of spacetime between arbitrary databases are not local to the customers of the database, so they can all see very different things unless a step by step equilibrium is maintained over time.

Leslie Lamport came up with an algorithm called 'Paxos' for doing just this in databases, allowing them to reach equilibrium, each according to its own clock, and limiting observability for others until the process was complete[170]. An array or 'vector' of all the clocks involved could be passed around the group without anyone being able to see the intermediate stages of disequilibrium before the data agreed. As long as different observers could trust one another's' data and update histories, observers could be promised a consistent view or consensus about the data for each allowed tick of their observer time. The relationship between these 'logical clocks' and the time outside the computer system is basically irrelevant.

In the last decade, many more approaches to the consistency problem have been proposed, and we are only just starting to come to terms the idea of a scaled equilibrium. This is not a problem that physics has ever needed to confront before, because we employ the phenomena of physics on such a relatively huge scale that equilibrium can be used as basic building block. That is not the case in the dataverse. Just look at figure 5.4, and suppose that a single update passes from one database to another, travelling from position 1 to 2, and then from 2 to 3. Assuming that they started out the same, the three databases might be in equilibrium again when the journeys are complete. But now consider what happens if an unrelated customer asks database 1 and database 2 at the same time according to its clock. Now the signals have to travel along paths unrelated to the paths used to update the data (and maintain equilibrium). The speed of the signals might be faster or slower along any of the links, depending on the relaying speeds of the points in between, so the observer could observe any state of affairs. It could see new data at 1, or at 1 and 2, or at 1, 2, and 3.

As godlike observers, with a simultaneous view of the entire situation, we could make the following observation: if a query were sent to databases 1 and 2 at exactly the moment database 1 was updated, and the query speed of the path between the observer and database 2 were faster or shorter than to database 1, the observer could see the change happen at database 2 before it happened at database 1! A godlike observer could explain why, but the observer trapped within the dataverse simply has no way to see what happened first other than by probing it with a query. That is its reality. We cannot call it an illusion, because within that limited spacetime that is the only available information. A godlike overview doesn't actually exist for any agent within the system, only in our thought experiment. This is exactly the situation with Einsteinian relativity too—only the process details are different.

Playing with time consistency

When time depends on the microscopic state of spacetime, without the room or duration for bulk equilibration, different observers may actually see time running backwards! The appearance of time running backwards is a version of what happens in movies where the spoked wheels of the wagons seem to run in the opposite direction to the movement of the wagon itself. This is because the sampling rate of the camera (observer) may be faster or slower than the changing rate at which the spokes of the wagon wheels pass the same positions. In a periodic process, like a wheel, the wheel revisits its own clock time every time the spokes line up indistinguishably. If they overshoot or undershoot relative the snapshots being sampled, then they appear to gradually move forwards or backwards, relative to the clock of the film frames. The spokes effectively discretize the locations promised by the wheels, making the locations appear to move. To godlike observers in the movie theatre, this is an illusion due to the cyclic nature of the wheel motion, but the observation is real to the camera.

In the synchronization of databases the same effect may lead data to fall behind relative to another. One way to avoid the muddle of inconsistent information is to prevent an observer from observing the intermediate states. If the camera motion were actually driven by the wheels of the wagon, drawn by the horses, the direction of the wheels could be made to look consistent—it might even appear stationary! Of course other motion in the movie might look weird if the rate of time were effectively driven by how fast the horses pulled the carriage. Databases are made to appear synchronized in a similar way, based on a discrete process, by freeze-framing observation until they have finished their updates to one another. This is called data locking in computing. It essentially freezes time for the observer, and by doing so creates a single superagent whose interior structure consists of three components in equilibrium.

It's quite common in computing and banking to use timestamps, based on real world clocks (which are just other computers connected to the same network) for ordering events. It's not hard to see that such an exterior source of clock time is only useful for comparisons relative to that same clock, so that we have to remember which clock gave us the timestamp. Even then, we have to assume that all clocks count monotonically, in only-increasing numbers. Clocks that count in finite cycles, like wagon wheels, may seem to run backwards. The order of events from different locations means nothing unless we assume something like a constant speed of light (which is an assumption we can make in a controlled way in physics, but not in computers). We can't really be sure whether the events actually happened in the same order when they don't come from the same source. Runners from twice as far away, running twice as fast, would appear to arrive at the same time—unless time itself ran differently along their paths. To the observer, there is only 'now' and 'before'. This is called a *partial ordering*, and the difference of opinion about it from place to place is another consequence of what locality means.

In the same year as Lamport's paper, physicist Jan Myrheim[171] showed that a partial ordering of events was the essential ingredient behind relativity that led to Einstein's field equations too.

> "It is a well known fact that in the standard continuum space-time geometry the causal ordering alone contains enough information for reconstructing the metric, except for an undetermined local scale factor..."[172].

There is no absolute scale in Einstein's theory of relativity, because the assumption of continuum spacetime has no scale, without symmetry breaking by masses, and therefore the scaling of locality is not well defined without matter to break its symmetries. In a discrete spacetime, like the dataverse however, that infinite freedom to count turtles is given a lower limit, potentially resolving the ambiguity about what local means. A definition of length amounts to counting hops between discrete locations. Myrheim pointed out, somewhat analogously to the construction of graph or promise theoretic spacetime, that if one starts with the assumption that there exists a number of points, and that these may be ordered in a certain manner, to which we attribute causality, or the unique *directionality* of events. This is enough to create our normal understanding of a metric space, compatible with Einstein's equations, except for the need to define an arbitrary volume scale. Interestingly, the number of spatial dimensions is not clearly defined, because it is related to arbitrary choices of coordinates, but Myrheim suggested that there might also be a statistical notion of dimension, analogous to thermodynamic direction.

Time ordering of spacetime points is the only concept required to generate a 'normal' spacetime in a discrete geometry. But the homogeneity required to explain the bulk properties can only be natural in a bulk statistical regime. None of this is a surprise today, as our sophistication has grown around these topics. We have to give up thinking about spacetime as one giant theatre that we watch with absolute knowledge. On one scale, it's more like a maze filled with twisty passages, where we can't see around the next bend—a network of processes providing channels of information to be sampled by each and every observer separately. Only on a bulk statistical scale do we recover the familiarity of Euclidean space.

INTERIOR AND EXTERIOR TIME—THE END OF PRIVILEGE

In this chapter, we've discussed time and motion and their relationship on a number of levels. In order to receive signal events, let alone experience velocity (different times for data to pass from one hop to the next), an observer needs to be able to sample at least twice as fast as the arrival rate of signals it receives back. In order to measure events independently each location needs a local clock. This simple point tells us that the fundamental changes that we call time, that promise the acceptance of signals has to originate from a process on the inside of the sampling entity's boundary, and—recalling Nyquist's law—be at least as fast as what happens on the exterior to avoid missing something.

The time an observer can use to measure information, and velocity is not the average thermodynamic time alluded to in the Arrow Of Time arguments. Countable change exists independently of large scale redistribution in space, in the fundamental ability for changes of state to take place. This realization—call it the idea of interior time—suggests that conventional ideas about relativity cannot be sustained on very small scales. Making quantum theory consistent with Einsteinian relativity is therefore likely meaningless beyond some minimum scale.

Within a tiny bounded region, the number of states must be small and there would not be enough memory to drive an arbitrary counting process, i.e. to make a clock to observe time as we understand it. A small region of space could not measure anything reliably, so how can we apply the classical concepts of observed time to it? Small scale time does not behave like large scale time. This begs a fascinating question. How big do you need to be to observe motion? Without a sufficiency of interior states, a region of space would have to experience time as circular or even random. The standard arguments based on dimensional analysis of continuum physics suggest that the scale at which this might happen would be the Planck scale, but this is, of course, only a speculation.

The idea that every point experiences its own time is therefore natural (in a counterintuitive kind of way) but also mysterious, because it implies that there must be internal states within the 'agents' that we call the points of spacetime, thus making them non-elementary. For example, one could imagine a clock by following the states of atomic energy levels. When an electron jumps into a higher energy level (orbit), its state has changed. By oscillating back and forth between the two, one can make a tick-tock pendulum. But a pendulum can't keep time, in the usual sense, because it has no memory. A third party could use it as a basis for an external clock (like a pendulum, as used in atomic clocks), but the atom itself cannot observe time itself as an entity—so its local role in processes is constrained by that experience. If we want to know how long something takes, or how fast it is moving, then we need to remember intervals of time, with many ticks, otherwise you can't do it. If your clock only knows tick or tock, then time goes forwards and backwards, forwards and backwards, because there are only two possible distinguishable states.

Without the idea of a pervasive universal theatre time, every isolated point is really a separate punching bag for random signals arriving, which it has no internal structure to remember. To make such a clock, you would assemble a set of locations believed to be regularly and easily observable to use as a clock. At first imagining, we might think that the points have to be close enough together to receive messages from them in the same amount of time, but what is more important is the ability to observe a uniquely counting number that changes regularly. What does regularly mean? Well, we have no clock to measure how regular it is without making another reference clock, so we can't know. The regularity of time must be a large scale phenomenon. It's turtles all the way down again!

We begin to see that, the idea of having a clock and being able to observe anything at all are deeply intertwined. Every signal received is a tick of a clock, but is it a distinguishable tick? This suggests a rule about observing remote events. We must not count events in a remote location as being part of the clock we measure time by. Recalling the supermarket checkout example, we should only measure time on the interior of an arbitrary spatial boundary that we refer to as 'local'. The trouble is that this rule cannot be complied with in all systems. In IT systems, there might not be sufficient interior infrastructure. In any truly fundamental system, there might simply be an insufficient number of states to represent a clock, which suggests that the concept of velocity is meaningless on the scale of the very small (why would such a small part of spacetime need to measure the rate of a process far away anyway?).

Motion is about the occupation of successive locations that fit a causal pattern. If there is no pattern, can we really call it motion? We sometimes use the term transition for those cases. The places a process can visit and not visit—what we might call 'access control' on a human level—must be determined in any system in order to understand what motion can be. Just how is adjacency really decided in a process, on any scale? In physics we have Equations Of Motion that express what trajectories might look relative to a fixed spacetime, under ideal conditions. But, as we examine phenomena on many scales, and observe the rich variety of consequences through the lens of 'give and take', the assumption that everything is neatly programmed into 'laws' of behaviour or deterministic algorithms starts to fray around the edges, even to the point where one begins to wonder whether we've been looking at things upside down. It's important to realize that physical law, as it has been handed down from the history of philosophy, is only a description of what happens relative to some coordinate map. Writing an equation that tells us which locations a process will visit doesn't really explain why they do so. We come a step closer to explaining motion by identifying properties (promises) that each location needs to make in order to open the lock and key combinations that generate observed patterns (like positive and negative charge, antigen and receptor, access control and credential, etc). Of course, that's a whole new level of explanation—one that still needs further explanation.

If velocity is a concept valid only for large scales, then it cannot be attributed at each point individually on a small scale, as we do today in the physics of bulk phenomena. The concept of velocity in classical mechanics doesn't really make any appearance in quantum mechanics. What's indirectly referred to as the velocity (actually momentum) in quantum theory is something implicit and more puzzling. It's fascinating to wonder whether these matters might be related to the troubles we have in understanding position and momentum in quantum systems, according to the Heisenberg uncertainty principle. There are certainly similar problems of definition. Imagine performing a search for a virtual process running in the dataverse. Due to our incomplete information about process management in the computing cloud, we don't know the exact position of a process in the cloud of computers. If we try to pin down the location of the process to a specific point or set of points, the smaller the region we look at, the smaller the probability that the process 'particle' will be running precisely there. The smaller the region we sample the less certainty we'll have about properties that depend on locations outside that region, like how it might be migrating from location to location. Issues that concern the simultaneous measurements of virtual processes are not particular to quantum mechanics.

Our ability to observe the motion of objects around us should be considered a privilege. An agent can't measure the speed of a single journey directly. It might be able to measure is a round trip time, relative to its own clock. If the agent has sufficient interior activity, its clock will be ticking away fast enough to count a length of time between sending and receiving a message. However, imagine the situation in which the agent has no internal clock, and is driven entirely by external events[173], like a database, or a security log, then that agent experiences no time in between sending and receiving. The round trip appears instantaneous, and whatever time was experienced by the other party is entirely unknown to the agent. For this reason, computers have to rely on independent timekeeping agents: builtin or exterior clocks that they have to trust as time keepers [174].

When humans look at the large and slow things that concern them (like ball games and traffic, etc), in the perceived space of our own universe, we effectively have such a view thanks to the dominant speed of light. But when our human privilege fails and we have to embed sensors and detectors down in the highly constrained worlds of atoms or networks, then we no longer have that privilege. Our detectors have to work within the confines of what information they can get from within the system itself (see figure 5.3). Any viable physics we develop to explain spacetime must work in this way[175]. Now all privilege of abundance is lost.

If you are part of the system and you experience no events, you experience no time. Others close by might experience events, even if you don't. The sum of both regions does therefore experience time, because things change on the interior of its boundary. When you sleep, you wake up as if nothing happened,

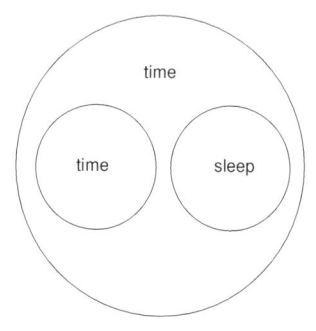

Fig. 5.17. Time depends on what events happen on the inside of some semantic boundary. Consider two regions, close together in some space. One has zero events and therefore experiences no time. The other does experience events and hence time passes. The scales region containing them both also experiences time because something changed within its boundary.

because you experienced no time. Your only evidence that time may have passed is that there are some changes to the states around you, that happened in a single tick of your clock[176]. Was it an alien abduction, or a mysterious quantum wavefunction collapse?

ONE CLOCK TO RULE THEM ALL?

Every day, the questions about Information Technology more and more closely resemble the questions we have previously had to ask about physics. Given the obvious universality of the concepts involving location and time in each, it's natural to ask whether there are principles that look like the relativity of motion for spacetime phenomena in computer science, analogous to Galileo's, Newton's, or Einstein's theories for physics? For there to be universal rules for motion, spacetime would have to be uniform, yet the low level inhomogeneity of the dataverse suggests that such principles will only be observable at very large scale. Before today's mega companies like Amazon, Google and Facebook built the largest datacentres on the planet, networks didn't look anything like Euclidean spacetime, and these questions were not even meaningful—but the rise of enormous cloud datacentres has greatly increased the uniformity of points, topologies, and clocks. Thus, we are gradually starting to see the quasi-continuum questions arise, across the cultural divide between physics and computer science.

At sufficient scale, we can start to define the relative motions of 'phenomena' between the virtual phenomena that correspond to matter and energy in our physical universe. We should not be too hasty in making these identifications. If we jump to think of the motion of data packets in the embedding space of the dataverse, we'll miss the point. Those phenomena have their place, but the spacetime issues are not *only* represented by how physical computers occupy our day to day idea of spacetime. As we've discussed, a comparable meaning for motion only begins to correspond to the way we view physics when we look to virtual machines and their virtualized data streams. Material bodies would seem to correspond to clusters of data processes rather than actual singular computer programs, as I'll discuss further in the chapter 7.

The analogues of 'matter' and 'energy' in the dataverse are scaled clusters of virtual processes atop a fixed and absolute infrastructure[177]. When the agents that make promises can themselves be transported, as for example, we see in the virtualization of cloud computing or the exchange of money tokens[178], then their ability to keep promises could depend on the speed of their motion, and we get 'relativity'. It must result from tradeoffs between the finite resources available to execute processes—a limit on their interior time.

Space and time are discrete in the dataverse. Each computer has an interior timer and a CPU which ticks at Gigahertz on current computers. The timer drives the CPU, so that the CPU can sample slower than the timer ticks. This determines the resolution of interior time for the computer, regardless of what happens on the exterior. If we are forced to include the ticks of a clock beyond our local horizon, then what happens to our notion of time? This is effectively the question that Einstein asked in formulating his special theory of relativity for astronomy. The result was that the consistency of queued events from exterior time forced an altered perception about remote moving processes. It's not hard to think of analogues of this effect in the dataverse too. When the speed of transport, for virtual processes, approaches the maximum rate of processing for the infrastructure—then, if the entire processing capacity of infrastructure is used up to transport virtual machine objects, those virtual locations will not be able to experience time, because there will be no room left in the interior time queue-processing resource to advance the execution of the processes. In that model, a finite speed of communication (analogous to the speed of light) would lead to effects like those we observe in special relativity, for analogous reasons: one reaches a resource limit.

Whether information is queued serially, or routed in parallel, there will always be a finite limit as long as spacetime has a finite smallest scale. The limits we observe suggest that spacetime must be fundamentally discrete so that it has such a cutoff. No physical process can ever have enough 'incoming phone lines' to accept arbitrarily high levels of arriving events. The ability to exchange information must, at some limit, become oversubscribed, with associated distortions to observed relativity[179].

However we look at it, whether in the dataverse or the physical universe, Newton's view of time as an independent continuum, with infinite resolution, has to be wrong. Time concerns changes that result in the effective processing of information. The region over which an observation spreads its information affects the ability to process the information. This is what makes data pipeline processing a challenge, and it's what Einstein rediscovered about observations of astronomical spacetime. Physics has been cursed by an infinity spell. We accept infinite or infinitesimal answers far too readily as a mathematical convenience. All observations are subject to limitations of finite information, so the real truth about relativity is both far stranger than the quaint stories we've come to tell, while at the same time being disappointingly mundane in its origins. It was surely Einstein's true genius so see through that obfuscation. The locality of spacetime processes takes away our power to know directly what happens beyond a certain horizon. What we can observe is snapshots of signals sampled over limited channels of information. Relativity is an information compression problem, and velocity distorts information processes—not space or time, but spacetime observation. We need to reset our expectations about having godlike powers of observation and start again, trying to understand how to measure space and time when its structure is discrete, small, and—most of all—when we are stuck inside it.

6
TURN BACK TIME

If I could turn back time
If I could find a way
I'd take back those words that hurt you

–Hit song by Cher (Dianne Warren)

The idea that each moment of history is like a place—somewhere we could visit repeatedly by dialling up a destination and undertaking a journey—became a common theme in science fiction after H.G. Wells published *The Time Machine* and Minkowski and Einstein popularized the notion of time as a dimension[180]. Do all those moments past still exist somewhere, or somewhen, not as stories recalled in books, but as experiences waiting to be relived? As a teenager—even before Cher's famous hit song entered the timeline—I used to watch reruns of a TV show from the 1960s called *The Time Tunnel*. In The Time Tunnel, a group of scientists used a device—a tunnel formed from spacetime bending rings—to open doorways to the past present or future. By running into the tunnel, they were able to travel back and forth in time. In each episode, the scientists were jaunted from place to place, and from time to time, as a convenient storytelling device. The view of time, put forward in the show, was that of a long corridor of events, in which places were conveniently restricted to events of history that would make for good stories! A similar view was portrayed in Star Trek's rather more moving episode *All Our Yesterdays*, in which the people of a planet, whose sun was about to burst into 'nova', escape into their own distant pasts to avoid destruction. Convicted criminals were even incarcerated in the past, in a bizarre form of temporal garbage collection (or perhaps recycling). Indeed, science fiction is replete with stories in which humans dip in and out of the process of time at different moments, as it suits them, with all the paradoxical consequences. Today, it has become tiresomely hard to find a science fiction movie in which the characters don't mess with history as an increasingly cheesy plot device.

Time as a tunnel sounds superficially compatible with the notion of trajectories, described earlier—an on-going storyline of the events belonging to all the processes of the world, accumulating all the moments and states visited as it goes. Couldn't you dial it back? In this chapter, I want to explore the limited ways in which we could (and do) and the ways in which we can't.

The problem with imagining time as a dimension, on a par with space, lies in a common misconception about how storylines (trajectories) can scale. Following a narrow event, like single particle, is one thing. Following a vast area like the planet Earth is another. The Earth is not a single static region: it is, itself, a collection of embedded processes, sitting atop a rotating planet that circles a star, circling a galaxy. Storylines remain local to an observer, at any scale, while events spread out in space beyond the boundary that separates local from non-local, as their influences propagate. To revisit the past, there would have to be some massively compressed cosmic memory that can be dialled up like a database, but the superficial similarity with a computer memory and the sum of all events in a closed system doesn't go very far—because processes are not simple and static.

The story of time as a dimension, analogous to space, has been oversold; it works in some parts of physics, but not in others. The notion of boundaries, described earlier in this book, helps to make it apparent how local time already conflicts with the assertion that time is just an expression of thermodynamic state, i.e. an accumulation of entropy. That will take some time or process to unpack, but it's not hard, if one can let go of familiar things.

In this chapter, I'll end up showing that—if we want to think of time as a dimension, with local achievable coordinates, then its scaling requires it to be not merely a single dimension, but in fact to be multi-dimensional, and granular across space. Time is different on the inside and on the outside of every boundary, and every interaction of processes mixes it over regions of increasing scale. The result is that space and time have to look very different on the small scale and on the large.

PICTURES AT A DATABASE EXHIBITION

The idea that you could *travel* in time is quite different from the idea that you can travel in space. Travel is a process that implicates both space and time, because it involves change, and distinguishable change is time. Spatial locations are all assumed to persist, in some frame of reference, but times are configurations of information located in space—on a particular scale, forming a particular clock. The ticks on a clock are states of information represented in that space, linked by a deeper process of change that we usually don't fully understand. The logical link between space and time is enabled by the *motion* of signals, by which we observe states beyond whatever immediate line of demarcation we use to define 'local', i.e. by processes in which information about observable moves along trajectories of adjacent locations. That summarizes the chapters up to this point; now, let's try to consider what it could mean to turn back time.

If times were places you could visit, they would all exist alongside one another as separate memory locations, like pictures at an exhibition, memories in a computer, or the pages of a book[181]. Some giant finger of god, or turtle flipper, in the embedding space, could be the master clock that tells which of those momentary states was the current one. But this picture doesn't make a lot of sense—if it were the case, then there would not be one universe, but an infinite number, kept in mothballs for no good reason. And what would cause the cosmic flipper to advance? If you went back in time, would you have a second private flipper, while the other kept going: would you be frozen there, or would you continue to live on while the other times continued? Time is a process of things. What would be the force that kept you moving from one moment to the next? It simply side-steps that question. Thinking of time as a sequence of pictures doesn't explain time, it only explains history.

The alternative to revisiting times, would be to turn back time by undoing the processes that brought about the current state of affairs. You can always try to retrace your steps—to undo the changes that led to the current configuration of time. Then you are faced with the confusing idea of whether you are going forwards in time to undo stuff, or whether that means that time is going backwards. Sometimes relativist physicists speculate on the existence of wormholes that would allow observers to join a remote location at a time that they could once observe as being in the past, relative to where they were[182]. But we have to be clear that what you see as the past, through a portal at one distance frame of reference, is not the same as what actually 'is' somewhere else, because it's only second hand information about the image of a particular remote process, necessarily distorted by the journey it's undertaken.

Information theoretically, the answer is simple. If the state of the entire world (including its on-going processes) is indistinguishable (in all aspects) from the past, then it is the past. So if you really could undo everything, you would indeed be back in the past. The difficulty lies in doing just that. Undoing processes gets harder and harder as they spread out and implicate more and more locations. This is what is meant by entropy[183]. Imagine trying to undo all the births and deaths of history, atom by atom. Could we undo every blink—pay back every purchase, in exact reverse? Could we unwalk every step? Of course, no one could do that even for their own narrow sphere. Now imagine trying to take care of all the influences those actions had on other people and places. The events and circumstances you want to undo may no longer exist—perhaps the locations may exist, in some static view of space, but even they cannot be fully recreated.

If this sounds like an artificial problem, it's actually not much different from the enormous challenge faced in trying to recycle the mounds of mixed garbage we produce each year. Trying to unmix and sort types of object to make the best use of them is an impossibly expensive task in most cases. Imagine running a non-local simulation of the entire universe alongside the present one, with infinite accuracy, and all the attendant locations and energy requirements[184]. No surprise then that the Time Tunnel scientists blew a few fuses in trying to do this.

Although I grew up on these stories of spacetime travel, with all the romance that attaches to them, and was trained in the ways of theoretical physics, I've slowly come to the view that I need to reverse my own programming. We all need to stop thinking of time as a dimension, and as a tunnel. The tunnel of time is our own vision of it: a set of stories about the processes we have intersected with, rather than a complete set of states for the universe. The stories are not objective realities. They are memories written in a compressed summarial form across locations of space that can only be accessed with the help of processes involving new time. There is a basic contradiction.

The mythology of time, as a corridor, may originate from historical tales, or even from the timezones marked out around the rotating Earth that calibrate the process of day and night with the processes that drive our wall clocks. If we fly fast enough, we can race the rotation of the Earth to get to one kind of process time, and effectively travel in time—but it's only one kind of time, very different from the time travel in science fiction. Even as we watch the sun go down, we know that somewhere else on the Earth there is a place where the sun is coming up. Our notion of time is split up into recurring phenomena and non-recurring phenomena. Time as a journey is not a number on a wall, but a collection of ordered ticks within a collection of interacting processes.

Undo undo undo

If you wanted to go back in time, and you hadn't taken the precaution of memorizing every snapshot in your cosmic database, you could try to reverse or undo all the differential changes across the entire universe since a certain calendar time (hit CTRL-Z repeatedly on your life computer and see what happens). On a computer, the 'undo' operation is achieved by remembering all of the deterministic and relative actions and reversing them, within a protected environment. In physics, we believe we have this information compressed in a form of memory about processes that we call 'Equations Of Motion' that form part of the 'laws of physics'. Many of these rules are reversible, as discussed earlier, so we can use them in reverse to predict backwards—in principle. It's not simply a case of throwing the universe into reverse gear of course—to do that you would need precise knowledge of every subatomic process in full detail, and have the technology and force to reverse them all deterministically. When you finally undid everything, you would have to set every particle on its trajectory again (except for yourself and your fellow time travellers), and then continue in a forward direction again as before.

The problem with this idea is that—in order to reverse those events—you would need to completely be outside the universe, otherwise you would also reverse yourself, your intent to reverse time, and your technology out of existence in one gigantic moment of regret. When you finally got back to the original state, you would then go forwards in time and arrive at exactly the same place and be doomed to live on a time yoyo for the rest of your deity's exterior time (all hail the second-, third-, actually uncountable-coming of Maxwell's daemon).

To be fair, something like this time reversal is certainly possible on a tiny scale in the dataverse. The virtual processes that run on top of computer systems can be suspended and even snapshotted, to the extent that they are isolated from contact with any other running process. By committing every detail of their execution histories to memory (like making a movie), time can be replayed for them (just like rewinding and replaying the movie). In that sense, one can go back in time, within an isolated system. It's all just information. However, we cannot do anything about the consequences of a process that breaks free of that isolation. The main difference between undoing processes local to you (the observer), and undoing the entire universe, is the size of the universe compared to the size of an elementary information system on the inside of your particular local boundary. The cost and likelihood of ever being able to return to precisely the same state, by undoing each change is not only prohibitive, it is not reconstructible because there is not enough memory to do it—that is the meaning of entropy.

Can time be reversed at the level of our universe? Initially, there might seem to be no compelling argument to refute this idea, but what we know about thermodynamics and entropy of information would put the cost of this at something prohibitively enormous. You would need a Maxwell's daemon typing CTRL-Z at every local region you interacted with, in parallel. The cost would diverge, and the information lost forever over event horizons (places that can no longer ever be observed) would require more superpowers for the daemon. So, the notion of time as either a journey or a reset is not really commensurate with the mechanics of time as a process of events unfolding. It's not too hard to claim your clock calibration can be going backwards relative to another separate clock, somewhere else, e.g. by travelling along parallel paths in spacetime with different latencies. This is what we call refraction. This would be like calling someone while they were moving through timezones[185]. Everyone's clocks would still be moving forward in their own frame of reference, by definition of what a clock is.

ROLLBACK, PLAYBACK, AND KERNEL PASSPORT CONTROL

One case where we can roll back time is in the movies—for a simple reason: movies are only one observer's local memories committed to an immutable reel of storage medium at one place. Once captured, this memory reel can live as a separate universe: one that doesn't interact with anything or anywhere else anymore. This is enough for a single observer's sense of reality, but it is not enough to capture the objective realities of all possible observers. It would be something like a private Westworld simulation, with just enough reality to fool you. When John Denver sang, 'You Fill Up My Senses' he was describing a process of saturating the cognitive inputs an observational process with playback, which I'll come back to discuss in chapter 8 (it doesn't sound quite as romantic when couched in those terms, yet this is what movies are designed to do for a limited amount of time). Dreams replace sensory input with random streams of regurgitated memories, crudely connected into unconscious storylines without the anchor of an exterior sensory reference.

Recordings that we make of the past, on audio, video, or in books, are immutable and isolated processes. We call such processes *modular*, i.e. they behave like separate modules. The recordings saturate the inputs of an observer, forming an isolated module, in order to supplant reality for that one observer for their duration. We can rewind those recordings and play them as many times as we like, in order to turn back time within that module, but we can't change them, because they are closed processes, with fixed boundary conditions and fixed rules. Turning back time inside a bubble is possible, but it's not what H.G. Wells wrote about (reversal outside the traveller's bubble)—it's a process of replay, subject to advanced or retarded boundary conditions like any other.

In computing, the idea of reversing time has been an integral part of database technologies every since their inception. 'Rollback', as it is known, is a way of managing isolated transactional changes, like bank payments between a specific and static set of locations. It's the CTRL-Z for database storage. A transaction, in computing, is defined as an exchange of information between fixed locations, that counts as a single tick of a clock, and can be confirmed immediately after it has taken place. By actually making information exchanges into a kind of clock—the clock that counts them—you really can undo time within your closed database universe, because you control everything about space (memory) and time (transactions). When transactions are important, e.g. in money matters, rollback is essential for fixing inevitable mistakes—but, it is only possible if you don't delete any information. For example, if you send money from your account to another's account, you don't want it to disappear en route. Either it should stay in your account, or it should move to the other account (with no denials). It should not fall between the cracks along the way.

Another way of saying this is that the total amount of money should be locally conserved, and money should never be usable at more than one location at the same moment time. Obviously, there is a lot to keep track of here. First of all, 'the same moment of time' implies that we are using the same clock—which means only on the interior of a boundary. This seems to contradict the idea of exchange, which can happen over great distances in the modern world. In the past, people met to exchange gold and silver, but now money is increasingly just numbers on a computer somewhere. A transaction requires two agents to come together with a common clock for the duration of an exchange. As soon as there is a possibility of forgetting information, or worse losing the distinguishing labels that tell us who owns what, you start to accumulate entropy. You can try to remember all information forever, with an escalating cost, but that becomes the cost of being able to turn back time, and reverse payments. In practice, we only need to reverse recent payments, so there is a limit to how much we *probably* need to remember. The financial time tunnel is quite short.

Scenarios like transaction reversal are common in the dataverse. The technology that enables this is called a monitor or *kernel*. A kernel is a spacetime (super)agent that behaves as a single local region—a kind of computer register. This kernel is isolated from all timelines, except the one it is currently engaged in. Its job is to add changes to one timeline at a time. In short, a kernel is a master timesharing process, which maintains temporary privacy for processes to prevent mixing, and the increase of entropy.

A kernel basically manages locality, by erecting a boundary between an interior protected region of time and an exterior set of states that are sampled in a controlled serial workflow. When a transaction happens, two processes give up their own interior time and trust the kernel's clock completely in order to drive the movement of information, without duplication or loss. Kernels are used to mediate the sharing of all important resources (like memory, CPU, disk, etc) between any number of different processes. By forcing access to these resources to be effectively interior to the kernel, the kernel wraps them and turns them into privileged services. A kernel also acts like a passport control or gatekeeper to the historical information, serializing access so as to preserve its integrity. A kernel is therefore the arbiter of the passage of time, through its porous boundary, temporarily joining or 'entangling itself' with semi-isolated processes. Anyone who has tried to work uninterrupted, in a private space (say writing a book), knows that you need protection from exterior events that 'steal your time'—otherwise you end up sampling events that are unrelated to the process you are trying to focus on. Modules offer protection from this, on whatever scale they are defined[186]. Modular kernel operations can be undone, within a protected boundary, as long as nothing external can alter the timelines in an uncontrolled (non-deterministic) way. However, because kernel processes are atomic—i.e. they only last a single tick of a clock—no one can observe what happens on the interior, from outside on the exterior. Just as the position of a pendulum can be undone, so we can undo transactions—but this does not extend to sequences of transactions. Transactions don't scale with the same semantics.

In the systems of infrastructure that reach across large and uncontrollable scales of the modern world, managing the details of everything that happens is an impossibility. On the other hand, many of our human processes are extremely simple and pendulum-like, i.e. transactional. You buy a tin of soup and you can unbuy the tin of soup. There was a simple exchange of things that is easily reversed. The laws of shopping are sometimes reversible, and within the microcosm of that purchase, time can be reversed. Nothing else of importance was implicated, except perhaps for some waste paper from the receipts—and, as long as not too much time passed and the soup didn't go bad, or nothing catastrophic happened in between, the reversal is innocent enough. We can manage events by erecting walls around them—by creating cells or modules that separate states causally, and make us the masters of interior time, even as we are the pawns of exterior time. These lessons are not specific to the example above; they concern information in any scenario from the physical universe to your shopping trips.

The analogues of a kernel, which mediate the give and take of information, must exist in other systems where there is conservation too. We just have to look for them. A surprisingly deep question lurks in this point: what is it that ensures that matter and energy, in the physical universe, move from one point to the next without getting lost? What kernel mediates that exchange? It's another turtle in the pile, of course, but not one that is explained by any physical theory to date. Physics, from the classical to the quantum mechanical, builds in conservation as an ad hoc constraint, preserving informational integrity[187], but no one knows why this should be the best model of the world. When spacetime agents are connected, by managing the boundaries of information to give and take a channel of directed information, they become temporarily co-dependent. The soup for the payment, and the payment for the soup. As long as this semantic binding of intent is not ruptured, the states can be reversed.

A bank payment can be reversed as long as nothing else spends the money in between the do or the undo, and as long as the payer or payee don't disappear. These changes are also ticks of time. A document edit can be undone as long as no one else edits the same text in between. The key issue is the ordering of events in 'time'. Of course, what we know is that each observer might potentially experience time differently, which is why locality is important. What Cher apparently didn't realize in her hit song was that she needed to not only store all of the events in her diary, from the moment of speaking the relevant words, she also needed to keep them private in order to maintain process state separation for the entire time she wanted to undo. If she'd never shared the words, their consequences could be easily undone. Then, indeed, all of the states would be available for a 'rollback'. It's a nice idea, but we can't even do this for simple systems over many changes.

The idea of 'rolling back' time to undo mistakes, is so tempting an idea that inexperienced engineers widely believe it to be doable, and even rely on it in technology every day. This leads to mistakes. Alas, the dream is a triumph of wishful thinking over common sense. In practice, rollback is really supplanted by a forward process of 'disaster recovery', and trying to restore systems from data backups. There is a cleanup operation to be done because of the spread of implications that radiate out from every event.

You can certainly try to reverse some processes locally by brute force, given sufficient memory and resources, but this may or may not be a meaningful operation, and could only be for an isolated transaction—not for the bigger picture. As soon as something else has depended causally on the erroneous state, reversing it will not automatically reverse all the domino consequences. What a kernel does is to prevent causal mixing, as far as possible—this means effectively preventing state-mixing entropy, from increasing. It maintains effective reversibility of states by preventing mixing of timelines. If timelines are recorded in memory, and retained with integrity, those (and only those) states can be freely undone. The process is said to be reversible. The mixing or change of entropy is an operation that effectively loses memory[188].

Information technologists have embraced modularity to such a degree that they often believe they are immune to the wider influences of processes on the exterior of those modules. In doing so, they forget that even those boundaries do not provide real kernel isolation, only an interior perspective. If the lessons of using a system on one scale to observe a system on another scale, e.g. in the weirdnesses of quantum theory or relativity, tell us anything, it's surely that you are going to see surprising and unpredictable consequences.

The desire to control time (and space) is a pervasive one and it intrudes onto processes in the dataverse at all levels. When mistakes are made, we want to be able to back out of commitments. The belief that one can replay more complex processes, in order to undo mistakes, is fundamentally wrong—as the growing understanding of 'complexity' in science has revealed.

Loops and multidimensional time

So far, we've implicitly thought of processes and their time as being a linear, directed stream of information events that increases in a single direction, as information is emitted and absorbed between a sender and a receiver. This cartoon view of time—the river of destiny—is a process known as a Directed Acyclic Graph in graph theory. It's simplistic and neglects some of the most important processes that take place in the world.

Think of any process that causes predictable change around us, on any scale, and you will find cyclic phenomena. Pendulums, waves, wheels, windmills, engines, central heating, the rotation of the Earth, the working week, and so on. All these phenomena that act as the driving force behind long term change convert constrained motion into a directed focused channel. Cyclic processes are what sustains all persistent dynamical phenomena. So processes are circular, but does this make time circular? Again, to answer this, we need to think about scale.

A cycle implies that some part of a system returns to its initial state. If you run around a field, on a circuit track, you come back to the same place after a while. Your position on the track is your clock and measure of interior time. If your only context were the track (say you are indoors and can only see the track), you would have no idea what exterior time there might be, but if you can see the sky, or you have a watch, then you can also see events beyond the track. As usual there is a boundary between interior and exterior: in this case, it's implicit in the cyclic process. Interior time is cyclic and reversible, due to its limited memory for clock counting. Exterior time is effectively irreversible and non-cyclic—it's open ended, unless it is part of a larger cycle that you can't yet perceive (say a seasonal year).

The time we know from our wall clocks is an arbitrary cyclic process that counts modulo 24 hours, returning to the same times again and again. Universal Time (UT) was created at the International Meridian Conference in 1884, and forms the basis for the 24-hour time zone system we know today (known as Zulu time in the military). Greenwich Mean Time (GMT) was chosen as the world's time standard, and its reference line, known as the Prime Meridian, was determined to be located at the Royal Observatory in Greenwich, London—that still houses a beautiful time museum. Daily calendar dates count independently, with different conventions around the world for Western solar calendars, or Eastern lunar calendars. They cycle around each year, and even have meta-cycles such as the animal years of the Chinese calendar and the cycles for Wood, Fire, Earth, Metal, and Water, but the numerical years count on and on forever. The time that matters to us today, on a small scale, is just a convenient counter.

Computers don't count time in the same way as we do. Since modern computer systems were invented circa 1970, computers count time according to Coordinated Universal Time (UTC), which began as a 'big bang' for IT in 1970. This is now known as Unix time (or POSIX timestamp from the standard by the same name). It counts the number of seconds that have elapsed since 00:00:00 Coordinated Universal Time (UTC), Thursday, 1 January 1970, minus various leap seconds that are needed to correct for the imperfectly countable orbit around the sun.

The processes and the times that matter to us form a hierarchy of independent causal processes, so it's peculiar from a mathematics, geometric viewpoint that we don't treat these processes as independent, and assign them different dimensions. Of course, we can do that. Space does not have the sole rights to multi-dimensionality. As already noted, it makes sense to think of time as a vector and as a scalar in different circumstances. To define a vector, we only need a sense of there being more than one independent process to count.

We could take the rooms in a hotel and label them from one to a maximum number per floor, and then add a number for the floor, making coordinate pairs (floor, room). We can do a similar thing with time. Indeed, we do exactly this when we make clocks. We make clocks as we make coordinate systems, but with even more dimensions that relate to the natural dimensionality of our environment. Compare the way we write coordinates on a calendar-clock to the way we write coordinates in space:

$$(x, y, z) \text{ or } (r, \theta, \phi) \leftrightarrow (\text{day,hour,minute,second}) \tag{6.1}$$

We treat hours, minutes, seconds as independent dimensions, only bounded, and we view time as a spiral that expands outwards, just like our number systems. Each column in our numbering acts like an independent dimension, with a basis vector equal to a different power of 10.

10^2	10^1	10^0		
		9	=	9
	1	0	=	10
	9	9	=	99
1	0	0	=	100

Instead of a simple cycle that can only count some of the interior processes, the complete wide area process of time acts as a *winding* model, with dimensions of limited size. It's a useful way of coordinatizing numbers.

I advocated this technique in the 1990s for coordinatizing computer monitoring, in order to facilitate machine learning of weekly patterns, because the most influential and important timescale for computer behaviour is the human working week[189]. By separating the processes into a hierarchy of interior and exterior regions, it was possible to control the tracing of distinguishable causal influences, without statistically losing essential information. Statisticians call this de-trending. If we then turn time into a kind of cylinder (figure 6.1), in which the main clock face retains exactly one week's worth of ticks, and the the distance along the long axis is the week number[190]. The result is a simple two dimensional time (see figure 6.1).

Our measure of time within a closed region depends on the states within it. If there is only a binary number, we can represent tick and tock, and then we're back to the tick again. There are only two possible times we can experience within this region. As we connect the region to exterior events, we might be able to use these to measure time too, but only if we can remember them, so we still need memory on the inside of the region to clock time for the region.

So does time really turn back at the stroke of 12, like Cinderella's pumpkin carriage, reverting to its former state? Over a local region, yes it does. The point is that in a small interior region, everything returns to the state it was in

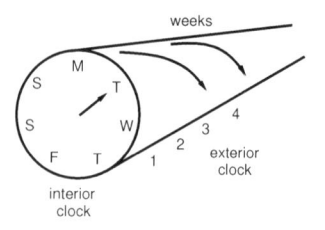

Fig. 6.1. Two dimensional time, with interior and exterior counting.

before, and is indistinguishable from what it was before. That is the meaning of time as a clock—in the sense that Einstein implicitly described. If the interior states are embedded in a larger region, and can interact causally with them, on the larger scale, exterior time can still accumulate changes. If any particular transition, anywhere within the spacetime memory of a region quickly reverts backwards through its counting representation, then it is entirely fair to call this a reversal of time. If all the states are back in the same positions, then the state is indistinguishable from the original time. The sum of all states is effectively the name of a distinguishable configuration spanning the system.

In the dataverse, we can stop computer programs and restart them from the beginning, resetting all their initial conditions to be identical. In that sense, we literally can turn back time within the limited scope of the computer process. But if an observer sees the results, coupling them to the outside world causing someone else's time to be affected then you can't say that their time went backwards. So the question is more like: how much time do you need to turn back? The more you ask, the more expensive and precarious it becomes. Modular isolation is key to this ability, because—as processes mix—undoing time may also require the unmixing of states. This is what we mean by *entropy*, and it's why the concept of reversibility receives so much attention in physics.

The largest scale process we are able to observe is the expansion of the universe. We have no way of knowing whether this is a directed acyclic process or a cyclic one. A majority of scientists probably believes this to be one way and irreversible, but several cosmologists have come up with cyclic models for the universe too[191]. To observe a cyclic process, we would need to be on the outside, with our own clock looking in. But any interior region of space knows nothing of this. It only has access to this local counting, and so from our perspective it will repeat its time cyclically, just like a pendulum clock, just like the hands of a clock every 12 minutes and 12 hours, and so on.

THE STIGMA OF STIGMERGY

Maintaining the distinguishability of timelines and trajectories requires memory. If processes move fast enough through space to lay out their changes in fresh ground, they can sweep out trajectories can leave behind a trace of their entire history just by writing on space. This is how processes 'blaze a trail'. This etching of a process image as a trail in space is called stigmergic memory. It is the basis of cooperative sharing in many non-linear phenomena that connect across the interior-exterior boundaries of component processes.

Stigmergy is the process of encoding trajectories as memory, or vice versa. The term is most widely used in cooperative biology. It's what ants do when they leave trails of pheromones to guide other ants, for instance[192]. It's a simple form of communication—of language—built on signalling. Ants don't have enough interior memory to recall trails through complex environments, so they can use exterior memory, hoping that it doesn't get overwritten by competing processes. Leaving a pheromone trail is not protected by a kernel, so it's non-private memory[193].

Ancient human geoglyphs around the world—figures drawn out on hillsides and open spaces, like the horses and men carved into chalk hillsides in the UK, and the Nazca lines in Peru depicting monkeys and birds, amongst many others—are pictorial memories written like pheromone trails but with rocks. In a sense, all memory is stigmergic. Processes have to leave a trace in something. That something is the available distinguishing features of space. If you speak, you leave a trace in the movements of the air. Others hear it by passing through it. Movies and audio recordings are also stigmergic memory.

Processes can get out of step with their memories, if semantics change along the way. When you save a spreadsheet or a file to your computer, you promise (+) the record of the process, and copy it into a memory space using a process designed for that. The space accepts it (-) and promises in turn to echo back the precise information as it was saved. Later, when you want to retrieve it, the memory space makes an identical promise to offer the data (+), but by now the software process that loads it might have changed to the point where it can no longer accept the data (-), breaking the integrity of the channel[194]. The inability to read back old data promises is a language problem, as we'll discuss in chapter

8. You can't always turn back time with impunity.

Processes that work in cycles, in space or time, are not Directed Acyclic Graphs (DAG), but Directed Cyclic Graphs. The loops they contain are responsible for feedback and form a critical part of our ability of reason about the world. Apart from the most elementary processes, it's hard to find processes of interest that don't feedback information to recycle it. When this happens, the use of space plays a key role in causal change, because processes can get stuck in loops that they never emerge from. The way out of that trap is to employ spatial memory privately. It really amounts to whether processes can maintain a separation of causal pathways, or whether there is mixing of states, and information gets lost. Databases overwrite old values with new ones, meaning that information is lost to the past. The universe does this too, with entropy[195]. By

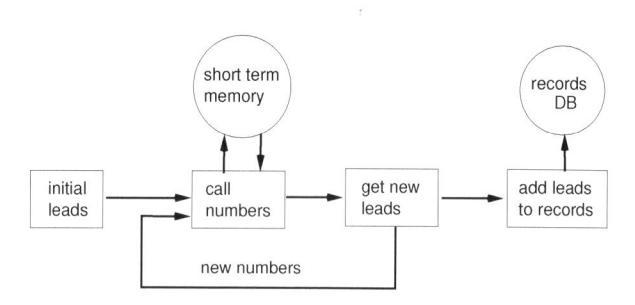

Fig. 6.2. A sales pipeline is a process starts with an 'initial boundary condition', i.e. a simple list (queue) of contacts to call, but hopefully expands iteratively as each new call picks up referrals that can be added to the list. Memory is needed to keep this process quasi-deterministic, and to avoid calling the same contacts too often.

using a database to look up and feed data back to an earlier part of the pipeline you can create a strange loop. This is where the relationships between time, causation, and space become complicated and muddled.

Suppose you are a salesperson, working on your sales pipeline (see figure 6.2), and you get a list of contacts. You start to process the list in order, by calling them up, and on each call you get more numbers to call. If the list is large, you might split the queue into multiple servers, like the supermarket checkout lanes discussed in the previous chapter. At that point, the calling process goes from having one time to having several parallel times, each with its own clock. Once any parallel search process ends, the list of new contacts acquired is appended to the original incoming queue of customers to call, mixing the past and present. It gets directed back to the beginning of the cycle, so 'now' looks like future, and past is the new 'now'. So information from the end of the process is fed back into the start of the process. There is a strict causal order, so in a sense you are sending information back in time relative to the processes—and it works because the process of serving a customer from the list of contacts is already a cyclic process. But now it gets interesting.

There are two ways you could communicate the information back: one is to publish the list of contacts to a database that's shared between all the salespeople, and is public. Or you could feed the new data back into the queue deliberately in order by a private channel. The main difference between these two is whether the information is mutable or immutable. By adding to a public database, you are not the sole source of change. Other changes might be happening that are not related to your private process: it's non-deterministic. Using a private pipeline, from the end to the beginning, you have a private (modular) channel that knows its source and is therefore deterministic. The two processes may or may not turn out the same, depending on whether chance favours a lucky separation of concerns, but this cannot be guaranteed. Just writing new leads into the same old database offers no protection between the independent sales agents, so agents might inadvertently call the same customer again. This is a common problem that search engines face in crawling the web, to give another example. If not careful, an agent could go around in a loop calling the same sequence of agents again and again (hopefully human sales agents are smarter than this, but machines certainly aren't). Loops are acausal, but still strongly ordered, and can more easily avoid repetition in a local region. Ultimately, the processes need shared memory to remember who they've already called.

Memory plays a role in the progression of time on a larger scale too. Indeed, the list of places visited acts as a clock, counting those accumulated states. Assuming the information provided by customers called doesn't change, or is immutable, then there is no harm in calling the customer again (unless they feel harassed, of course) but it means that no long term progress is advanced. You are stuck in time, by not spreading out in space. Thus the database represents a version of the thermodynamic time that physicists call the Arrow of Time: the informational entropy in the database reflects the spreading out of state.

Real circuitry is not purely deterministic and timelike in its behaviour, especially when it is scaled. Think of how an orchestra plays. The instruments play synchronized 'in time', on a collective scale, but that is a gross oversimplification of timing in the orchestra. Each instrument has a very different sound process: a string starts more quietly and rises to a crescendo. A pipe or brass instrument makes a sharp sudden sound. Drums also have sharp transients, while cymbals have a very complex sound shape. Each note is a process that lasts for many sampled moments for a listener. Each instrument 'hits the note' at a slightly different time, and yet our scaled perception of a good orchestra is that it plays together. That is the human skill of the players, accommodating their nuanced phrasing into a broader timeline. When modern electronic instruments try to simulate playing with synthesizers, playing the notes exactly as written in the sheet music, the result sounds awful, because a computer cannot make play notes to predictive targets based on listening to what's going on. Feedback breaks the stricture of causal reasoning, as a larger scale 'thermodynamic' or statistical state intrudes to absorb corrective information. What directed private channels between process agents can do, to a certain extent, is to maintain a single record of deterministic progress that's calibrated to a single clock. That modularity brings what physicists would call reversibility to the description of the process, if not truly to the process itself: by never throwing away information (assuming you can afford to keep it all) you maintain the ability to go revisit every transaction independently—to see the entire timeline again. But you might not truly be able to reproduce it.

In some processes, like in neurobiology and like in the software maintenance processing model CFEngine, loops that get processes stuck are prevented from recalling old customers, playing repeat performances, or revisiting previously visited places, by locking them for a 'dead time'. Switching off a causal pathway, by blocking it, turns out to be another cheap stigmergic way of remembering information. In brains, neurons that fired recently are blocked, and database locations that were stored, can be effectively blocked or switched off for a key time period in order to form a simple barrier between process information. While a strict branching of states is a reasoning strategy, a prototypical form of logic, blocking information is a complementary process, and one of great importance that I'll come back to discuss in chapter 8.

Modular time—seeing the wood and the trees

Some processes turn out to be single-episode stories. Some are spacetime trails that may seem completely unique on first inspection, but, on a larger scale, they turn out to have something in common with neighbouring episodes. In space there are all the similar blades of grass in a lawn, or trees in a wood. In time, there are all the recurring events like the sun coming up each morning. Anyone—any process—with the capacity to learn and recognize phenomena, does so by remembering the distinguishing 'promises' made by the superagents formed from similar episodes—with memory. No two episodes are truly similar if you go out of your way to distinguish them—we do this by adding data like timestamps or names to phenomena—but we also do the opposite, stripping away labels to make phenomena comparable. We look for a wood by aggregating trees. Throwing away uninteresting distinctions is the basis for pattern recognition.

The scientific method is also based on the idea of revisiting evidence formed from multiple similar fragmentary episodes, with the intent to compare them and calibrate them to a common standard. In mundane terms, science votes on the weight of similar experiences. In accident investigation, the data recorded by a black box flight recorder represent a timeline encountered during the private universe of a plane formed by each modular flight. The recording is specific to each episode and has no a priori meaning outside of that context. However, most commercial flights are repeated events—and they are repeated with a high degree of consistency. If we strip away information that labels a particular execution of a flight as being unique, then we can gather multiple executions of the same flight and compare them for the purpose of learning. Just as biology creates multiple instances of a species and subjects them to different contexts, leaving those survivors to reproduce and retain the memory of what context made them fit for purpose—so one can use memory to make judgements about the variations in experience, by overlaying multiple flights onto the same timeline. Normal calendar time of the flights is then essentially redefined into a two dimensional form, from

$$\text{date} \rightarrow (\text{flight episode, time into flight})$$

The absolute context of the flight is stripped away by this dislocation, leaving only a relative significance—a kind of semantic relativity principle. In this view of superposed realities, the master sequence of flight episodes is presumed not to matter, only the net offset time into the flight, for that limited 'interior' duration. The states of all the aircrafts' private flight spacetimes can thus be superposed to create an ensemble average, and a range of variability that acts as a statistical envelope of expectation for the future. This is an example of how statistics can be used to meddle with time in the interpretation of observations. We can overlay or superpose ordered sequences to realign boundary information, then measure the variability within the bounded regions and use this to predict something about a probable outcome for the future. In a sense, this is an exact complement to the process of predicting outcomes theoretically, by first determining equations of motion independently of context based on general principles, and discovering a general solution of those equations to be a superposition of all possible solutions, to be later pinned down by specific boundary conditions. In a theory first approach, the sequence is:

$$\text{principles} \rightarrow \text{equations of motion} \rightarrow \\ \text{general solutions} \rightarrow \text{pin boundary info}$$

And by observation in repeated experiments:

$$\text{remove boundary info} \rightarrow \text{general superposition} \rightarrow \text{trends} \rightarrow \text{principles}$$

The reasoning here is similar to the chain of interpretation used in probabilistic Quantum Mechanics too.

Losing your mind to logic

The dataverse builds its reputation on being right, not on being random. We trust computers because they don't make mistakes like humans do—or so we think. However, with a proper spacetime understanding of what a computation is, and how it scales, this belief can be exploded, and I want to explain why. So far, we've been able to get away with believing that computers had no uncertainties in their computations, because we've pinned our idea of computation to a single scale: bits in computer chips.

Computers work by distinguishing two states that we call 1 and 0 and sometimes even 'true' and 'false'. The technology deliberately avoids phenomena where we have to distinguish between overlapping states, though this is impossible as computers deal with large aggregate amounts of information. Computers reason using a form of logic that was inspired by the work of George Boole and others (Boolean logic). It's based on a maximal distinguishability of states: 1 and 0. We use this distinguishability to build 'logical' processes, and logic has acquired an undeserved reputation for its correctness—as compared to, say, the apparently fickle nature of human reasoning. In logic, true and false cannot overlap. But then there are statistics. Statistical aggregation looks at collections of states, with possibly mixed values, and in sufficient number to count the weight of evidence for each independent value. It's a form of voting to determine what the collective state of truth should be. These techniques were quite clear to Boole, who had no particular trouble with them. However, the way we've built our processes in the modern world has placed these two interpretations of truth on a collision course.

To get a sense of the artificial limitations of dealing with concepts like true and false, and the flawed thinking that results, we only need to look at a larger scale process that is built on the same ideas: democracy.

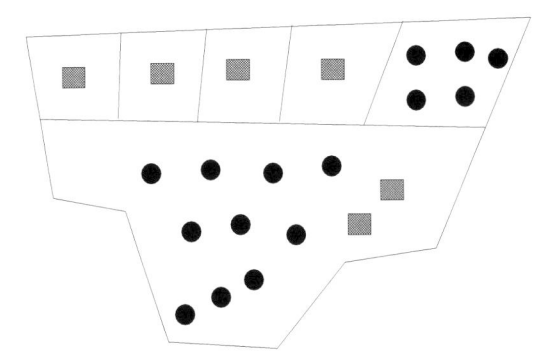

Fig. 6.3. Naive Boolean logic doesn't scale with impunity. Without boundaries, the circles win the election by counting heads. But if we first average all the different regions, so that a region majority counts as 1, then take the vote based on the averages (like seats in a parliamentary government), then the squares win, because the population is quite different in the different regions. By rescaling, one ends up comparing apples and oranges to get a total reversal of logic.

In a democracy, at least with a two party system, voters decide for one party or the other. The party with most votes wins. Suppose we call the parties True and False (that might cause some controversy, but we can indulge in a little relativism here). The problem is to define the counting process. If we count individual voters, we get one answer. However, this is not what happens in elections. We vote by regions of independent governance: modular, cellular, bounded regions with their own interior states. Each region votes and decides a majority, then the results may count as one vote—so that the largest number of votes by regions wins.

The results of a voting process are unstable under rescaling. Regions 1 to 4 have a majority of squares, whereas regions 5 and 6 have a majority of circles. If we vote by region then squares win. But the total number of circles is greater than the total number of squares, so if we count by population rather than by regional votes the circles must win. These are different semantic priority interpretations. Local concentration of information makes decision logic unstable to categorization. The scaled version by region is a completely different process, as shown in figure 6.3. By rescaling the boundaries of observation, you can literally rescale true into false and vice versa. Any statistical weight is a vote based on collected data (see figure 6.3).

Computers are subject to the same instabilities of reasoning as humans are, as we'll explore in depth in chapter 8. Clear distinctions do not scale with impunity. The reason IT gets away with its reputation for determinism is because, up until recently, it has dealt with tiny amounts of information. When unexpected results emerge, we chalk them up to 'bugs' in the system[196]. A lot of computer algorithms treat the matter of certainty naively, by asking for a 'quorum' of identical promises. Computer systems use quite small sets (typically 3 to 7), because agreement is costly to implement; this makes decisions quite unstable to small changes. For larger sets (like national elections) the results are still unstable at the cusp of indecision, but a large statistical base of voters makes the outcomes less sensitive to perturbations[197].

Statistics is an obvious case where conclusions, based on multiple replays of a process, with overlapping significance, are defined based on a level of comfortable searchlight of resolution—on which the promises made collectively by coarse observations can be observed to be broadly stable—or, in which there is a voting majority. The semantics of a state depend on scale. In a rescaling process that involves aggregation and discarding the distinguishing labels of individual details, we can't undo time, or even reverse engineer the reasoning, because the loss of information by basing a decision on a counting of indistinguishable states is irreversible. From a count, we cannot say what each voter voted—this is useful for privacy and to avoid retribution, but it means that the clock cannot be turned back on certain events, as soon as a process is subjected to distortions of scale.

OVERLAPPING TIMES, WHEN PASTS COLLIDE

Statistical aggregation leads to a related confusion. What if the states that tick for time are not fully distinct states but aggregate states actually overlap with one another? This sounds like a peculiar idea, but it's more common than you might think in data processing.

Think of a video camera sweeping out an area, back and forth. It collects observational data from a finite region, not from a 'single point' at a time, because—even if it were efficient to do so—the camera doesn't know what a single point is. It samples according to its promise to receive (-), no matter what might have been revealed (+). Pictures are snapshotted over the whole region as it moves. So as the camera pans slowly, it samples overlapping regions containing the same locations multiple times—collecting potentially different numbers of samples from each location and leading to an unfair weighting of the data. In the days of optical cameras, this might have lead to unfair exposure of the film. Only the edges of the image change on each movement, so apart from the edges all the middle points get sampled many more times than the edges. This would bias a statistical vote. We usually think of new samples as fresh events, but some events are cumulative, like the image built up by a searchlight or radar when scanning in the dark, gradually revealing patches and building up a picture. This is called a sliding window measurement (see figure 6.4). It's a technique used in computing to smooth out measurements and eliminate noise from signals.

The technique can be applied on many scales. Recall the black box flight recorder. Each strictly partitioned recording is different, but episodes can be stored and compared by superposing and comparing them. A smart car journey may collect data to analyses engine efficiency, but how does it know when a journey starts and ends? Boundary conditions, like the starting of the engine are one way to mark out the interior region of an episode. Exterior events can be kept separate, such as whether the journey was made in winter or summer. Whether we accept (-) or revoke distinguishing traits (+) plays a major role in causal process.

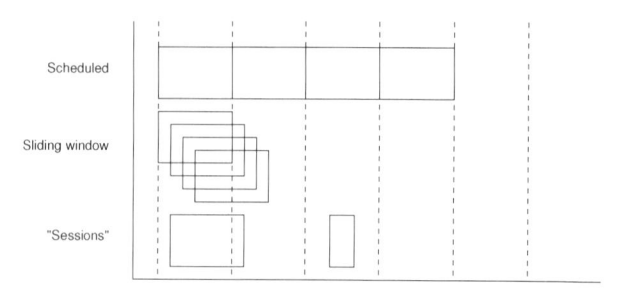

Fig. 6.4. Different ways of building up a statistical picture from samples of different received size. This can play a role in selecting an outcome—like the voting example, but now voting on a smaller but more frequent scale for every measurement. You can be systematic and continuous with a scheduled plan, slide your searchlight over the stream, or snatch uneven glimpses of the signal with "sessions".

Repetition is crucially important in many phenomena of interest, but reproducibility is even more important. Industrial business processes are repeated in the hope of achieving a consistency of customer experience. Business promises imply consistent semantics, often with guarantees. We speak of commodity goods, i.e. those that are highly reproducible and easily comparable to similar goods from competing sources. Occasionally, production anomalies result in certain promises of regularity (patterns) not being kept. Industry wants to correct its mistakes—'to take back the transactions that hurt'. That might involve undoing a sale, revisiting the stages of a repeated process to find out where the irregularity occurred, and altering it so that future transitions—from previous stage to next stage, in a production line—follow the desired trajectory. When trying to undo mistakes, previously accepted transactions might need to be undone and redone, going back in process time and re-engineering the future, leading to entirely new pathways. Whether we think of this as going forwards or backwards in time is really a semantic issue. In practical terms, the distinction is not really relevant, because outside of the relatively isolated modular process the rest of the exterior world continues to advance without any knowledge or influence of this virtual time travel, even after trying to correct the interior mistake. Time, and local modularity, are inseparable concepts.

The consistency of accumulating data over aggregate regions is becoming a major problem in the dataverse, as we deal with an increasingly globalized world. Not only are many business locations separated by relatively long latencies, leading to very different interpretation of time, they may have to travel through intermediate agents in unfamiliar or untrusted political regions whose influence of the information is basically unknown. It's hard to say that data are comparable when there is no clear causal basis on which to make that decision. The field of modern statistics was designed primarily for 'normal' processes of independent events, in which independence could be assumed by virtue of a particular model of spacetime. Information can mean something completely different in different regions.

Statistics is an obvious case where conclusions, based on multiple replays of a process (different windows onto the past), are defined according to a baseline level of resolution on which the phenomenon being measured can promise to be countable and broadly stable—in which there is a repeatable voting majority. Modularity—or kernel formation—plays a central role in semantics and distinguishability, but it's not the panacea that many computer scientists believe either.

Imagine data arriving from remote sources (perhaps all the taxis in the city) over some channels, and being aggregated into a single database. This is a very common scenario in IT. When streams of data feed into a database, some labels become meaningless. A database is like a reservoir for data—it mixes data, eliminating the order in which the queries arrived. Software developers may try to at timestamps to data records, to preserve that order. But a timestamp is not real time, because it's only a label attached to a process. It's what lawyers may object to as 'hearsay' in a court of law. Databases may attach a label or timestamp from some clock, but which one? Who gets to provide the label? The source of the data or the database receiving it? As we already showed, the order of data might be reordered in transit between source and receiver, so the source time order is different from the receiver time order. In fact the receiver's time is the only active process that represents a real clock so it's the only meaningful choice when data are being aggregated from different sources. The problem then arises when someone wants to query the database for its data.

It's quite common for programmers to abuse time and say 'tell me all the latest data, since I last queried'. That works okay in a linear process, like a queue (first in first out), because it preserves time in a single linear channel. But a database doesn't. By 'latest' a database query means 'data with timestamps later than the last timestamp I last asked for'. However, when we put data into a reservoir and eliminate the process ordering, we only have labels to go by.

We can miss data arrivals because they were labelled incorrectly, or because they arrived with labels marked before the last query that arrived after the query completed. Moreover, retrieving the data from the database, by checking labels and outputting a new ordered list, is itself a linear process that can miss data as they are arriving (interleaved processes). Latest is a moving target, not a fixed boundary condition.

A simple solution to this issue is to use only specific queries, never open ended times like 'now' or 'latest'—but humans tend to think in these terms, so we often make mistakes. We have to trust that all of the intended data have arrived, and that the data in the database are in a quiescent state before asking for answers based on it. This is a non-trivial assumption too. It's why banks leave a grace period when processing transactions for payments to 'clear'—to allow any latent processes to catch up with possible queries. Determining answers from aggregated sets of similar data is fraught with issues of scale!

It's not hard to see that placing an aggregation point—a data reservoir or 'database'—as an intermediate stage in any process chain muddies the concept of time by replacing a clock with an untrusted stigmergic collection of labels, leading to problems in both reproducibility and reasoning. This is why we can never fully trust intermediate agents, or points of space, as relays. This is a major issue in information technology, and particularly in financial services. As for physical spacetime, we can only speculate on what this might mean for processes at the lowest sub-quantum levels.

The catch is that modules act like databases too. So whenever we use a separate module to accumulate private results of a process, where the data it collects are expected to be similar to other modules, they may be completely unrelated. Patterns happen on a large scale. Isolation prevents patterns from forming. If the modules open up and share data, then they become subject to non-deterministic change[198]. In physical spacetime, we call chalk this up to entropy, or to twins paradoxes. In biology, it's rare to find ordering problems, because evolution has generally selected fixed point mechanisms at scale, but the lack of a well defined fixed point bedevils many systems in the dataverse. Computation as a process, with relatively few degrees of freedom for ballast, is fragile but clear with sharp edges. As the degrees of freedom grow larger, the sharpness of its conclusive states ebbs away and results seem increasingly ad hoc. This is the curse of scale: the more information you have, the more it seems like noise.

PROTECTING TIME WITH CELLS AND MODULES

Boundaries play a clear role in defining outcomes of measurement and inference. A boundary itself is simply a pattern in the matching of (+) and (-) promises made by spacetime agents. Processes may possess boundaries (as borders in space, or as episodes in time) that protect the integrity of their information, and these will always lead to a distinction between interior and an exterior state, and interior-exterior observer view. Exterior observations are observations of the relative past, according to interior time, provided interior time can count to a high enough number to tell the difference. Combining observed states with interior states imposes a 'many worlds' separation of causal influence from the sources. When they meet there is a confluence of wormholes into the past (see figure 6.5), something like Star Trek's *All Our Yesterdays*.

In the dataverse, the problem of parallel timelines has become quite acute since computer systems have become distributed, both in terms of geography and in terms of design. Monitoring and tracing systems that track software behaviour have not kept up well with the distributed nature of software systems. For instance, most computer systems have legacy systems for recording messages, like 'syslog', which serialize all messages into a single timeline. Multiple processes can fire unstructured messages, at will, at a single receiver process, leading to an unpredictable ordering of interleaved entries from causally distinct processes into a single timeline. The mixing entropy of the timeline is thus maximized; the result is a stream of irregular messages with unclear semantics that may or may not contain valuable information. Moreover, the representations of time used in the messages are usually local origin clock time, represented in some irregular format, and without any promises about calibration of the clocks to a common standard. The received order of the messages implies a local destination clock time, but there is no way to infer or even define message latency, so there is no consistent scalable meaning to the timestamps. Software developers have come up with a vast number of search and organize software projects to try to unmix the messages and find order in the chaos. This involves a high overhead in computational resources, and in general is not impossible, since most of the information is already disconnected from its contextual origins, leading to irreversible primordial soup.

Aggregated events, sampled from multiple independent sources, present us with a mixed sense of time—a perception of events that depends on several sources and can end up seeming quite inconsistent and even unpredictable. If

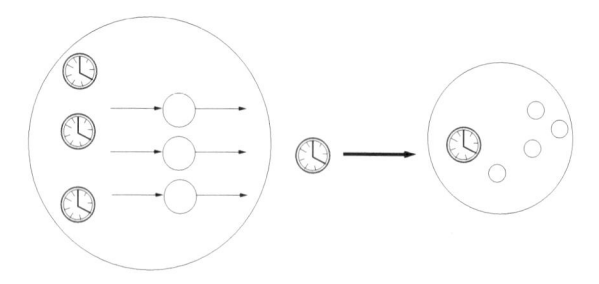

Fig. 6.5. Aggregation of data along DAG pipelines, from independent sources, involves different interior clocks whose signals are integrated according to each receiver's clock, in a pipeline.

tunnels along several routes to the same destination arrive at a single location, what one observes along each may even contribute differently and contend or interfere, effectively mixing the outcomes of multiple processes. The same phenomenon might be observed through different portals, with different outcomes. This relativity of discrete differences is weirder than Einstein's large scale relativity.

The give and take of information is always at the root of these phenomena. Data streams (+) flow together and mix different sources, along different paths. The result accepted (-) at a single confluence may see completely different pictures of reality as a result of small variations, with events reordered. The average web-page you browse on your computer is basically a collection of portals to multiple past times, or separate universes. Different portions of the page might stem from different times, and therefore be out of synchronization with respect to wall clock time. This might have consequences when we need to base a decision on the sum of the information. Similar inconsistencies have been claimed in information transfers in quantum mechanics too, e.g. when causally ordered events are unable to preserve their ordering due to the uncertainties of observation[199].

A more mundane example helps to make this obvious. Shops get deliveries from several suppliers at different times. The shop experiences this stream of deliveries as tick events, and samples them into its modular world. As long as the arrivals remain unchanged by the module, this has no semantic significance. If the shop simply sells the same goods on to customers, no problem, the delivery ticks are passed on individually, but if the different sources are combined into an aggregate result, there is a synchronization of events and waiting reduces the total number of ticks. A bakery builds the cakes before selling, so it aggregates the ingredients before sales. Its operation has to wait for all the parts to arrive. Similarly, a car on a production line is aggregated and assembled from a confluence of several processes: one that makes engines, one that makes bodies, another makes wheels, and so on. These processes occurred on timelines that are completely irrelevant to the car process, and can be discarded in favour of the local clock. The information and configurations of state that comprise these components. To make a car you need to wait for not one but four wheels, one after the other, two front seats, one body, and so on. In other words, we have to aggregate arrivals of information from the different channels. Whether or not you consider the production of a car to be an atomic process depends only on the scale at which you can observe it.

When should we keep processes separate, and when should they merge back into a single world view? Advanced data processing lies at the heart of most business processes today, from accounting to self-driving cars. These questions are essential to making decisions when voting with data. Without a proper understanding of how spacetime scales in the region over which data are processed, no one can hope to fully understand the answers they get, yet we increasingly place our lives at the mercy of poorly understood information scaling. Whether exterior data arrivals should automatically define temporal progress in a process is not a cut and dried issue on any scale. Processes sometimes need to wait for certain minimum requirements to arrive before a larger process can proceed. This affects the coupling of process clocks[200]. Industrial processes need to reproduce outcomes reliably within a controlled context. Empirical science is also based on such patterns of observation, relative to different regions and contexts. It's not *reversibility* of behaviours that matters in general, but reproducibility: not the ability not to turn back time, but to replay time. This is connected with modularity, because the separability of paths to avoid a mixing or conflict of interests between agents can be minimized with private modularity, i.e. by an interior-exterior boundary placement.

We have to think of time in terms of the dual roles of give and take: as processes that change and processes that measure. When a change occurs we need to access the causal graph of spacetime linkage to compute a wave for causal updating, i.e. the consequences of the change. As usual, there may be advanced and retarded perspectives on this (see figure 6.6). These are two kinds

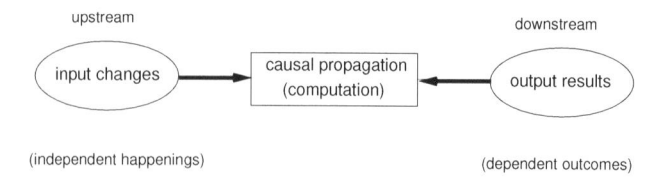

Fig. 6.6. We are used to thinking of time from left to right, but recalculations of causal influence may be initiated from advanced or retarded perspectives. Imagine a data pipeline or a manufacturing production line. New components at the inputs (left) can trigger the manufacturing process to bring about new outputs based on causally independent or 'random' happenings (right); but, new design (re)combinations may also constrain the instigation of new process channels, rerouting inputs to new desired outputs (right); thus outcomes may also be considered a trigger of change in process sources. In biology, the retarded process is initiated by random species couplings (sex), and the advanced process is selected as DNA recombinations by environmental conditions. We anthropomorphize the latter as 'intent'.

of process: unintended receipts of information from self-illuminated exterior sources (information pushed), or 'intentionally' observed data of sources—shining a light on to process dependencies, like radar (information pulled). It's also worth mentioning the link between modular thinking and the scattering model in physics (see figure 4.4).

From a Newtonian perspective, the causation involved in processes like figure 6.6 seems strange. We are used to understanding things from absolute past to absolute future, which is a throwback to the days of a deterministic clockwork universe. For a data pipeline, input sources are polled (observed repeatedly) for changes, and a promise to propagate from the sources makes the information available non-locally to the computational spacetime process. But the same component input information could also be 'fetched' (re-observed) by the process of linkage to be recombined in a new way, e.g. to make a television instead of a smartphone. Observers, as 'middle-men' can also drive processes to transmute passive information into future change. This kind of 'decision' to recombine to a new future outcome implicitly connects causal predecessors to causal ancestors—process past to process future—together backwards. Although I'm using intentional human language, this is not an argument against—rather, it's a definition of how a concept like intent comes about (see chapter 8). It brings up a new point: there is a semantic difference between these interpretations of time. There is a *causal past*, which is a predecessor along the proper timeline of a process, and there is an *observational* past, which is the sampling of a signal arriving at a sensor after some journey. The two are only the same if a receiver uses the information from the observation to drive a process from its past to its future state. To a classical physicist, this sounds a bit cheesy, because one is used to describing things that happen involuntarily—as if there were an unwritten obligation to respond to every signal sent. What computer science teaches us is that this is a belief one simply has to shake off. The consequences of indeterminism go far deeper than implicit probabilities.

Cellular structure, or modularity, is a hierarchical phenomenon. Boundaries can subsume other boundaries to create super-cells—or superagents. Atoms scale to molecules, molecules to cells, cells to organisms, organisms to species, etc. Each layer of separation has a different channel separation purpose[201]. The dichotomy between separation and aggregation strongly affects predictability of outcome The gestation of offspring is the obvious example in biology, wherein a wide open process of sexual mixing of DNA is shut off completely in order to stave off interruptions: placenta or eggshell isolates the offspring from exterior time. Semantic correctness (promise kept) is maintained within the shell by a (e.g. nine month) time lock on the process. This protection also affects our ability to reverse aspects of a process—to turn back time—because it becomes more like a protected causal DAG, when shut off from the world with an effective 'desired end state'—rendering advanced and retarded boundary descriptions as much the same as possible over a noise-free channel.

In an open world, reversals are more impractical: separating garbage into glass, metal, paper, plastic, etc, is almost meaningless. Separation of these major categories is the tip of an iceberg of information. Some plastic can be reused, some can only be burnt. Much paper and plastic is combined with other materials making separation impractical. Separating garbage for recycling is going to require an entropy reversing technology that acts on a microscopic scale, not a macroscopic one, larger than the garbage itself: the information mismatch is just unworkable. With sufficient access to small scale process details, we can go back in the time ticks of the process: strip the engine, replace the cylinders, etc. There is no change in local entropy, but there will always be consequences on a larger scale: entropy increases elsewhere—due to scope creep. A widening spatial net gets implicated in the wave of influence spreading out from the localized car process, and forming a larger scale clock of its own.

A QUIET BELIEF IN ANGLES

The idea that we can simply choose to separate processes from one another, as if making the world a collection of orthogonal vectors and alternative directions, is surely one of those great lies that we tell ourselves. Industry has a deep seated and unrealistic belief in the superpowers of modularity, as a simple everyday example reveals. Figure 6.7 shows a workflow associated with a

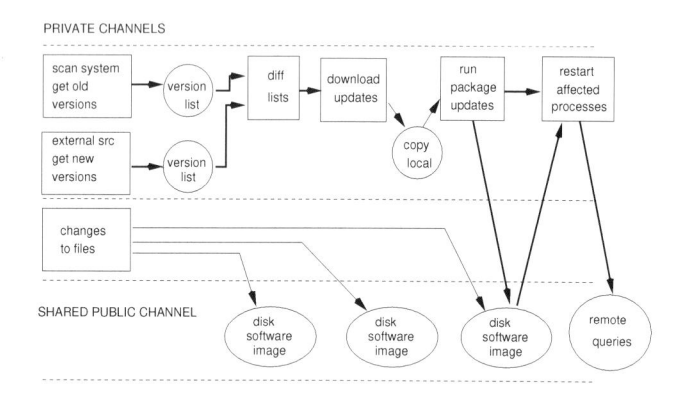

Fig. 6.7. A computer workflow pipeline process, for updating software on a computer. The process mixes changes to private (local) and public (global) state, leading to potential indeterminism, meaning that rolling back to earlier versions is not possible in general.

pretty common task: updating the software on a computer. Initially, computer operating systems were installed lock, stock, and barrel as if they a single entity (a single modular location). Today computer software is sub-modularized into many different packages that can change more or less independently, as they come from different providers. Instead of mixing together all the software in a big public space, software is now broken up into different regions that promise independent responsibility for being up to date and consistent with their surroundings. This is not totally successful. Packing software into bundles is analogous to what biology does in packaging its processes in different cells, organs, organisms, and so on, within an overall functioning ecosystem. The difference is that software packages depend on one another more for services, so they can't be completely autonomous. This is true in biology too, for oxygen supply and so on, but in software the tower of dependencies is much higher and more fragile. For instance, your web browser needs access to the screen and the keyboard, as well as the network, so updates to the screen or keyboard drivers will affect your browser too. A simple mistake in one of those packages might render your computer unusable.

In the software update process, the computer scans the space of the computer as quickly as it can for a list of which versions of software are currently installed (assuming that nothing much will change over the time it takes). Versions are what passes for time relative to the update process. Each new version is a tick of the clock used by the updating process. The updater then downloads a list of current intended versions by the software manufacturers, and compares these lists—again, assuming that nothing changes while it is doing that. Any new packages are then downloaded from a remote source to the local storage of the computer for installation.

As updates happen and rewrite the timeline as a set of uncoordinated changes, sometimes with imperfect boundaries, partial updates (even just downloading some packages) can make changes that interfere with the ongoing processes. Time is not a straightforward process with a single clock, it is a vector time, with processes advancing according to one clock, processes being updated on a different clock, and interactions between processes happening on a clock formed from the entanglement of processes locked in step through dependencies.

When changes do occur, modularity offers some protection to processes from semantic inconsistency—as long as the boundary is impermeable to interference. You shouldn't be able to get half a change to a software package—updates should be 'atomic'—meaning that only complete approved changes can be 'observable' to users. There is a promise of discreteness implicit in the superagency of the packages. Sometimes this works, but sometimes it doesn't. When updates occur, they may not be fully separated even on the disk. Since they interact with one another, and with users local and remote, there is occasionally disruptive changes to the dynamics and semantics of give and take between the agents involved. The packages maintain their interior integrity because they are bounded, but they then violate that integrity by unpacking their contents into a public space—like a shipment container being unpacked. As soon as this happens, they mix together leading to entropy, and interfering with any other sources of change. Say, for instance, the user of the software typed in some software preferences, entered credit card details, and account information. The software update might erase all that information—or more scarily replace it with something else.

Any kind of shared memory is susceptible to this. This has led a lot of software developers to argue for complete separation of data—the elimination of global data. Of course, this is not possible. Databases are inherently global. We actually want data from outside sources. But we also want to know when changes take place, and different processes might need to preserve certain parts of the data for their own use. All of this comes quickly into conflict with the narratives of order and predictability demanded in civilized society. Business processes should not be ad hoc, or customers will find a competitor who can do better. The amount of effort we put into creating the experience of predictable repeatability The quiet belief in separation of concerns and deterministic outcomes is a pernicious flaw across information technology. It leads to unseen conflicts.

No U turns from equilibrium!

Now that computers are scattered all across the planet, and connected to neighbours by pathways that are very inhomogeneous, one of the problems information technology faces in designing distributed and virtualized data storage lies creating stories with a decisive timeline. It doesn't have to be unique, but it has to be able to define a source of arbitrated truth. Traditional databases and memory systems are not engineered to locate memories, with an unambiguous universal notion of time as a key, for the reasons we've discussed up to now. Subsequent events can rewrite the memories of earlier events, which may in turn influence future events, databases (or any other kind of memory storage) are unreliable sources of truth, and act as a distorting mirror of the past.

Following the financial crisis of 2008, a new kind of database was designed to replace timestamps as a source of truth with an explicit vote on the acceptance of what constitutes a process tick from an exterior world. In the borderlands between databases and data encryption, this database basically emulates a kind of time tunnel with a steadily growing and immutable history of events, based on wide area consensus voting. The database is now referred to as a *blockchain* (so named for its interior structure as a linked chain of data blocks), and it shot to fame around the cryptocurrency Bitcoin.

Blockchains are a process of wide area equilibration, designed to create many exterior copies of the interior chain structure. There will be no time paradoxes within a blockchain time-tunnel. You can go back in time to look, but never be seen, never to intervene or change what's passed or past. This sounds attractive, and many believe it to be a new panacea to all the world's ills. The catch however is that it's also incredibly costly. Voting costs resources, validation of individual identity across space costs energy, and distributing the uniform indistinguishable copy of the chain to all locations in the blockchainverse also costs energy. The energy used to maintain Bitcoin now exceeds the electricity consumption of major nation states, in spite of supporting a tiny fraction of their population.

New data arriving at a blockchain are sent as transactions, like payments. They are duplicated and broadcast to the interior of a closed network of agents, each acting as spacetime locations—and collectively acting as a single scaled point. Requests are published like 'letters of intent'. Intentions (+) are collected (-) by all participating agents that aggregate such transactions until there is a bulk amount called a block. The agents all promise a clever interior process, which makes the validity of the transactions dependent upon the integrity of the entire prior history of other transactions from the entire network. Invalidate something, invalidate everything. This has the effect of making the present database valid only for an immutable version of the past. The effect is something like a building a tower. You couldn't change history without climbing the whole tower of interlocked steps, or without tearing it all down—but there are too many copies to tear them all down. The blockchain is both supremely well validated and supremely fragile at the same time. The effect is to so entangle the fates of past, present, and future memories that it is entirely impractical to tamper with them.

In a sense, a blockchain is something like an inefficient combustion engine: a transaction spark ignites a transactional explosion of data copying that fills its space, and is contained. Work is done along the way (much of it wasted) to contain the explosion, and reducing it back to a homogeneous equilibrium across the region. It drives the blockchain forward, one block at a time. The autonomous nature of the nodes in a blockchain network means that what's offered (+) may not be accepted (-) by all locations, but if the changes are absorbed by the entire network, then the transaction can be considered carried out. Meanwhile more events are arriving from around the network. What the blockchain design emphasizes, once again, is that *functional* characteristics, including the dimensionality and adjacency of nodes, depend only on the logical connections between agents—not on whatever underlying physical computer system it may run on.

Blockchains and other replicated databases try to engineer an illusion of simultaneous agreement for observers across disparate regions—what Einstein called global *simultaneity*—but it can only be done on a thermodynamic scale. Einstein pointed out that simultaneity is an illusion of scale—a major part of his theory of relativity. We manage it by drawing a boundary between local and non-local. On the interior, we can think of time as global and simultaneous to a fair approximation. On the exterior, we will be wrong about that. To pull off the illusion of consistency across a wide area, we have to effectively shield the observability of the cluster (or put a lock on it), decoupling interior and exterior time over a region of space. We stop the clock for observers until the magician is ready to unveil the illusion. Each agent's interior time continues during this cover up, but the exterior state of the collective is frozen until there is interior consensus, just as in Lesley Lamport's Paxos mechanism.

In the mid 1990s, before the big Internet explosion of the early 21st century, when the dataverse was still growing, I developed a very different model for managing computers by equilibrium—some autonomous 'robotic' software called CFEngine[202]. Computers go wrong quite a lot, and they also need updating, garbage collection and all kinds of maintenance. CFEngine tried to build an equilibrium state for computers, so that users experienced a highly stable system. Instead of allowing time to progress through changes, CFEngine would restore a predefined state before anyone could observe any change—like making a course correction with a satellite navigation (GPS). I called this an artificial immune system, because this is essentially what the homeostatic and biological immune system does. When changes occur, it tries to restore balance on a number of timescales. If you get out of breath, you pant to restore a stable oxygen level. If you are affected by a process that opposes the desired state of your body (either by a rogue interior process like cancer, or a rogue exterior process like a virus or bacterial infection) the immune system instigates countermeasures: processes that detect the molecular signatures (+) of these agents and oppose them with an array of immune cells that lock onto (-) the signatures and neutralize them. Both biological and technological immune systems don't try to turn back microscopic time, they try to halt time, by meeting the invasion of changes with a counter wind and keep it in check, until the next infection comes to deliver a new tick on the balance of time.

Before the software revolution on the Internet really took hold, this seemed like a good way to build stability in computer infrastructure, but as the World Wide Web took hold, there was an explosion of activity, bringing new ideas, and a desire to evolve systems much more quickly. An immune model like CFEngine was too slow to stabilize systems on very short timescales, and it underestimated the desire for human interventions. Today, the computing cloud had recycled many of the principles of CFEngine, but has re-engineered them to work relative to different scales, and allow for human activity. The result is self-healing platforms that also try to maintain a dynamic equilibrium on a much faster timescale.

Give and take in the unfolding of events

Many processes benefit from modular spacetime isolation. Computer science uses modularity to isolate human reasoning as well as to protect against external change, but it occasionally uses it naively, promoting instability. Today, 'microservices' are a popular and extreme form of modular design for IT systems, that have to achieve consistency by costly manual equilibration, somewhat analogously to the robotized blockchain. Politics also divides and conquers using national and regional modules to avoid human conflict. Biology exploits modularity at all scales: an ecotone is a region of transition between two biomes, habitats, or biological ecosystems[203]. In other words it is a virtual boundary in the processes of life that maintains a certain stability. Biological phenomena have found ways to form cells or modules with different local semantics on all scales. Even the weather exhibits the spontaneous formation of convection cells that isolate wind patterns and regions of temperature. Through these boundaries, the regional agents of spacetime can keep the entropy in one modular region from appropriating the states of another. Modularity is how scaling preserves its semantic integrity.

In quasi-intentional language, one might say that the decision to form a cell leads to the promise of stable functional behaviour. We can also reverse the causation to say that the functional result was an emergent side effect of a confluence of unintended causal promises (+) that found a matching (-) in a complementary process. The directional flows of give (+) and take (-) are not only expressions of orientation—call it causation—they also address the increasing capacity for distinguishing direction with increasing scale, by expanding into new available locations[204]. Entropy is often presented as disorder and chaos, but Shannon showed that it is information itself—we need it to make large scale progress. Managing and maintaining irreversible persistent IT processes, with a unique historical record (what are called 'stateful applications' in IT) in a virtualized cloud environment, where rules of data relativity operate, is a notoriously difficult engineering challenge, precisely because the price of virtualization is to give up absolute coordinates on the interior space.

When information has to migrate through spacetime without loss, it requires a suspension of exterior time (a process lock) to prevent contention by other processes. Time itself—as an accumulation of changes to state—has to be counted as a distributed (non-local) phenomenon, which is expensive. To be inside overlapping clocks, belonging to different processes, makes consistency of observation highly challenging. When we treat non-local phenomena as if they were local, paradoxes ensue. And when events are interleaved from different causal origins, without sufficient memory, it becomes impossible to reverse state and roll back time.

Non-local processes imply the spread of second hand information, subject to distortions as they travel through time and space. The crucial question of determinism and relativity in systems is thus not really about relative motion, as we imagine in a Newtonian view, but about where we can draw the boundary between interior and exterior: local and non-local, observer and observed.

7

DIAL mv FOR MYSTERY

'Well, in our country,' said Alice, still panting a little,
'you'd generally get to somewhere else if you run very fast
for a long time, as we've been doing.'

'A slow sort of country!' said the Queen. 'Now, here,
you see, it takes all the running you can do, to keep
in the same place. If you want to get somewhere else,
you must run at least twice as fast as that!'

–Lewis Carroll, *Through the Looking Glass*

All is not as it seems in physics. On the one hand, Newton's three laws of motion summarize what are probably the most familiar aspects of classical physics to ordinary folk. They explain the day to day movements and collisions of bodies around us, albeit in an idealized way, and thereby form the basis of what every school child can be expected to learn about the subject. Newton captured the essence of daily experience about motion and the resistance to changes in motion (inertia), and turned it all into a simple yet precise mathematical formulation. Newton's laws have been used to predict the trajectories of bullets, cannon balls, rockets, and other projectiles, of all sizes and scales, from human scales all the way up to planetary scales. Few theories can claim to be so successful—and yet there is something quite odd about Newton's laws with regard to the story we've been outlining in this book. The laws conceal deep anomalies in physical behaviour that need to be explained. There is nothing else like them. As laws with ostensibly universal significance, they are oddly unique in the world of phenomenology.

THE THREE LAWS

Let's review them briefly. The first law, originally due to Galileo[205], states that: any body will remain at rest, or continue in a state of 'uniform motion in a straight line', unless acted upon by an external force. In other words, it's hard to stop a moving train, or to get one started.

The second, more quantitative rule, tells us that forces are the influences that start and alter processes associated with motion: specifically, the force F on a body is equal to the rate of change of a quantity called *momentum* (denoted p), defined by the strange combination of a mass and a velocity $p = mv$. The velocity v in, in turn, is the rate of change of position x with time. The rate of change of momentum can be written as a mass m multiplied by its acceleration a, when the mass is fixed.

$$F = \frac{dp}{dt} = \frac{d(mv)}{dt} \simeq ma \qquad (7.1)$$

So the key variables in motion are mass, location, and time—though physicists normally condense this into just two: location and momentum, as mass only has significance during dynamical processes.

The third law is usually expressed by saying that for every action, there is an equal and opposite reaction. This is equivalent to saying that all forces or influences come in pairs that sum to zero, and is more formally expressed by saying that the total momentum in a closed system is constant.

The fingerprints of space and time are clear to see in the laws. They refer to motion, position, time, velocity, and so on, but then the laws go on to introduce the concept of 'force' F, and mass m, without explanation. Most of us will take the idea of force for granted, as we are used to the idea of applying force in daily life to push open doors, or open bottles, though probably no one can explain exactly what it is. We are most familiar with 'contact forces' like pushing a door open, but we now know that this idea of force is an illusion of scale. It's really the electromagnetic repulsion of atoms close together that transfers force from one process to another. There is no contact, only a kind of communication. Physicists can't explain force any better than that either, at least not without referring to the other quantities mentioned in the laws, like momentum—so the entire mystery around these quantities is really swept under the rug.

Then there is the other mysterious quantity, which seemingly comes out of nowhere: the mass m of a body, which seems to be quite unrelated to space and time. Momentum is the mass of the body[206] multiplied by its velocity mv, which is an odd composition to be sure: it takes something already scale-dependent and non-local (the velocity) and combines it with something equally unknown in order to form a quantity that's treated as local. This might lead one to suspect that mass must have an explanation in terms of space and time too. All we can truly say, from the laws themselves, is that Newton's laws are a self-consistent set of rules for generating stories about the particular set of variables it sets down on paper—most of which are related to spacetime. The stories generated match our experience of physical phenomena well—but with some explanations missing, what exactly do they say about spacetime processes?

Newton's achievement was to construct a simple formulation of behaviours, starting with the assumption of a Euclidean theatre and the three laws. The result has been so influential that the principal variables of position and momentum find their way transposed into most other descriptions of phenomena too. Quantum Mechanics and Einsteinian Relativity both have quantities that stand in place of the momentum, but are not semantically similar—and their presence there doesn't have a strong motivation. Position and momentum are called 'canonical variables'—almost an open admission that they are more a belief system than natural phenomena. There is an entire branch of mathematics is dedicated to the study of 'canonical systems' described by position and momentum. However, the peculiarities of the laws with respect to an information based description make it worth questioning whether that imposed transference is truly appropriate.

In this chapter, I want to try to unravel some of these issues, and consider what force and momentum might correspond to in other kinds of spacetime, on other scales—like the dataverse, biology, and economics. If momentum is so key to explaining the motion of objects in physics, might there be corresponding quantities for these other spacetimes, allowing us to make predictions for them too? The answer is, probably not, yet there is a profound lesson to be learned by asking the question, and naturally it is directly related to scale.

Recoil and conservation of momentum

Why should a body continue in a state of motion in a straight line, without stopping? Why should it be hard to stop a moving train? Perhaps it seems like an odd question, but this behaviour doesn't really occur in any other area of experience than those where Newton's laws apply. If I send you money, you receive it and you pocket it. It's yours. The money does not continue moving on to the next person and then the next. Nor do you feel any force impacting on you (setting aside metaphors of greed and the uncontrollable urge to shop). Files on the Internet don't move from computer to computer until captured by a stopping force; and, if you receive a file on your computer, the computer doesn't recoil from the impact, even if it fills the entire memory of the device. Conversely, when I throw a ball, why doesn't it simply disappear from my hand and reappear at its destination, like Captain Kirk beaming from one location to another—like an information transaction? Most other elementary phenomena involving exchanges are transactional[207]. Something about motion, in the Newtonian sense, is coherent and self-sustaining, and leaves a smooth persistent trace along its trajectory, like a process loop in a computer program rather than a transaction. That tells us that motion and momentum cannot be elementary processes.

One of the important consequences of the 'law' of conservation of momentum is that masses seem to carry a strictly accounted amount of motion with them—which is what the momentum measures. How does a simple point carry this momentum—a property we associate with bulk? An agent can partially transfer its own transport process on to another when they collide, forcing an impact onto each other. When a mass strikes another mass head on (e.g. as in billiards), both masses will generally continue to move in some direction: an initially stationary mass will recoil from the collision, and the incoming mass might even change direction and go back the way it came (depending on the sizes and elasticity of the bodies). This is a very strange idea from the perspective of a computer scientist. If we send a message to a computer, the computer doesn't recoil in horror at the message. The message does not get reflected back from the computer back in the direction it came. Information does not behave in this way on an elementary scale.

In fact, nothing behaves in this way except massive bodies on the approximate scale of our human world. This tells us that there may be something special about mass—it could be a phenomenon primarily associated with aggregate scale. Most elementary interactions are transactional. There is one exception to that, which is when processes are *statistical* descriptions of deeper transactional processes, or bulk aggregations, where there is larger pattern of behaviour that can't easily be stopped by a local interaction because it's non-local. Conserved motion, in the Newtonian sense, could therefore be a result of scale. In looking for a computer analogy we could simply be thinking too literally, in terms of the embedding space instead of the actual space of the dataverse, but actually there is much to learn from doing this.

It seems reasonable that if we leave a process alone, it will continue to do whatever it was doing before. However, that sentence is filled with many assumptions about space and time, and its processes. We don't normally think of a stationary object as a process. For a 'thing' (an agent, or a state of space) to be 'doing' anything (a spacetime process) it already implies a cooperative interaction of agents. The continuity of a process seems to involve non-elementary scales. Why would a process continue its pattern of behaviour? What's so special about moving bodies in physics? An analogy might help. Suppose you are sitting still, but deeply involved in a conversation, or writing of a book. The effort to free yourself from those processes (which are not related to exterior motion, but still involve interior change) might cost some effort. Moreover, to shift the conversations and processes to the new location would also cost some effort—unplugging from local services, and reconnecting with equivalent services at each location along the way, in order to sustain the activity. So an accumulation of on-going dependent interactions could easily be a hidden source of 'mass'. This is an idea I'll explore some more below.

A clue that this scaled view is significant comes from looking on a small scale. On a quantum mechanical level, particles don't seem to obey Newtonian laws of motion in any obvious way. The Heisenberg's Uncertainty Principle is often touted as implying that particles jump around in a quite random manner, as observed by exterior measurements, though that is probably overstated. But there is at least a possibility that uniform motion is only something that can happen on a large scale, in an average sense.

An information process view

On an elementary informational level we have to think about Nyquist's sampling law. In order for a property like momentum to be passed on, from agent to agent, from location to location, along a chain of receivers, each agent has to be sampling for it twice as fast as the process by which the property is signalled. Clearly each downstream agent can't be twice as fast as the next one, so—if spacetime were merely homogeneous, and the locations were all alike with an interior time—then the next point may only have an even chance of receiving the message. If we could imagine some third party godlike observer looking down on two points about to exchange momentum, calibrating their interior times by its own clock, it might see a lot of failed attempts to absorb the momentum message, perhaps because its own clock sees the two other points' clocks as running at slightly different or even erratic rates.

Of course, as far as the physical universe is concerned all this is speculative. We have absolutely no idea about what fundamentally drives change or how the states are arranged at the fundamental level in physical spacetime. But this is absolutely clear for the dataverse, where we can measure such processes and see the issues take place. When large amounts of data enter a region of space in the dataverse (i.e. a cluster of computers through a load sharing lens), the cluster of computers it interacts with—as a collective superagent—has to be able to sample the data about twice as fast as it arrives. This might not be possible as a serial process, but it can be also be parallelized, if we allow for the superagent's interior structure. A lens effect, discussed earlier, would be involved in this process.

Recoil, on arrival of the data, would look analogous to the impact of a sudden transition of local interactions to a new location. Think of moving house, and transferring all family processes: school, work, shopping, etc to the new location. Once you've lived in a place for a few years you start to grow roots: regular interactions with local agents and processes. You have a job, a house, friends, a pension, etc. All of these incur a cost to moving: to move involves a lot of effort. In other words, there is an inertia because of the network you are attached to. If you overcome these hindrances, and you can finally move, they are still there, replanted at the next location. If you keep moving, with the same encumbrances, your mass remains constant. It will continue to cost to move these. But if you can shed some of them, the mass goes away.

An impact only makes sense if the movement carries active processes with it. Recoil is thus a scale phenomenon. Momentum is, like velocity, a non-local phenomenon. In the Newtonian mechanics, the variables ascribe recoil to a change in motion, but we also know that inelastic collisions of bodies result in crushing and breaking up, and other transformations of the bodies that can only be explained in terms of their composite structure from atoms. The dataverse is a discrete space, so there can be no smooth motion with recoil, in a Newtonian sense, unless an existing bulk process were capable of relocating on average and being handed on, piece by piece, like a package along a production line. This kind of effect can actually happen in the computing cloud, where a job gets relocated because there is no longer room for it in a certain region. The recoil from a message would not be in the same computer, but rather in the form of similar promises a software entity could hand over to others to keep. This is a little mind boggling, but again it makes perfect sense as long as we can think of the bigger picture, rather than being obsessive about locality. The key is to think of mass and motion as processes, not as intrinsic and inert properties.

We perceive the passing of momentum to balls in billiards as being instantaneous, but by now we realize that we have to ask what processes are responsible for sampling and absorbing those properties, and whose clock is driving that process? The assumption above is that what is sent must be received, like an obligation. Classically, this kind of transfer by mandatory obligation has simply been assumed as divine right. However, this is not the true picture in most cases, if any. Sometimes that which is offered (+) is not accepted (-), as we saw in chapter 4. Information offered may be refused. Atoms fail to absorb photons. There are 'forbidden transitions'. Letters may be returned to sender. Some messages in the dataverse are simply dropped.

Understanding conservation of momentum and recoil, in Newtonian terms is a non-local, non-elementary phenomenon. If we imagine the arrival of a huge amount of data to a *region* in the dataverse, like say a cluster of computers viewed from a very high level, we might indeed see data impacting as transferred motion, when single locations exceed their local memory and have to offload to others. This spreading over space could perhaps be viewed as a kind of load sharing that exceeds some quantum of storage, as a maximum threshold. The conservation of momentum then looks like a kind of lens process that reorganizes local processes when they become too concentrated.

The non-locality of mass, inertia, and exterior time

The concept of inertia nagged at philosopher Ernst Mach (1838-1916), during his lifetime, and his views on relativity were highly influential to Einstein. His position on inertial acceleration is now known as Mach's principle. Bemoaning the absence of an absolute frame of reference against which to measure motion, he noticed that the only effectively absolute measurables in the universe are the fixed stars: by virtue of their being so far away that they appear static, they provide a surrogate for an absolute frame of reference. Since all reference frames experience them as effectively static, inertia must be a measure of the effect of the distant stars dragging us back. His principle is bizarre and not quite convincing, but what Mach was getting at was that inertial mass has a non-local aspect to its behaviour—which turns out to lead to the modern view of mass as a process encumbrance.

We don't need the span of the entire universe to explain mass, as long as we don't get too caught up in Einstein's principle of non-absolute motion. As mentioned in earlier chapters, you can have relativity as a virtual principle without it implying the non-existence of an absolute substrate[208]. We apply this principle directly in the dataverse. Mass is only relevant for *processes* in space and time. Statically, mass has a negligible dynamical significance, with respect to Newton's laws of motion: stationary masses also produce a gravitational field that induces a very weak acceleration. This tells us something about the nature of mass as an interior process: it can't be a purely local phenomenon either, else it would not reach out through space to induce accelerations. A massive superagent is actively sampling surrounding superagent regions to grow connections that we think of as the gravitational field. If we stop time, that field has no meaning either. Its the curious thing about spacetime, which makes it tricky to think about—it's hard to imagine space and time without each other.

The gravitational mass, also identified by Newton, plays an almost totally disconnected role from the inertial mass in Newton's physics—linked only by the concept of force. Why would a dynamical view of mass lead to gravitational attraction? It was Einstein's great accomplishment to show that gravity could be viewed as a distortion of spacetime processes—the very meaning of a force. In popular science, Einstein's gravity is depicted as a curvature of space, but this is wrong. It leads to a curvature of spacetime (or processes). If the bindings that lead to mass could also reach out to neighbouring bindings and join with them, like plant root systems linking up, then new adjacencies could also form and draw locations closer by rewiring the adjacencies of space. This would likely be a very weak influence. It happens in the dataverse that processes effectively attract one another by sharing mutual attractions, as a form of covalent bonding

7 Dial mv for Mystery

that I'll discuss below.

There is no immediately apparent analogue of Mach's principle in the dataverse, because distant locations are not observable in the same way as the distant stars, so there is no obvious set of signals by which to measure motion. Perhaps the closest things to the fixed stars would be the thirteen global DNS root name servers that hold the Internet namespace together[209]. One could plausibly measure changes against the information in these fixed locations. Virtual processes in the computing cloud treat these as constants, but the dataverse also has an absolute substrate, formed by its physical infrastructure—analogous to the luminiferous aether. We would therefore expect the inertial mass to be related to the processing delay experienced by passage through certain regions of absolute space.

In spite of the rejection of an absolute space in physics, there is a precedent for the view of mass as a quasi-local process in physics too. Starting with the dynamics of materials, there is the so-called *polaron mass*. This is an effective mass that an electron or hole experiences as it moves around a crystal, induced by its own interaction with neighbouring atoms. Freely moving conduction electrons polarize the outer electron charges in the fixed lattice of atoms, inducing a shadow with a backreaction—like the wake of a ship moving the water. This produces drag on the electrons as they move around. It's like trying to fight your way through a bustling crowd versus dancing freely like Julie Andrews across the open vistas with a song and a spring in your step. The interaction with the neighbouring space gives motion an effective inertial mass to drag around with it. The mover feels heavier because of its involvement in the process of engaging and disengaging with the crowd.

The same basic idea is used in the Standard Model of particle physics to explain how the Higgs field gives mass to the families of fermionic particles. The Higgs mechanism in particle physics is rather similar to the polaron effect: particles induce a cloud of interactions with a scalar field that surrounds everything, like a virtual material and those interactions generate an effective mass for the other particle-field promises, like a tax on change. In a sense, the Higgs field plays the role of the luminiferous aether that Michelson and Morley searched for, but could not find, because they were looking on entirely the wrong scale. The principle is the same, even if the specific promises and agents are different. The Higgs field was detected, indirectly, by CERN in 2013; it is a quantum field unlike any other because it is a boson rather than a fermion. Its principal role is to give mass to the other fermionic particles in the standard model, as well as to block the long reach of some of the bosonic signalling particles, due to its interactions with them. Those fields behave like polarons, moving through the stationary Higgs field, with its uniform local processes playing the role of a crowd.

The Higgs field, in the Standard Model, is interesting because it removes the need for inertial mass based on Mach's principle of fixed stars. The Higgs field—a featureless constant background—becomes the effective absolute of spacetime by the assumption of its uniformity. It plays the role of a luminiferous aether on an agent level that remains unaffected by large scale movement of superagent clusters, because motion is virtual and on a larger scale than the processes that generate it. The interaction picture doesn't fully explain why Einstein would predict the mass of a moving body to grow as a function of its speed relative to light speed though. Interestingly, this is where Mach's principle comes back in a more local sense.

Einstein underlined the importance of observation. Can something actually happen if we can't observe it? Einstein showed that a remote observer sees the effective mass of a body increase to prevent it from reaching the speed of light. In fact, to prevent the observer from ever seeing it reach the speed of light. Does that mean that it really does reach the speed of light, but we can't see it? Or does it mean that the very act of observation puts the breaks on the acceleration by adding to its mass? It's observation by light itself that leads to a distortion of dynamics. So is observation itself mass? We need to be careful here. It's certainly the channel information rate of photons that leads to the observation of increased mass and finite speed, not human consciousness that limits motion. And it's not just humans that observe—we aren't making humans the arbiters of spacetime. Every interaction, including light from the body scattering off space dust counts as an observation—an entanglement of separated locations. Something is always 'watching' a body, unless it's completely dark. And if dark, if nothing is watching, is it really moving? These are unusual ideas, but also straightforward ones. Is the cost of apples rising is no one is buying them? We can set a fictitious price (+), but it no one accepts it (-) there is no cost. In the interests of having a stable picture of the world we tend to say 'yes' the price of apples or the mass of a body increases, but the answer is really irrelevant. It's just a story that never happens. Of course, photons should not have a privileged status either. If we understand the encumbrance of mass in terms of process, then simply 'being next to' a location through some adjacency channel may be enough to distort space and time. This has interesting implications for dark matter, alas beyond the scope of this book.

MASS AND THE DATAVERSE

Ironically, the Higgs phenomenon is becoming relevant in the dataverse too, imbuing computing processes with a tax, or mass, on an entirely different scale. In the dataverse, modern cloud infrastructure has grown additional layers of virtualization (modular process wrapping) in recent years. The many strata in systems, built layer upon layer, implies that the overhead of percolating through that network of dependent layers is becoming a problem. There is also a shift away from designing simple monolithic processes that are reliable and long-lived, within their own skins, to using clusters of many unreliable redundant processes in parallel in order to take away some of the risk of relying on a single localized process. The effective mass presented by these extra overheads is becoming an acute issue, because the sharing of underlying resources, through layers of protocols and software systems, is beginning to stretch the tolerances of technology design. What used to take seconds directly on slower physical infrastructure may now take ten on much faster hardware. Obviously this slows down dynamical performance, but it can also lead to faults and software failures, because most communication protocols are based on 'timeouts'—when a communication takes too long according to its own clock, the expectant receiver gives up—regardless of the non-local reason for the latent delay. The time delay has an effect of forming a 'potential well' analogous to the energy barriers in Quantum Mechanics, which prevent electrons from visiting neighbouring locations. But, instead of redesigning all the layers of software to account for virtualization more efficiently, the old systems are simply run on top of the new layers, adding more and more overhead—adding effective mass. To make matters worse, software engineers have recently begun to adopt a model of programming known as 'microservices', in which software processes are broken up into many small agent parts, which are spread out in virtual space, and held together by interactions called a service mesh. This forms complex molecules of agents that interact across their length and breadth. The cloud of interactions with surrounding services means that, when a process has to change its physical or virtual location, those relationships have to be moved too. The result is a bureaucratic burden—of cancelling and reforming process relationships. It's ironic indeed that the effective mass of cloud infrastructure is beginning to stifle the very mobility it was designed to enable.

Mass plays a prominent role in consensus determination for decision-making systems in the dataverse too, though obviously computer scientists would not use this language. Decisions frequently have to be made in distributed data systems, to understand whether all copies of a particular record are updated: this is the problem of data consistency, particularly important for version tracking and backups. In scenarios where data copies may be inconsistent, we settle the question of the 'correct' answer by insisting on there being a significant quorum of the participants to vote on what the correct value is. That's just like a meeting protocol for boards and governments, used to document, trust, and act as a tie-breaker in accepting outcomes of votes. By engaging a cluster of agents in a decision process, one adds a mass encumbrance for each voter in the quorum, which adds a significant performance cost for distributed computer systems: a dynamical mass that takes time to process. It slows down computer systems considerably and thus it remains a contentious issue in computing.

In biology, too, there are varying extreme forms of entanglement with surroundings: organs that are tied rigidly and permanently to their functional contexts and cannot be moved without being transplanted between patients (a very expensive process) are totally confining potential barriers. They have effectively infinite mass. There are also phenomena like tumors that borrow local resources in order to grow—they can move, but only by detaching and re-attaching those dependency relationships with local infrastructure at a new location. Their mobility depends on what they depend on.

Plants obviously set down roots and can't easily move. However, biology also has its share of virtual phenomena, like slime moulds, that have no fixed adjacency structure. They form weakly coupled superagents that can dissociate their bindings when expedient, and move to a different location to set down new bindings. With a wider scale perspective, one could also call animals low mass virtual processes: they form temporary bindings to locations where they feed or drink, but can more easily transplant themselves due to their weaker binding. As always, the multiscale richness of biological processes leads to great complexity and rich semantic behaviours.

Whether it is the addition of processes consuming time that accounts for increased mass, or whether it is a slowing of interior processes (a latency) that yields the exterior effect, may be a question of relativity. Both mechanisms are possible, and the lesson of relativity is always that the observer is the ultimate arbiter of experience.

Physics defined mass historically as a fundamental property of matter, rather than a process, but quantum theory has already shown that to be untenable in the need to renormalize the masses of particle fields to make sense of the quantum theory. The mass we observe is a dynamical quantity in any consistent view of spacetime.

MASS AND MOTION TO ALIEN OBSERVERS

Scale is a powerful illusionist. If we look over a long enough timescale, say by using stop motion photography—sampling snapshots of clouds or opening flowers at regular intervals, and aggregating the tiny moments into a smooth perception of reality—and then speed everything up, then the appearance of continuous, deterministic change to coherent bodies has an almost Newtonian signature. The mysterious qualities of momentum and mass can similarly be understood by zooming out to a kind of stop motion observation.

Let's conduct a thought experiment to illustrate just how perceptions can be scale dependent. Recall that observers who sample slow moving changes see them as smooth, whereas observers who change on the same or slower timescale as the phenomenon, may miss them altogether! To see a change on the scale of our human lifetimes, we would need an observer who could record images over many generations. Imagine, then, an alien from a distant planet looking down on the Earth (through envious eyes, as H.G. Wells put it, but not with a telescope but a fashionscope, specially adapted to see clothing fashion brands). Imagine too that there are two clothing chains of shops in a city where people live. Initially, the alien sees a mixture of the two brands of shop, almost identical, surrounded by a cloud of messenger particles (people).

As the alien observes, a force appears to act between the brands causing them to repel one another, like charged particles. The alien postulates that like-products repel. Since the clothing stores promise the same kind of product, they tend to repel one another, or diffuse outwards, because they want to move into areas where they have exclusive access. They begin to move apart unless there is something to counterbalance that. The alien sees that the shops belonging to the same brand diffuse apart over many years, but they never move too far from each other, while the different brands have the same behaviour. So the brands behave like coherent bodies formed from many individual stores—but two distinct bodies that repel one another.

He postulates a kind of nuclear attraction that keeps the stores of the same brand together. A clothing chain is bound together by ongoing relationship through its promises to belong to a single corporate identity, so the shops belonging to the same brand will be held together by mutual interest, a central office, factory distribution network, etc. They cannot move too far apart because they are attracted to a 'centre of mass' distribution point. The cluster of shops is not a location in the normal sense, it's a superagent cluster. The mutual reliance on maintaining the processes of the brand keeps them together, while the inefficiency of being too close together tends to drive them apart. The overall effect will then be that all the clothing shops tend to move away from one another, but the mutual dependence will keep each brand from getting too far apart, so they remain coherent but separate entities. Each brand will each be held together as a 'body' by the co-dependence of their distribution networks, but they will repel one another, because they don't have any shared resources. The alien reflects wistfully that, from the great distance, the pinpricks on the Earth look much like the cells and atoms under his microscope.

The shopping brand story resembles how charged particles appear to us. Bodies with the same charge (promises) repel one another, based just on the distinctions we call promises. They could coexist by changing their promise so they no longer perform exactly the same process. To diffusion of types and affinities could be one way to imagine the illusion of force. But what about inertial mass? How fast do brands move apart? What causes them to accelerate quickly or slowly away from each other?

Think again of a celebrity who tries to walk through a crowd of fans. He or she is slowed down by all the interactions with fans and onlookers, talking or holding on to them. The presence of the celebrity 'polarizes' the cloud of people: fans are attracted and non-fans are repelled. In order to move, the celebrity has to effectively drag the cloud of fans along for the ride. This is how light is slowed down in materials, and the aforementioned electrons swim through a kind of treacle of displaced electric charge in materials. If the celebrity does not try to move, the mass has no role to play in slowing them down. It's only when they try to leave that the cloud of interactions plays a role. Going back to the clothing brands, the exterior interactions each brand binds it to processes from which it can't easily escape: leases on factories, shops, employees, contracts, and so on. These are not things that can be quickly or effortlessly transferred from one place to another, because they depend on the interior processes of other agents, whose clocks may be running at an unknown rate. As long as the brand depends on them to keep its own promises, they will hold it back. These bonds are called 'dependencies' in software systems, and 'ecology' in biology.

To observers on Earth, none of these large scale effects are apparent, because we sample too often and live too short lives to notice the trends, but our alien fashionista sees all this clearly. The clothing brands experience a preference to move away from one another, with a sluggishness formed by its interactions with the cloud of nearby agents. One by one, the individual shops will begin to move, and the collective cluster begins to move off, but the alien can't see these small changes. It costs a lot and takes time, according to our busy clocks, but the alien has time to kill. The acceleration and speed depend on the strength of the ties that bind them to their original location, and the pressure exerted by the environment of shoppers. First, they might move down the street, then across the city, perhaps to the neighbouring city. Finally, they might move to another country, and to Mars, where their brand promises find a more willing acceptance in the local environment. Now the movement is noticeable to the alien.

Later, the alien might see two other such brand bodies approach one another, attracted by the promise of a new shopping district. Looking like blobs formed from smaller particles, they collide, and then slowly slink away from one another like two amoebae. The masses don't look like rigid bodies, but they are coherent entities, bound together by inner forces the alien cannot observe. All the alien sees is that macroscopic average migration of two bodies, colliding and repelling one another. The speed and directions they take might be regular enough for the alien to codify them as 'laws of motion', knowing that the reasons and the details were hidden from view. In order to maintain motion in a straight line, we would have to postulate a process that could remember the direction it's going in, which itself has an inertia or resistance to change. Motion is an exterior process, but it must depend on an interior process that remembers a non-local vector, pointing in a direction that it couldn't otherwise know anything about.

This is not exactly how we think about uniform motion in a straight line, as Newton described, but this is the effect of scale. We are never asked to confront the enormous differences in scales in our daily experience of the world. The difference between the brand stores, the subatomic scale, and a billiard ball is truly staggering: far greater than anything we could even begin to approach in cities or information systems on Earth, So we are stuck with trying to understand the principles over rather smaller scales, and extrapolating to perceive a connection.

From mass m to momentum's process p

The picture we have of mass is something like the "baggage" we carry around with us in life. It's the ties and relationships can't easily be escaped, because they are on-going dependencies. They are necessary to a process, and therefore they slow us down if we need to move on. Although that sounds a bit cheesy and clichéd, I think its an accurately scaled picture of what inertial mass represents.

In Newtonian terms, momentum is more complicated. Momentum represents the impact felt when one is struck by a moving mass. With our microscopic view of mass as a crowd of relationships, 'striking' or 'colliding' starts to sound unclear. We can no longer think automatically of billiard balls—a slow motion event, like watching two crowds mingle, could be a more accurate depiction. But let's return to the idea of processes as being promised properties of spacetime rather than being a completely separate 'matter' phenomenon.

As far as we know, 'empty space' does not feel an impact when a mass ploughs through it[210]. The Michelson-Morley experiment discounted this, even though we also know that what we call 'empty space' is actually far from being empty. So mass apparently only strikes other masses in the Newtonian sense. This also needs to be explained. Does it imply that mass has something like a 'charge', i.e. a promise or a semantic label that is recognized by other locations with mass? Something which establishes a special handshake between masses, allowing them to interact, but which is absent from empty space? Well, if mass is an on-going collective process, rather than an inert property, why would this be? How does momentum get transmitted from one collective location to the next? The obvious answer is the very processes that lead to mass could themselves interact to couple the binding processes. Unlike electric or fashion charges, masses attract one another.

If we think of mass as the process burden of interacting with an environment, carried along as a collective moving influence, then we need to think quite carefully about what we mean by the motion of a mass. Does an electron drag its mass along with it (like a turtle with heavy shopping), or is the mass more like the wake of the turtle as it fights its way through spacetime infrastructure? In other words, are we witnessing motion of the first, second, or third kinds?

For a regional process to move, relative to a set of spacetime agents that can form the attachments, the process needs to have an inner integrity of its own, even as it moves. It needs to look like a virtual process in the dataverse. This probably rules out the explicit gaseous motion of the first kind (see chapter 2) from certain laws of mechanics. It might also explain why bodies of gas don't collectively impart momentum on other bodies of gas: they only exert pressure on their boundaries. Einstein showed that gravity could be understood as the shape of spacetime processes, generated by mass—in other words, he showed that the connectivity of spacetime was identical to the force of gravity. By implication, states of space which exhibit mass, by binding to neighbouring properties with mass, have potentially two kinds of binding: the mass binding that causes drag, and the adjacency binding that says 'I am next to you'. Are these the same, or different?

According to the Standard Model, the Higgs field (whatever it might be) is responsible for the masses of most particles in the model, but is it also responsible for gravitation? Is it responsible for inertia? For that to be true, the Higgs field would have to cause spacetime to form superagent clusters representing the drag on normal fermionic matter, and also distort the adjacencies of spacetime agents non-locally to weakly shape them according to the observed patterns of gravitation. The Higgs field's role seems to be a link between gravity and the other forces, but it has a different character. Unlike the direct attraction of labelled charges, gravity has more of an indirect form of attraction through a third party.

This leads to an interesting question about the nature of gravity versus the nature of other forces in physics. If any force is an ability to impart a change in a process that binds it—i.e. the effort imparted when uprooting any localized process—then how should we understand the other forces like electromagnetism, and the strong and weak nuclear forces? Do they form their own bindings different from spacetime adjacency? Do they interact with gravity too? Do they ride on top of gravity as they ride on top of spacetime? I find it remarkable that these questions remain hidden behind layers of mathematical technocracy, despite a century of looking for ways to understand them.

IONIC AND COVALENT MASS: WHY GRAVITY IS NOT LIKE THE OTHER FORCES

Everything going on within the universes we've been discussing revolves around the connectivity of agents and the messages they pass to one another. There's a small variety of forces, accounted for by labels we call charge in physics, or protein complexes in biology[211]. In physics, intermediate bosons, shadowy polarons, and Higgs fields play the role of intermediaries that communicate binding processes for all the spacetime influences. Yet these signals are not all passed in the same way: for some are directed intentional signals, while others are passive stigmergic go-betweens.

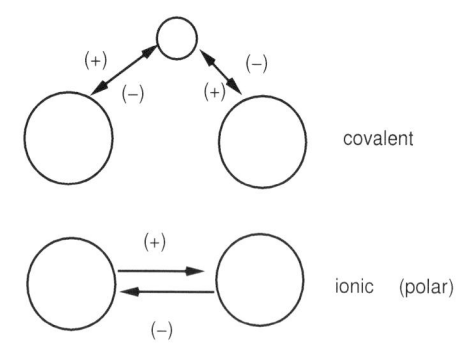

Fig. 7.1. Bonding is a generic spacetime relationship, whose structural patterns are recurrent across many scales and phenomena. Ionic bonds promise like plug and socket, while covalent bonds work by mutual attraction to a third party. From an information perspective, gravity's signalling channels represent the adjacency of spatial agents to agent. Other influences or forces ride piggyback over the channels of adjacency, via covalent messenger particles.

In chemistry, students learn that there are two ways to bind atoms together (see figure 7.1): one is called an *ionic bond*, in which an atom promises a spare electron or two and the counterpart promises to that it can accept the spares. This forms a natural directed lock and key 'promise' that forges a direct adjacency, as discussed in chapter 4. The other method is called a *covalent bond*, and involves a mutual attraction to a third party—like when two people know each other through the intermediary of work, or because they belong to the same club, not because they offer one another something directly. This geometry is universal in character, not merely an analogy to something that happens in chemistry. Covalence in chemistry is a special case of the general interaction geometry.

In the physics of the Standard Model, the seeding of mass around the Higgs fields is a form of covalent bond. Whatever labels spacetime agents as massive clusters, binds them through the intermediary of the Higgs fields—as if the Higgs fields are universal fans to the celebrity particles. The seed acts as a catalyst to push reluctant agents into a triangle interaction, which attaches a cloud of interactions or a network of promises to congeal around it, and form the tangle of interactions that leads to its dynamical mass. Particle detectors observe these clusters as particles. The Higgs field in particle physics plays this role, as does the phonon field in material physics and superconductivity. The appearance of an effective mass, in turn, hinders certain processes partially or completely, distorting other processes and leading to effects like superconductivity, semi-conductivity, Bose-Einstein condensation, and more. They are examples of feedback phenomena, muddying the waters of time and space with eddies of interaction, leading to all kinds of physical phenomena.

Not all processes in a spacetime can have mass. It makes sense that any processes playing the role of signals would be massless, and neutral with respect to the promises they communicate. If that were not the case, massive superagents would never hold together, and charged particles would be deflected by themselves! That would not lead to a stable universe. In physics, this seems to be true: neutral and massless intermediate bosons play the role of messengers for the semantic information, and thus remain unencumbered by the need to pass messages of their own. Think of a postman who constantly had to mail himself messages to remind himself about what he was doing.

Applying a dataverse perspective to physics, we might say the following: being able to separate process information (spacetime) from the functional modular channels (private for electricity, magnetism, strong force, and weak force) is the way one would maintain locality, because The separation of interior from exterior processes, which suggests the possibility of there being 'particles', needs separability. Somethings need to *not* interact. Too many messages in all directions would lead to a maximal entropy soup. However, one would expect this separation to be a virtual process on top of underlying exterior spacetime connectivity, because the elementary structure of spacetime has to support both the interior and exterior structures. Could it be that the Standard Model should be seen as a kind of virtual chemistry layer on top of gravity's spacetime infrastructure? This is implicit in the semantics of spacetime processes, but it is not the received view amongst theorists today.

Intermediate messenger bosons in physics, like the photon and gluon, represent directed ionic interactions. The Higgs field is a covalent interaction, mediated by an ambient background field. Spacetime may not need a distinct intermediary for the gravitational force, like a graviton. It was only assumed because of the model of forces. If there were intermediaries for gravity they would also feel gravity and there would seem to be an infinite positive feedback of gravitational activity. Indeed, quantum gravity is non-renormalizable (pinned at a certain scale) and plagued with infinities when one tries to imagine messages between the messages. It would make more sense if gravity's adjacency bindings were of a directly ionic character without the need for further message turtles[212].

Just as time is distinct from the other dimensions of space, spacetime adjacency (which Einstein told us is gravity) is singled out as a special information channel amongst the other virtual ionic channels commanded by the Standard Model forces. The phenomenon of mass is not like those at all: it's a covalent association that leads to process distortions, like drag and latency, for all other processes. Spacetime adjacency remains a privileged binding, as it is with respect to *that* embedding space that we measure position and time, and therefore motion. Whatever other dimensions or degrees of freedom that might carry messages for the other forces, their effect is observed thanks to the promises of spacetime, including inhomogeneous gravity. That's why gravity may never be unified with the other forces, in the way we've imagined, and why time will never be a proper dimension like space, except as a mathematical artwork.

Mass is a concept invented during an age of Enlightenment. It explained observable phenomena with amazing simplicity. Today, we know that the underlying explanation of mass is much more complex and general than the attribution of an effective number to a body. It's a bit like attaching a price to something in a market: the price is the result of a history of interactions that determine it, but the final role it plays is quite independent of that. Process encumbrances can happen on any scale, from the bottom to the top. Imagine if every packet of data sent across the Internet had to be approved and signed off manually by each computer it passed through, in a nightmarish form of bureaucracy. Well, that's what we call a firewall proxy server. Imagine if every sperm cell had to call a taxi to reach its egg, or be carried on the back of a bee to fertilize its mate? If we think in terms of processes and the promises they keep, those modular phenomena that try to make the world look diverse and separated seem less mysterious after all.

CONSERVATION OF MOMENTUM

According to Newton, when a billiard ball strikes another, the two balls transfer some of their energy and momentum to one another, sharing it out so that the total is conserved. What does this momentum conservation mean from the viewpoint of information? The law of conservation of momentum applies for idealized rigid bodies or particles, in a continuum spacetime, but it is approximately true also for macroscopic objects like squishy balls and compressible cars too, as long as they don't conceal momentum by converting it into interior state, e.g. when a car is crushed by a collision. Then the law fails a bit because the crushing of the material disperses the processes of change differently.

Momentum, in the sense of uniform motion in a straight line, is a strange concept indeed. How do bodies even know there are straight lines? A force that forms a field through space might give persistent influence at a location within the field space, and mass may give persistence to velocity (spacetime), but these characterizations explain nothing. If classical Newtonian motion is the scaling of a collection of transitions, then an interesting question arises: why should the locations of space along a trajectory cooperate to enable such long trajectories? This is a question that conventional thinking in physics and computer science would prefer brushed under the rug, but promise theory exposes questions like this to the harsh light of scrutiny, not least because they are of huge importance in the dataverse.

Taking a dataverse perspective, we can reexamine our assumptions. a natural question to ask is: why do moving bodies have permission to move in space? Internet packets don't automatically have the right to go anywhere. They are carried by voluntary cooperation. Letter and packages don't force a postman to deliver them So what authority does a process trajectory have over the resources of spacetime, or is a trajectory really an emergent 'guest' property of the cooperating spacetime agents along its path? In the dataverse, the latter is explicitly true: packets are received and processed at every location along their path. They could be stopped at any moment, even be discarded. In biology, cells move either by repelling others or being pushed around by other processes—say, a heartbeat. What makes physical processes so special?

When a ball starts rolling, does it have an unfettered right of access to its future path, or does it have to apply in writing to pass through checkpoints along the way? The locations forming a moving body's trajectory may be 'unable' or 'unwilling' to accept the passage of information belonging to the process. When you throw a ball, how do you know that each spacetime location from start to finish will hand the ball over to the next location? If we believe in locality, i.e. autonomy, this can't be taken for granted. The question sounds silly on the scale of a ball, but only because it defies experience.

Let's try a different process: when you send money, how do you know that each bank will forward the money without siphoning off all or part of it for itself? When you buy soup from the supermarket, how do you know that it is what the factory produced, i.e. is not diluted or poisoned? The answer is that we don't know. The promises made by sources of information are no longer valid once they are handed over to another agent. It's what we refer to as trust on a human level, but that trust is an expression of our belief in locality—the *a priori* independence of locations.

How trustworthy is your universe? The problem with intermediate agents is that we have no idea what they really do with our information. We can't really observe them directly. The meaning of locality is that each location only promises what it 'chooses to'. This has huge consequences for propagation. Every time messages have to travel from location to location, across a number of agent boundaries, the promises for give and take conceal interior memory and processes, whose information could be altered locally. How can we trust that the processes will preserve the integrity of information? This is a pretty important question in banking, for instance. Payments are made to banks, and the records are distributed around the world to different databases. How do we know the money won't get altered along the way?[213] It's worth emphasizing these matters, because we take them very much for granted in the world of physical processes, but we can't take them for granted anywhere else.

FORCE AND ACCELERATION

We understand a force to be an influence that can lead to a change in the velocity of a moving body, on Newtonian scales—that means it could be a change in speed or a change in direction. Mass is the intermediary that converts force into an acceleration. Without mass as force seems impotent. That means that electrical and nuclear charges are dependent on mass in some way, which also explains why there is no force without inertia.

Contact forces—pushing and pulling by touch—are an illusion of bulk matter. There is no actual contact between atoms: there is only the electromagnetic repulsion, which involves action at a distance by the electric field—and a force field a simply a region of space (empty or not) over which a force can be experienced by charge. The concept of force is all quite circular. In the alien fashionista example above, I suggested that a force might be a kind of process entropy, shaped by the matching give and take promises; but, the conundrum of how an empty space can exert a force is another stack of turtles that physicists try to avoid thinking about.

The concept of message channels and messenger particles, which inform charged bodies about forces have been invented to give a kind of material substance to what this means, but the picture isn't fully consistent because it doesn't explain why mass and charge go together. The labels we call charges are just a tautology to conceal our ignorance about forces—labels that we have no explanation for. A charge associated with a location in space means that it both generates and experiences a force, i.e. it is both a sender and a receiver of information with that label. The information guides a process. The picture of force as an irresistible imposition is already broken in this view. Force can't be both the push and the acceptance of the push: the force must lie in the binding of the two.

When a force is applied to say an electric charge, there is a channel of influence sent by a source and accepted by a receiver. The source must also be a charge, so this is just an information channel. The charges become 'next to' one another, i.e. adjacent with respect to that particular field process. The two processes might not agree on which locations are next to each other at all. At this point, the underlying adjacency we think of a gravitational geometry plays a peripheral role, but once the charges start to move, those gravitational mass adjacencies are implicated in the process picture. Figure 7.2 shows a common arrangement of a parallel plate capacitor, with a net positive electric charge on one plate and a net negative charge on the other. The charge imbalance leads to a region in which there is a force field—a region in which any charge in between the plates would be accelerated. This is a long standing model, which is more descriptive than explanatory.

In quantum field theory, the channels are represented by an intermediate boson, like the photon for electromagnetism, or the gluon for the strong nuclear force. All agents that carry charge also carry mass, so all particles in a particle-physics sense are charged with gravity. Einstein, and those who followed him, taught us that a force is a relative effect that can be understood as a distortion in the channels of adjacency between spacetime agents. This seems reasonable, because a force is an acceleration via a mass, and an acceleration is a distortion of spacetime in which a non-local process sees reduced clock time across consecutive locations of its trajectory.

But Einstein also showed that the distortions of adjacency channels between points *are* gravity. So the other forces create an effective gravity on an amplified scale. This only makes sense if the other forces operate on a scale much larger than the elementary units of spacetime. Spacetime has its weak covalent influence, just as atoms can feel direct ionic attraction and weaker covalent influences in chemistry. The semantic roles of the different forces on matter are also somewhat different. Some forces influence the motion (electromagnetism), some effectively constrain the distances between charges (strong force), while others lead to transformations of state (the weak force). Our concept of force, in modern physics, is quite different from Newton's concept.

It's truly hard to talk about force and motion without using the concepts of force and motion—those ideas have been drilled into us our entire lives. Yet our understanding of them should be immune to redefinitions of scale and circumstance, as we zoom in and out of processes. Information offers a sort of neutral measuring stick for both quantitative and qualitative (metric and semantic) by which to compare processes. That's why it's so important to rethink what we use as the foundations for those concepts. As always, other spacetimes help us to see more clearly.

In the bioverse, cells are an obvious example of the kind of cluster of processes that I've associated with mass. Cells are explicitly collective objects that interact as a cluster of parts with close surroundings, as they move through tissue or blood plasma. Cell walls mark a boundary around cells, preventing the interaction of interior and exterior processes, except through special channels and receptors on the exterior. Those receptor signatures act as a form of semantically rich, ionic charge labels.

There are several kinds of motion in the bioverse, but they basically amount to

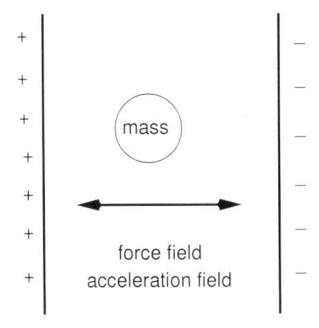

Fig. 7.2. No one really knows why forces cause accelerations. The geometry of force's semantics is shown here. Labels called charges tell us which locations in spacetime can generate a 'field' through an empty space, i.e. a non-local region in which other charged locations will experience a force. Mass plays a dual role as a charge that associated with a gravitational field and as an inertial encumbrance to resist other forces.

either swimming through a fluid, i.e. repelling and propelling themselves against some kind of background, or crawling along a static background scaffolding of other structures. These are the biological equivalents of jet engine and car transport—it's motion of the first kind. Amoeboid movement is the most common mode of movement in nuclear cells: a small leading edge, a main cell body, and a posterior protrusion cooperate in redefining the boundary around the cell in a coordinated way. The directed locomotion is based on a structural polarization of cells, primarily through the interior localization of proteins to specific areas of the cell membrane. Protrusions extend to reach out to adjacent substructure. The cell thus 'renormalizes' its length span along the axis of motion, relative to what it crawls on, stretching and pulling back together to keep its boundary integrity. As a unit of spacetime, a cell is not elementary—it fluctuates in both space and time, from an exterior point of view. It has the role of a virtual process, crawling along an underlying granular structure. It's interesting to ponder whether all motion has to be virtual, or whether there really are elementary agents that can support the necessary and sufficient processes. Biological motion is not restricted to individual cells—we can continue up the hierarchy of scales to find it building on layers of bounded structure. Trees, for example, are now believed to pass messages to and from one another, in forest networks, using spores and fungi as 'messenger bosons'—thus adding another scale to the picture of spacetimes made from agents and promises[214].

What is significant and interesting about cellular motion is that the geometry of cellular motion is encoded in the structure of the cells themselves. There is no need to idealize spacetime as pointlike particles and idealized vector fields; the finite size of everything is just part of the story. Cells are not spherically symmetrical balls; they are polarized units, meaning that they make different promises along different directions. Cells remember direction, like a vector field. This is why they can move autonomously: they have built in non-locality on a microscopic scale. If we treat cells as the unit of spacetime, then motion appears mysterious, but if we accept interior structure, whether visible or invisible, it is understandable. Cells that move, or exhibit directional characteristics, have a front and a back, like a car or a 'plane. The spatial symmetry is broken explicitly to select the direction of their motion. Memory is the only informational requirement for this to work, in any kind of spacetime agent. Once an agent, connected to neighbours along a number of links has received information along that link, it can remember it and retain a preference for that direction as a memory process. That tends to preserve motion in the same direction again, so there is a mechanism for continued motion in a straight line even from an autonomous agent. The trouble with that argument, however, is that—for most graphs—the incoming and outgoing links are not the same, nor are they related geometrically along the same axial direction, without some deeper reason to connect forwards and backwards (see figure 7.3); so, the problem of maintaining directional stability, on a large scale, is not solved by local polarization. Only the non-local polarization of spacetime, on a sufficiently large scale could lead to the Euclidean lattice structure we experience—like the field between charged plates above (see figure 7.2).

The spontaneous symmetry breaking involved in programming directions into spacetime can be provided by a number of mechanisms. Mathematician Alan Turing initially attempted to explain pattern formation in multicellular processes by chemical interactions with asymmetric molecules[215], fulfilling that promise of interior directional structure. If a network of interacting chemicals (in this case proteins) interacts, their fluctuations can be amplified into large-scale stable patterns, bridging a molecular length scale to a cellular or even tissue scale. This is a form of phase transition—one based on another game of Turtles, of course. One broken symmetry propagates to another scale, but what caused the original? As discussed earlier, there is ultimately no explanation

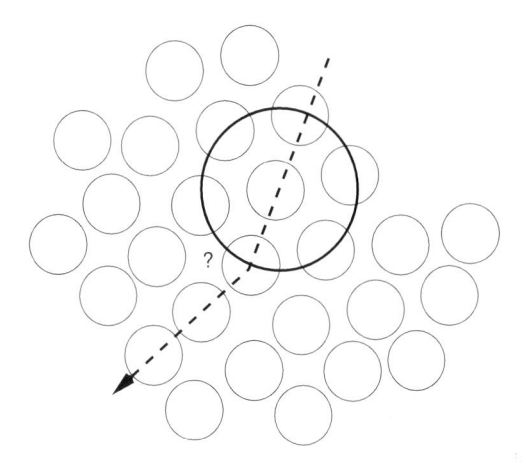

Fig. 7.3. A non-local structure that remembers the dimensions of an underlying spacetime may not find its memory consistent with the local spacetime structure, if that structure is not 'homogeneous and isotopic', i.e. the same everywhere. Local variations can be smoothed over by the size of a cellular agent.

for symmetry breaking and polarity of agents without the external input of information.

At larger scales in biology, a similar principle of polarization applies to organisms, known as 'cephalization'—meaning the formation of a head. Most organisms have a head or a mouth, which is associated with an opening for eating. It leads to a head-tail axis along the body of the organism. This polarization, on a larger scale, starts with the evolutionary adaptation to the exterior process of *feeding*. A non-local field thus applies a bias to the process over long times. For mobile organisms, that feeding axis is also associated with their motion, as they move towards prey. It's quite hard to imagine any meaning to motion without some prerequisite symmetry breaking. In this case, it's a give and take (prey-predator) relationship.

The idea of spacetime as a collection of cells might feel specious (no pun intended) and artificial to some readers. Why should we need to understand cells and their motion in terms of space and time? After all, we can now see processes and scales above and below those of cells, even though it was not always the case. Part of the answer to the question lies in forcing us to rethink spacetime in terms of processes, and to confront the many faces of processes head on. We don't know for sure that any picture we have of space and time is fundamental, so we need to compare all possible representations, on all possible scales, to fully open our minds to the subject. In that spirit, a natural bridge between the worlds of the bioverse and the more rigid dataverse is provided by a form of algorithmic behaviours known as Cellular Automata—computer programs that simulate populations of cellular biological activity inside computer memory.

Cellular Automata are models of spacetime in which the locations form a fixed lattice of cells, like a Euclidean grid or honeycomb structure. A process clock drives time along by updating the state of each cell in the grid, based on its probing of the state of neighbouring cells. For this to work as a cooperative tissue, each spacetime agent location promises to reveal its state to its neighbours, as if there were a Shannon information channel going in both directions. Similarly, each agent samples its neighbours on each clock tick. It then updates its own state based on a simple rule based on the sum of that information.

Cellular Automata were discussed initially by John von Neumann as a model of living things, and were developed significantly by Stephen Wolfram and others[216]. The interesting point here is that they are really only a model of spacetime: there are locations and there are changes that are computed according to some clock. Spatially, they are arrangements of agents (drawn as cells), usually in a regular pattern, each of which evolves according to entirely interior processes, but which can sense the state of the cells around it. Each spacetime agent thus has information channels to its neighbours and interior processes that sample neighbours and bring about change. Because of the implicit linkage, and the interior memory, large scale, non-local states form as structures over longer timescales than a single tick of a cell clock (see figure 7.4), from entirely local rules, as information spreads to influence more and more cells. In Cellular Automata, the virtual shapes and entities appear, they move around, and they can be destroyed, all within these simulations. They are mini-universes without any specific rules about motion. The motion is an entirely virtual process, which uses the scaffolding of rigidly positioned cells as a kind of operating infrastructure. Certain cells typically have to be activated at the beginning, as a boundary condition—which represents the input of information. The evolution of that initial input can then sustain the cell universe over multiple generations. Each cell has an interior time and there is normally a coordinated global time which represents generations of the cells. The *Game of Life* is a very well known example (see figure 7.4), with many videos available online. Even this simple model exhibits processes that are self-sustaining and which move at a constant velocity in a constant direction[217].

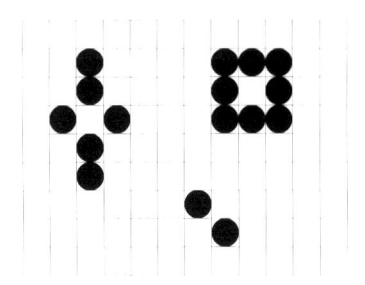

Fig. 7.4. The Game of Life is a cellular automaton. Cellular automata do not have to be put into regular lattices. Computation on generalized networks is a natural generalization, more suitable for information infrastructure. Nonetheless, combined with the notion of autonomous promises, cellular computation offers a powerful perspective on reasoning and stability, semantics and dynamics.

Like any discrete spacetime, the local adjacencies that transport information limit the speed of movement to one cell per iteration of the global clock. Moreover, because the cells are laid out in a grid, the space has a well defined (actually prearranged) notion of distance, as well as time. There are no processes that can tunnel from one cell to another without passing through the intermediate cells. Thus the space is totally ordered in each identifiable direction. Constellations of cells can move relative to that absolute grid, just like amoeboid motion converting transactional changes into something looking like a smooth movement in a straight line. As in figure 7.3, the straightness of the lines is a direct inheritance from the background adjacency structure of the grid. The homogeneous and isotropic grid is what allows uniform motion in a straight line. The total ordering of the points is what maintains a single measure of distance. These properties are not independent features—and this tells us something about the large scale structure of a spacetime that can support uniform motion in a straight line. It doesn't tell us too much about the small scale structure.

We see from Cellular Automata that scaled information models can simulate Newtonian motion in a straight line. However, the concept of a straight line, and other directional information is implicit in the conserved geometry of the agents or cells—so it's baked into the simulations from the beginning. The constancy of the motion is implicit in the conserved rule sets. Both the conservation of motion and direction are processes based entirely on information: the geometry and the interactions are simply constraints applied to states updating according to an interior clock. The Game of Life has a single global clock, but this isn't a necessity. The Internet itself is effectively a cellular automaton with many local clocks.

Cellular automata models are not just models for the study of cells and biology, they are also used as models of ordered collections of processes, such as those found in biology on a wider scale: tissues, forests, swarms, etc. They can also model material substances, like crystals and gases, chemical processes, crowds of people or herds of animals, and much more[218]. The successes of Cellular Automata acknowledge explicitly that a vast number of phenomena we separate into convenient modules are all children of basic relationships between space and time, with universal characteristics.

Having crossed the bridge from biology to the world of technology, we can compare the behaviours of more general processes too. Motion in the dataverse is an artificial accomplishment, built on top of electronic circuitry, and layers of virtualization, but we shouldn't discount that as cheating in any way. Motion is motion; processes are processes. The most common kind of motion, in the dataverse, is the intentional transfer of information over the Internet, driven by the background processes that run on the computers all over the world. Unlike many of the mobile cells in the bioverse, data packets in the dataverse can't normally move by themselves—they are carried by conveyor belt processes, like cells carried by the blood stream. Those processes guide the motion of packets along pathways in the dataverse, based on an *address*, or desired end state location—a feature of spacetime that's absent from Newtonian ideas.

In a real sense, there is a force field that guides packets as they move, quite analogous to the parallel plate force field in figure 7.2. The geometry of the Internet force field is not like the plates though—it matches the geometry of the Internet: a hierarchical and tree-like network, with loops and cycles that need to be avoided (recall the possibility of Internet black holes from chapter 2). In other words, there is a pattern to adjacencies induced to guide motion. The way it works is through routing tables and hierarchically ordered addresses. An Internet packet can be carried to a predetermined destination by setting up a special kind of charge relative to a fixed 'field' of influence. The charge is a specific semantic label, given by an IP address. Just like a postal delivery address—which guides a postman in a field of signposts and roads—a data packet, dropped into the force field of the Internet, is ushered along by routers and switches (computers that specialize in delivering packets) which acts as crossroads. When a packet arrives, each router knows only which direction in which to send the packet based on its address, because all addresses of a particular address are clustered together, and are therefore best reached by taking only one of the pathways out from the router. In the physical world, there are only three or four kinds of force, with three of four kinds of charge, depending on your point of view, but in spacetimes with more distinctions there could be any number. Just as biological cells can label themselves according to a vast range of protein labels, computer data can be labelled with any number of special tags that allow it to be custom-handled by the agents it passes through. The only assumption is that all the agents would be able to recognize and act on the labels in the same manner.

There is nothing corresponding to the idea of address labels in physical spacetime—no way by which to determine delivery to a particular location, because most locations don't have addresses at all: they are indistinguishable. What we mean by empty space is really just 'indistinguishable space'. Assigning explicit addresses is not too scalable: it's expensive. The Internet (version 4) has already run out of addresses using its original scheme, and has had to introduce a new set of labels and routers (version 6) to extend them. But there is another approach to addressing too, that forms the basis of the next chapter: semantic addressing. This is where we locate and measure spacetime processes relative to 'signature events', like landmarks (mountains) and catastrophic occasions (the year of the flood), etc. The universe is homogeneous for the most part, but we can still sometimes 'program' spacetime processes to deliver matter to a virtual location, by managing the constraints on a trajectory with forces at different times and places. This is how we send a rocket, say, to the Moon, or to Mars.

When we have numbers to crunch or records to process, batches of data are moved from process to process along a 'data pipeline', with each location transforming what it samples into something else. Each stage in the pipeline is a local agent that absorbs data and emits new data, performing a transformation step in a directed graph, but the entire process is non-local in total and involves the data moving in something like a straight path through the pipeline. Sometimes data might even circle back and be reused iteratively by the stages of a process. This is what eventually leads to computation.

The process is bounded with inputs and output, like scattering region (see chapter 4), and can therefore be initiated as either an advanced or as a retarded process. A system designer can choose whether the input triggers a series of cascading events that retain forward momentum, or whether a desired endstate reaches back and samples a chain of prerequisite stages, in a recursive backwards cascade, leading to the final output. In the first case, the pipeline acts like a conveyor belt—or in the worst case as an explosion of cascading triggers. In the second case, the process orders its dependencies 'Just In Time', riding a wave of backwards travelling recursion.

One kind of 'Just In Time' process that decouples the presumption of triggering by obligation, separating them into offer (+) and sample (-) is known as a Publish-Subscribe process (or PubSub for short). It's a stigmergic form of cooperation based on publishing and distributing data to be picked up by a receiver. Upstream agents in a process make their outputs available to downstream agents, but don't transmit them in a directed channel. The downstream agents can then sample those published data by probing for them. The result is a 'voluntary' or 'autonomous' transmission of a process that views momentum not as an absolute conserved imperative, but as an emergent side effect of a cooperative process.

Conveyor belts carry information around the Internet for the most part, but there are processes like amoeboid cells that can effectively move themselves too. Programs can teleport, by copying themselves and deleting the older copy, just as long as there is available infrastructure to move into. Computer viruses move by hijacking existing exterior processes to hitch a ride, just as biological viruses hitch a ride inside cells. These processes spread by attaching themselves to agents, hooking into existing mechanisms with unspecific acceptance promises (-), intended for something else. Viruses are not be recognized as foreign entities by those processes, because they make the same initial promise (+) as the usual process. Labels clearly play a major role in the routing of information by external forces.

In the dataverse, a destination address is like a force label, but one with a far higher granularity than we find for physical forces. This is what enables the enormous menagerie of distinguishable forms in both the dataverse and the bioverse. The exterior processes that pass the labels along can treat different labels differently, enabling a greater breadth of functional behaviours, with high degree of complexity. Label specification is the essence of how path selection, or routing can work—whether at the post office, in high speed Internet routers, or along Darwinian selection channels that evolve one step per generation. The features of motion that present themselves in biology and computing are: specific polarization of agents, and the interaction of interior with exterior processes. To create locomotion, you need to build a motor, which involves processes not inert parts. This is a conundrum indeed.

PROMISING THE DETAILS

The model of spacetime I've been using to compare different scenarios to a common standard is the promise model: a collection of agents that makes promises based on information to one another. Promises offered may or may not be accepted by a recipient. The result is like a network, in which spacetime could be arranged like a gas or like a solid, depending on those promises. An advantage to this model is that it lays bare the information relationships at work explicitly. In modern physics we've abstracted away the concept of processes to a large extent behind the veil of the field—a continuum model of a smarter spacetime. This sleight of hand is a deliberate concealment of our ignorance about the processes responsible—rendering them to the level of what Einstein called 'spooky action at a distance'. Since fields embed their own Equations Of Motion, they conceal interior processes, whose interior states are 'elsewhere'. However, this was not always the case.

James Clerk Maxwell (1831-1879), the physicist who unified our pictures of electricity and magnetism, built his first theory based entirely on spinning wheels: not on forces and points but on spacetime processes (see figure 7.5)[219]. The action of the magnetic field was driven by rotating currents and fields. Later physicists dropped this mechanical facsimile of the field, even though it worked, eventually disparaging the mechanical analogy in favour of abstract fields and equations. But let's be clear—the change was not real progress, only a change in viewpoint. We could even say that it was a deliberate mystification of the causal processes. In fact, it is part of pattern in theoretical formulations ever since: replacing processes and actions with a *generator* for action, by adding a veil of abstraction. This is what we would call a layer or virtualization in the dataverse.

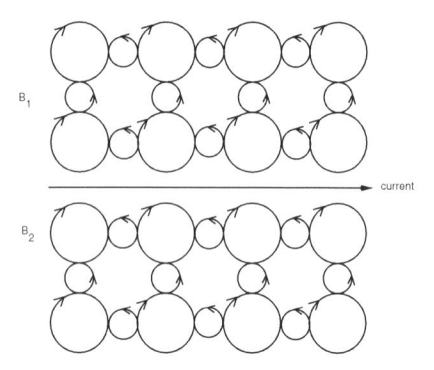

Fig. 7.5. Maxwell explained electromagnetism with a spacetime model based on rotating vortices before eliminating the picture from his equations.

Yet, one can't wriggle out of the idea of processes altogether by speaking of abstract symbols. What we can say is that the specific representation of Maxwell's vortices need not be exactly true. The question is whether progress is made by obfuscating process with abstractions. That path has led us to extreme mathematizations of physics that has contributed more to mathematics than to physics. There must indeed be a process at work, but it needn't be a tiny copy of a spacetime with rotations, even though the effect may be the same. An analogy might help to swallow this. We know that computers can generate transformation processes that appear to be rotations on their screens, but which look like entirely different strings of bits in their internal memory representations. The lesson is that appearances are in the eye of the beholder, but processes are the unavoidable reality of information changing. Almost a century later, the well-known physicist Richard Feynman developed a diagrammatic causal view of virtual processes for Quantum Mechanics too, now called Feynman diagrams. This was later shown to be equivalent to a field formulation developed by others—for many the diagrams are only representative of 'fake processes' generated by a procedure for calculation, but can we really say that they are more or less physical than the total abstractions that eliminate networks of interactions in favour of inscrutable symbols? I think that is hard to argue[220].

A similar case of representation also confounds the issue of a quantum property known as 'spin'. Spin is an interior property, predicted and observed for electrons, photons, and other 'particles'. It was discovered from quantum mechanical theory, where it arose almost in a form virtually identical to the representation of exterior angular momentum (like the orbit of a planet around a sun): yet it appears to be entirely interior to electrons. In Quantum Mechanics, there is thus orbital angular momentum of electrons moving around a nucleus, but also an intrinsic spin of an electron about its 'own axis' (a special axis that can only be 'up' or 'down'). If you believe particles are points, an axis of spin doesn't make too much sense—but, if you believe in particles as spacetime processes, then it can. The obvious answer to this paradox is that i) the electron is not a mathematical point in a Euclidean continuum, but an agent with interior processes—a *spinor* whose size is undefined, and that ii) those processes detect and influence the geometry of the exterior environment. On this point, I can't resist mentioning the subject of *twistors*. In the 1960s, mathematician and physicist Roger Penrose, came up with a model of spacetime based on a complex number representations, in order to solve certain questions in quantum theory. His primitive agents of spacetime were not simply inert points, but spinors (complex degrees of freedom concealing a spin-like process) that he called twistors. His spacetime agents could have an interior orientation of up or down.

The simple binary duality of spin states is tantalizing: up or down, zero or one, give or take. I confess that I've often wondered if there is a connection with the Promise Theory view of spacetime as matching 'give (+) and take (-)', or with the transactional interpretation of Quantum Mechanics, with its offer (+) and acceptance (-) waves. For now, that question remains unanswered as far as I can tell, but I suspect we haven't heard the last of the issue.

Charge to the rescue?

In classical Newtonian physics, mass and electric charge play similar roles, with respect to different background fields. During my days as a physics student, it was not uncommon to hear people compare mass with electric charge. The reason for this comparison is the force laws for electrostatic force, and gravitational force:

$$F_q \propto \frac{q_1 q_2}{d^2}$$
$$F_m \propto \frac{m_1 m_2}{d^2}.$$

Apart from the name given to the constants of proportionality, the force between two charges is proportional to the product of the two charges (multiplied together) and inversely proportional to the square of the difference between their centres. Similarly, the force of attraction between two masses, is proportional to the product of the masses and also the inversely proportional to the square of the distance.

But, while mass and charge play the same role in these force laws, there are obvious differences too. First, charge comes in two flavours: positive and negative. Opposite signed charges attract, while like-signed charges repel one another. Masses are always positive and always attract. Indeed, there is no negative mass, gravity is always attractive.

The suggestion is that gravitational force is not really a force in the same way that electricity is. Rather it's a figment of the tendency to bond *covalently* rather than *ionically* through intermediate glue agents, like the Higgs field, or scaled phonon feedback. The similarity in force laws is fortuitous, but as Einstein showed, the force of gravity is absent in free-fall, and the force we feel on our feet is the electromagnetic force preventing us from falling into the spacetime distortion produced by the mass of the Earth. In other words, gravity is a relative illusion of the geometry of space and time, which is related to the relative shape of location adjacencies and the sampling rates of their interior clocks. A bend in spacetime (as in figure 1.2) accounts for the dynamics of acceleration and thus has the characteristics of a classical force. The encumbrance of the mass distorts time.

ACCELERATION BY SLOWING YOUR CLOCK

To better imagine how forces can relate to the interior characteristics of process, we need to revisit the observer view of motion. Who can measure force and an acceleration? Going back to the picture we used to define velocity in chapter 5, suppose we try to define acceleration in the language of information and graphs. In figure 7.6, we see the geometry needed to measure an acceleration. There are three agent nodes: S a sender of some information, an intermediate node I, and a receiver R. Since the distance between all points in a graph is 1 hop, we are interested in the time Δt_1 it takes for the first hop from S to I, and then the time Δt_2 from I to R. If Δt_1 and Δt_2 are different, then there is an acceleration (or deceleration). Now, the observer, at the bottom of the figure can only observe these events if the agents are kind enough to oblige by sending a signal back to the observer indicating some time at which they had the information.

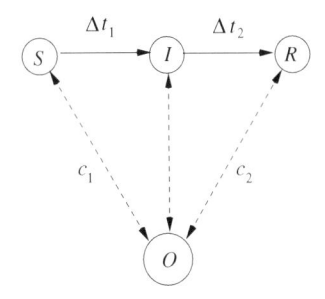

Fig. 7.6. Acceleration in a discrete graphical spacetime.

By Nyquist's theorem, we would normally expect to miss that event altogether unless the sampling rate, at which signals are sent from all the nodes along the path to the observer, were at least twice as fast as the transition rate or rates between the agents along the path—*and* the observer was able to receive and process these signals in the same order as the events happening along the path. This seems like a lot of caveats, and that's true. By Nyquist's rule, an observer might only observe such a transition if capable of sampling at twice that rate, but that assumes the interpolation of a continuum of points, which we don't need at this microscopic level. So there are two choices for observing an acceleration. Either the observer and its motion is part of the clock, in which case every change advances the observer's clock by one tick. In this case the observer can only measure one speed: 1 unit hop per 1 unit tick of time. So the observer sees a constant rate, like the constant speed of light in a vacuum. This is a consequence of the detailed balance of changes happening in space and time, with limited degrees of freedom. There can be no acceleration. Alternatively, the observer has its own interior time source and does not rely on the ticks from the motion to detect the motion. It can sample at least twice as fast as the changes of state occur somehow (this must be fortuitous, since there is no causal connection between the non-local remote velocity and the interior clock of the observer: the only connection between the two is the very signals that transmit the position of the state we are tracking). The second case allows for the possibility of measuring a speed, which is not precisely equal to the natural velocity. Since the observer has interior structure, this observance of non-standard speed can only happen above some critical scale, when the observer's clock can tell the difference.

But what about if the observer is being accelerated? Then every signal around it is part of its clock. The more acute the acceleration, the greater the differential rates on each consecutive location along the axis of the acceleration. The non-local effects result in the sense of compression or feeling of a force.

In computer networks, it is these details that make building a consistent picture of causation very difficult. In a computer network, acceleration can happen as information passes through routers and switches, which are the analogues of the spacetime points. There, congestion mass (too much signal traffic to deal with) causes delays, because the nodes have a fixed rate at which they can pass signals along. Queues build up and interior time slows down per exterior process. In a computer network, the times Δt_1 and Δt_2 can be different. If the observer is constant, then the only reason for measuring a different time is that the agent nodes along the path withheld the information for different durations, relative to the observer's clock, i.e. if the information had different 'inertia' or 'latency' at S and I. In computer science we can only accelerate a dataflow by speeding up time, or increasing the number of ticks. We cannot alter space, and transitions are effectively instantaneous across single wires. Transitional motion is fundamentally unlike velocity in a Newtonian sense, but on a large scale the effect is equivalent.

WAVE-PARTICLE MOMENTUM FROM THE BOTTOM UP

Apart from the very specially constructed cellular automata, what we don't find in the dataverse is any analogue to the Newtonian rule about continued motion in a straight line. Only waves possess an obvious process clock by which to sustain their motion unabated. Waves are processes that disturb a medium—either a material field, or a force field, like the electromagnetic field. Waves carry their disturbances with them as they go, storing energy in a potential field and releasing it into a kinetic field on each crest or cycle of the wave. The wave moves slightly as this happens due to the lag and the physical extension of that oscillation. This tells us that waves use spacetime non-locally in order to move. Point particles can have no such mechanism, so from a process perspective point particles seem to be ruled out.

In popular—and even professional—writings about quantum physics, the conception of particles, waves, and quanta is mystified and muddled in an appalling false narrative. We continually ask: Is matter particle or wave? We could equally ask: is water a wave or is it made of atoms? What a ridiculous and artificial issue. Waves are large scale (non-local) information about an underlying variable; particles are localized information about singular observations. And quanta? They are countable processes. There are three utterly distinct ideas–is it really so hard to get this straight? Do we still teach students this Harry Potter version of physics? Is it time for physics to grow up a bit? The entire notion of wave particle duality in Quantum Mechanics is an appalling mystification of the facts, held onto to sustain an almost religious belief in locality. Particles need some kind of self-sustaining process in order to move through spacetime too—a process that crawls along the scaffolding of the gravitational adjacency matrix.

Einstein effectively showed that the gravitational channel is identical to the spacetime adjacency channel: the shape of spacetime is identical to gravity. But that doesn't tell us whether the other force channels are separated from the gravitational channels, or whether they ride on top of it—either would be possible. However, the fact that they are mediated by photons and gluons that travel at the speed of light, and which are affected by the shape of spacetime, does imply that they travel on top of gravitational spacetime. This means that gravity has a privileged role as infrastructure underneath the charge-based forces, and thus it seems unlikely that it could be unified with them in the way that Grand Unified Theorists have expected. Just as time and space are not the same and are not truly unified by Minkowski or Riemannian descriptions of spacetime, because time is different, so it seems less and less likely that gravity should be on an equal footing with the other forces. Gravity and time are the odd ones out—or the odd one out?

THE ACTION PRINCIPLE FOR BOUNDARY CONTINUITY

One of the reasons we seldom confront the deeper meaning of Newton's laws of force and momentum is that momentum conservation has been elevated to a high principle in physics. This was initially a supposition. However, in 1915, the German mathematician Emmy Noether (1882-1935) proved a theorem which showed that there is a direct connection between the assumption of continuity in space and time and the conservation of dynamical quantities like momentum and energy. In other words, the assumption that the processes of spacetime do not change erratically, from location to adjacent location, is compatible with the assumptions that energy and momentum will be conserved in processes that span those locations. Similar arguments can be applied to the conservation of other process variables like electric and nuclear charge.

It was perhaps this result that shifted the focus from the Ptolemaic central observer focus on atoms and central phenomena, to the idea of translational symmetry as the important truth in physics. Ironically, however, the translational invariance so hallowed by theoreticians does not imply conservation by itself—only the absence of boundary information does. The discovery of variational methods as a formulation for quantum theory no doubt also played a role in this shift in thinking. Noether used variational methods to show the result, and variational methods have also become one of the most powerful and widely used techniques in modern physics. The variational method is sometimes called the Principle Of Least Action, and it's appealing because it plays into our received beliefs about economics of energy and its role in making things happen[221].

The 'action', as it is known[222] in physics, is an expression related to the total energy of a process, and is expressed in terms of the variables we expect to describe to process. The action principle plays a major role in quantum theory, electrodynamics, and all theories where the belief in the conservation of energy dominates our formulation of the 'laws' of behaviour. It's a beautiful construction, which has seduced many (including me): to use what we expect to persist about a system to show us how it can change without violating those expectations. However, the action principle's formal role is that of a generator—one that uses the differential continuity property to derive equations and conditions for motion under the assumption that nothing is created or destroyed on balance, As such, it is one of those process obfuscations—like the very concept of a field—that offers a beautiful economy of expression at the expense of detailed causal understanding[223]. It's a bit like describing the behaviour of a person by maximizing the happiness: it might not be wrong, it might be attractive to take that view, but it doesn't actually explain the reason for the happiness or the behaviour.

In his writings, Richard Feynman used the action principle as the basis of his 'path integral' formulation of Quantum Mechanics. He interpreted the variations in paths by saying that electrons did not only take different paths, they actually took all possible paths at the same time! The time he was speaking of was Newtonian time—that of a godlike observer, as is usual in physics. But, if we re-couch his thinking into a scaled view, then it only emphasizes his picture—making it more plausible. What we call a particle trajectory can be the result of not one process, but many overlapping processes. Particles are not elementary observations, but superagents, just as non-local as the trajectories they undertake. Because the clocks themselves interfere, for each agent along the path, so the result that finally persists, over measurable timescales and for an exterior observer, is the paths that are common to all processes: those that don't change during the variation.

Technically, the real meaning of variations is to expose the large scale consequences of spacetime continuity. If you don't know the details, they aren't important. If you do: when we try to vary the action of a process with respect to different parameters, we probe its invariance under those perturbations, but always under the shadow of space and time, and at boundaries. By varying the action with respect to a chosen parameter, and asking when its variation can be zero, we ask: under what restricted set of circumstances would the process action not leak information? The variation yields a set of equations that determine when the continuity is maintained on the interior of the processes described by the field equations (i.e. Equations Of Motion). This is the part that varies the least amongst the paths on average, so—as a spacetime process, rather than a pointlike entity—classical trajectories find an explanation in terms of a renormalization of what a path means. The separate condition for continuity when crossing an exterior interface or boundary is, by construction, the complement of the condition for conservation of process.

The action principle is another example of the duality of interior and exterior promises being kept independently, relative to some kind of boundary line in space. The effect of the general principle is to predict the usefulness of assuming the conservation of what we account as energy, by choosing a description of spacetime that is continuous in its properties from neighbour to neighbour. There is a subtle difference between the success of these harmonious assumptions and saying that they are a direct and precise match to what is actually going on in the structure of spacetime. In a sense, the hallowed principle of energy conservation is a book-keeping trick that works because all its processes have been designed to make it work. This was part of the genius of the enlightenment philosophers, culminating in Newton's accomplishments. Most of all, it's remarkable that this accounting scheme really does match our observations, which might simply be an indication of how far away we are from probing the truly discontinuous nature of spacetime in physics.

In the dataverse, and in the bioverse, nothing is necessarily conserved unless we work quite hard to directly impose constraints on the local promises kept by its agents. For some, this means that computer science and biology are unruly and simply non-fundamental, but, in practice, this might be more a question of scale and the complexity implied.

From continuity to productivity

We've covered enough basic theoretical background now to be able to see how to apply cumulative information processes, with their give and take interactions, to scale the heights of possibility—to see what might emerge from processes through combinatoric complexity. Modular locality, with its interior and exterior processes, is a pattern we see repeated at all scales. It is particularly common in the worlds of technology and biology, where it offers a rather different perspective on what makes processes tick than mainstream physics does. Conventional Newtonian mathematical formulations don't seem to incorporate locality at all: solutions get 'imposed' onto a passive spacetime background, with a purely exterior concept of time. That viewpoint fails to explain the world on both large and small scales. Einstein's Relativity corrects for the mismatch of sampling for information delivered by finite speed observer channels. Quantum Mechanics begins to separate 'sampling' and 'process' from the global Newtonian theatre at the level of the very small (and thus very limited), but there is a strong sense that it hasn't yet found its proper form of expression. Each of these models feels attached to specific scales and fails in its applicability beyond them.

A key mistake one could easily make would be to assume that there is only one channel of spacetime active in the evolving dynamics of phenomena. There is more to be said on this matter in the next chapter. Information can be perceived in simultaneous channels, both dynamically by parallelism or superposition, and semantically using different labels and tags. Semantic information channels are the subject for the next chapter. Whether it's quantum superposition, photons measurements versus quantum teleportation, proteins and signal blockers, antigens versus hormones, or data and metadata, there are abundant examples of the processes whose full understanding is a superposition of processes across different information channels.

The question that links motion to more complicated and creative processes is: where does the boundary information that steers a process come from, and how is it passed along the graph of causal development? This too seems to be a scaling question: below a certain scale, there is insufficient complexity to do more than pass information from agent to agent, with errors and uncertainties (see figure 7.7). We can increase the complexity of information both for agents' interior processes and for the messages sent between them. Information injected

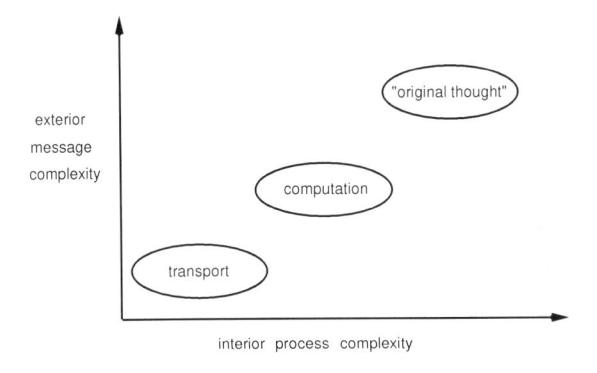

Fig. 7.7. Sustained cooperative processes spans a range of levels of sophistication from simple transport to 'original thought', depending on the complexity of the agents and the messages passed between them.

at the input as 'boundary conditions' gets passed along the unfolding chain of agents, through the passive message channels that are exchanged between active agents—as intermediate messenger bosons, transmitted data packets, signalling molecules, spores and fungi, and so on. The agents emit (+) and receive (-) these messages, and it is the combination of the signal and the agent that yields an effect. Primitive agents support primitive processes, like simple persistent motion. Once we reach a certain level of complexity the interior programming can transform inputs, and then transformations dominate over mere transport. This is what turns transport into computation, and reasoning— it's a straightforward story but one that is no less amazing for that.

We study the basic processes of space and time not only to uncover the mysteries of fundamental physics, but to apply them to technology and to unify our understandings of dynamical phenomena everywhere. Whatever the truth is about our physical universe on the smallest scales, we may never know for sure.

It's clear though that our imagination may have become temporarily stunted by the unreasonable successes of models by Newton, Einstein, Schrödinger, and the Standard Model—and may even have become caught in the more baroque mathematical talent contests spawned by the latter. The fact that we need a certain minimum amount of spacetime structure, and thus scale, in order to be able to detect any process, places a lower limit on how small we can get before it makes sense to even speak of the classical variables of observed motion, mass, and momentum. I believe this throws into question the usefulness of the entire dynamical formulation of quantum theories as fundamental theories. As long as there is no agent capable of observing classically inherited dynamical variables, they can play no role in the system of information on a small scale. That has to be true in our universe, in the bioverse, and in the dataverse.

But, doesn't the idea of interior-exterior concept of process dynamics already kick the can of spacetime's origins down the street? Simply put: yes, but this seems inevitable, and no worse than before. Our task is to make the arrangement of roles and concepts more plausible by extending the most elementary things we know to reach all phenomena—and the key to that lies squarely with information.

8
SPACES THAT SEE AND THINK

> He spoke of lands not far
> Or lands they were in his mind
> Of fusion captured high
> Where reason captured his time
> In no time at all he took me to the gate
> In haste I quickly checked the time
> If I was late I had to leave to hear your wondrous stories...
>
> –Jon Anderson, Wondrous Stories

We've reached a stage, in the description of space and time, at which we have a plausible—if unorthodox—model of bounded spacetime processes: one that applies across a wide range of scales, even across phenomena that no one would have dared to compare to one another in the past. It paints the basic element of any universe as a process (finite and discrete) rather than as a mathematical point or a smooth trajectory, and in the scope of that picture we can account for just about anything. I now want to move on from describing the themes and phenomena behind those processes to consider the role of their *semantics*. What can processes, and their scaling, mean for our understanding of materials, chemistry, machinery, life, technology, economics, and even society? Clearly, information and semantics have played a huge role in filling space with amazing phenomena. The evolution of life on Earth is perhaps the apex of that journey, and information continues to play an ever-increasing role in our lives at the level of human society too. Today, as we exploit information enriched technologies and create the illusion of smart adaptive infrastructure, so we interact with tailored artificial processes in nearly every encounter. The mechanics of those processes will set the limits for what our species may achieve in the future, so it seems worth studying in more depth. But even more surprising is the idea that the themes we've explored up to now could lay the basic principles one of the greatest mysteries of the natural world: consciousness—how we perceive the world around us, and think about it. As we begin to understand and develop the components of Artificial Intelligence, we also find the core themes of space, time, geometry, and information underpinning every technique there too. Naturally, as you'll have come to expect, the existence of modular boundaries plays a central role in this picture.

No agent is an island

As we look around the natural world, and even the technologies we fashion from it, a recurring pattern of behaviour stands out: information arrives from a wider environment, to a localized border—a ring-fenced location or a spacetime agent, like a place, a cell, a building, a brain, a measuring device. The agent samples the information at its boundary and alters its own interior state according to its interpretation of what it perceives (it alters its memory), as in figure 8.1. This is how atoms absorb light, how cells detect proteins, how you apply for a driver's license from a bureaucratic institution, and how you see and recognize things around you. In physics, we call this the 'measurement problem', or 'observer relativity'. In computer science it's called 'modularity'. In biology, it's the emergence of 'organisms'.

On any given scale, the complexity of processes, on the interior of such an enclosed boundary, may be lesser or greater than on its immediate exterior—because the boundary marks not only a separation of processes, but also a separation of scales. Up to this point, I've assumed that the insides of those boundaries were elementary and primitive, but now we are going to flip that around to see what happens when an interior process is much more complex than its immediate exterior. The results, very naturally, lead us to intelligence.

Where there are dense localized processes, on a small scale, and sparse distributed processes, on a larger scale—relatively speaking—processes can not only transmit but also transform information. The structure of an atom is busy

8 SPACES THAT SEE AND THINK

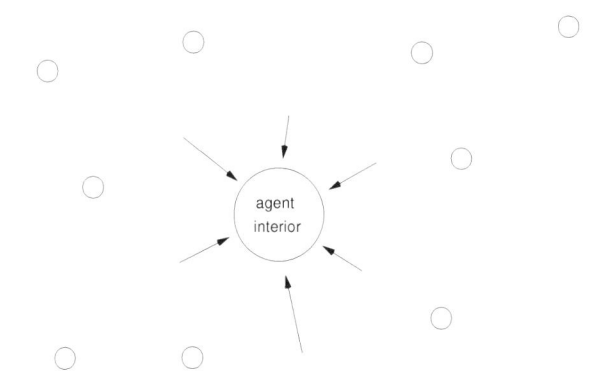

Fig. 8.1. No agent is an island—it's a receiver with some internal memory, which must be greater than zero, else energy could not be conserved.

with subatomic details, compared to the empty space around it. The processes in plants and animals are concentrated inside their skins, compared to the air of fluid around them. The complexity of a planet is much higher than the empty space around it—and so on. This is a pattern—separating interior from exterior promises—that localizes information for its stability and for the integrity of its processing. We say that spacetime memory is *sparsely populated*.

The *semantics* of the local behaviours are based on the give and take between the receiving agent and its pre-existing state or interior memory, and the external stimuli from senders. The interaction happens on the scale of those agents, but the significance of that interaction might lead far beyond the simple change of state, when viewed on a larger scale. Meaning propagates and expands hierarchically. The joining of a sperm and egg, for example, is a simple docking process and molecular interaction—but it is also the start of a much longer process of gestation, forming a baby, whose significance over longer times will affect kin, tribes, and society at large. Time passes on the interior through a combination of interior ticks and exterior tocks, and those scales interact strongly or weakly across the boundary depending on the relative significance of each. This simple model system, which has played a central role throughout the book, is the prototype for what we can think of as a *cognitive* system.

8 SPACES THAT SEE AND THINK

LIFTING COGNITION BY ITS BOOTSTRAPS

When we talk about cognition, in a day to day sense, we think of it as a human capability—an ability to perceive, reason, conclude, and respond. In other words, something much fancier and more sophisticated than the description above. It turns out, however, that cognition is just a process too, yet one whose more sophisticated aspects all have reasonable explanations in terms of that simple toy model above, scaled up.

The Oxford English Dictionary defines cognition as the process of acquiring knowledge and understanding through thought, experience, and the senses. By assuming that information arrives at a boundary, we are assuming a sensory capability, and by assuming memory, we assume the ability to store experiences. But what about thought? What I want to argue in this chapter is that the process of cognition—and by extension *thought*—have their deepest roots in the simple process of Shannon's information channel, scaled up through layers of complexity, with only elementary spacetime notions at their core. This is yet another twist on familiar theme of sampling information from an observer viewpoint, across a local boundary. From an evolutionary perspective it makes good sense, and potentially answers the conundrum of learning: how to get started without any prior knowledge[224].

The question that nags from an evolution point of view is: how did humans' advanced cognitive abilities, recognition, reasoning skills, and our capacity for abstraction pull themselves up by their bootstraps from nothing? If you don't know about faces, how can you learn to recognize them? If you don't know what food is, how can you learn where to find food? If we dial evolution back to the very beginning, the only stimuli were from phenomena that changed across space and time. The most basic interactive process is one that receives some input and changes as a result. From that elementary process, we can keep building until we arrive at complex thought. Even today, the most basic issue any living organism has to deal with is to respond to distinctions in space and time: to remember patterns and locations, to repeat journeys to important resources, and so on. For animals that can move around, mobility requires them to learn about relative motion; for plants, whose sources come to them by carrier, they need to recognize random arrivals of data. Both kinds of organism need to recognize concepts like 'helpful' and 'harmful'.

In the foregoing chapters, I've focused on what happens on the exterior of this type of bounded region—i.e. when a region plays the role of observer from the inside looking out. We've discussed how information can be connected by processes, how it's transported, and how its details are aggregated and summarized, with the help of a variety of lenses. When an observer examines exterior regions of increasing scale, details get dropped or washed out, in favour of aggregate summaries or averages—called 'coarse graining'. It involves the elimination of detail, or the approximation of information that leads to increasing entropy as details are mixed away into indistinguishable grey mush—which physicists tend to attribute to an arrow of large scale time. These perspectives can also enable the decoupling of renormalized behaviours at different scales, which leads to all important stability and persistent longevity of processes. Yet all this is the opposite of everything that cognition is.

In this chapter, I want to ask the complementary question: what happens on the interior of a boundary, when distinguishable information arrives from the outside, and is not merely accumulated in a formless grey stew, but is separated and archived in distinct 'piles' for later use? Rather than equilibrating away the distinctions into a grey oblivion, the answer turns out to lead to processes rather closer to what happens in our grey matter. The essential leap of faith we have to make, in going from the simple nuts and bolts of spacetime processes to the level of organisms, is to leave behind that basic prejudice described at the start of the book: that spacetime is just about outer space and Einstein— a more or less Euclidean continuum. We need to take on board the lessons of the foregoing chapters and think of the universe as being more like layers of entangled circuitry, then the mysteries of cognition melt away into a very plausible set of processes that range over scales from microscopic regions of space to vast expanses, and respond over timescales from nanoseconds to aeons.

"AI"

We couldn't discuss systems that think without immediately mentioning the field of Artificial Intelligence (now usually called simply 'AI'). AI has a long history, from science fiction to engineering, and has had something of a renaissance over the past decade, thanks to the computing cloud and technological abstractions it enables. The basic idea is to simulate intelligent behaviours with computers. Different people have different goals for AI, but it's fair to say that one of its goals is to mimic and scale up human reasoning, in the way that industrial machinery mimics and scales human activity. Of course, that presupposes that we understand how to scale reasoning—certainly a topic I certainly want to address here. Our view of artificial intelligence is based mainly certain prejudices and myths that have grown up about human intelligence, as it applies to a human scale—like, the idea of a brain being 'just a computer'. This often misrepresents the science we know. For instance, it's sometimes argued that because a computer (actually a Turing machine) can simulate any computable process, the brain must be equivalent to a Turing machine. This is a lazy inference. A Turing machine might indeed be able to simulate a brain eventually, over millions of years, if every detail were actually known. What the result doesn't imply is that the brain is made up of simple flowchart algorithms like a computer program, or that a computer could mimic a brain on the same timescale. The brain is certainly a machine, but it needn't be a Turing Machine.

The successes AI has had in playing certain games, like Chess or Go, are based on a mixmaster approach to human mimicry, not on developing an innate process of learning and thinking. Nevertheless, we are homing in on what key processes are at work in thinking and recognizing the world around us. Mainstream AI focuses on pattern recognition, based on a few techniques that researchers have stumbled upon. It typically neglects collective reasoning, such as Swarm Intelligence[225] that we see in ant and other insect colonies, as well as the reasoning processes that are implicit in evolutionary selection—just to mention a couple of counterexamples.

As AI has grown and developed, various metaphors have fallen in and out of favour to explain the human thinking. Probably the most influential, and perhaps damning, is that of the mind as a computer. The association of thinking with mathematical computation may turn out to be one of the great embarrassments of AI. It has led to many bold assertions and careless pronouncements about what is possible and impossible—analogous to the rumours about how it's impossible for bumblebees to fly, that circulated in the 20th century. Computing, as we understand it today, builds on the simplified idea of formalized thought expressed as *algorithms*. They are simple machine-like processes, based on small numbers of distinct states or 'symbols', together with a model of decision-making called *first order logic*.[226] Computers mimic human decisions—by trying to take short cuts with information in isolation. They don't really arrive at them in the same way natural processes do[227]. However, modular isolation plays a key role in computation—so there are clearly matters we need to put into perspective. A vexing example of this happened to me with my so-called intelligent 'smart home' agent, which can control lighting as well as perform some other tricks. Following a scheduled power outage, intelligently scheduled to cause minimal disruption at night, the system was unwittingly rebooted and was revived before its Internet connection. Its response was to panic, switch on all the lights, and cry out loudly that it couldn't talk to the Internet—as if this were an emergency worth waking the house for. Clearly, someone (beset with an absence of context) decided that this would be proper default behaviour, and coded this behaviour. A human might have reasoned with more consideration for the sleeping household. Indeed, a programmer could have used more consideration too. Today, AI is little more than a set of circus tricks that build on stimulus-response tables—quite difficult to call 'smart'. Our threshold for promoting simple machinery to the ranks of intelligent entities is built more on surprise and novelty than careful judgement.

In practice, what makes Artificial Intelligence an interesting area of research, in the context of this book, is that it specializes in understanding how space and time can be used as computational methods. Not everyone would express it like that, but I've found that insight to a fair representation of what goes on, and an extremely valuable for understanding thinking processes. We are nowhere close to being able to understand or mimic human thought—though we are closer to mimicking human behaviours. Today, researchers use AI's techniques to augment human senses, to give early warning of cancers from distinctions too small for our perceptions to resolve, to plot the practically invisible signs of disease in cells, or to see into atomic interactions of superconductors to identify the mechanisms at work. We are excited enough about the ability to see across the vast range of scales in our expanding world of interests, and to relieve the tedium of banal but demanding work, to avoid fully confronting the idea of a thinking machine and what it might mean.

Spacetime has a central role to play in artificial intelligence: reasoning needs to understand the relativity of spacetime because intelligence, as we understand it, takes that for granted. But it also needs to understand how processes themselves emerge to represent *thought* over time and space. Today AI's methods are based on absolute pattern recognition. Animal senses detect patterns like "Motion ahead" which is relative, but a home security system detects "motion in the fixed garden", based on a pre-learned map of absolute locations. Smart cars need to adapt to changing circumstances, and relative contexts, but they are currently unable to do that. AI is currently a bit stuck on the paradigm of sequential processing algorithms, because of computing's obsession with certainty. I'll argue that AI's current ideas about thinking, cognition and intelligence are far too narrow minded to explain animal intelligence, and that the real key to understanding these phenomena is to embrace a more sophisticated model of spacetime processes, built on all the phenomena we've discussed up to now. Without a framework of cognitive systems to underpin it, AI is little more than a collection of pattern recognition tricks, glued together with some recipes. To get to grips with the idea of spacetimes that think, we need to dig deeper into what cognition really means.

Cognitive systems

Cognitive systems may yet prove to be the culmination of everything we know about data processing, scale, and memory. Cognition is an idea that transcends spacetimes: from brains, to smart technological spaces, to organisms, and ecosystems. There is no doubt that phenomena within our physical universe experience cognition (we belong to that category), but it's more subtle and daring to proclaim that anything that qualifies as a spacetime might itself be intrinsically able to form a reasoning system, at some level of complexity and on some scale. Figure 8.2 shows a schematic diagram of how information

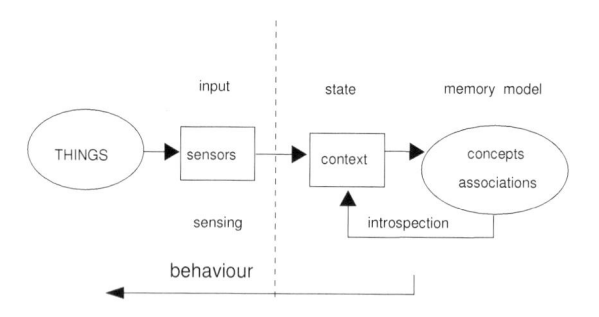

Fig. 8.2. The functional roles involved in continual sampling of observations, transmuting into interior states that represent concepts (memory model), from left to right. Feedback from sequences of scaled memories, from the right, can also cycle back around as 'thoughts', and be placed on a par with direct observation.

flows in a cognitive system. Information is sampled through sensors from an exterior, across a boundary represented by the dashed line, into some a set of states that track the exterior changes—I call this 'short term memory'. It represents snapshots of immediate experience, on the timescale of the sampling, and eventually gets used to 'name' phenomena and address them in the future. That short term context gets condensed and turned into long term representations of memory that build up an archival picture of persistent experiences over time, including the objects and things we learn to recognize. There is thus a *separation of scales*. Although the roles are drawn as separate entities, they might not be conveniently separable in the space they occupy. The picture is only schematic. Learning is the repeated experiencing and recalling of phenomena over time, creating a stabilization of memories, and connecting them to other memories by building overlapping connections that wire them together. What happens in learning is that one spacetime pattern representation—sampled from an exterior channel—gets copied into a new spacetime representation, on the interior, which is a version of the patterns it can perceive through its sensors.

In a cognitive system, interior processes that build on long term memories are continuously feeding back impulses alongside the incoming channel of sensory information. A cognitive system is polarized along an observation axis, as we discussed in the last chapter. Short term memory thus consists of exterior sampling and pre-processing of compressed summaries, alongside recycled storylines from long term memories, in a kind of 'stream of consciousness'— both steered by a short term model of recent context. So the short term processes of a cognitive system are a mixture of snapshots of the exterior and whatever 'experience' it happens to be thinking about, recalled from the archive. In other words, there is—from the very beginning—a mixture of processes on two different clocks: a sampling process and a recall process that are only weakly coupled. If they were strongly coupled, we would be entirely reactive organisms, like plants.

What distinguishes a cognitive system from a simple receiver of information, is that it has interior processes, which are active in interpreting and classifying the information—scaling it and archiving it, without too much loss, forming a *retrievable* representation. It combines memories associatively through the keys triggered by that awareness of a current state, based on the mixture of senses and introspection. Storing memories is one thing, using the memories to modify behaviour is another, but recognizing similar phenomena and being able to reproduce them from memory is an entirely different level of process.

THE SCALING OF ADAPTATION AND MEMORY

In the modern age, we've come to think of memory by the various technologies we use to assist us: computers and the Internet. Add to that our own failing wetware: brains and even books. But, we don't need computer chips or even grey matter to remind ourselves of the past. Memories of the past are all around us, in the places we visit and in the familiar objects we fashion and keep. Memory is everywhere. Memory is what is expressed by space.

Every time we change the state of a space—say the content of our homes—in response to external events or interior deliberation, the space learns something about the exterior and represents it. It remains there, as a reminder, until the trace is lost to other changes. Memories are exactly what I've been calling interior and exterior state—as long as we can observe the traces of past processes, we have a memory. Many processes are involved in memory: in storing and retrieving. Our memories are within us, and around us, and we the memory of other processes, from our DNA to our scars and injuries that document past events.

Memory encodes information on across many scales. For example, cell DNA and messenger RNA, along with other proteins, keep the long term memory of our species—as a branch of biological processes on Earth. The contents of those data are selected by the survival characteristics they endow to the end results of a complex building process, which in turn is based on a generational-social process that biases our chances of survival. DNA is memory passed on from generation to generation, at the level of an individual. It is sustained at the level of a species. A species is a superagent of organisms, expressed over space and time, that changes and evolves by a clock based on birth and death. Short term memory, on the other hand, lies in fortuitous matches of protein bindings in cellular biochemistry, as well as chemical signals, such as hormones, that switch those long term adaptations on and off. Together they promise the functions adapted to the cell's survival. The cell walls prevent the structures from decaying and the signals from being washed out by surrounding chemicals, providing a neat modular boundary to pin a scale of operation.

Biology is hard to imagine, because it acts on such a huge scale, but our homes are easier to comprehend. The colours and decorations in our homes reflect the outside world. When we look at old photographs from the 1970s, say,

8 SPACES THAT SEE AND THINK   339

we immediately remember the time and all its associations. More exactly, the short term state of a home also changes in response to the exterior: a sensor may turn on heating when it's cold outside, and accumulate food during cold weather. The thermometer is a form of short term memory. The peeling wallpaper is a longer reminder of the excess dry heat over long winter. The cans of food in your larder and the contents of your refrigerator represent memory processes too. Memories are also represented in the possessions that document family life. You hang a picture on the wall that was a gift from a friend. The year we bought chicken soup in bulk was the year of the flu epidemic. The household boundary protects the interior state from leaking away and becoming someone else's entropy.

Cities gradually adapt to the processes around them too: the pretentious Greek columns on many public buildings remember the history of ancient Greece. More practically, roles become ingrained. Ports are towns that grow out of trading processes, and are fashioned into districts to support different activities and cultures. Memories, in the form of slowly changing building structures, and rapidly changing populations with their social norms, reflect the outcomes of past processes, in the way they support the visitors and the residents, Towns build up around railways and rivers, shaped by their specialized promises to outside agents. The modularity of cities prevents them from diffusing away into mere countryside, and allows them to contain sub-agents with their own specializations—just as organisms contain organs. Memory is hierarchical—outer memories affect inner memories, and vice versa.

Clearly memory is complicated, and surrounds us at all levels. All of space is memory of processes past. Memories can be encoded anywhere, but interior memories are carried with an agent, and are available locally. Memories encoded in other agents on its exterior become unavailable if an agent changes its relationship to nearby agents changes, or they become unobservable. Cognitive processes are the basic template for how memories are encoded. We can use tools like Promise Theory as an organizing principle to unify all the apparently different mechanisms that lead to memories being stored and retrieved, across their many different scales[228]. The interesting conclusion of that work was that the concepts to build up a knowledge representation of an exterior space turn out to be rescalings and embellishments of four simple spacetime concepts[229]. In other words, there are only four kinds of connection needed on the interior of an agent, to support the encoding of information as memories that recall spacetime events on its exterior. But before we get to the four types, we need to get a feel for how connections of any type can not only form memories, but actually represent spacetime processes—because these processes are at work in biology and in the dataverse too. This is the story of how special structures become machines, with functions.

Simple cognitive systems, say like a camera, just record an image of what they observe directly by rearranging some data on the inside. The more impressive achievements of advanced cognitive systems involve abilities like learning, to generalize and recognize similar things, to form concepts, to refine concepts as we learn more, the ability to piece together stories from relationships between them, and so on. The challenge to understanding those cognitive systems lies in describing how 'concepts' form from observations. Concepts are some kind of ('invariant') patterns that are stable to all the minor changes and variations we experience through possibly unreliable and limited sensors. You can't define anything if the basic elements are shifting and changing faster than you can define them. Stability is the bedrock of memory, and stability is a local phenomenon that is closely associated with the separation of scales. The separation of scales—what happens on the small and the large, trends from variations, long term from short term changes, intentional from unintentional—this is a topic I discussed at length in my book *In Search of Certainty*[230]. To recap briefly, it's helpful to think of concepts as being a kind of chemistry formed from information.

Let's pursue that analogy just for a moment. There are only a hundred or so chemical elements from which all molecules can be built. These are approximate invariants on the timescale of the planet (isotope numbers are not important to the chemistry, to a first approximation). They combine to build a much larger set of molecules from a fundamentally bounded set of bricks. In other words, the total set of derived concepts possible in chemistry (the set of all molecules) is large, but countable. We need the set of underlying components to be finite in a memory system, otherwise the evaluation of observed state or context, taken from sensory inputs, would not lead to repeatable outcomes. This is important, even though it's counter intuitive. We seem to be able to imagine any number of concepts, but a single agent might only be able to know a finite number of them, in its limited access to memory. The finiteness means that it's possible to revisit the same concept multiple times, put it into context alongside other memories, and to recall it with the same meaning each time. If everything changes its meaning all the time, memory would be useless. The process of observation needs repeated outcomes, else nothing could be remembered—it would be entropy.

Humans have vision, sound, smell, touch, etc. (at least the lucky ones). They are the bases on which we recognize patterns, and form the elements of our representations about the world. We combine the concepts we observe into new concepts: two arms, two legs, and a head make an animal, for instance. It all starts with the transformation of patterns into tokens—the essence of data compression, i.e. how we take a complicated observation and turn it into a simple symbolic memory.

Pattern recognition by spatial imprint (ANN)

The present face of artificial intelligence is a spacetime process known as an Artificial Neural Network (ANN)—or just a neural net, for short. Neural nets go back to the 1940s and the work of Warren McCulloch (1898-1969) and Walter Pitts (1923-1969) who were trying to emulate the work of Donald Hebb (1904-1985), a model of learning in the brain[231]. The story is the subject of many books and articles, so I won't repeat it here.

Neural nets have advanced significantly since their inception, yet after over half a century of study no one has a clear understanding of how they work. On a simple level, we can think of them as something like a smart sieve for finding gold in dirty inputs (see figure 8.3). Data are poured into a wide end of a network, and come out of a narrow end, sorted and labelled with some kind of name for what went in—a bit like the coin sorting machines that banks use (see figure 8.3). For years, researchers lost interest in the idea of using highly connected arrays of processing elements as smart filters, for a variety of reasons. Although they are adaptable, they are not particularly efficient in one sense of the term. The revival of interest over the past decade occurred as companies with massive cloud infrastructure, like Google and Facebook, were able to apply brute force computation to these neural nets, in a technique called now called 'Deep Learning'. With enough processing power, neural nets proved to be an effective approach for recognizing images and patterns (see figure 8.3). They have transformed the fields of speech recognition, text recognition, and all forms of visual recognition. They are pattern processes, not merely pattern filters.

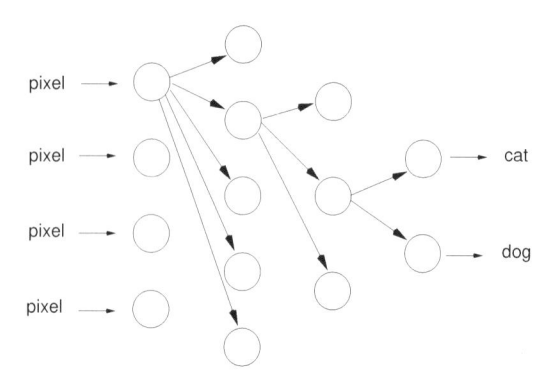

Fig. 8.3. An example process for an artificial feed-forward neural network. The inputs on the left hand side are larger in number than the outputs on the right hand side, allowing 'dimensional reduction', or classification of patterns on the input into distinct exclusive categories at the output.

Modern neural nets are still approximately 'one dimensional' processes, on isotropic and homogeneous arrays (see figure 8.3). The homogeneity of the spacetime structure in a neural network is broken to some extent by applying different strengths of weights on the wiring of its adjacencies. The efficiency of the links is tuned by a process called 'training' or Machine Learning. For ANN, training is supervised. A supervisor collects a lot of inputs that are known to contain examples of the concepts you are trying to recognize. Then a process of 'back propagation' is used, where on receiving the known inputs and fixing the known output, an algorithm can tune the links in between to make the input generate the output. It's a bit like what mathematicians do to solve an equation by varying the coefficients of series of possible terms until the equation gives the right answer. Backpropagation is an advanced boundary condition method: the desired outcome is used to work backwards to the Equation of Motion for the network. The result is a custom tuned machine that's hardwired by the training process to recognize what it's been taught to detect.

The neural nets are interesting arrays of locations connected by links. They are graph spacetimes, and they have variable give and take—which is where their 'intelligence' lies. The variable strength of the links is not naturally interpretable as variable distances, but rather as signalling differences, i.e. semantics. Different pathways are weighted differently, a bit like quantum mechanical path integral with far fewer degrees of freedom.

We now know that this is not a good model of how the human brain works, though it is a reasonable model of how eyes and ears work, i.e. how our senses have adapted to processing[232]. There is no evidence of backpropagation in animal brains, perhaps because the conceptual elements we employ to understand the world are not simple named buckets, but rather evolving processes that don't need to be driven by an explicit exterior training clock.

The spacetime analogy might seem to be of limited use in understanding neural nets. If you think in a Newtonian sense, then it has little to offer, though a gravity expert might argue that the variable weights in a neural net are a bit like some representations of a curved spacetime, in the Einsteinian sense. Either way, the phenomena involved in thinking networks are not well characterized by 'motion', in the sense of chapter 5. The changes are more like chemical interactions, more like biological networks. As mentioned earlier, the importance of 'translational invariance' or uniform motion in a straight line is supplanted by the attention on the importance of give and take, and locality. Motion, as a sense of flow, is something that happens recognizably when a space is mostly homogeneous, or featureless. But information is exactly what destroys that translational symmetry. So the more a space learns and adapts, the less spacelike is will appear to us. It will look more like a gravitational distortion field.

Of course, from the usual scaling laws, we know that can often restore the appearance of approximate translation symmetries at a large enough scale, just by zooming out enough—assuming the underlying thing is large enough to have large scale behaviour, because zooming out involves aggregating space into coarser grains[233]. The more we zoom out and brush over details, the more alike different regions seem. Whether or not a phenomenon identifiable as motion occurs on that scale depends on the extent to which new information at that scale can break the large scale symmetry. Artificial Neural Networks are quite small structures, even when used on large data sets, so they don't have properties on 'large scales'. That means they are more like chemistry than like Newtonian physics. There is no motion in the way we understand it in an ANN—at least during normal operation. The structure of the space is fixed, and programmed like a filter, through which data pass. It's similar to a series of sieves, where each layer has pattern of holes fixed by training. A decision process works like wave flowing through the structure. The distorting lens of the weights, or the 'curvature of the local space', i.e. the relative timings and weights affect the results coming out transforms the outcome and compresses it. Artificial Neural

Network recognition is not motion, then, in the sense of particle motion, it is an unfolding of states more like a wavefront passing through uneven shallows. Imagine waves from the sea that come to shore in a particular pattern, when the wind is from a certain direction, or when a particular ship passes by—the waves might just conspire to hit one particular rock on the coastline. That rock then becomes a surrogate for the incoming pattern—the pattern is recognized when the water hits that rock.

Even that picture is not accurate, because unlike a smooth Euclidean space, the spatial locations in neural networks are not an unbroken continuum of elementary points along a beach, they are localized strongholds that have their own interior computational processes, each with their own sense of time. Each processing location makes use of its individual ability to take part in a mesh of 'give and take'—to accept different amounts of what is handed to it from its hardwired neighbours. It's more like a subway interchange, with several different train lines converging on a station. When a lot of traffic comes from certain directions at the same time, it might trigger passengers to leave by certain exits (perhaps to void congestion). An observer, on the outside of the station, might look at the flood of passengers coming from a certain exit and say: 'Ah, I'll bet the 7:45 from Oxford was late again', because exactly those conditions are known to cause a flood of passengers (at this time) who are forced to leave by exactly those exits. The analogy is crude, but it conveys the basic idea.

So now take that idea, and imagine further that you were in a position to manage the traffic through the station, setting up barriers and opening extra doorways to route passengers. Then you could actually train those directional routes to handle each situation and *learn* the optimal routing of passengers. That would make the behaviour at the inputs match the behaviour at the exits uniquely—and you would have achieved deterministic *pattern recognition*. A coin sorting machine routes a particular shaped coin along a particular path, but a neural network would take certain combinations of coins and result in a particular light flashing at its output—so you could train it to learn the size of a tip.

The patterns, by which a neural net accepts relative amounts from different incoming contributions, are determined by an independent process of 'training'. Training imprints an effective geometry onto the space, forming a kind of lens, through initial signal is viewed. By altering their structure in this way, neural nets are able to arrange for certain incoming patterns to produce an output that points to a particular place on the exit layer. Those exit points then represent the recognized types. In spite of the analogies I've presented, the details of how this happens are still not well understood, which makes neural networks both fascinating and somewhat questionable. In a world where we favour predictability, neural networks currently work at the level of curated hope. In essence, neural networks are formed by hardwiring the spacetime connections of a network into a directed pipeline with certain lensing properties (see figure 8.3). A neural net is designed along an axis, pointing in one direction. As data are fed into the network, they are propagated along the axis, at a rate determined by the individual nodes, and are split up along all the forward pointing directions, i.e. distributed to all the nodes in the next line of nodes.

If we want to use more Newtonian or quantum language to describe this, we might say that the signals move from each location forwards into multiple 'parallel worlds', where they experience different rates of time and different effective masses. The rate at which data move through the network may not be constant—that depends on how the networks are implemented. It would be more or less constant if implemented on the interior of an agent with a single clock; however, modern networks are too large to fit into the limited memory capacity of a single computer, so they get spread out across multiple machines in 'the cloud'. There are thus several independent notions of time, along each story path, which are synchronized only by what is received from the next layer of space. Synchronicity is in the eye of the beholder. Each input state goes through a number of cycles of 'mix then separate', with a final projection layer playing the role of classifier. After each layer of lensing, the results are mixed back together, to layers with fewer points. The hope is that, by training the link weights appropriately, the signals will come together again at only one of several output locations[234]. A neural net is a bit like the Feynman path integral or action principle realized as a technology.

In normal operation, each input path of a neural network could correspond to a pixel in an image, to a word in a stream of language, or to any other variable in a data source (see figure 8.3). A neural network is not a precisely homogeneous spacetime, but it has controlled homogeneity, like a lens. Some researchers have studied the implications of approximating decisions as smooth spaces called *perceptual manifolds*, thus deepening the mathematical likeness of decision processes to spacetime[235]. I think it's important to keep a perspective on the relationship between these deep mathematical formulations and the

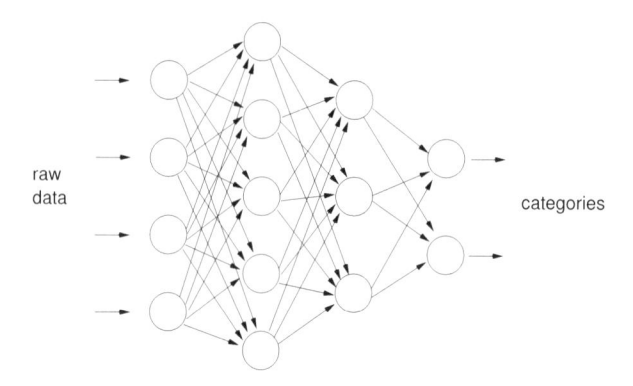

Fig. 8.4. A completed example structure for a small artificial neural network in figure 8.3. Unclassified data left are mixed and projected out as classified outputs right.

important essence of what goes on in processes. Everyone is impressed by algebra, but ultimately mathematics is only a tool for defining how to measure a process by counting. That might either help or obscure the understanding of the mechanisms at work, depending on your interest.

PARALLEL DIMENSIONS AND SUPERPOSITION

The dimensionality of neural networks is interesting, because they are deliberately inhomogeneous and anisotropic—different by distance and direction. The pathways information take through their space get scrunched up in places, or spread out in others. Remember that the effective dimension of a spacetime is not that of its embedding space (i.e. the network is not two dimensional like figures 8.3 and 8.4), but depends on the number of independent arrows at each location along the paths. Traditional spacetime ideas from physics are none too helpful in understanding those degrees of freedom, as neural nets are neither translationally invariant nor rotationally invariant as it expects.

On the large scale, ANN processes are almost one dimensional as drawn—dominated by left-to-right propagation—but the inputs and outputs at each location fan out into differentiated alternatives, each of which represents independent freedoms. They bring other dimensions to the space. The alternative branchings of each path lead to 'many worlds' that are superposed, and eventually collapse back onto intended classification agents further down the line. As the signals get bent, mixed, and otherwise distorted along their journeys, it's tempting to liken the distortions of processing paths to the distortions of spacetime caused by gravity, and to see certain pathways as being time dilated relative to others. That isn't too helpful for understanding pattern recognition in the dataverse, but perhaps it helps to demystify the gravitational distortions of spacetime in physics just a little.

The process of backpropagation, by which the routing policy for information travelling along different pathways is determined, is beyond the scope of this book, and still remains something of a dark art although there are now industrialized techniques for training them. What is interesting is that the symmetry breaking, involved in making the pathways unequal, allows spacetime to calculate an answer to a question. For some researchers, the layers of the neural net simply approximate a representation of a function or transformation of the input points, and they are satisfied with this explanation. But this is not much more profound than saying it behaves like a different shower head.

Give and take in ANN—from discrimination to reasoning

The mixing of information entering at the inputs, passing through several staged layers of a neural net, makes it act like an information aggregator and discriminator—which turns it effectively into a parallel statistical analyzer. Each location is *potentially*—but not necessarily—linked to every other location down the line. Moreover, because the different paths are effectively weighted, and some weights might even be zero[236], each local agent promises only to accept some or all of the information fed to it, and will assign a weight to each input it receives from an adjacent neighbour, according to its interior capabilities. This is reminiscent of the interactions between offered information and acceptance information in the Transactional Interpretation of Quantum Mechanics, separated by operator filters—or a prototype path integral—with the subsequent analogy between definite outcome states and probabilistic frequencies (see figure 8.5). Again, this likeness is interesting and helps to show that the peculiarities of Quantum Mechanics are not unique to that realm, but doesn't inform the process of pattern recognition too much. On the other hand, what seems noteworthy is the ability for such spacetime structures to aggregate separate channels of information and to differentiate or discriminate between different information. That is the basis of decision-making—of basic *reasoning*. It takes us one step closer to understanding how a system with the geometry of a cognitive agent might acquire the necessary processes to 'think'.

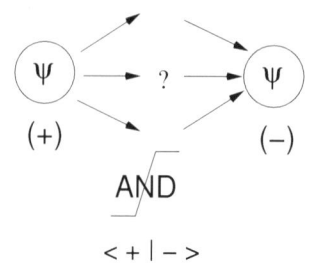

$$\psi (+) \quad ? \quad \psi (-)$$
$$\text{AND}$$
$$< + \mid - >$$

Fig. 8.5. A neural network is a directional causal fabric, something like a digital form of an astronomical diffraction grating. Each layer promises an incoming pattern of rays from the source, and each subsequent layer may or may not accept and integrate this. The result is an interference pattern that may sharpen or blur the inputs to some end. Blurring is useful for eliminating specifics for general categorization, while sharpening is useful to distinguish different cases.

The special 'magic' required of a spacetime, in order for it to exhibit computation and reasoning is thus accomplished by manipulating what's accepted and what is promised in response. If there is sufficient weight, or probability, accumulated, a certain output becomes activated at a critical threshold. This is exactly what happens in a transistor switch or logic gate, as a superposition of signals (typically with one biasing another) and leads to *switching* from 'on' to 'off'. Acceptance channels are mixed in different amounts, and then digitized or truncated at the output, by collapsing them across thresholds. When there is enough matching information flowing through a point, it gets 'switched on' from 0 to 1 (see figure 8.6). Spacetime acts like a carrier wave for a discrimination process, based on observer semantics. Each point in an ANN looks superficially like a load balancer, that spreads its load evenly across the spacetime channels of all the checkout servers—but the receiving points don't necessarily promise to accept all of that load. Also, their sense of time might not match—some might be deliberately slower than others. Think of a passport control checkpoint at an airport: passports are usually separated into parallel semantic channels: local residents and foreigners from different locations, America and Canada, European Union, etc., and then miscellaneous. If you are able to think of the

8 SPACES THAT SEE AND THINK

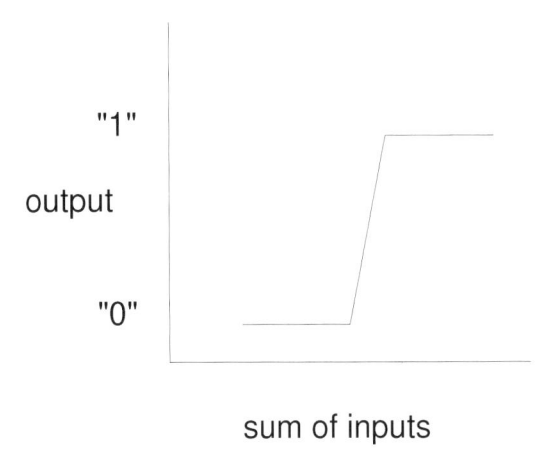

Fig. 8.6. A 'logistic' or 'sigmoid' function is a classic way to perform switching. A process that behaves like this is bad for HiFi signal amplification, but great for computers and heavy rock guitarists looking to distort themselves. The digital cutoff, capping activation strength, was originally motivated by a model of neuron activation in the brain.

passport line as a neural network, then you could ask it questions, like how to discriminate between a planeload of passengers: those with valid tickets and passports, and those with other combinations. Adding more dimensions to the semantic pathways: one passport officer might speak French, another Spanish, another Chinese, etc, so there will be a natural match to feeding certain passengers through certain channels. Not all passengers would be accepted by every channel, and that could be seen as a different effective speed and mass.

If a network is not properly trained, the discrimination process will be inefficient, but if the nodes are all trained, then the right passengers will end up in the right lanes. Now a manager standing at the exit could look at how many people come through the different exits and be able to guess which plane it was that landed, or at least what part of the world it came from. This is the basic idea of a classifying network. The term 'neural' is a throwback to a time when it was believed that they could be models of a brain. But, we can think of an artificial neural network as part of a much larger class of smart spacetimes that makes use of broken symmetries to identify and classify signals. This is the basis of pattern recognition.

Scaling patterns into concepts

Artificial Neural Networks can only do so much. They typically form the first sensory layer of a cognitive system, because their processing is geometric in nature, and is locked in by a training process that can't be repeated too often—training data take a long time to acquire. Each neural net is basically a one-trick pony, and the tricks have to be compatible with the symmetries of the network array. If we look for processes like neural nets in biology, we don't find brains. Rather, we find arrays of cells, such as those specialized in vision. There are different kinds of polarized cells for detecting motion, direction, and movement, as one would expect if eyes were designed by spacetime processes. That's not to say that networks cannot be bent to deal with other issues, but it would likely be inefficient to employ a programmable regular array for a process that was not regular in terms of its input channels. Then, we come back to virtual layers of scaled processes as a more likely candidate.

Recent studies of brain activity, using Magnetic Resonance Imaging (MRI) to scan the brains of listeners during English language readings, showed how brain activations were grouped into to broadly cohesive cortical regions, with similarities even across different subjects[237]. The authors posited from this that English words were mapped by human brains into non-local but clustered meanings that were labelled by the following categories: visual, tactile, outdoors, place, time, social, mental, person, violence, body part, number. In current parlance, these could be also called spacetime, state, self, senses, relationships, and danger. There is some evidence then that concepts are represented by modular clusters of locations in the brain. This certainly fits with the geometry of the local processes we've discussed throughout the book.

How then can we store memories that represent concepts, in a way that captures the meaning of the concept without having to rethink the entire history of events that led up to it? We all know what dinner is, for instance, without having to go back to the initial feeling of hunger, nutrition, prey, hunting, cooking, social occasions, and so on, to build up the concept from its components. It would be like having to synthesize the food, from basic elements, in a chemical factory every time we wanted to eat. Clearly, the outcomes of recognition need to be remembered and they need to be stable to the arrival of new information. The timescales for sensing and for learning and for concept formation must all be very different.

Without an initial selection principle to seed concepts, deciding which con cepts built on others, and in which context—memories would end up with the structure of a flat archive, ordered according to time of arrival, or even mixed amorphously in the absence of an arrival number. That's what a computer database is like. Spacetime forms a useful blank canvas onto which we can impose short sequences of memory, but it can't organize memories effectively. To scale understanding, a cognitive agent needs to be able to *discriminate* observed phenomena and maintain the distinctions is can discriminate over time. If all memories were dumped into a formless database, they would never become smart. Rather, memories need to self-organize in such a way that the same input leads to *approximately* the same recognition from memory each time. If the map were too precise, it would apply to only a single context and its ability to recognize variations on past experience would be poor. The relativity of the map's encoded representation is thus what determines its eventual usefulness. One of the skills a cognitive system has to deal with in an exterior environment is to identify when phenomena are similar, in spite of their possibly being on a completely different scale: a giant dinosaur and a small gecko are both recognizably lizards, but on very different scales. The problem is that our eyes and ears are scale dependent, so we need a way to scale the way we address concepts in memory, to recognize things that are similar only in terms of their relative semantics (legs, scales, tongue, tail, etc), not in terms of quantitative coordinates. Notice how these are all spacetime concepts. It's a very interesting irony how an initial critique about the importance of invariant properties, at the start of this book, has led to a simple theory of cognition, in which we finally uncover the reason why such invariances have value: they minimize the amount of information needed to maintain the integrity of a stable cognitive process.

SEPARATION OF TIMESCALES

The arguments I've discussed concerning symmetry, scale, and observation lead to a plausible story for the general structure of cognitive processing, based around the principle of *separation of timescales*. I believe that this principle—well known in physics—is more important than (and leads indirectly to) the preferred approach taught in every Computer Science class, which is the *separation of concerns* (semantic concepts). Skipping the details for brevity[238], one ends up with a picture summarized roughly by figure 8.7. The quick version is this:

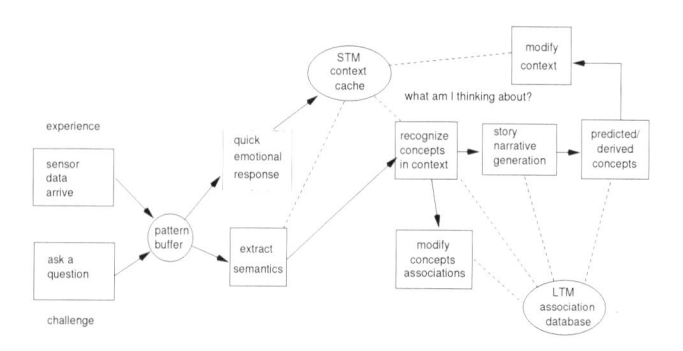

Fig. 8.7. A data pipeline or feedback workflow for cognitive reasoning, with long (LTM) and short (STM) term memory caches, based on the model in reference [4]. There is explicit mixing between public (global) and private (local) state leading to non-deterministic recognition and reasoning, making the process non-repeatable—time cannot be replayed.

- Scale dependent sensors recognize patterns in exterior space, projected onto reduced dimensional structures, analogous to neural nets.

- Quick, secondary responses from the senses can also detect important situations, like 'danger', 'stress', and 'happiness', which tag memories with important low-grade reasoning. It's quick and dirty, but can save a lot of expensive time trying to reason in terms of stories.

- The neural nets 'tokenize' the complicated patterns into simple episodes which are encoded like names, like turning complicated descriptions of, say, cooking into numbered menu items. The menu of names describing the episode fragments, contributes to a part of what we can call *context*— a list of concept fragments, a kind of situational genes.

- The concept names that can be produced by senses, lead to a basic vocabulary of concepts (hot, cold, moving, etc). Clusters of these names that are activated simultaneously lead to superagent concepts. This principle of 'co-activation' is seen as a detection discrimination and confirmation mechanism, in many natural processes, including the biological immune system.

- Long term memory can keep short sequences of events and changes. If a memory could not record time, it could not imagine anything dynamical, and could not model the exterior world. So memory cannot be static, as in a database.

- As clusters of these basic names form, they don't just accumulate, and reinforce long term memory over time, but also feed back as the other half of context, as the agent is 'thinking about' those concepts. In other words, when a concept is activated, the agent is thinking about it, and this feeds back to the context, eventually dominating over the sensory inputs. The sensory channel becomes more like the bias one feeds a transistor, while the rest of the thoughts are the amplification of ideas from recycling old concepts (when we dream, we lost the bias of exterior senses, un-anchoring memory from exterior context).

- When the cognitive agent wants to recall something, it needs a trigger to find the memory. The trigger is based on the contextual tokens, rather than on an address, like a computer archive. Once a long term episode fragment is found, related fragments get activated if they are adjacent to it, and thus a number of possible stories can be generated by following the network of concepts.

Storytelling is the way in which one gets from simple sensory fragments to complex time-dependent memories and reasoning. Chains of events are streams of sensory observations. This stream is distinguished by a variety of hardwired semantics, whose origins are still unclear, but are likely defined self-consistently by evolutionary selection and through stability of feedback. We don't need to describe exactly how memory is implemented—there are many possibilities. Nor do we have to explain how connections between memories are made. These are 'virtual questions' for every technology or phenomenon to find on its own scale. I'll content myself with the spacetime principles here.

Initially this stream seems to have no scaling structure on which to seed a hierarchical process, but the answer to this could be the biases from emotions. Emotions play the role of a blunt classification scheme for an agent's exterior situation. They don't promise any specificity, but the lack of semantic precision is what allows them to switch quickly and pervasively. Emotions can trigger on certain patterns of events quickly and flood the system with a loud and blunt response. The effect is to form a signpost in the landscape. These signature markers break the symmetry of the sensory stream. Emotions are also great outcome selectors, helping us to decide when a story is satisfactory. If we feel satisfied, we stop, otherwise we keep digging into a story. Think, for example, of the bizarre dreams we have. Our brains put together some random memories and ideas into an utterly weird spacetime experience, but they feel quite normal because they trigger the usual emotional responses, and we feel satisfied or disturbed.

A short term memory keeps a snapshot of the current state observed on the exterior of the cognitive system. This is temporary because the cognitive system is evolved to respond to changes in the exterior at a rate driven by the exterior clock (see figure 8.7). Sensory combinations, combined with feedback from the existing interior processes, to 'shine a light on' larger concept fragments, thus opening us networks of possible 'stories' formed by following associations. The associations themselves are encoded by co-activation of pathways. Long term memories thus form by repeated co-activation, leading to a long term structure which is reinforced by repeated activation of the contextual combinations, based on the hardwired processes.

The picture emerging is one in which evolution plays a key role in bootstrapping concepts, from the only phenomena available in the beginning: distinctions in space and time. Random mutation and selection is not a fast process, but once it gets going its efficiency can grow exponentially. In biological organisms, hardwired adaptations detect basic concepts through senses, as developed through evolutionary through selection. In the dataverse, they are hardwired by the experiences of programmers and expert designers. In both cases, there is a relatively long timescale of learning the basics, over which elemental processes behaviours get set, and determine the constraints (analogous to the Equations of Motion) for the realtime dynamical process by which information is sampled and stored in the cognitive agent.

By laying down memories that feedback on the process itself, a cognitive system employs a stigmergic mode of memory recycling to cooperate with itself. This is the trick of a cognitive system: it cooperates with itself by looping back its recorded states to influence its future behaviours. If the exterior were constant, this might settle into an equilibrium and 'die', but as long as there is information on the exterior to drive and modify the interior's states, it will be unlikely to reach equilibrium. Once this works on the interior of an agent, it can also extend its stigmergic droppings to the environment, through pheromone trails, forest paths, or book writing. Then the social memory comes into play to sustain tokenized concepts, sustaining them as tokens, without having to recompute them from first principles[239]. So, just as long as the social species of cognitive agent survives, with all its layers of memory, its intelligence—in terms of recognition and storytelling is likely to grow exponentially too. This is not just supposition. We see this process in action, and the evidence for it lies in the evolution of human languages. Languages change as concepts change, and as fragments of concepts get recycled with other meanings (through metaphor, and so on). I'll come back to this below.

Static concepts are formed, in a sense, by 'voting' on what the senses and memories report—i.e. by a statistical process of averaging, or normalizing. However, as more and more samples are accumulated, the average result may be renormalized to a different effective value—and this is how we understand values at scale. Recall the voting problem discussed in chapter 6, in which redefinition of boundaries completely reversed the outcome of a process. It might take only a small difference of experience to end up with a very different conclusion—a very different conceptualization. If you think this sounds like a precarious process, you would be right. Averaging can be unstable to small changes. This is a big problem in reasoning—and, speaking of reasoning, we need to explain the concept of an explanation!

The cognitive process model suggests that reasoning is a process of storytelling, formed by navigating paths between existing concepts. That isn't what many people imagine when we speak of reasoning, so why not? Our ideas about reasoning have been biased by several centuries of mathematical idealizations about reasoning that were developed to try to understand the limits of certainty—by *logic*. Logic is a mathematical model of reasoning, which is like an over-constrained version of storytelling that is impractical and too specialized for general purpose process adaptation. In mathematical logic, certain concepts can be true or false, but nothing in between. There is no room for uncertainty. Probabilities add back some uncertainty and provide a middle ground for calculating a likelihood of getting a true or a false, but the mathematical reasoning never quite escapes from this straightjacket. Philosophers didn't always think this way—indeed this has taken over the discussion of reasoning mainly since the invention of computers, and the idea that the brain might be a kind of computer. In science fiction, as well as science, logic and rationality have been idolized as reliable truth, and human emotions or leaps of intuition vilified as an unreliable weakness to be contrasted with rational reasoning. In fact, this is probably exactly wrong, as I'll explain below. The main capability that leads to reasoning is the ability to make up stories from experiences (+); the criteria for constraining them or selecting the best ones (-) is of secondary importance.

There is something intrinsic about being able to connect the dots between ideas into a well-formed story that's central to our ability to plan and understand things. Some days it's harder than others—and when we struggle to put two and two together we feel quite stupid. Other days, we might forge ahead and take great strides in our narrative. My experience, as a teacher at the university for twenty years, taught me that helping students to combine incremental steps of understanding into a larger storyline is one of the most important ways of helping them over hurdles they experience—like tunnelling through a potential barrier. This is what gives us confidence, because it's intrinsically tied to trust and its emotional basis. This ability to chain together causal steps without losing one's way feels like a major limiting factor in how smart we are in a given context. And it does depend on context: what feels obvious to one person, say in woodworking, may be an insurmountable barrier to someone experienced in gardening.

As experiences arrive at the cognitive agent's sensory boundary—ordered by exterior time, and classified by specialized networks that discriminate patterns— concepts are somehow acquired, modified, recombined, and further built upon, leading to a network of memories. The ability to separate observations into well defined concepts is by no means assured. Contextualization of exterior situation is the obvious key that one would like to organize observations with respect to, and this is where spacetime plays a central role. After all, the observations come from the exterior in the first place. The ability to combine concepts into larger derivative concepts suggests that a cluster of smaller concepts might be unified under a single umbrella concept, acting as a connective hub. We know from studying networks in Promise Theory that the role of centralization is for calibration—which is exactly what we mean by the clarification of semantics. A single hub concept can then act like a network switch. However, this suggests that there must be an innate or pre-arranged bias in the processes of observation that acquire and store the memories in the first place in a predictable pattern. Why else would memories be kept separated, and tagged for recollection in durable ways? Obviously that would be a highly beneficial adaptation, so it may be no surprise that it might emerge under the right evolutionary conditions, but we still need a plausible mechanism to explain it.

Semantic names and addresses

It's not enough to be able to store memories—we also have to be able to find them again. In a simple warehouse model of storage, as we often use for computers, we can use a numbered map of spacetime to store and retrieve items. Computers, file archives, shopping malls, and libraries all use location-based addresses formed from numbers and coordinates. Databases and other indexing methods go a step further and allow us to look up items based on related keywords, which are then mapped to a coordinate location by index—transforming words into, say, page numbers, or locations on a warehouse shelf. But what if there is no underlying numbered coordinate system to rely on?

An alternative to numerical addressing is to use concept-based addressing, such as the aisle labelling one finds in supermarkets. Hierarchical classification plays a role in the 'namespacing' or categorization of concepts[240] Computer file systems, web documents, markup languages, or document data formats, like HTML, XML, and JSON, all have a regular hierarchical structure from the root of each document. This may be used to define a coordinate systems, based on the names and ordered numberings of identifiable objects within the document. Locations are specified by URI, URN, URL path identifiers, as discussed in chapter 3. Path locators provide nominal semantic coordinates based on a 'spanning tree' formed from the elements of the journey. It's like giving directions to find someone's house.

In order for similar semantics or meanings to map to similar locations, the properties that form the key must promise to be similar too. Names, in human language, are not too reliable. The keywords 'car' and 'lorry' would not be likely to end up close together in a common neighbourhood that we could assert as a category, but if both were associated with wheels and transport, then they could be. A further generalization of a path address is the idea of taxonomy. Taxonomies were popular in the 18th century as systems of knowledge. A taxonomy is basically a tree of names, forming an ordered hierarchy of concepts, separating conceptual branches into 'namespaces'. The result is a path-like coordinate system, commonly used in computer file systems that use folders or directories for each nested item.

In biology, for instance, taxonomies have been used to divide the world into categories like plants and animals, and then animals into say mammals, birds, and reptiles. Each category can be divided further into smaller and smaller categories. The idea is much like the way we try to tidy up information into categories in archives, libraries, and other filing systems. There are two problems with taxonomies: the first is that they assume that everything already has a name and that the name fits into exactly one category. If you have an animal (like the Duck-Billed Platypus—an egg-laying mammal, or a flying fish) that defies the arguments for classification then you are already in trouble. Taxonomies don't work for semantic categories, because semantics are in the eye of the beholder, and thus arbitrary. Moreover, taxonomies are meant to be standards that everyone is supposed to agree on—but rarely do. Any addressing scheme needs to be approximately unique, so that it gets you to the right 'ball park'. Tree structures are unstable because they apply true-false dichotomies with sharp edges to discrimination criterion, hoping to divide and conquer items into mutually exclusive branches. The problem with that is, quite like the true-false approach of logic, the smallest mistake sends you down the wrong alley, and then you are lost.

A generalization of taxonomy, as used in computer science, is called an *ontology*. Instead of a tree, it uses a web or semantic network of names that represent ideas, along with metadata to explain relationships, using a multitude of labelled associations. Ontologies try to do more than simply put names and ideas in order—they try to understand them at the level of relationships. This sounds like a nice idea, but it also has its own problems that I discussed in *In Search of Certainty*. Briefly, the assignment of associations is equally arbitrary and quickly gets out of control. It also has no innate basis. Categories and associations are named according to human trainers' whims, not by an approximately deterministic process, so—once again—they are fragile and non-reproducible.

Ontology techniques have evolved in recent times, since the widespread use of machine learning, using neural nets, for instance. They've become more like deterministic key hashing, thanks to the separation of a learning phase which pre-categorizes items by dimensional reduction. As an approach to addressing, taxonomies and name spaces are be inefficient, especially when they become too deep, because divide-and-conquer methods need to be parsed serially, and executed recursively, using a relatively complex process to traverse the entire space from a known starting point to come up with the result. Worse than that, they assume an invariance that isn't there. In computer programs, data can be stored in locations determined by a 'hash', or sorting code. A sorting code is quick to compute and reduces the dimensionality of a space for approximate match, much like a neural network.

For instance, to make a very simple hash, you could take the first letter of a name, and store items alphabetically or by subject code, as libraries do. Then when you need to retrieve the information next time, you look at the name and go the room for items beginning with that letter. It narrows the search, but it doesn't get you all the way, and it assumes that you already have a symbolic representation to work with. If you keep on subdividing the name by letters, then you end up back to an effective taxonomy that's based on spelling. For the human world, knowledge is already a sophisticated business, and we can't get something like a hash without knowing the spelling of names or about complex mathematics. That's why a cognitive agent needs to bootstrap itself by learning. That's what cognitive agents are adapted to do. Of course, a single cell in biology can perhaps remember concepts like 'the right protein', but it can't memorize the names of animals or the phone book. Simple cognitive agents have much simpler tasks. What is truly extraordinary about the cognitive process model is that as the scale of the agent grows bigger, its capabilities grow rapidly and significantly.

CONTEXT ADDRESSING

The way observable qualities are detected by sensors and then aggregated into memory clusters suggests a way for determining types and categories of knowledge that emerge automatically. We want to end up with a token, which can then be sorted quickly and reliably, using a simple mechanical process. In other words, the same set of sensory triggers has to lead to the same concept every time it is observed or thought about. Suppose, then, that the sensory representations could be used as the addresses themselves; then ideas would be built up around 'pure' exterior concepts based on spacetime similarity—which are the same for all observers.

Assuming that the process of observation is constant and reliable—and discounting issues related to exotic phenomena of Einstein's relativity or the quantum level, which few organisms will have to experience—those invariants will then lead to seeding of 'roles'—see the schematic figure 8.8. The first concepts are patterns, then patterns cluster around similar locations, by network routing, based on which sets of patterns are co-activated. As the number of stable concepts grows, learned concepts can also become part of context—through

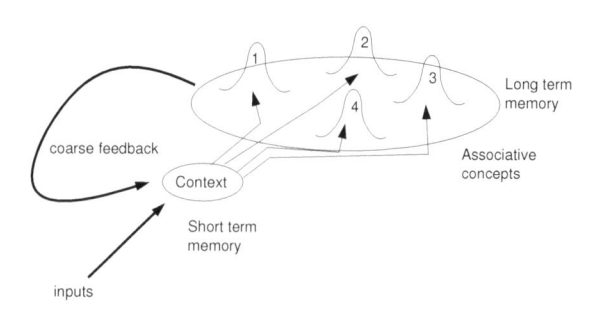

Fig. 8.8. Semantic indexing: sensors receive data, patterns are recognized, and contexts are discriminated by building on the presence or absence of invariants that are detected by sensors. These can be mapped to clusters of concepts, creating seed roles. The seeds might not be physically localized, but they must be virtually localized, and therefore work as a lookup key.

introspection, or 'what the agent was thinking about'. Derived concepts that build on different clusters can be defined by adding in the feedback from pre-existing concepts to the sensory channel. Having a feedback loop adds a new source of stability: self-consistency or 'eigenstates' like those that characterize quantum systems[241].

The innovation that cognitive agents bring is to compute a semantic generalization of spatial patterns that measure context. It is symbolic processing, a bit like the neural nets described earlier—but going much further. The neural net approach is inefficient beyond a certain scale, because it is dense in its expensive connections, while concepts become much more rarefied or sparsely separated. A closer look at context is thus warranted. The definition of context as an accumulation of observed patterns, from the recent history of an observer, gives us the keys to store and retrieve long term memories. Long term memory emerges, on a separate timescale, as a summary of tokens, organized as a structured network of concepts, and tokenized by the contextualization through spacetime patterns, stored as process fragments or episodes, and associated by a kind of adjacency links. Long term memory is yet another virtual layer of spacetime, within the dataverse, connected by associations. Figure 8.9 illustrates how the accumulation of short term patterns can lead to long term stable concepts.

The accumulation of concepts from smaller concepts, where concepts can be sequential 'film clips' that include motion, suggests the existence of a concept hierarchy—but not just any kind. It is not a taxonomy. Researchers who have

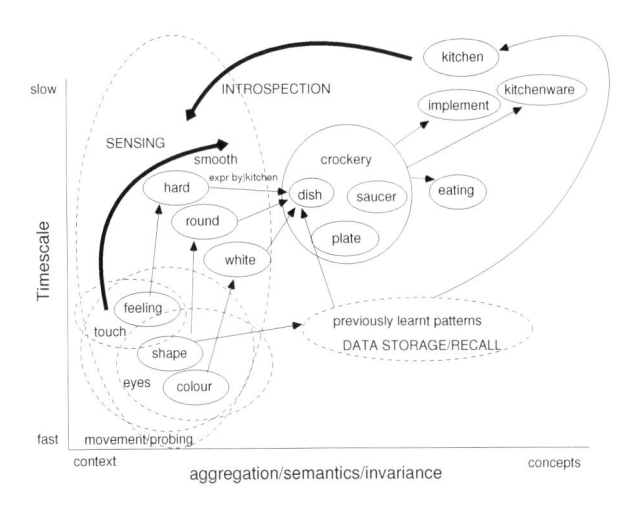

Fig. 8.9. Sensory patterns cluster into roles, which form anchor points to build concepts, through little storyline pathways.

studied knowledge representations have used hierarchies of concepts in the past but their models were static. Is there a structure that can support dynamics as well as semantics? The answer lies in the stability of aggregation rather than instability of discrimination—the latter being the mistake of using logic as the basis of reason.

The representation of patterns in a neural network is quite different to the representation of concepts in a semantic network, so I now want to discuss semantic networks in more detail. It means that there are at least two independent mechanisms available for recognition and reasoning—both methods are believed to exist in the biological brain, though no one understands exactly how; but the corresponding processes have certainly been used in the dataverse to seed the formation of hierarchical graphs from feedback processes[242]. This is the essence of AI today.

Semantic networks provide a kind of emergent and porous modularity, which is curated by long term process semantics rooted in spacetime phenomena. The result is a collection of stable but overlapping modules whose boundaries are context dependent. For the reasons above, it seems clear to me that Artificial Intelligence (AI) cannot be about monolithic structures, like Deep Learning networks with regular translational invariance. Moreover it can't be about linearized algorithms that conjure single pathways. Such specific processes are too fragile and too restricted in exploring the possibility spaces. Quantum computers that can explore parallel paths simultaneously can solve problems that take linear systems exponential time to solve, but we don't need quantum computation. Any parallelizable system can do the same, in principle. So reasoning must be possible based on parallel paths too. The integrated storyline that we experience as consciousness is only natural from the viewpoint of centrist observer structures—the boundaries we've been discussing. So spacetime enters into the picture at every level: classifying the semantics of associations, and in the dynamics of their aggregation into compound and modular structures. Moreover, this picture scales. The formation of social networks has the same basic structure as the knowledge networks alluded to above. A social network may be considered a smart space, and indeed an intrinsic knowledge representation. Links between members of a social network lead to clustering of cliques and regions[243].

THE FOUR PROMISES OF SPACETIME PROCESSES

If we begin from the hypothesis that spacetime—or more correctly, the hierarchy of spacetimes we've been discussing—is the basic fabric on which to build all phenomena, then it makes sense that all conceptual relationships would be reducible either to scalings or relabellings of combinations of basic spacetime properties—because we have to start from something to build a tower of concepts, and the only thing present at the start of the process would be space and time. That includes the existence of matter, which can be treated as one of those spacetime processes. Those notable properties become symbols in an alphabet that recombines to form the words of a new expressive language. This is not an allusion—the combination of parts really is a language, according to the classification of languages dues to Noam Chomsky (1928-), at least in one dimension.

There is something profound in this idea that follows from the previous chapters too. If the forces of nature, as we understand them, are also just proxies for spacetime structure, based on different flavours of adjacency, then all we have left is to describe the semantics of space and time. This idea is similar, but not identical, to the use of hidden dimensions imagined by Kaluza and inherited by string theories. By taking a promise theoretic point of view of a cognitive system, one comes up with a set of basic promises on top of which all subsequent phenomena might be built. The hypothesis becomes that four elementary classes of promise would necessary and sufficient to describe the semantics of spacetime (see table 8.1), and allow the construction of processes as combinatoric 'stories' based on this alphabet. The promises are:

- To be inside or outside (to be part of)—*math:* a polar centric vector.

- To follow or precede—*math:* a translation vector.

- To be next to another location—*math:* a primitive vector.

- To expresses information—*math:* a scalar property.

Table 8.1 shows how these four classified types relate to more familiar concepts. It remains to be shown how all relationships can be reduced to embellishments based on these four semantic elements. To make the interpretation work, one has to bend the simple meanings quite far from their simple interpretations, to encompass what we would think of as metaphors in language—but actually, that is the key to understanding scale and meaning. Bending meanings, by allowing associative metaphor, is exactly what scaling and accumulated complexity allows you to do.

Stating the four types in this way seems straightforward and innocent enough on casual inspection. They can all be realized using an even more primitive level of connection: promises, in the sense of Promise Theory. The real power of the idea lies in what happens when we scale them: when being next to something turns into 'being semantically similar something else'; when preceding something turns into directing or causing something; when containing turns into generalizing the meaning of something, carrying or owning something; and when expressing a local property becomes a promise and a representation of information itself. Then, using these four distinct kinds of semantic link to connect points in a space—forming a promise graph—allows us to construct anything.

Scaling is a very difficult thing to imagine—as the foregoing chapters will have made clear. Placing yourself in a frame of mind to think about geological time, or the influence of DNA on an ecosystem is mind boggling. Yet that is where the major insights come from about how processes work. It seems particularly significant that we have to assume that spacetime is discrete in order to make the scaling of distinctions additive, rather than allowing them

	FORWARD	BACKWARD	INTERPRETATION
1	is close to approximates is connected to is adjacent to is correlated with	is close to is equivalent to is connected to is adjacent to is correlated with	contiguity PROXIMITY "near" Semantic symmetrization similarity
2	depends on is caused by follows	enables causes precedes	ordering DIRECTION "follows"
3	contains surrounds generalizes	is a part of / occupies inside exemplifies	boundary perimeter LOCALITY "contains" / coarse graining
4	has name or value characterizes represents/expresses promises	is the value of property is a property of is expressed by	qualitative attribute DISTINGUISHABILITY "expresses" Asymmetrizer

Table 8.1. Examples of the four irreducible association types, characterized by their spacetime origins, from reference [4].

to succumb to entropy—because in a continuum spacetime you can't make or retain sharp distinctions. The mathematical points of zero size, apparently allowed by calculus, may be consistent in mathematics—using the prescriptions devised by Newton and Leibniz—but that should not be taken to imply that they can be interpreted literally. The real meaning of those limits is to enable very accurate counting estimates, when you know exactly what to count and you have a large number of items. The continuum limit is a scale renormalization trick: what calculus tells us is only the behaviour from a large scale perspective, where small intervals appear like a smooth continuum. There is no implication that spacetime itself has to be smooth.

To understand how these four concepts suffice to build up concepts of arbitrary complexity, we need to make a digression from the raw spacetimes we've looked at up to this point, to consider virtual spaces that express semantics. It is the scaling of such conceptual networks wherein lies the possibility for a spacetime process to think on a sophisticated level. Human cognitive abilities build on vast arrays of patterns, concepts, and complicated associative skills. The ability to make connections and analogies by association is no simple feat

THE SCALING OF MEANING

During his fifty or so year reign around 1500 B.C., the Egyptian Pharaoh Thothmes III erected two seventy foot columns by royal decree, before the great temple of Heliopolis, near what is today Cairo[244]. These two exclamatory masts were powerful symbols of the Egyptian spiritual and technological supremacy, singular and immutable signals, in contrast with North Africa's seething and inconstant desert sands. For almost three and a half thousand years, they adorned the entrance to the impressive temple as a symbolic gateway, with little competition from their surroundings. Tall, thin spikes sticking out of the flat Earth, making a statement.

The first of the needles changed hands, as a gift to the British people, in 1819 in recognition of Nelson's Victory over the French fleet, at the Battle of the Nile in 1798. An odd gift perhaps, a lump of stone so heavy that it took years even to muster the effort to move it—but the significance lay more in the singularity of the conquest, than the original significance of its shape, or its composition. The symbolic gesture was repeated for the United States in 1881, when the Khedive of Egypt, hoping to stimulate economic investment in his country, gave the twin monument to the city of New York. Since then, the obelisks have been emulated in the monuments of several large cities, including Paris and Washington. They have nothing at all to do with Cleopatra (preceding her by some 1500 years), nor are they made of any valuable substance. It is their singularity of form—standing out like material substance against the bland background of 'space'—that conveys a powerful symbolism, recognized by all cultures. Singularity—the act of defying formless entropy—plays a recurrent role in symbolism.

In Shannon's language, an obelisk is a low entropy structure—a distinct signal that has not succumbed to mixing with the states around it. It contrasts with a background that is relatively featureless, like sand dunes or flat Earth, which we might call 'maximal entropy' structures. This is a bit confusing in relation to the physics view of entropy, which accounts for it differently. Shannon's entropy is a measure of variability along the timelike trajectory of a message (a signal is a sudden bump versus a flat road along a path). The thermodynamic entropy, discussed in physics, is a usually a measure of variability between 'equivalent' slices of a phenomenon, laid alongside one another 'spatially' (a distinction is a brown egg in a packet of white eggs with no particular order).

It's curious but no accident that information and meaning turn out as opposites. When nothing happens, we attach no particular significance to that. It's the anomalous or unexpected changes that we attach meaning to. Of course, this is a human convention, but it makes sense. The more information you have, the less meaningful it is—the more like noise is it. This is where our sense of value for novelty and uniqueness plays into our value judgement. Information needs contrasts in order to be distinct and observable. Symbols are precisely contrasting distinguishable patterns in a process. The more information captures the 'attention' of a process, the more meaningful it is. This gives us a simple interpretation of what attention is, without referring to brains, consciousness, or thought. It's plausible that unusual signals might be danger signals so any process (say an organism) would be interested to pay attention to them. Persistent repetition of the same information is sleep-inducing.

Crucially, symbols and information are formed from discrete patterns—regardless of whether their embedding spacetime is continuous or discrete, the active phenomena of information are discrete. The paradox about information is that having too much of it is like not having any at all. When you accumulate distinct, discrete data, sharp with clarity, into a single grey mush of aggregate averaging, then the effect of accumulating and mixing renders it featureless and insignificant. Meaning is depth rather than breadth. Breadth is what we mean by entropy.

The *significance* or *meaning* of information is not the information itself, but what weight and associations we attach to it—how well it stands out. Discrete symbolism and variable weight work hand in hand to define at least what humans consider to be meaningful. To make a symbol, we attach meaning to patterns etched in space or traced out in time. A symbol becomes a surrogate *representation* of an intended meaning—an icon to stand in place of the real thing—to promise its intended meaning, in promise theory parlance. A symbol can be much smaller than the thing it represents. This is the basis for compression of information: it's how we form abbreviations and writing from sounds. With an alphabet of different symbols, we can express complex relationships. One obelisk by itself has a limited range of meanings. Obelisks dotted around the world all signify the same basic idea, wherever they are found. If there were a collection of different shaped forms: obelisks, pyramids, sphinxes, and so on, then suddenly, we have a larger vocabulary to express ideas with. Two pyramids and a sphinx in close proximity could signify a royal burial, while a sphinx on its own could just be a warning to stay away.

Symbolverse

Accepting that there is an equivalence between symbols (meaning, semantics, etc) and phenomenological patterns is the very essence of what mathematics calls *algebra* (from the Arabic word that means the convergence of ideas). It is the basis of practically all mathematical reasoning, and all of our formal descriptions of the world. In a sense, algebra is the language of processes. It feels as though there is a deep connection between symbolism and spacetime.

Every pattern made up of discrete characters that we call states. Every particle, or field, expressible as measurable characteristics of space, can be thought of as a symbol in an unconventional alphabet of such states. So the 'particles' so often mentioned in physics are really representations of symbolic information. As we sweep out paths through these characteristics, by stepping through space and time, we create messages: the variations in properties at different points, over regions of space, express patterns that tell stories. If those stories are based on symbols that were previously assigned meanings inside a cognitive process, we can call them 'thoughts', because they refer back to the same overarching process that created them. In this way, a cognitive process sews together past and present, with the help of memory. The symbols or patterns of memory can be associated with forces, colours, shapes, standard characters, words, and phrases, which can then be used to communicate from past to present, and on into the future. Some messages will be gibberish, while others may be ingenious, taken in context.

This is an analogy obviously, but it is not as spurious or fanciful as it might sound. It's natural to extend insights on one scale with others. Spacetime, which promises a variety of semantic labels on a variety of scales, certainly has all the necessary structure and more. It turns out to be quite a fruitful exercise to relate spacetime to language too. The strings, words, and sentences of language organize into messages, and the messages exhibit all the variety we see around us: they are places, atoms, DNA, buildings, living organisms. The universe, in this point of view, is surely the most ambitious story ever told. What other story can rewrite itself, even as the characters look at themselves from within?

Symbolism bridges what is simply promised by spacetime (+) with what is interpreted (-) from it. It invokes qualitative semantics, and semantic relativity, rather than Newtonian or Einsteinian quantitative relativity—but it builds on physical representations that can be expressed in space, so it is slave to physical relativity too. By standardizing measures, we can compare what other variations happen within the measures. Within each square on this page, you find different alphabetic characters. This is how we encode language.

A basic alphabet of symbols—even from a periodic table of elements in chemistry—describes a vocabulary for communicating concepts. And by concepts we mean: structures that recur in a process—phenomena we meet again and again. In physics, the vocabulary of phenomena consists of different forces, their agents, and their manifestations. In biology, the vocabulary is built from chemical elements arranged into amino acids characters, and from there proteins that act as phrases or whole sentences. In computer science, the vocabulary begins with bits and standard measures of bits (bytes, words, etc). These allow representations of numbers, images, and human alphabetic characters that we use to describe all the wonders of the Internet. In linguistics, it's patterns of sounds or symbols that trigger associations to experience.

From the bottom to the top, we express meaning in small or large packages by emphasizing reliable distinguishability. The trick to scaling the meanings of small things therefore lies in the way that concepts can be associated in order to reinforce those meanings, and build new ones. It has to be stable, so it has to be done in a way that cannot be dispersed by mixing. Relationships between concepts will be the new adjacencies at this level of virtualization. Yes, as expected, we are looking at yet another virtual layer of spacetime: conceptual locations connected by associative links. The connectivity expressed by a space of symbols, joined by meaningful associations forms a network space: the symbolverse, if you will. The boundaries of meaning can be modular and hierarchical too, just like any other space. A collection of parts becomes a car, a collection of cars becomes traffic, and so on. So, to connect the dots between simple spacetime processes and high level concepts, we just need to understand how scaling can occur, without turning distinct patterns into mush.

SCALING LAYERS OF COMMUNICATION

Claude Shannon defined the concept of communication in terms of symbols, because he realized that communicating intent is about a simple process of passing on symbolic meaning. By transferring intended meaning to a collection of easily recognizable proxy patterns, the mechanics of the process transforms meaning back into information. Turing too based his theory of computation, and Turing Machines, on symbols written on a tape, which could be manipulated with rules for associating one symbol with another. Those transitions, known as 'state machines', look a bit like the scattering processes of quantum theory, yet they can be made to exhibit simple processes like addition, subtraction, and elementary logic.

Let's consider how we get from elementary information to more complex concepts by scaling symbolism. Concepts are ideas, but symbols are just distinguishable information: patterns in spacetime. Clearly, we can make patterns on any scale, but ideas go beyond information. Take the idea of the number "5" as an example (see figure 8.10). Five is not a single clear idea. There is a whole network of related concepts that cluster together to create 'fiveness'. There is the idea of five things, which might be represented as five fingers, or a tally with four bars and a stroke through it. Five is associated with counting. Five precedes six, and follows four in a sequence. On a computer, five may be represented by the font image we see on the screen, which is a collection of dots (agents collectively promising 5 as a 'superagent'), and in software it's represented as the ASCII code (53) used by programs to label the font pattern, while in memory it's a binary pattern 101. The binary patterns are the ones used by the computer's CPU process. The decimal value is the one used by the programmer, and the font pattern is the picture seen, by the user, on the computer's screen.

The concept of five is clearly an accumulation of patterns, not any single pattern—which is one reason why computers aren't too smart at understanding: as passive memory, they deal in one thing at a time; only an active process can seek out multiple relationships by linking regions close together as a measure of similarity. At present, we only have technologies that can recognize some of these patterns at a time, not all of them. The basic alphabetic building blocks for

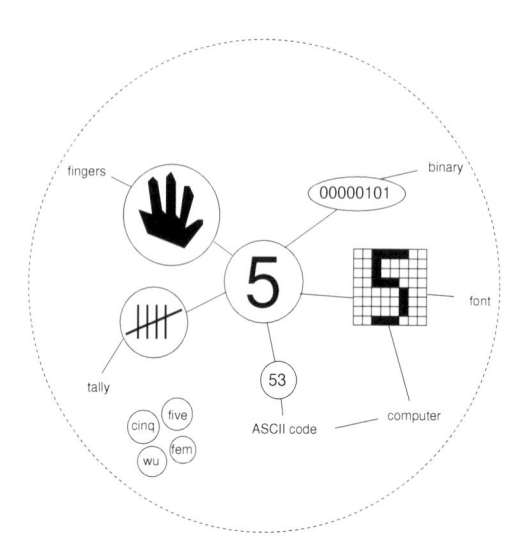

Fig. 8.10. Associated representations of the concept of five. There is a hierarchy of layers that overlap with other concepts.

all of the concepts can always be called 'agents', in Promise Theory parlance, and clusters of those form superagents with their own implicit boundaries. They are bounded spacetime processes, starting with very simple and passive processes that maintain a stable state, and accumulating complexity as they are combined into ever more intricate stories. Euclidean spaces have no natural way making boundaries between different concepts, because they are scale free—but networks do: gateway nodes, like the hubs in figure 2.18(a) and (b) and in figure 8.10 act as natural borders, just as gateways form natural boundaries for other semantic structures: houses, cities, countries. Roughly speaking, the links, or associations, promised (+) and accepted (-) between superagent clusters, is what gives them all meaning in different contexts—but we need to go further to put meat on these bones. We need to be able to say what kinds of relationships the lines represent in figure 8.10. It's easy to wave one's hands and claim a relationship, but mathematical sciences aim to be more precise than that. The four classes of relationship indicated in table 8.1 turn out to be all we need to perform basic storytelling feats, and therefore to describe processes and to 'reason' about them.

Figure 8.11 shows the structure of an 'event' characterized in terms of

8 SPACES THAT SEE AND THINK

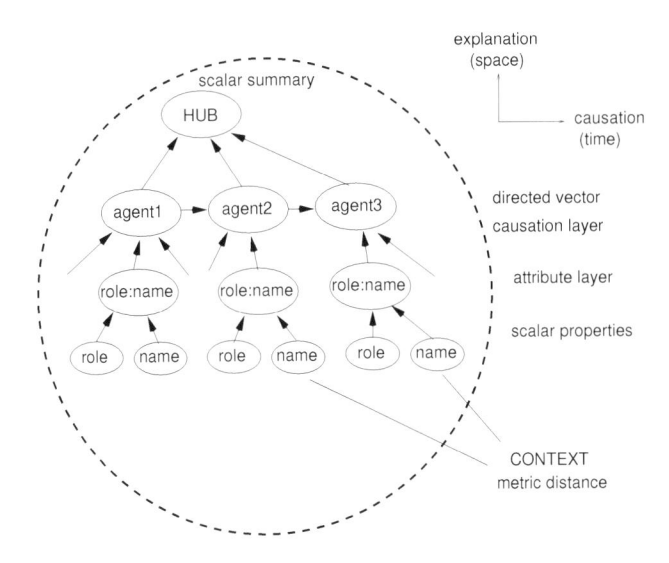

Fig. 8.11. If the types of adjacency are not homogeneous and isotropic as in Euclidean space, but are polarized to capture ordered sequences, and are weakly typed, like the fundamental forces in physics, to capture the four types of spacetime role—then adjacency can perform reasoning, and form a representation of episodic memory. As one scales up concepts based on deeper concepts and episodes, there is the possibility to form complex cognitive processes and advanced 'intelligence'.

high level concepts. An event is a process atom on an advanced conceptual level—a unit of 'happening', as a person might experience it. The structure is hierarchical and it is polarized like a layer of vector like dependencies—a bit like a blockchain, or chain of evidence. Chains of relationships cannot form concepts without being bound together by something that names them as a single meta-concept, so there is a hub that binds the parts together 'covalently'. The layer at which events are connected is also far from the layer at which their detailed properties are explained and disambiguated. For example, suppose we take a short episode or process fragment (see figure 8.12):

> Professor Plumb murders Miss Scarlet in the library, with a bread knife, because she refused to marry him.

This sentence, which humans have little difficulty in parsing and forming a picture about, is already a surprisingly complex representation of concepts,

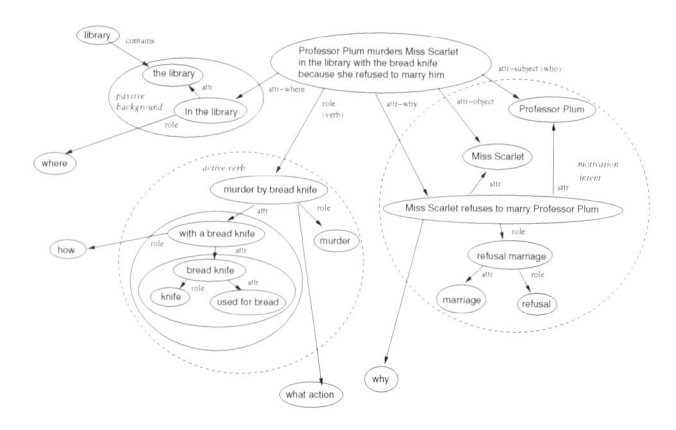

Fig. 8.12. A structured reasoning map, with semantic scaling, based on the four basic spacetime associations.

ultimately related to sensory perceptions of space and time. We have to be careful to separate our linguistic inferences from the actual process, in this simple case, to understand why the two levels of understanding can't always be separated. Let's try at least for this case.

The concept of 'Professor Plumb', 'Miss Scarlet', 'Library', and so on, are active agents in this scenario. They are 'nouns' in language parlance. They interact using named interactions 'murder by bread-knife', and 'refused to marry', that represent channels of communication connecting the agents together. The channels use other agents as proxies, co-opting and virtualizing their roles into mere messengers. We understand 'knife' as a weapon rather than caring about its shape, or its manufacturer, or whether it was a butter knife, a fish knife, or a cheese knife, etc. So the agents in play have roles that disambiguate their meanings. In general things have names and roles. In computer science, the roles are called 'object types'—they represent a kind of promise made by an agent, in a particular context.

The concept of knife, in this context, is a descriptive attribute of 'murder by knife'. Murder is also a descriptive attribute of 'murder by knife'. That common fragment binds to two concepts together and becomes a new atom[245]. This hierarchical structure avoids an explosion of different rules for different types, allowing all concepts to be expressed in terms of the four basic spacetime relationships. Every episodic concept can be represented and referenced by its hub, in this way. The hub is the single point of calibration that represents it. If we only link to that, we 'inherit' all the sub-connections that explain it.

Neuroscientist Jeff Hawkins, of Numenta, has developed a form of semantic memory called Hierarchical Temporal Memory, which provides a model of how this kind of structural semantics can be learned from a purely dynamical spacetime process on a homogeneous background, using only symmetry breaking at its inputs. The model uses superagent hubs to pool related ideas as a network of hubs. When following well-trodden paths of contextual memory, storylines are clearly illuminated by closely linked context. When searching or learning blindly, the hubs broadcast to all their neighbours like the flooding algorithms used in network science—a kind of spacetime loudhailer[246].

Note carefully the arrows in the diagram: adjacency or 'being next to', in this space, is not a mutual property: the relationships are one-way streets that point towards the summation of knowledge. It's as if the sum of fragments of information vote for one outcome by pointing to it. That outcome might be a shapeless blob, a blank stone, or a calligraphic symbol—but by having the support of all the contributing concepts pointing to it, it assumes the meaning of the full sentence description.

The order of the words, in the example, is a linguistic convention only, for convenience: the sum of the elements generates the meaning. The names, like Miss Scarlet are also hubs that expand into meanings. The fragment 'Miss' is really 'Miss in the role of name title' (not miss in the role of failing to hit a target). The concept 'Scarlet' is 'Scarlet in the role of name' (like Captain Scarlet, as opposed to the colour scarlet). Each fragment boils down to a collection of blobs whose only meaning lies in their occurring in a context. That context comes from the sensory information a cognitive agent collects and processes. So it seems clear that context plays a large part in addressing and naming locations in a memory, and that the polarized structure of the memory connections determines an order of processing. The order actually has two dimensions: extent or space, which feeds into the disambiguation or *distinguishability* of the entire superagent, and sequence or time, which expresses the order of events in a story fragment.

Once we have a fragment like the one above, we can build on it further:

> That time when (Professor Plumb murdered Miss Scarlet in the library, with a bread knife, because she refused to marry him) was just after (we watched Star Wars in the new movie theatre in New York).

The figure explaining that concept would already be too large to draw on a single page, but the principle is the same. As usual, scale confounds our ability to imagine it easily—but we can imagine the principle behind it, and even implement it in machinery, like computer software.

In practice, there are layers of abstraction formed by hubs—just like the switching hubs we find in communications networks (your home WiFi router), but on the virtual level of semantic adjacency. Because abstraction is a received interpretation (-), it doesn't matter too much what physical structures underpin (+) the agents and promises in this picture. It feels futile to separate hard physical processes from abstract virtual processes—there probably is little meaning to the distinction, which throws into question what is truly elementary in our understanding of physics. The nature of *interaction*—with the binding of give (+) and take (-)—contains all the degrees of freedom necessary to explain complex phenomena and meanings without the need to invoke any veil of mysticism.

Hierarchy of meaning

Having represented ideas as networks of symbols, there are different ways to reason about them. Reasoning has come to be a loaded term in the various compartments of science and technology, and it means different things to different people. I take in perhaps its purest form, as a word that has its origins in the 'consideration of alternatives'. The simplest kind of reasoning may therefore be to treat symbols in a network as indivisible atoms and to trace out paths that combine symbols in order to tell a story. Reasoning, in this fashion, is about being joined together as neighbours in the semantic network. Each location is an idea, and each finite path becomes a fragment of a story.

Locations in a semantic network scale from single atomic agents to superagents, and each new level of aggregation can make its own promises about what other concepts are next to it. For example, combinations of binary patterns can represent alphabetic letters or images, images may represent faces, characters aggregate into words, words into compound words, and so on. We can also introduce different channels, like different 'forces' that have different significance—these are the four basic associations described earlier. In the previous chapter, I described how electromagnetic, nuclear, and gravitational forces are effectively different types of adjacency that guide the evolution of changes in physical spacetime. In the dataverse, there is a similar channel separation of channels between passive and active process in computers. The reason we need to go from purely metric or dynamical spacetime to semantic spacetime is that the property of agents 'being next to' one another has no invariant meaning outside of a clearly defined process.

But defining thoughts and ideas as stories formed from connected concepts is not a widely accepted hypothesis in computer science or in neuroscience—I suspect that's probably because most researchers don't understand how to think about the processes that operate on multiple scales. Certainly, Computer Science has no such body of work on scaling of semantics or interpretations. However, having worked on Artificial Reasoning systems for a number of years, I've come to believe that traditional ideas about reasoning have become distorted by an inappropriate belief in mathematical idealization. Turing's computation is an extremely limited form of reasoning, and one that's not representative of human thought. It doesn't contain any of the key ideas, like context dependence, fast prioritization, and slow verification, etc, even if it could eventually simulate them. Computers operate on computational 'logic'. As mentioned at the start of this chapter, mathematical logic is a deliberately over-constrained version of rational reasoning, whose goal is to eliminate uncertainty. The trouble is that you can't eliminate uncertainty, unless you eliminate most of possibility—and that's not what intelligent reasoning is about.

When AI tries to reproduce general intelligence, researchers have mistakenly tried to create a rational form of intelligence, which is not what humans identify with. Rational methods use mathematical ideas and numerical argumentation that is supposedly unclouded by foolish human emotion, but I believe that this attitude to logic makes a serious error of reasoning—and that emotion is, in fact, a rational optimization of reasoning, which is far from foolish. Probably, a more useful way to think about reasoning is in terms of stories, in an analogue of the way Richard Feynman pictured process histories. That needs further explanation.

A computer program is also a process that depends on stories—but formal stories formed from symbols that belong to a single alphabet—i.e. agents belonging to a single semantic scale. In a computer, there are already two kinds of symbol: pure data symbols and meta-symbols, which form 'command instructions', that guide the behaviour of the computer. The meta-symbols form actionable instructions relative to the process that interacts with the data. On a simple level, a command is any special combination of symbols that changes the trajectory of a process, but commands are patterns of symbols at different scales or behaviour: bits, bytes, words, documents, images, movies, etc, all play different roles in the world of information. A program is only a template for translating symbols into behaviours, encoded in the memory of the computer—just as chemical processes are templates for activities of attractive forces in chemistry and biology.

Any form of space that can adopt different recognizable states can be used as memory or to issue instructions, and can be programmed to command (+) instructions to other processes that accept (-) them. This is how biology forms an ecosystem of communicating neighbours. As long as we can tell the difference between a 1 and a 0, a yes and a no, we can cluster locations together to make any kind of symbol. There just have to be enough locations to capture all the details. As the scale of processes increases, agents can incorporate more interior complexity, and more complex processes without exterior help. This is how biology can create such amazing contraptions from its application of the processes. That's one reason why we shouldn't expect methods used by Artificial Intelligence in the dataverse to compare easily or directly to processes performed by active cells in the bioverse. The process clock in the dataverse is exterior to its raw memory locations, whereas the memory cells in the bioverse are living things, inside and outside the arbitrary boundaries we name as entities.

In his seminal work on the theory of computation, Alan Turing represented memory as a tape, because that was the level of available technology at the time. Memory in a Turing machine thus became a single one dimensional space, but not a one dimensional timeline. Even so, time was not a unidirectional matter in a Turing machine: processes could run the tape back and forth to alter previously written values, turning back time locally. A one dimensional space is effectively equivalent to a purely timelike process, if it is unidirectional, say in a one dimensional memory, so a computation that aggregates more than one input can only be carried out with spacetime reversals. Even quite simple computations involve an interaction between data in space and pathways that define a time. The virtue of having a multi-dimensional memory, such as we find today in three dimensional universe, or in a parallel cluster of computers, is that it allows a data to express symbols of any kind with respect to any process, without needed unnecessary reversals. This is how we arrive at our familiar unidirectional idea of time.

Turing proved that his definition of what we now call a Turing Machine could simulate any computable process—yet this result remains widely misunderstood. Some have claimed it to mean that everything computable is a Turing Machine. That is not the case. Although you can simulate any process with Lego, it doesn't mean everything is Lego. The role of multiple dimensions and parallel paths of communication is an obvious case in point. The relatively new field of Quantum Computing makes clear use of this extra dimensionality, as does biology and even cloud computing.

The locality of processes, on different scales is what leads directly to the special importance of a central observer interpretation of space and time. Physicists and relativists may have been obsessing over the wrong symmetries, unable to see what phenomena of interest lie in information science—and that's what a cognitive system ultimately derives from: observation from a single vantage point. That translates into point-to-point channels of information, and hence a sequential or *linguistic* form of communication, as a single stream of information. A single localized cognitive process can easily transmit and accept a single stream of information from a single source—from mouth to ear or from hand to eye[247]. It has to queue up information in order to process it. That doesn't mean there cannot be forms of communication that are not serialized, but those might not be suited to the geometry of a singular localized organism.

AGENTS AND PROMISES OF LANGUAGE MECHANICS

Spoken language doesn't seem much like the spacetime of Newton or Einstein, but it is spacetime in the sense of being a process that traces out a changing symbolic structure in sequence. The agents of writing are spatial symbols; the agents of speech are phonetic patterns, which could be observed as wave patterns if we had advanced vision. The sounds and shapes of languages form something like an alphabet that we draw on again and again. We divide the sounds into vowels and consonants, and in the alphabetic forms of writing the alphabets are closely related to the sounds used. The particular set of sounds varies greatly between different languages, just as the writing systems vary: from the written Semitic languages that seem to have no vowels, to the spoken Danish language that seems to have no consonants.

The spacetime agents may be phonemes or letters or words, at one end of the spectrum. They might be sentences (modules) or paragraphs. They could even be entire stories that can be wrapped neatly and packaged. The natural boundary of an agent depends on its exterior promises, i.e. on its semantics. This is in stark contrast to the Euclidean view that points and locations are universal and independent of any interpretational concerns. That is surely no more than humans projecting their beliefs onto an objective reality. The encoding of information in the dataverse can take on a variety of representations, of which natural languages are the most complex forms. In information technology, it has become the norm to speak simply of 'data', as a stream of symbolism for encoding by a plethora of underlying binary formats (ASCII, EBCDIC, UNICODE, least significant byte first, binary-coded decimal, big-endian, little-endian, etc) is taken for granted, precisely because it is of fundamental importance. But of course, before any of these digital forms existed there were physical patterns of sound, gesture, and sequences of events ongoing in the environment of early organisms, whose recognition would confer a survival advantage. Symbols can take any form, and do not lead to language by themselves[248].

The innovation of language lies in compressing symbols into definite or bounded events that can be strung together to scale meaning from single events into stories. There two urgent questions immediately come to mind: where do the symbols of our human languages originate from, and how do we combine them into something that becomes words and ideas? One natural source is simply spacetime.

If we look beyond the apparent one dimensional stream of symbols, whether written or spoken or growled or mimed, there are certain universal structures in language: including dimensionality, scalar and vector promises. The encoding of the information trajectory varies considerably, but the specific symbols might not matter as much as their semantics. The causality is also encoded as a universal set of precedence relationships. This is what we call representation in mathematics. It is a step towards incorporating semantics into a purely dynamical pattern. It's usually much easier for us to look at a document of text and see something the looks like space, divided up into regions and sections (unless you look at Arabic script), because writing and pages have a simple metric coordinate grid, with certain locations playing special roles. Words act like cells, forming a tissue of sentences (the expression a tissue of lies is still used). Parenthetic digressions allow us to venture into other dimensions—to form superagent cells—in order to add embellishment using interior structure within a flow. These process structures are what we call grammar, and linguistic grammars were identified and classified as computational processes by linguist Noam Chomsky.

In 1959, a young Chomsky wrote a seminal paper[249] that classified the mathematical properties of formal language models according to the kind of computational process that was needed to comprehend them[250,251]. He identified four levels of grammar, but only in the idealized sense of formal or pure languages. He showed that certain features of patterns required more sophisticated processing than others. For example, the simplest 'regular' languages are just streams of symbols that form repeated patterns, according to basic rules of combination and repetition. However, if we introduce parenthetic remarks (like this) then the interpreter would need to have some memory in order to recall its position in the main sentence, while making sense of the phrase in parentheses, and later attach that sense to the entire meaning as a model of the concepts being discussed. In a spacetime sense, a parenthesis is a hidden dimension that feeds into the sentence like a digression in the road (see chapter 2). Physicists would recognize it as playing the role of a loop in a Feynman diagram of particle physics. Chomsky associated the four levels of linguistic complexity with four kinds of 'cognitive machine' (in my terminology) that could model them. Simple regular languages can be followed by so-called Finite State Machines, that have no memory at all. Increasing amounts of memory and sophistication are needed to parse each layer of language, until—in the final instance—the most general language requires a Turing Machine, i.e. what we understand to be a full computer.

Grammars are nothing more than the coding of repeated structural patterns that allow the scaling of meaning from a stream of symbols. They are multidimensional expressions of symbolic aggregation, representative of processes that are effectively multidimensional, but which need to be packaged into a single dimensional queue in order to be carried over a single channel. Grammar is not a set of rules to follow, as we sometimes learn in school, but rather a set of navigational markers that help us to recognize the patterns of intent in messages. The linguaverse is a squashed superset of the symbolverse. Patterns of symbols are first collapsed into sequential (one dimensional) stream, using markers like spaces, commas, and parentheses, to express 'inner dimensionality'. Sequential streams are, by their nature, temporal when communicated, but may be stored in a spatial memory[252].

Human languages don't follow precise rules of Chomsky's formal languages:

they are 'stochastic', meaning that they vary in the use of different constructions, parroted and assembled from memory rather than necessarily being expressible through mathematically precise rules. In scaling terms, human language has an entire chemistry of symbols. We learn and attach linguistic meaning not only to the basic atoms, but also to quite complex molecular patterns. Every individual speaker of a language shares the same basic atoms (characters, sounds, and words), but we have cultural extensions that are private on an individual level, e.g. within a cultural group a particular phrase can mean something unique. The alphabet is not constant—it's extended via scaled superagent clusters. There is therefore more information in human language than in a formal grammar, which is almost certainly a key reason why machine learning methods of AI have shown themselves to be superior to model-based methods for language recognition. Humans simply learn certain phrases as patterns of words, just as we learn words as patterns of symbols, and songs as patterns of phrases. Layer upon layer of patterns at different scales dominate out cognitive processes—learning by rote, inconsistently across the hierarchy of process scales, in order to form new meta-symbols. These get repeated, without reference to any original meaning. For example, the ability to classify patterns does not necessarily lead to an ability to find meaning in new patterns, as Chomsky's own example of legal grammar indicates:

Colorless green ideas sleep furiously

The coarse classification of word types in grammar can only capture the most common instances of language. The extensible freedoms a cognitive learning network extend those rigid rules—the ability to add contextual meaning by brute learning—is crucial to scaling an adaptive living language. Our parrot-learning of phrases plays a more prominent role in human language than is strictly necessary in formal systems. One might argue that human languages are not mathematically efficient as we rely on memory instead of precision to match concepts approximately. This is a long and fascinating discussion that, unfortunately, will not fit into this margin. Nonetheless, this is what one expects of a network of concepts coupled strongly through a rigid hierarchical process.

Chomsky, and his many disciples, made inferences from his work that proved to be controversial, and which went beyond what was immediately implied by the model. Their school of thought posited that the capabilities at work in a human mind were of such a general and invariant nature that humans must

8 SPACES THAT SEE AND THINK 381

be genetically endowed with a kind of universal ability to parse grammars, and some even proposed that a single innate layer of grammar must span all human languages[253], and he offered some evidence to support the claim. Many linguists have since refuted the strength of the assertion, offering their own counter-evidence. Some even rejected it entirely, claiming that language is learnt entirely from birth with no role to be played by innate ability. A spacetime perspective is thus interesting to think about.

In order for such a cognitive system to evolve, one would expect the root stimuli that lead to concepts to arise directly from sensory experiences alone, i.e. from the environment. Being alive or dead is one pretty fundamental and binary sensory input, so selection based on survival is also the basis of a cognitive process on a species level. We know that animals have evolved sensors for light, motion, shape, direction, sound, etc, and these are triggered in different combinations to discriminate between different phenomena. These basic patterns already form a basic genetic vocabulary on which to base a conceptual hierarchy, but they are very far from the kind of advanced concepts we experience today[254]. There are two other sources of memory that can play a role in modern language, each of different timescales: brain memory from birth, and societal memory that is passed on from person to person in real time, and deterministically from generation to generation in social groups. Genetics can learn to adapt to fundamental properties, on very long generational timescales if they confer a survival or selection advantage, but societal memory retains a library of patterns that a newborn baby can absorb wholesale in a short time, without having to rely on a very long term process of probabilistic selection. Over a single lifetime individuals can grow a model of language(s), which is very likely closely related to the way we connect and embellish the concepts in our brains. The basic hardware mechanisms need to have a genetic factor there in the linkage, but that occurs on only one of a hierarchy of scales involved. The question is whether that genetic influence survives scaling, or is washed away by other dominating influences.

Promise Theory suggests that it's likely that the structure of the four universal relationships is encoded in memory and is innate. It is based in spacetime processes that derive from our sensory apparatus. But it also suggests that the details of languages' patterns are a memory process, where memory can be anything from interior memory to exterior memory, social memory, genetic memory, and so on. So the two claims may both be right about different levels of learning.

Navigating language with dimension and coordinates

Chomsky's work identifies language as a process, and therefore as a spacetime. The dimension of the spacetime—like any other—is partly a matter of definition. This book is itself a semantic spacetime: a series of modules formed from symbols, connected in a structure that is somewhere between one and three dimensional. Dynamically, it has ordering dependencies that make a timelike trajectory follow a single string of characters in one dimension. More artificially, it has chapters and sections, that break up the text into coordinate tuples, which we can choose according to our viewpoint:

```
(CHARACTER)
(PAGE,LINE)
(CHAPTER, SECTION, LINE, WORD) ETC
```

Text editors often count line numbers and word counts—those metric values are sometimes useful in structured documents, like computer programs. The embedding spacetime of the page is only one representation. If we read the book on an electronic reader, then page numbers become an irrelevance. The dimension of spacetime theatre, as we see is somewhat arbitrary.

The dynamical trajectories—the pathways—are stories. A book tells a narrative, in either a single stream, or as an anthology of small stories. We try to separate fact from fiction, believing one to be more fundamental than the other, but linguistically (impartially) they have the same form. The property of telling stories is a linguistic spacetime structure. Space corresponds to the varieties of words, sounds, or ideas, we can visit along a path that represents an intentional utterance. As we continue to scale these utterances, they may lead nowhere, or they may take us on a very specific semantic journey, which we call a narrative. It takes some people longer than others to reach a destination!

There is scaling here too. Words or sounds scale into ideas. Ideas scale into larger ideas, buy composition, and context comes out of this. The Chomsky hierarchy underlined the deep connection between linguistic (serialized) representations of symbols and finite state machines. The implication of this connection, for spacetime, is that the nested structures of agents being next to one another, or inside superagents, are all derivable from the behaviours of simple causal machinery that follow invariant rules. They are, indeed, bits of machinery that can perform different levels of computation. Language is a process: it is transmitted and received by agents. It is sampled causing clocks to tick at the endpoints, one character at a time, but it is a clock in its own right too. Each symbol character is an event that moves the shared clock of sender and receiver forwards.

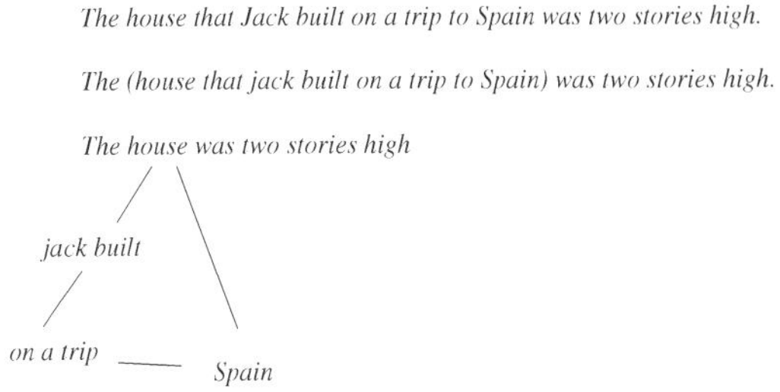

The house that Jack built on a trip to Spain was two stories high.

The (house that jack built on a trip to Spain) was two stories high.

The house was two stories high
    /   \
jack built
  /
on a trip ——— *Spain*

Fig. 8.13. Multi-dimensional grammar. The parenthetical remarks in adjectival and adverb phrases are dimensional extensions of the basic symbol for a house.

STORIES, HISTORIES, AND PATHS ARE SEMANTIC SPACETIME

The role of spacetime in the process of communication of intent is pervasive, at all levels, and refers to several different virtual layers. Sensors adapt to process exterior spacetime phenomena, in the environment around an agent. This is a physical reality. The sense of exterior time from this environment would affect ordering of memories, else we would not be able to recall sequences and stories. Our own interior sense of time must be used for recalling memories, and playing them back to tell stories in different representations. This is an interesting prediction, implying the existence of some kind of storytelling organ that takes scattered memories and replays them into what we perceive as sensible storyline. Interestingly, when we are asleep, cut off from the anchors of sensory inputs, our dreams construct quite unhinged stories from random places that memory.

Within the physical network spacetime of our brains, there is a virtual spacetime of concepts and associations between them, featuring extended patterns. These could only be formed from aggregation of smaller parts into larger parts (the formation of superagents from agents, in promise theory language). The interior structure is most likely related to way we form grammar in our spoken language (and vice versa) else it would cost too much to perform complex computations every time we wanted to say something. The scaling principle tells us that, as these conceptual superagents grow, they become entangled with all kinds of contextual references that embellish them, leading to quite new promises being made by the superagent itself. So a cluster of words like 'son of a gun' can make quite different promises than the words it consists of. Eventually, these phrases become habitual and their new interpretations become reinforced[255].

Although this spacetime model of language in relation to consciousness is speculative, I would find it surprising if our extensive facility for metaphor and metonym in language were not a simple consequence of the scaling of the only concepts available to evolution for bootstrapping a representation of conceptual symbolism. Basic spacetime distinctions are the only initial inputs we have to go on, and from there we would need to rely more and more on memory to define and embellish concepts.

When we finally come to pass on ideas, through language, a receiver needs a common set of concepts, sounds, and symbols (emphasizing the role of shared societal memory) in order to recognize what a sender is talking about. Here again, experiments with semantic representations in the virtual spacetime of concept networks, show that the scaling of concepts, by forming bounded clusters of concepts held together through interior relationships, must be quite close to the structure of the language and its 'rules'. So the question remains: do the innate aspects of our brains play a strong role in determining how language concepts are represented in memory? A network spacetime model suggests they would only do so in quite simple languages. The more sophisticated the language, or the greater the number of levels of virtualization, the less likely it would be that we could see any evidence of it.

Using simple promise theoretic considerations, I was able to show that this story is plausible for the generation of readable sentences in one version of English, even on the tiny scale of a model, and using only the four basic spacetime associations discussed in chapter 8. The scale of a real brain would make such a toy model seem ridiculous, but it confirms that the mechanism could contain some truth—very far from being proof, of course, but I think it shows that processes, whether organisms or machines, don't need anything more than basic spacetime knowledge in order to develop a rich language, provided they have sufficient memory to model their historical experiences.

On top of all this, Chomsky's grammatical classification forms a kind of map over the one dimensional streams, and therefore implicitly over the fragmentary representations in memory from which sentences and stories are formed. The Turing machine is the pinnacle of this serial version, and cellular automata general that the spacetime fabrics. John von Neumann and later Stephen Wolfram argued the importance of cellular automata.

The separation of (time)scales is a difficult concept, easily forgotten even by experienced scientists, but it must play a vital role in complex systems. Language evolution fits that mold, and remains the summation of nature and nurture over successive generations. Genetic memory is not really different from brain memory except that it is learnt very slowly, over many ticks of the reproductive generational clock, while brain memory is learned much faster over ticks of the experiential clock. Chomsky's work implicitly forms a description of the modularity in both human communication and computer programs, as well as in other algorithmic processes like business process.

Stories play a central role because they are a literal representation of processes in the symbolverse, so they represent spacetime at that level. Creating stories by tracing out concepts, in the symbolverse, is how a cognitive agent both communicates on its interior and exterior, as well as how it generates ticks of interior time. It's quite important to realize that language and stories are not concepts exclusive to human imagination and communication. Stories are simply another name for processes. Admittedly, not all stories may be immediately recognizable to us, as humans, but our ability to decipher the stories that make the universe work grows every day that we study the sciences and the arts.

Nature tells stories of its own, through its many processes and timelines. Feynman called his version of Quantum Mechanics a sum over histories. I call processes in semantic spacetime a collection of stories. Both are representations of information along directed paths, but stories are symbolic, and that means they are effectively digital. At the blunt end of the spectrum, we could mention the layers of geological deposition around the Earth, which contain a timeline of history. From this history we have extracted murder mysteries, ancient treasures, and even dinosaurs. The layers are uneven two dimensional shells, deposited on land and seafloor; the radial dimension plays proxy to the role of time, as each new layer of memory is accumulated. The deposition of the memories (+) is a completely independent process to the geology, archaeology, and palaeontology process that retrieve and interpret them (-). Neither were planned, yet both come together to enable a transaction of information, just as if there were a geological wavefunction across the surface of the planet. Across the modular regions of the planet, there are parallel stories ongoing within different spatial boundaries, usually carved out and bounded by the topography, as mountains and rivers form a natural boundary to isolate cells. The surface dimensions express variations that have differentiated due to local conditions in those random positions across the erupting landmasses, the unequal effects of the sun, and local chemistry.

In the animal kingdom, stories and labels are written in scent trails, or enacted with sounds and gestures, to explain the semantically recognizable features of a journey, perhaps to find food. Simple honeybees, for instance, have a symbolic language' that allows them to communicate the precise coordinates of food sources to one another. In this dance language', a scout bee performs a repetitive sequence of movements on a vertical part of the hive. The bee moves forward in a straight line for a few centimetres, then in a half circle to the left, back to her starting point, straight again, and then circles to the right. The duration of the straight path tells other bees the distance to the food source. The direction of this path relative to gravity encodes the direction relative to the sun, whereas down' means fly in the opposite direction of the sun'[256]. So bees transform one set of symbolic information (topography) into another (gravity and asymmetry). Not only do many organisms seek out and explore space, either intentionally or at random, they manage memory processes to follow repeated trajectories to map out regions. They transform journeys into new virtual representations based on signalling, and can explain their experiences to another organism.

The invention of *language*, in the animal kingdom is a direct representation of spacetime. It seems very likely to be a direct response to the need to recall journeys through complex topography and to pass on the memory to navigate paths. Language provides another perspective on how memory plays a role in our understanding of spacetime. The two dimensional maps discussed in chapter 2 were only one possible representation of spacetime, usually as two dimensional pictorial representations from a god's eye view. Nature found other representations too: in language, a map that takes the form of a number of actions to carry out from a well defined starting point. It is exactly what I've called a process with a retarded boundary condition, encoded as a one-dimensional stream of symbols, and interwoven with layers of parentheses and orderings.

Evidence that bees can memorize and reproduce in order tells us that they have the ability to serialize experiences in memory and retrieval, and thus they have the capacity to understand stories at some level. This is the first spacetime necessity for consciousness, so—while we might debate the depth of their experiences—we can't debate the fact that they have experiences, however simple-minded. The degree is once again a matter of scale.

LINGUAVERSE PHENOMENOLOGY

So language itself is another spacetime process: a sequential process that takes place, using symbols, over a communications channel. In other words, one that incorporates memory explicitly to learn meanings, associating semantics to clusters of symbols. Underneath that process is the simpler dynamical process by which the atoms of meaning are transported from one agent to another, as a string of symbols: either phonemes of sound, or symbols in writing. Shannon himself wanted to separate these two layers of process—distinguishing the meaning of symbols from the pure dynamical process of communicating them, even just on the scale of a single communications channel[257]. That fundamental insight effectively opened the doorway to modern computing. Underneath that again, is a representation of the process of communication using forces and agents of physical transportation—which is a language of its own, simpler and more limited in its expressive powers, but whose words are ultimately reducible to the four spacetime types of associative meaning.

There are two other ways in which language invokes spacetime structure: in the encoding of words and meanings within the stream of sounds and writing, and the rescalings of meaning that come from the combinations of those words into sentences, phrases, and metaphorical associations. If we look at how the meanings of words are inherited from earlier words and previous usage, concepts overlap in different ways in different languages and regions. For example, in the UK, a goal scored in one's own goal (for the opposing team) is called a home goal, attaching to the use of 'home' as part of a self vs non-self narrative, but sounding like a domestic aspiration. In Cantonese, southern Chinese will use a term 'oolong', meaning 'mistake'—and sounding like a kind of tea. So the concepts overlap with quite different associations along different dimensions of meaning, because of accidents of history. Another example, in the UK, the political parties for 'left' and 'right' are called Labour and Conservative parties: there is an association with Labour as done by workers, and Conservative in the sense of religious values (rather than, say, parties for giving birth and pickling vegetables). In the United States the names are Democratic and Republican, but in terms of values they align with 'left' and 'right'. So we have names relating to opposing directions, Marxism, self-restraint, voting, and rejection of monarchy. None of these associations are plausible in any sense today, but the names stick, and continue to be built on. This is what you would expect from a cognitive model of language, based on memory rather than logical reasoning. The particular metaphors we use to build terminology depend on where we are in our thinking at a particular moment. This is easy to see in the way words and concepts are built from different combinations of cultural ancestry. Language evolves quite analogously to genetic evolution, though on a different timescale by a different process of mutation and selection[258].

Is that plausible? It's known from neuroscience that human senses are directly linkable to spacetime properties like place, direction, motion, common patterns like faces, and so on. The question is: does that seeding of conceptual atoms continue to more complex ideas. That's a speculative idea certainly, but several linguists have discussed ideas of this kind[259]. Very often, common linguistic expressions are nothing more than proxies for spacetime relationships. Consider just a few examples of common phrases that use concepts about distance, speed, time, mass:

> Christmas is fast approaching, has arrived
> We are getting close to Christmas
> We're approaching a result
> The conclusion is just around the corner
> It's a far far better thing...
> A sufficiently large contribution
> Time flies when you're having fun
> The moment of truth approaches
> The subject is heavy reading
> Just North of 2000 dollars
> Within the bounds of reason
> What makes it tick?
> Can I point out...
> At this point of the proceedings...
> Took a left turn

8 Spaces that see and think

> A long time
> far and away
> Not remotely like
> Close enough
> A long running process
> On some level
> Just in time (i.e. within the boundary of acceptable time)
> How long will you be there? etc.

See also the quotation at the beginning of chapter 1, by Heisenberg for a sentence replete with geometrical content! When businesses talk about the space of possibilities, or the journey they are undertaking, these metaphors are largely built on spacetime notions.

> We are moving ahead, but there were setbacks.
> Our position in the market is strategic.

Phrases exploit the relational ideas of space and time to explain associations and relationships, because spatial relationships are the ones we have evolved to think about. Even words can often trace their histories back to ideas about space and time. Think of the word "situation". It still means placement or location, but we rarely use it in that sense anymore. We use it to imply our context or condition. We say "currently", "At the same time,...., meanwhile" as drop in phrases that punctuate our expressions. All these phrases are transposed from simple spacetime relationships.

Search me!

There is circumstantial evidence then that language is an adaptation resulting from a cognitive process that builds concepts in the image of spacetime phenomena. Language itself is a process, which exploits space and time, and is composed from an accumulation of junk stimuli, associations, and baggage, whose contextual significance may have long since passed. After all, the purpose of language is to act as a kind of map of things that happen in the world. We should be clear though, this doesn't mean that we should expect to find language in all cognitive processes. The processes are a probably a necessary but certainly not a sufficient criterion. Nor should we expect to find recognizable spacetime concepts in all the words of every existing language. The origins of words are deep and long, and are distorted beyond recognition over generations. Compare the English language a thousand years ago with modern English, and there is little to recognize. But that doesn't invalidate the proposal that language can be seeded in this way.

What we can do today, thanks to the tool of computers, is to try to demonstrate parts of the process, using a model that constructs fragments of communicated meaning—from words to sentences to stories using this spacetime approach—to see if the result is plausible. Words are arbitrary patterns, so their origins are not the most interesting part of the problem, but sentences involving words convey accumulated meaning. One can try to assemble concepts using aggregates of words, mimicking human language, and then try to generate reasonable The outcome of a cognitive process should be a form of Artificial Reasoning or auto-explanation based on whatever words we associate with basic concepts[260].

Let me try to describe a few of the key steps. First, you pick your cognitive system, by deploying sensors that are adapted to collect data from an exterior process, then detect and discriminate different patterns. The process could be a human process, a factory process, a logistical process, even a computer game. The key is to analyze the process, i.e. to recognize spacetime phenomena on the exterior of the cognitive agent, by transforming them into a spacetime process on the interior of the agent, but on a different level. Sensors promise to tokenize— or assign single name identities to—the important states that characterize the sensor input. The techniques for this could involve Deep Learning[261], or even be short-cut by human expertise, like so-called expert systems[262]. Using human concepts might sound like cheating, but it's a fair compromise, because that adaptation is the longest timescale part of a cognitive process—the part where biology had to wait for billions of years for suitable adaptations to evolve. One can skip those parts and build on patterns we already know. The tokens from sensors form contexts, just as in the figure 8.7, that lead to concepts being activated, and result in a semantic map[263]. What we want to try to do is enable a constrained set of processes that allow a semantic network to condense around the concepts seeded initially by sensory inputs and the context they measure. Once that semantic network has formed—and grows large enough—pathways can be followed from any point in order to generate fragments of stories that describe the processes in that outer world.

This sounds quite straightforward, and indeed it is, but it's not quite the full story (no pun intended). Generating stories about the exterior world is easy, but generating stories that describe interesting happenings leads to a curious, but fascinating, new issue. Storing memories is easy, but retrieving them is much harder. Moreover, retrieving relevant memories is an even harder problem. How is it that you know which of all the things you know are going to help you in a given situation? This is something our human minds do so effortlessly that it's almost hard to imagine why it would be an issue. Yet, it is one of the major issues of Artificial Intelligence.

Think for a moment about how you store files on your home computer, or your work archives. You might have a perfect system for storing data according to a particular system, but then along comes a situation in which you are asked about a process that you worked on several years ago. How would your computer know where you put it, and be able to retrieve exactly the relevant parts without your help? Because our present generation has been spoiled by search engines, like Google and Bing, your answer might be that we look up the answer by searching. That just kicks the can along the street though—how do Google and Bing find data? Search engines run processes that crawl around all the information they can reach, by scanning the space of concepts, a bit like the way a harvester reaps crops in a field—by brute force. They then index the information using search phrases to make it quicker to find. When you type in a search phrase, the search engines basically pull up a list of index entries for the topics you search for. There could be literally millions of possible entries on the Internet for given search terms, so the entries need to be ranked according to relevance. So what is relevance?

Relevance is the extent to which your current state of mind overlaps with the state of mind that led to the information being stored. On the Internet, it's not just one person's mind, but many different people, so how could we possibly know? There is no simple answer to that question—there are many ways to assess relevance. One is to rank information by what id the most commonly sought after information. To do that, you have to gather statistics from all the other people who asked a similar question. That's not something your brain can ever do, and yet it is able to retrieve relevant memories, so we can probably discount that as a useful method for autonomous reasoning.

Another approach is to look for the concepts that are best connected to other concepts. That's an approach known as network centrality. It's like asking who is the most important person in a social network? The answer is similar to a statistical voting process, except that you collect the votes implicitly by learning over long times. An important person is a person who is deemed important by other important people. If you have a hundred friends who are kindergarten kids, you aren't too important, but if you have a hundred friends who are world leaders, then you are. The same applies to them. There is a self-consistent way of measuring importance from networks based on the links locations have. So a search engine can rank information by learned importance. All that tells you where to start looking, but then a human being has to rummage through the information to use their own intelligence to find what they want.

This gives us some idea of how to start indexing and locating memories, but it assumes many things that are far from clear. First, what does the index point to? How does it address locations in memory? Are memories in cognitive agents always numbered, like the memory in computers? That's certainly not true in brains, for instance. Are memories stored in taxonomic trees, perfectly sorted in advance? No, that can't be true either. In short, no one really knows how memory retrieval works in brains. However, it is possible to use a system of lookup based on short term memory to index long term memory. Short term memory is a characterization of a cognitive agent's current state—the things it is seeing and hearing from sensors, combined with the things it has been thinking about recently. By using a short term memory with some stickiness (what physicists might call hysteresis, or computer scientists might call a Time To Live), token concepts are switched on by virtue of being recently activated. Those concepts effective 'shine a light on' parts of the long term memory and can provide points of entry that make certain locations more 'lit up' than others. The intricate details are beyond the scope of a book like this, but it's not hard to imagine how—by virtue of interacting with an environment, and having internal feedback that keeps an agent recycling recent concepts—one has a natural indexing ranking system for a network.

So, suppose we can use an approach to memory retrieval, based on something like the method described—we still need to retrieve the stories and explanations relating to those concepts from the network of links. Again, the details of this could be intricate, but essentially we do that by following the connective links, or spacetime adjacencies. Now, if all links were the same, as they are on the Internet, then stories would end up being complete nonsense, like data dumped from the hard-disk of your computer, in no particular order. Some kind of meaning needs to be encoded. Semantic networks try to encode human meanings, but these are arbitrary and incomprehensible to a dumb process. Unless humans

8 SPACES THAT SEE AND THINK

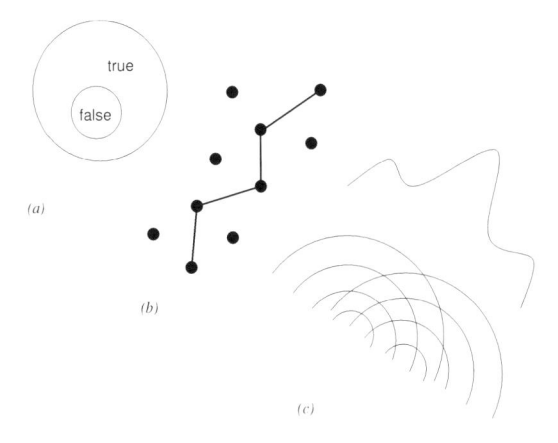

Fig. 8.14. How can we highlight or block certain paths during reasoning? Reasoning is about the selection of a spacetime path through a network of ideas—the accumulation of a story from incremental relationships between spatially localized concepts: (a) logic can only put a fence around a presumed outcome, (b) a semantic story as a DAG, and (c) interfering wave processes highlight certain points by positive and negative voting.

write specific algorithms to behave according to specific meanings, scanning through human engineered relationships is not possible for a process that can't already think like a human. It's catch 22. Moreover, when ontologies make too many distinctions between similar concepts, the aggregate meanings are lost and one finds the network of connections is over-constrained. Over constrained logic gives few or no answers because situations never recur. Everything is new, so memories are ineffective—like a many worlds interpretation of memory, in which every memory is in a separate universe, unrelated to all others. Strict first order mathematical logic turns out to be a liability, not a help. What else could we do? The most natural answer seems to be to use spacetime processes to induce meanings (see figure 8.14)[264].

The answer seems to lie in the four basic types of spacetime relationship described earlier. The four types represent the foundations of all kinds of fancy human concepts, using only the situational relationships intrinsic to spacetime. If you were going to base a system on something elementary and intrinsic, that evolution could build on, then spacetime concepts would be a natural starting point: is something close to something else? Does something follow or precede something else? Is something inside something else? Or is something an attribute that describes something else?

In the experiments done over ten years with different sources of data, I've been able to show that it is possible to use a spacetime approach to generate meaningful stories from networked concepts, with no prior knowledge except what is learned by from the evolved pattern recognitions of sensory inputs. Others have found similar results using other approaches. That's is a pretty encouraging result—but, of course it's only the beginning. The problem with such simplicity is that it generates too many stories. If ontologies are too picky, then generic spacetime categories are too relaxed. The key to relevance lies somewhere in between. Provided there's a large enough pool of data gathered by experiential learning, then there are usually too many paths that can be followed, and the interesting ones get lost amongst a dearth of completely uninteresting stories. All the stories may be factually correct, but they lack a sense of relevance to context that makes them seem absurd to a human judge. Clearly the issue of contextualization of knowledge is more important and more subtle than any one technique.

Artificial reason

To understand what's currently wrong with the reasoning based on story generation, we need to take a step back and describe the structure that allowed concepts to be combined in the first place. It's almost the opposite of the structure advocated by taxonomy or ontology. The taxonomic top-down approach tries to define concepts in human terms and then neatly file away instances of those concepts into separate folders underneath. This approach quickly leads to inconsistencies and arbitrariness. But the hierarchical approach clearly has an important property: it provides umbrella terms for similar concepts. Promise Theory predicts a solution that takes care of this, as a bottom up approach. That solution has a lot in common with structures well known from linguistics, which is surely interesting.

In the story model of reasoning, hub concepts unify component concepts, just as hubs unify network locations that a 'next to' one another. They describe inclusion rather than division, or category sorting. A reasoning-friendly spacetime builds hubs to link unique combinations into *instances* of phenomena, not generalizations. Ironically, this turns the generalization hierarchy upside down, without sacrificing the idea of categories. Indeed, it opens it up to enable emergent generalization, rather than generalization by design.

Let me emphasize that story-based reasoning is a model of Artificial Reasoning in an AI sense—not intentionally a model of the brain. No one currently knows the structure of concepts in human brains (the only natural reference point we currently have to advanced thinking), so this is necessarily speculative in a wider sense. However, from a computational or even physical mechanistic point of view, it feels natural to seek an explanation, which is both compatible with what we know about language, but which can be turned into a higher generalization of that linear and sequential mode of description. Language plays a role in some of the ways in which we think—in particular, how we explain things. It isn't everything of course, we also know that we can jump to conclusions without having to use language. Flashes of insight happen as parallel processes, as pictures and sounds in our minds. It's only if we want to explain the result—to ourselves or to others—that language becomes essential. Perhaps that's why animals that don't have language act more instinctively than rationally, and by inference why emotion plays a key role in non-linguistic reasoning.

Nevertheless, I believe that, in order to understand reasoning, we have to understand the compatibility between language and the parallel processing that we call intuition. It seems that science is still some distance away from a full explanation of this topic, but I can relate what I've learned from applying Promise Theory to semantic spaces.

Pathways of explanation

There are clues to be found, in the way human language evolves. Language can point to what a reasoning network would need to look like, in order to make the process of explanation simple and natural. The study of compound words is one place to start. For linguists, this a large topic that I am in no position to treat fairly in this book, but I can offer a flavour. Compound words leave clues about their origins, in the way words get joined together (recall the word geometry); the histories of the words hint about how humans think about concepts—and, more importantly, the clustering smaller words into longer ones also tells us something more trivial, yet more fundamental: how 'adjacency' of concepts, in memory, allows for a process of refinement rather than novelty.

Concepts, whether primitive or derived, behave a bit like 'lexical fragments', or 'syntax fragments' of language—atoms of meaning that can be reused like a table of elements. This is a spacetime issue in the following sense: two locations next to one another can be considered parts of a single location, rather than two separate locations. Similarly, two concepts, as words next to one another, may be part of the same concept, not two different concepts, depending on how their distance is measured. Some relativity may be involved in those measurements. This is how we understand scaling. The scope of concepts in language is often expanded, in this way, by this form of 'superagent' scaling—it's a kind of semantic renormalization. Root concepts become coarse grains of related ideas, surrounded by a cloud of explanatory context, conceptually similar to the way physicists think of electrons as being surrounded by a cloud of interactions processes that modify its essential properties. The concepts are like a generalization of charge in elementary particles: specific signatures for the give and take of information.

There are big ideas here, but we can start—more humbly—with paired words to illustrate the technique. A compound term is then, in essence, an extended name, which contains some qualifying context about the role of the word, in a functional sense. Consider a few examples of word pairings:

> Drinking water
> Washing machine
> Newspaper
> Coffee table
> Table leg

These compound words are fragments of meaning, represented as symbolic strings. The strings are superagents in the symbolverse; they play a role, in language, something like that a subroutine plays in a computer program. Indeed, the analogy between language and computation is borne out in a precise sense by Chomsky's analysis. Drinking water is a kind of water, but more than just water. A washing machine is a machine, but one specifically for washing things. The adjectives supplement the meaning of one string of symbols with another string of symbols. To see this in a spacetime way, you could imagine recognizing locations based on certain characteristics: a string of houses is a terrace. A string of houses near a road is a street—a slightly different concept. A string of houses with a road and a shop is a neighbourhood, and so on. In this example a house acts like a letter in a word. Having a road adjacent to a string of houses

8 Spaces that see and think

forming a terrace is like placing an adjective next to a string of letters that forms a noun: from **HHHHH** to **RRRRR HHHHH** to **HHHHH RRRRR S**. A house can act as a letter in a kind of language, on a larger scale. Who is going to speak this language? The region in which the houses are built speaks the words, but—in this case—it's not clear whether any other process is listening. Perhaps town planners see these expressions and think about them. The most likely answer is perhaps also the most surprising: the agent listening is the same agent that's speaking: a region of spacetime that acts as a cognitive agent can be talking to itself, by recycling the information it remembers (as houses and shops), and placing it in the context of influences from its surrounding interactions. If this were happening in a brain, we would call it thinking. This is a bold assertion. Does a body of language think? Certainly, it can, if the right process connects its fragmentary memories together. This book has been doing some thinking, thanks to the process of reading and connecting together ideas t tell a story. There is only one story on the page, but it probably induces all kinds of side stories in your mind, because the cognitive agent that can make sense of language is not fixed—it's a dynamical process. So let's get back to the process.

The structural part, about how compound ideas extend one another, teaches us something about representing meaning. For example, the leg of a table can be referred to by a single compound word or phrase 'table leg', in which the compound has the role of 'leg' (by analogy to an animal leg, which plays a similar role) and a contextual attribute 'table'.

$$\left(\text{leg} \xleftarrow{\text{role}} (\text{table-leg}) \xrightarrow{\text{qualifier}} \text{table} \right)$$

Or, a coffee table is a type of table:

$$\left(\text{table} \xleftarrow{\text{role}} (\text{coffee table}) \xrightarrow{\text{qualifier}} \text{coffee} \right)$$

These examples are straightforward, and fairly unambiguous. A name reflects the role of the word, and additional elements promise qualifying context, that clarifies their usage. Given that the usage is normally observed or communicated, in the first instance, the clarification is based on the process origins leading to the context.

Compound concepts don't need to directly reflect their origins, because processes forget details once a pattern is learned well. When you don't know a word well, you have to think about the stories that explain it. Eventually, you use it so often that its meaning becomes secondary to its role as a simple name. Familiarity breeds contempt for origin. Consider 'window' (which is a distortion of the Norwegian compound 'vindauge', meaning eye for the wind, clearly before the invention of glass). A window is neither literally wind, nor an eye, yet the word serves its purpose with a whiff of the metaphysical. An eye has the role of letting in light, so there is a plausible analogy. Why not wind mouth? That would be another possibility. Clearly there is some arbitrariness in selections, which lead to metaphor. Eventually, the choice of metaphor becomes irrelevant, and the pattern of window gets renormalized into the 'sensory token' part of an agent's processing. Not everything needs a story to have a meaning, but perhaps everything has a role—or makes a kind of promise.

Terms can use *metaphors* in language to associate one idea with an idea that seems to have a similar structure. We have to come up with names for things, and there are limited things on which to build communication, so we have to combine smaller parts—but referring to all of them every time is cumbersome, so we assign a name to the whole collective. It's a form of semantic approximation. Another way in which we approximate concepts is through the use of abbreviation: a *metonym* is a word used as a substitute for a related concept. For instance, when we say 'I love The Beatles', we mean 'I love the music of the band called The Beatles', not a collection of insects. The association is implicit, because the most common usage of Beatles is in that context. The type of a word is often left implicit, and the logic of it replaced by the size of the vote made by its experiential footprint—its frequency of usage. The implication seems to be that types are not as important as simple names. All the discriminatory labels that we might pile upon concepts to pin them down are a hindrance—a mass that isolates them. It is the simple and arbitrary names that correspond to contextual anchor points. Once we've found the names, the additional attributes can be used to disambiguate. Metaphors allow us to build new concepts from similarities or analogies to other processes we already know[265]. Metonyms allow us to approximate away the origins and replace them with a kind of commonality index.

Compounds remind us about the simple lessons of scaling, extended with informational semantics for distinguishability: parts of concepts may represent names, while others may represent functional or contextual roles. The meaningful attributes are contextual attributes. Meaning can start out being explained, as stories, generating exterior paths relative to the word fragments:

> The (leg like protrusions that support the surface) called table.

Eventually, process repetition joins the story path together and it forms a single superagent, treated as one thing. As a cognitive agent learns and selects, its experiences expand, morph, and prune parts of the conceptual relationships, clusters of independent words (or word agents) form a superagent cluster, and can then be collapsed into a single concept whose origins become divorced from its original conception.

> (story description) The LEG of the table.
> (compound cluster) The TABLE-LEG.
> (renormalize to word) The taleg

The relationship between names and meanings can be altered by rescaling concepts graphically. If we take a story as a superagent, the interior agents can be grouped and collapsed into more singular agents, by effectively putting brackets around them, to mark out boundaries. Everything inside a parenthesis boundary can later forget its origins—or the opposite can happen: a vague high level term can be supplemented by accreting details that build interior structure, to distinguish a special instance from equivalent cases:

> the ball landed
> (the ball that flew) landed
> (the ball that (flew through the air)) landed
> B landed.

The outer parentheses show regions with interior details that can be renamed to reduce the dimensionality of the locations.

The structure of this explication follows a simple pattern. Think of each concept in these fragments as a spacetime location, expressing a scalar property from the alphabet, and each word as a cluster of spacetime locations, making each parenthesis a cluster of clusters. Different superagents can be formed around a unifying name, with different roles or disambiguations. The roles play the roles of namespaces for terms that are imbued with private meanings (see chapter 3). Think of the concept of 'Doctor' may refer to a medical doctor, a general practitioner (GP), a surgeon, or someone with a PhD, etc. We could simply find unique names for all of these, but that is an expensive and

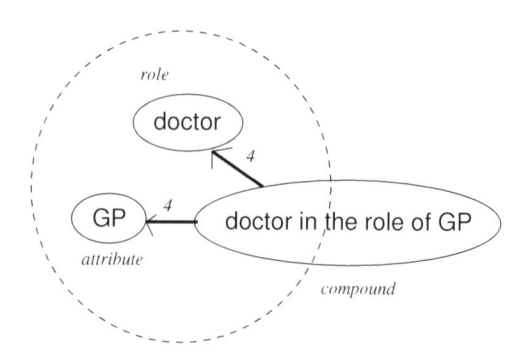

Fig. 8.15. Compound concept formation can be reduced to a functional pattern that has an unambiguous spatial structure. A representative name linked to partial concepts (type -1 association) one of which is a main subject header (like doctor) This is a dimensional reduction strategy, because we can now refer only to the compound name and drag along all the conceptual baggage of its associations accordingly. Note, however, that the interpretation of a privileged role for is sometimes ambiguous.

even inefficient approach, which does not easily scale; hence, it is normal to describe 'namespaces' as contextual constraints. This can be done by forming a compound name as a phrase 'doctor in the role of GP' for instance (see figure 8.15). This implicit compound relationship might not be spoken out loud, but its meaning is clarified by a spatial relationship in the symbolverse. The concepts are next to their roles in the semantic memory. Spacetime comes to the rescue, through its connectivity. A spacetime process will select one kind of doctor by referring to a hub 'doctor in the role of...'. Meanwhile, thinking about the concept of 'doctor' in general, points to all the possible roles adjacent to it.

It's ambiguous which part of a compound is the role and which part is the qualifier in the construction. This really doesn't matter. The roles are interchangeable. It could be argued that 'GP' is the role and doctor is redundant, or that 'doctor' is the role and 'GP' is the qualifier. The distinction is somewhat arbitrary from a computational viewpoint (this is the danger of 'schema' thinking), but it is sometimes helpful to be able to distinguish a privileged component that describes the behaviour or function of the term. This is the power of the approach—no one needs to make an arbitrary decision up front. The answer will emerge from the process that reads the memory.

Compounds are everywhere in human languages. In the following sentence:

> For he invented a nominal compound, most efficacious in every way.

the words *in-vented*, *com-pound*, and *eff-icacious* are all words whose origins as compound concepts have withered and succumbed to some entropy, leaving familiar words with altered meanings.

I don't want to get too carried away with human ideas. In the dataverse, no recognizable language is needed to name phenomena—names can be assigned by equivalent processes that distinguish uniquely[266]. From those basic attributions, one can add supplementary data, using data-types, subroutines, virtual packaging, functions, or macros. It is evidence of the deep relationship between meaning, process, and spacetime.

Numbering items is a simple approach to naming, if you have memory to remember which numbers you've already assigned. Computers rely on numbering because they are engineered to label memory locations according to a numerical coordinate system. We often forget that numbers are also concepts that need to be named. Integers are, in a sense, just names that are arranged in a prescribed order, what is called a precedence relation in mathematics. Similarly the real numbers are just naming patterns that satisfy a similar precedence relation.

$$(\text{Surname}, \text{Given name}) = (10004, 3056) = (x, y)$$

Our tendency to lay out concepts in a kind of empty space is very likely related to the fact that our brains have evolved to process phenomena about space and time, so those are the tools we have at easily our disposal. We should be aware of this bias. It takes far more effort to force abstractions based on other criteria, describing more complicated disciplined ideas, like groups, groupoids, categories, and the whole pantheon of mathematical families. When using spacetime as an analogy, we think simply and intuitively in terms of degrees of freedom and constraints.

Scientists of the so-called exact sciences today are sometimes guilty of abusing numbers to tell their stories. For example, they assign pointless numbers to things instead of giving them names, when the order of the numbers has no significance. In information technology, it is also common to abuse integers because they are convenient ways of managing and keeping track of uniqueness. When trying to come up with a 'user name' for a new account, we might want to choose 'mark', but this is already in use. So mark1, mark2, etc are offered in turn. By using the simple space-filling idea of counting, we avoid having to search for a name that is not currently in use, just by maintaining a single memory or the high water mark.

Words don't have automatic meanings. They need to be calibrated. The same is true of numbers, actually; the calibration of numbers is just a bit more standardized, thanks to mathematics. Numbers can be considered universal in their context. But remember that Euclid described geometry entirely in terms of words, long before Descartes was able to bring the modern language of numbers of bear on the problem. The distinction between a name (coordinate) and a functional role is blurred by rescalings of symbolic clusters. It's just another in a long line of examples of how spacetime regions play different roles on different scales, and how renormalizations of scale often reveal entirely different process interpretations for interior and exterior regions. We draw rings around related items and call them sets or groups[267]. The ring boundary marks certain elements as being on the interior of the set or the exterior of the set. Venn diagrams are spatial aggregations that represent relationships. They form a link between classes of things and the things themselves. We often use the term high level concepts and low level concepts to describe the whole sets and their members, respectively—or even 'parent' for the whole bounded region and 'children' for the members inside. The implicit directions 'up' and 'down' thus correspond to conceptual generalization and conceptual instantiation (realization, reification, etc) of concepts from their parent classes.

Is language intelligence?

Philosophers have long speculated whether language and intelligence are related[268]. We humans talk to ourselves a lot, in order to 'reason' about ideas—especially when we are trying to come up with a convincing story. An explanation is a stream of thoughts that generates a process. Its clock runs by counting idea after idea joined together. This construction of storylines is also an integral part of what we think of as consciousness—the serialized storytelling, not just of talking to ourselves, but of perceiving the world as streams of sensory experiences. Those experiences happen in a certain order, which doesn't have to be spoken. Our sense of time comes from that ordering, and that has to come from connections between ideas. This perceived storyline operates quite like language. How this works is beyond the scope of this book. Nevertheless, we reconstruct memories by walking through these stories, piecing together stored memories that may be scattered around at random in space. We also have flashes of unconscious inspiration that have no obvious timeline. Those flashes of insight are not linguistic reasoning; they show that thought is more than language to thought. We turn to language when we want to form a convincing reason—to tell a story about something that makes it seem a consistent process.

The French writer Jean Cocteau said, of history and myth, that history is made up of truths that gradually become lies, while mythology is made up of lies that become truth. This is how memory systems work. Increasing symbolism comes to dominate memory through simple-minded naming, as memories are renormalized by processes of entropy. Details fade away and are replaced by the stories we like. During sleep, when memories are not anchored to context by the senses, we dream by forming storylines based on random memories—often selected preferentially from recent events or major important events in our memories. That is what you would expect of memory as a network: central memories and recent memories stand out the most. What is extraordinary, and feels largely overlooked in neuroscience, is what process is responsible for joining memories together as stories?

I would not claim that language is necessary for agents to form conceptual thoughts, but rather the other way around: language has a structure parallel to that of conscious thinking. Both are ordered storylines. Both manufacture a timeline from non-ordered concepts, as we speak, write, or when we dream, or become aware of our surroundings. Our stream of consciousness has the ordered structure of linguistic grammar, just far more complex. Any timelike process that separates ideas in a storage network, and assembles scaled concepts into trains of thought, would seem to have an essentially linguistic character, when represented as a stream of symbols. Viewed in terms of processes, the relationship between language and thought does not seem so mysterious.

BREAKING TIME SYMMETRY IN THE THINKING PROCESS

A question that linguists don't seem to address, too often, is why our streams of communicated language actually start and stop. Why don't we just burble a continuous stream of nonsense from our mouths, on and on without end? Perhaps you feel that some people do! Yet the fact remains that sentences have a beginning and an end. It is evidence of the discreteness of meaning, but also of the discreteness of stories.

Part of the answer is surely that language is based on symbolic distinctions. If communication were a continuum that never ended, its significance would not be symbolic, and could be lost as noise. Noise is basically a signal that offers (+) more information than a receiver (-) is inclined to receive. A common admonition of poor writing is to say that it reads like 'a stream of consciousness', which is to call it is a meaningless sequence of utterances, lacking sufficient structure to attach meaning to. It lacks a final (advanced) boundary condition: a point, a goal, a completion of its message. Indeed, I started the discussion of symbolism by talking about significant structures, like needles, that stand out discretely (not discreetly!) against the continuum and act as anchor points for bounding meaning. Clearly, discreteness or episodic partitioning of ideas is important to meaning.

Why would cognitive processes separate concepts into fragments of meaning, in their internal representations? Why don't our brains simply record our lives like a single long running movie? One answer is the scaling of effort. Cognitive processing is expensive, so it doesn't pay to maintain a state of high alert at all times. But the deeper reason is symbolism itself. Without the discreteness (the boundedness) of symbolism, semantics have no protection to diffusing away into their surroundings. The scaling of cognitive processing involves sampling at different rates and coupling of inputs to certain anchors in memory. Those anchors need to be rooted and invariant. Without separation and protection, they can drift. This also offers a further clue about how memories might be retrieved from a semantic network. Cognitive coordinate systems are unlikely to be numerical, like Euclidean coordinates (x, y, z)—after all, nothing in nature builds such a regular grid-like coordinate system, except perhaps for a spider. More common is to navigate space using signposts, which stand out like Cleopatra's needles, to mark out significance starting places.

Cognitive spacetime relies on the ability to control precision. It builds structure from information, rather than dissipating it as motion. That's why physics—mainly concerned with motion—has little to say about it. Cognitive learning is not like the weakly coupled processes of thermodynamics that lead to diffusion and entropy. If information diffused away, its precision could not be controlled, and we would not be able to tell the difference between a tree and a forest, or even a country or an ocean. Different scales of space and time interact with one another, quite strongly, making connections with an identity which is preserved. The role of Shannon's error correction processes is, in a sense, to preserve the integrity of these scales that are interpreted as symbols. Unlike Shannon's picture, the language of meaning is not only a single scale. It's a hierarchy. Short term memories form a boundary condition (a context) that helps to shape the way long term changes are organized, across a multitude of scales. For short term characterizations of meaning we use the term *context*—a network of short term relationships, that amounts to what states a system is currently expressing—or, more daringly, what a system is 'thinking about'.

Recording events captured by sensors is different from recording concepts. Scale matters here too. Events are short-lived. Concepts are approximately constant or time-invariant on a much longer timescale—they are formed and they evolve only slowly, whereas events are a stream of sudden arrivals, some of which might be missed by an observer. But it's not that simple either. A diligent database can remember sequential events using timestamps, from a reference clock that counts continuously as a non-interruptible service, but humans have no natural numerical counter to keep memories in order. A computer trades numerical reliability for no special significance to any time. Humans recall timelines episodically, relative to some especially notable event—just as we record places relative to landmarks of greater significance (e.g. the year of the flood). Figure 8.11 illustrated how the four relational types could record short sequences, without sacrificing explainability and semantics. The trick seems to lie in the finiteness of episodes.

There is no doubt that these two forms of recall are complementary, which is why archives of external memory, such as books that are numbered, and indexed administrative records, play such an important role in our working lives. Remembering all details in a meaningful order is expensive, and it can't continue forever else the indexing becomes useless—eventually information needs to be pruned and garbage collected, as the significance of memories decays by the 'passage of time' or as old events become superceded by new events. The key to understanding searchability, as well as the boundaries of thought processes, is to associate the beginnings of episodes with such significant signpost events—for instance, "The Year of the Flood", "The Day of the Triffids", "The Tsunami", "The Financial Crisis", and so on. Clearly, these events weren't the beginning of time, or of any particular process. What they did was to mix in new information that changed the interpretation of events significantly, starting from that moment. By attaching new significance to injected information, a cognitive process can renormalize the scale of attention, when storing and recalling memory. A cognitive agent can be inattentive as long as events have low significance, and then ramp up sampling activity, from sparse information to dense information, on detecting a significant event. Time perceived by the agent is thus not a purely numerical scale, the names and addresses of events play a role too: we might say "After The Financial Crisis of 2008...X happened, then Y happened". The process can thus be reduced to thus three events:

$$C_{2008} \to X \to Y \tag{8.1}$$

It's inefficient to try to remember everything that happened from birth to death in a single stream. These landmark events act as index markers for looking up memories, allowing us to dive into the stream at any point, or reconstruct the stream from matching fragments.

When we ask for the cause of an event, we typically look backwards through memories for an index event on which to blame the occurrence. The Tsunami was caused by the earthquake, not by the continuous movements of plate tectonics over the past billion years. Our interpretations are ruled by significance, not by the calm reasoning of unbiased causation. An index point is a semantic marker—a significant event, like an earthquake or a birthday that places us in a certain semantic 'frame of mind'. We measure interior time relative to the start of this timeslice boundary. When we ask a why question, we are really asking about what prior events (especially index events) happened before the event of interest. Some of those events might trace back to different locations, others may be local. What matters is their influence on the here and now.

Processes don't always have simple coordinate systems, especially when they are dominated by change, rather than being laid out across a static spatial memory. Timelike (change-dominated) processes need recognition of events to be guided one-off signposts. If you miss the page number in a book, you can go back. But if you miss your train, you can't. So cognitive systems, which have adapted to realtime changes, have to be able to prioritize and even anticipate important events. The problem lies in determining how and when to pay attention.

If we look at biology, the immune systems found in vertebrates form an interesting cognitive subsystem in organisms. An immune system is dormant most of the time, otherwise we would be feeling ill most of the time. How does it know when to wake up? The question is still debated, but the answer that

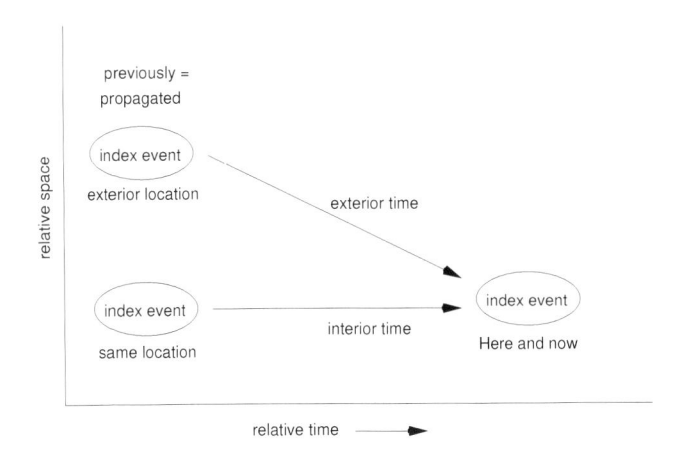

Fig. 8.16. Understanding why an event occurs here and now involves relating prior index events or anchor points in a storyline formed from the history of the here and now. Causation can propagate locally through the passage of interior time, or non-locally as influences reach the 'here and now' from 'then and there'. There is no ultimate reason for any event, only a chain of happenings that contributed to the current one. It's happenings all the way down.

everyone agrees on is that immune systems have sensors that detect trouble. Sometimes trouble comes from the exterior (like microbes, bacteria, parasites, etc), and sometimes it comes from the interior (genetic disorders, cancer, etc). Over long (evolutionary) times, sensors which detect trouble have evolved, and they convert that 'trouble' into well known signals that start a process of response to neutralize the trouble. The detectors and the responses are all cells programmed with particular receptors (-) that match the characteristics (+) of the trouble. When the cells bind to those trouble sources, the trouble is neutralized and the combined agent (trouble plus 'antibody') make a new recognizable promise. Other cells recognize those neutralized cells and eat them. Eventually, the garbage is collected and expelled through our lungs and renal system, which is why we end up coughing a lot at the end of a protracted illness[269].

The immune system is a classic cognitive system. It has a separation of timescales: the long term evolution of sensors to 'index' the patterns of observations into 'concepts'. The fast process of responding, and the storage of memory through—yes, memory cells that form our acquired immunity. In computer systems, techniques are currently far less sophisticated for detecting trouble.

Computer systems may view the starting or stopping of processes as significant events, as well as faults and failures to keep promises. Software engineers create certain checkpoints for monitoring processes: firewalls, gateways, etc. Detectors signal the detection of events by logging of messages to a special alert channel. It's still very much left up to humans to make a decision about what to do—though increasingly, self-protecting responses are being automated. The key events that mark significant symbolic changes are what I call Index Events, because they play the role of an index reference against which to look up historical happenings. They are a kind of 'semantic address' or signpost that marks the entry point to a subordinate episode amongst all the stories in memory process—like a new chapter in a book. Changes of context certainly signal semantic changes to humans. As we enter a particular context, arriving home, or entering an airport, we passing through a semantic boundary that changes our mode of conceptualization. These are cognitive index markers.

When a key event happens, in a cognitive system, the state of 'awareness', or information processing, ratchets up a gear, and it begins to probe more actively for information that may otherwise be too deemed insignificant. If you are resampling the same information a lot, then you are sampling too often. The dynamical sampling rate means that the clock rate of the process changes too, along with its notion of relativistic time. Whatever organ or process performs this task of integrating samples into a timeline, through short term buffering in memory, plays a key role in what we experience as consciousness. After all, the intelligent recognition of concepts, as we perceive a serialized story about the external world, and even when we dream, is what creates the integrated awareness of our experience.

Many neuroscientists focus on consciousness simply as a response to senses, but I believe this misses the point. The running movie we experience through our senses is not created by senses, but through a process that creates the illusion of a timeline, even when we dream, cut off from senses. This is as much about memory retrieval as stimulus and response. And the serialization or memories requires a localized queue—so what local part of a human brain is responsible for this process? That 'organ' would be the seat of our consciousness[270].

WORKING BACKWARDS FROM THE ANSWER

This picture of reasoning, as a process of potentially incoherent and even competitive storytelling, through the linkage of separable concepts by a hierarchy of bounded aggregations (superagency), is not as clean and precise as the pristine view of logic once assumed by computer scientists and philosophers. It's a picture that makes a lot more sense from the perspective of processes however—something more like a digital representation of the interference of waves. All our experiences from elementary physics to the heights of biology suggest that processes extending in time and space mix information in ways that are not well modelled by mathematical logic and its rigid attention to efficiency[271]. But that needn't be the end of logical reasoning. There are many instances where determinism still plays a structural role. Data compression is one.

The reason we can often compress certain data is that the same outcomes can be obtained by different processes. Instead of speaking every word of a story, we can build rules that take a certain input and expand the data into a predictable outcome. The actual information necessary to do that is smaller than the final outcome. For instance, suppose you were to send a circular form letter or mass email to a list of people. The inputs are the list of names and addresses and the text of the letter. By substituting the names and repeating the letter contents, we generate huge volumes of mail from just a small amount of data. This is mostly replication: there is nothing new, but there is a 'lens' process of magnifying and combining the information in a smart way. The inputs completely determine the outputs, but they are not one-to-one. The reverse process is also possible: data compression works by factoring out the redundant copied information—taking the mass mail and outputting the form letter and the list to result in the smallest amount of information that can reconstruct the mass volume.

This ability to trace a process outcome back to a smaller set of significant information is the key to processes of diagnostics and analysis. We might know the answer, but be interested in knowing how we reached it. Most people believe in the chain of cause and effect, i.e. that for every effect there is some cause that precedes it. Though philosophers and statisticians argue about this, a simple punch on the nose helps to dispel any doubt. However, some take the idea too literally, as if to mean that everything that happens is the fault of one unique cause. Now we can understand that there may be no single reason or 'root cause' behind a given happening, only a choice of 'index events' to blame the occurrence on. In practice, we consider satisfactory explanations to be those that *end* with the event we are trying to explain, and *begin* with some index events that we are willing to take on trust as the cause. Explanations never end, they only give up reason to faith. At that point, we are simply happy to stop asking questions. And those who never ask any questions, are satisfied with no explanation at all. Note the role of emotional satisfaction in that decision, rather than a rational criterion.

Reasoning backwards from a fixed point is a theme I've visited several times, in the preceding chapters, in connection with advanced and retarded boundary information. Usually we consider processes in terms of simpler dynamical changes, but there is nearly always a semantic equivalent for dynamical phenomena—and recognizing that is one of the reasons for writing this book. The semantic version of retracing is a process we use in order to seek explanations for outcomes in symbolic processes. Having been quite critical of logic up to now, it's tempting, here, to mention one of the great accomplishments of logical reasoning that have allowed humanity to solve a vast array of problems that would otherwise be impossible: algebra. Algebra is a process too—one that is driven by *advanced* boundary conditions. We use it to solve equations in order to answer questions. In mathematics, algebra begins with symbols—proxies for numbers or concepts that are written down in an organized way to form a language. If you don't know what number is the right one in a particular role, you give it a name and write down the constraints it has to obey. Suppose the constraint is that: the buses alternate between two destinations, so take the next bus after two cycles have passed. Which bus arrival should we take? We can write it as an equation.

$$2x + 1 = 11$$

Then by a simple process of sliding numbers around and adding them, according to simple patterns, we can isolate x and find the answer. Algebra is nothing more than a storytelling device. When we solve an equation, like the one above, we are really saying that there is only one possible story that leads to an equation being true—in this case, given the outcome 11, which paths x keep the promise that $2x + 1$ gives that outcome? In this case, there is only a single story that keeps that promise:

$$x = 5.$$

The goal is to figure out which possible pathways give the right answer, or how we trace back the 'time steps' along only key paths that lead to the origins of the equation: the input values, or boundary conditions that make it consistent.

No other value will make the result 'true', i.e. consistent with the rules of the path, given the meanings of the other symbols.

The brute force search for a unique answer—trying every possible value—is cumbersome and expensive, luckily the structural rules of manipulating algebra are basically variants of simple spacetime processes, so we can solve equations using pattern recognition. The expense of the process to find the answer is known as its 'complexity' in computer science. The expense is no doubt why we tend to oversimplify algebraic approaches to reasoning in order to find an approximate answer quickly. That is not a fault of algebra, but it is a symptom of the expense, and thus a weakness of the strictures logic as a concept. I discussed this in more detail in my book, *In Search of Certainty*.

Reasoning pathways, like these selected stories undoubtedly exist in all spacetimes, at all scales—there is certainly a minimum level of complexity needed to differentiate pathways of sufficient length to find the narratives that best represent symbolic outcomes. Most likely, the level of complexity at which recognizable human ideas emerge is exactly that scale at which biology begins to distinguish itself from molecular chemistry. That's a speculation, but perhaps not an unreasonable one.

The 'give and take' of epigenetic reasoning

Recall that generating stories from semantic concept networks is a process that leads to large numbers of stories, most of which don't make sense on further inspection. The reason is that stories—generated naively from a memory that accumulates information uncritically over all possible times, events, and contexts—tend to overlap and wash out any modular sorting or contextual relevance. The result of a search is usually a mixture of wisdom and junk. How can a cognitive process learn to understand relevance?

At first it might seem that the solution to this problem would be to find a way to make a smarter algorithm that can select only relevant stories. But that's top down thinking, and the world is built bottom up. The context information lies in the memories, not in the process that runs on top of them, so it doesn't make sense to try to extract context from memory to try to decide which memories to access. Instead, we might try to block out certain memories to avoid seeing the context information. In the natural world, processes are governed by the give and take of equilibria: forces and counterforces keeping processes in balance, not on ensuring precise and perfect behaviour. This is the essence of Promise Theory. The secret of intelligence itself may also lie in the full implications of switching, using give and take. Evidence for this comes from an unexpected source.

The field of genetics is, by now, quite well known, even to a popular audience—but the version we hear most, in the news, is the idea that genes are directly responsible for particular traits, one at a time. This is not really what genetics implies. It gives scientists and business opportunists a simple task: to search for genes that switch on or off certain characteristics. In the language of Promise Theory, genes are assumed to make certain promises. The dream of therapists is that adding or subtracting the effects of a gene would allow us to cure problems that lead to illness and suffering. This view is both simplistic, and incompatible with a spacetime view of processes. Genes act as promises to cellular machinery to make recipes (+), which can then be read (-) and used to trigger new cellular processes. What genes promise is to select certain alleles, which switch on variants of protein processes. Those processes ultimately lead to outcomes that can be associated with the growth and functioning of an organism.

We know that genes are read—and their intentions interpreted—by a class of molecules known as RNA, which function as messaging channels, printing presses, and apparently also as signal blockers. The thing is, there are only some tens of thousands of genes in organisms (humans are not significantly more complex genetically than organisms we consider to be significantly more primitive). If we look at the stability of natural organisms, i.e. how little actually changes about them on each new generation or how uniform they are across their species, then we have a conundrum. There is both too much variation and diversity to be encoded by these genes matching characteristics one for one, and being switched on or off. At the same time, there is far too little variation expressed compared to the full number of variations composable from every possible combinatoric combination. Initially it was thought that genetics was all about positive addition of attributes, but more recent discoveries have identified the competition between give and take in genetics too.

Genetic memory is a long term memory that encodes processes using retarded boundary conditions. Why isn't genetic determination a simple factory production process then? How can adaptations be controlled during the process of gestation of life, and during an organism's lifetime? Such adaptations point to the need for cognitive processes, but that would require both a short and a long term memory. The contextual short term memory for genetics turns out to have been largely overlooked until quite recently. From earlier chapters, we've seen how the give and take of interacting parts plays a crucial role in process phenomena. Epigenetics is about the interplay between an environment that expresses certain promises and a fixed string of promises encoded as DNA, which represents genetic characteristics[272]. The chemistry and combinatorics of those bindings are more complex than was initially realized, and can shape behaviours in real time, meaning that the initial picture of genetics as a purely retarded process is wrong. Epigenetics is a virtual process, so genetic determinism involves networks that are both physical and virtual. Spatial connections between molecules are only the beginning. Spacetime processes lead to evolving networks between genes and proteins, which complex boundary shapes. The amount of information in these molecular processes has defied all attempts to estimate it, but scientists continue to learn more about how introns and extrons interact with multiple forms of RNA, proteins, and enzymes.

How is it that we achieve such stability in the basic adaptational patterns, while at the same time allowing for such variety in personal appearance and characteristics. The transmission of characteristics from parents to children is often eerie if not scary in its specificity, to the chagrin and humiliation of children who might try to resist at all costs. Some of this might be attributed to simple brainwashing through 'nurture', i.e. social transmission, out of band of the genetic channels. However, this is not entirely convincing. Different siblings or friends, who are nurtured in the same way, can grow up with very different characteristics, and kids who never met a parent can ape their behavioural characteristics quite closely. There is plenty we do not understand about how genetic transmission works, but it seems to be more powerful than we have realized in the past.

Today, the study of epigenetics has brought to light a new possibility: namely that genes are not the main signalling route. They form only a baseline, whose encodings are modified by another layer of signalling, known as epigenetic signalling. This is analogous to a switching-routing network in the dataverse. It connects or blocks certain channels of cooperation between genes and proteins. Each location on a genetic string (chromosome) expresses a promise, but if all the locations were active at the same time, they could lead to all processes being active at the same time. The result would be chaos, like a lack of town planning. When scales do not separate, and the expressions of upper scales are not simpler than the degrees of freedom in the lower scales, chaos results: a lack of semantic stability. There can be contention and loss of stability. The ability to simplify—to make a clear selection—is the key to stability, and this is as true in genetics as in any other kind of system.

The combinatorics of genes, through complex protein pathways could have been a classic statistical mechanics of combinations, leading to probable outcome, were it not for a counter fact: the stability of a few strong themes with minor variations suggests a perturbation or modulation by one or more secondary information channels. This would be hard to explain by the combinations of positive attributes alone, but with a secondary layer that can switch on or off aspects of the baseline, an enormous number of new variations becomes controllable or *addressable*. Most combinations of genes will lead to nonsense, just as most combinations of words lead to nonsense. Only certain stories end up having meaning, and this observation can be translated directly into switching. Genes generate digital diversity.

Darwin's counterforce, for stable adaptation, was *selection*, i.e. abstaining from certain pathways by killing them off, on the basis of their being ill-suited to particular contexts. If the attributes offered by genes (+) match an environment where they can be accepted (-) that can bring about a propagation of the attributes and maybe the genes too. Genes remember diversity (+), but what is it that remembers selection (-)? It would be inefficient indeed if both these processes we not standardized and commoditized for the markets. Now, the inspiring and remarkable experimental work[] of scientists has discovered new classes of information message channels, which don't connect but rather block connections in the normal DNA to RNA to protein manufacturing. There seems to be a highly complex and refined language of pruning of the combinatoric explosion, which acts as a digital mechanism, far more efficient than waiting for random births and deaths to apply statistical sorting[273]. Such an information-based selection mechanism that can be applied at any scale, simply by labelling special locations in space to reveal or block their special properties.

Epigenetics may hold the key to the story problem elsewhere too. The way that complex gene pathways control the expression of characteristics, in a context sensitive way, may be analogous to the way context enables or blocks pathways in a reasoning network. To have full control over adaptive processes, one needs to be able to select a particular pathway, or to construct roadblocks to prevent it, even add street-lighting to illuminate preferred pathways in a given context. Many fascinating developments are happening around epigenetics. From a Promise Theory perspective, this is interesting. Genes generate the possible pathways (+), and the switching by blocking of paths (-) has to have a much greater specificity, which suggests that the bulk of the smartness lies in the selection of what is passed on from parents, not in the mutations themselves.

We come back to the switching on and off of adjacent information channels as a fundamental principle of spacetime processes. Communication is the key to information transmission, and its channels are the links that connect space together. Information theorist Claude Shannon, himself, speculated on the role of information as an interaction region of spacetime in his early work, while still under the guidance of Vannevar Bush. He has worked on algebraic approaches to electrical circuits, allowing logical reasoning with electrical circuits to be written down within a mathematical form. Such an approach was revolutionary in electrical engineering, but it was quite unheard of in biology.

Ironically, one of the early projects for which Shannon wrote a dissertation was *An Algebra for Theoretical Genetics*[274] He attempted to construct a causal circuitry for genetic behaviours in cells, by parameterizing the spacetime of the processes like a kind of tensor calculus (analogous to Einstein's representation of spacetime in General Relativity[275]). At the time, the details of genetics were in a fledgling state, and unconfirmed by experiments. Shannon based it on straightforward reasoning. In any genetic process, there would a source (a gene, say) and a receiver that bound to it (a kind of protein that could read the gene). Combinations of these interactions would then flow through a number of transformations, like circuitry.

Shannon's formulation was to create a square discrete space of genetic promises, with one dimension representing genes, and the orthogonal direction the set of chromosomes. The tensor-like objects would have to match up, with an offer (+) of a gene and a receptor (-) to complete a unidirectional binding. At the time, there was no knowledge of how this worked, but mathematically, the matching of promises falls into the familiar pattern seen again and again in science: senders and receivers, matching. No one begins engineering a system, by planning for balls thrown from (+) to (-) to be dropped along the way, or even be deliberately blocked, but that is effectively what epigenetics points to. It's like trying to draw with shadows. Today we know that the (-) epigenetic screening channel is not merely a passive connector that accepts everything offered by the (+) DNA channel; it has an independent dynamics of its own. For every (+) tensor in Shannon's work, there must also be a (-) channel, predicted, if you like, by a Promise Theoretic view of transmission. In brains too, there are two kinds of neuron: type I excitory neurons, and type II inhibitory neurons, further suggesting the importance of signal blocking to adaptive reasoning. More recently, a new kind of neuron—called a 'rosehip neuron' (for its particular bushy shape) has been identified as a signal inhibitor in the human brain. It is not present in mouse brains, which poses many questions about the role of inhibition in scaling and intelligence—and what might be special about humans.

Cognition and causal reasoning

The quest to understand Artificial Intelligence is well underway, but it is currently based on only a few techniques, limited in scope. The understanding of cognition in systems on different scales will, I believe, change current attitudes to intelligent reasoning, and help to remove some of the deep-seated prejudices about the nature of intelligence. The processes I've discussed in this chapter are passive adaptive systems that do not try to modify their exteriors, though they can modify their interiors. Robotic systems that can extend their own territorial interiors, by manipulating the world around them to become a part of their memory resources, have the capability to grow and evolve. Manipulation of surroundings can blur the boundary between interior and exterior. A smarter agent may want to renormalize its boundary to expand its territorial influence, when its interior world seems insufficient.

Many researchers have argued that pattern recognition is insufficient to qualify as intelligent behaviour, in spite of grand claims made by some working in the field of AI. Even making probabilistic predictions, based on learned data is insufficient—after all, even Quantum Mechanical systems can make probabilistic inferences. What does seem to be a worthy hallmark of intelligence, however, is so-called *conditional* or *counterfactual* reasoning.

> If it hadn't rained, it wouldn't be wet.

> If the professor had not asked her to marry him,
> Miss Scarlet would still be alive.

If artificial systems can one day perform advanced reasoning based on conditionals, we might call them reasonable, if not necessarily smart. I believe that the way to scale reason is through story-based explanation, using semantic networks, fed by cognitive systems. The studies on this have been ongoing for a decade or more, but are still in their infancy today. Nevertheless, there are clear pathways for exploration and development.

Recognition of things and behaviours is not reasoning, but neither is the kind of rigid mathematical logic employed to make decisions and inferences in computer programming. Human reasoning is not often logical, it's based on trust. We tell stories about the reasons for events until we reach a state of emotional satisfaction. Stories don't always make strict causal sense, and they are truncated when we can say 'because I believe X', where 'X' is anything that could plausibly be used to start chain of reasoning. Think of how reasonable we are when voting for politicians: the answers we come up with are invariably nonsense: 'because he's my guy', 'because I feel I trust her', and so on.

We need to be able to tell stories that explain what, when, where, how, and why things happened, but ultimately the stories can't go on forever. We need an understanding of both spacetime and likelihoods computed by statistical or probabilistic methods, if we are trying to be rational, but rationality too is only chosen if that is what the storyteller trusts. Rationally, causal reasoning has to be separate from more familiar relationships like correlations and associations. Causation leads to an ordering of relationships and inferences. Statistics can only express the idea that two things are similar, not whether one caused the other, or whether one is an attribute of another. Statistical reasoning cannot embody differentiated semantics. For that, we need to be able to distinguish different relationships—actually four basic types.

Computer Scientist Judea Pearl, who has studied the meld between statistical inference and causal reasoning, writes of a causal hierarchy with three levels of graphical reasoning—i.e. three kinds of spacetime process that can be considered thinking[276]. The first, which he calls 'associative', is a conditional promise in Promise Theory language. It's the crux of the observer problem: an agent can measure (-) a source (+) if it exists. This is a simple promise binding combined with a conditional inference. Even this level of reasoning is more advanced than the assumptions of deterministic predictions in Newtonian views of the world. What does a symptom tell us about a disease? What does a survey tell us about beliefs? The second level Pearl calls 'intervention', in which a known change is imposed or promised as a given boundary condition, to add a measure of certain knowledge to a process. I would call this the injection of information. What if I take the medicine, will my headache be cured? Then the third level is 'counterfactual' reasoning: Was it really the medicine that stopped my headache? Would Kennedy be alive had Oswald not shot him?

Conditional 'what if' reasoning is filled with potential ambiguities and vulnerabilities due to incomplete information, but it plays a major role in modern science. Conditional promises, like Pearl's three levelled inference are a kind of middle ground between formal reasoning and spacetime interference. They have the flexibility to cope with more than idealized propositional calculus. Symbolic logic could never address a question like : what are the stories told by the processes of chemical interactions on the scale of a planet, in the context of a certain range of temperatures.

From the surprising evidence of epigenetics, we confirm that suppression switching mechanisms, or negative pathways, seem to be more important than many have realized in fitting or selecting outcomes. We are usually concerned with generating only the right answer, but nature seems to do the opposite: it generates far too many answers and then eliminates them, like counterfactuals[277].

Flawed reasoning or poor taste?

One of the mistakes that we humans make, in our eagerness to attribute semantics to all things, is that we take too literally (or at least too seriously) the stories we make up as the connective tissue between ideas. More generally, our stories about natural processes may be equally flawed. If we believe the simple model described in this chapter, then connecting ideas through memory processes has a virtue regardless of any sense of 'literal truth' or 'optimum truth' to characterize the wiring. Phrases like 'right answer, wrong explanation', 'right for the wrong reasons', 'the end justifies the means', and so on, epitomize the way in which getting to a certain place has an intrinsic virtue, even when certain connections don't stand up to scrutiny. Nearly all explanations in science end up being 'wrong' in the light of new understanding—or, more correctly, they end up being deprecated in favour of new ones. It's difficult to call Newton's laws wrong, given how successful they are—yet some sticklers would insist that they are. The question then is: is anything ever truly correct?

Take the role of metaphor, which we've already established plays a central role in language. In ancient—and still in modern—times peoples have sought to represent ideas in terms of well known 'elements' of our world: Earth, Air, Fire, Water, Gold. This is another classic case of attributing properties to static concepts of material things, places, or the properties of locations rather than to the broader processes that underpin them. These elements were deified and still hold a descriptive significance in philosophies Feng Shui and Eastern philosophies, for example. We now know that there is no literal sense in which these ideas play a causal role in explaining connections between illnesses and patterns of fortune, but that doesn't really matter. By using the ideas, we might still make connections—like a self-fulfilling prophecy—by analogous (or 'covalent') reasoning. What we choose to call a connection is unimportant, if the connection turns out to be useful. Sticklers for a version of correctness will complain about the use of such concepts in Chinese medicine, for instance, but this is a bit like complaining that evolution is incorrect because a certain creature was too ugly to be the missing link in an evolutionary history. The aesthetics are subjective; the utility of the link is the only causally important issue.

Sometimes Western literalism strives to place great faith in the representations that connect chains of reasoning, out of reverence for the greatness of science or religions—i.e. by elevating tribal authority over pragmatic criteria. If Chinese medicine cures your stomach ache, or your malaria, complaining that its explanation used the wrong words ('heat' instead of 'Histocompatibility' or some Latin excruciation) in basically just unhelpful. It does not mean there is no truth in the reasoning. The network may be real, although the names are not. Reasoning works by connecting ideas together into stories. Not all the stories will use the best paths, but they might still work on some level: our puritanical ideas about truth and falsehood are likely just misplaced. We just don't know how far down the stack of turtles that argument may hold sway. A bridge that connects parts can be made of many materials to support a leap of faith. As our trust in various bridges grows, we favour them above others, but are they better? Mystics and charlatans are those who push form over effect, in order to take ownership of the power that goes with apparent knowledge. Some stories become powerful instruments of manipulation. The unique power of the scientific method lies in its constant questioning of the validity of its own stories. This is why counterfactual reasoning is so important to the success of modern science.

Stories may not have to be inscrutably correct to play a role in the world. Eventually certain explanations become accepted and normalized—or *trusted*. In *In Search of Certainty*, I discussed how trust can be explained game theoretically as a cost-saving mechanism, built on learned familiarity. Trust allows us to truncate investigations that would go on forever, curtailing Zeno's paradox for explanations. In the long run, stable reasoning processes acquire normalized explanations, in the form of well-trodden pathways—the locations play the role of concepts. This fits closely what we know about classical physics. Physical 'law' can be represented as a 'principle of least action', or a 'path integral' in Quantum Mechanics: the average effects of fluctuations around a norm that eventually conserves energy is to cancel out non-classical paths, and leave the familiar classical paths which persist over multiple process trials. It's not an inviolable law, but rather a pattern formed from persistent traces, leading to a simple story to be handed down.

THE LONG NIGHT

Our ability to sense the phenomena which unfold just beyond the perimeter of our senses, is limited ultimately limited by the speed at which we can sample the world (the clock that drives cognitive processes in our brains), as well as the memory we can muster to build up a model of the world. This might seem fantastic and unique, but it is a straightforward scaling of an interaction like the absorption of light by an atom—scaled both dynamically and semantically to a new level of complexity. We accumulate patterns and concepts over a lifetime of learning, becoming more advanced and more nuanced as we gather more and more experience. With no prior knowledge to guide the process of adaptation, bootstrapping the recognition of spacetime phenomena takes evolutionary time, leading to a few hardwired capabilities, a highly adaptable set of learning techniques, and plenty of limitations too. We now know that both nature and nurture are involved in learning, recognizing, and reasoning about the world: eyes and senses are not just cameras, but actually discriminators that intrinsically encode concepts like direction, motion, location, and even faces[278]. But learning can only help us to process what can perceive, and that is a tiny fraction of what goes on in the universe.

Trapped within the three dimensional realm of the universe, and all of its embedded phenomena, we are ultimately limited by the ability for electromagnetic signals to discriminate patterns for us. In the dataverse, we are limited by the ability to collect and recall what happens in an environment, where we have no senses, artificially. Biological systems rely on chemical diffusion and even weather patterns to carry messages and etch them into memory. Economic systems rely on polls, data collection, and yearly accounting to gauge the state of the local and global patterns of social development. We have to rely on information from somewhere. This is what it means to understand processes.

Phenomena that are too fast, too small, or too far away, will always be challenging to observe with raw senses. The limitations of network pathways, scale, and memory may lead to incorrigible distortions during observation that we characterize as 'weird' or 'counter-intuitive'. Yet humans show great ingenuity in manipulating the phenomena around us, to tease out information from every available channel. We do this by manipulating time and space using memory. Information technology allows us to record events and play them back, faster, slower, rescale them to fit inside our heads, to see and hear frequencies of light and sound that our natural senses were not designed to hear. By mapping extended sensory observations onto our own experiences, we can do a lot to understand a world that is more complex and vaster than any of us can imagine, all in one go. The analogue of scaling, applied to learning processes, in what happens when stories are built on other stories. We see this happening through language: from symbols we get words (bigger symbols). From words we get expressions as sentences, and actions. From actions we get episodes. From collected episodes, we form stories and explanations. From those explanations, we build new explanations—on and on.

> 'when infested with harmful insects that start to eat the plant, the stressed plant produces chemicals that attract other insects to the plant, and these feed on the parasites. Those carry with them fungi that cover the plant and form a signalling layer....'

> 'The atom emits a photon, and the electron jumps to a lower energy level whose electron cloud has a different symmetry....'

> 'The markets responded to rumours about the government's plan, and the price of the dollar fell allowing foreign buyers to deplete the reserves of the commodity...'

These fabulous stories! How do we come up with them? None of those above could be observed at first hand, from beginning to end; the stories are sewn together from a patchwork quilt of many fragments, joined by a satisfaction in beliefs formed by learning in other contexts. We say that we understand something when we have a story that ends in questions that we are ready to answer based on trust or fatigue, without proof. These are what satisfies our curiosity—because satisfaction is an emotional response that has evolved to cut short the need for the complex reasoning known as system 2. It would, after all, be exhausting to question every turtle in the pile.

The ability to chain together concepts and patterns into storylines may turn out to be the elusive basis for reasoning and consciousness[279]. Something has to generate our sense of sequential time from non-ordered networks of memorized concepts, and episodic fragments. It may only take a proper understanding of *common scaling*—the simplest principle of the processes described in this book— to bridge the plausibility gap. I suspect that, when we understand brain function more fully, we'll discover a process (perhaps even a localized organ) in the brain responsible for extracting stories from stored memory. When that mixes with contextual sensory inputs, the result may be what we call consciousness—a sense of things happening in time[280].

In his book, *The Emperor's New Mind*, Roger Penrose wrote an extensive survey of ideas about space, time, and computation, as a commentary on the problem of Artificial Intelligence. The book is dense and delves deep into the flaws in both physics and computer science. It claimed, amongst other things, that quantum theory—including ideas from quantum gravity—might be needed to understand how ideas may emerge during intelligent behaviour. This feels very unlikely, in its literal sense—and I'm not sure that Penrose did either—but pointed out valid inconsistencies in the state of quantum theory and computer science. I find it more likely that we'll find the answers to quantum theory by studying the scaling of discrete systems—information networks without getting caught up in the belief that existing formalisms are an immutable starting point. Experiments in artificial reasoning show that, for a thinking system, ideas are not hard to come by: what's harder is the elimination of ideas that are not relevant— essentially counterfactual reasoning. The problems and answers lie in negative space, just as the resolutions to causal reasoning lie in the counterfactuals. It was always the goal of science to understand how an empty universe, that erupted from a singularity could result in its own introspection: organisms and processes that can think. We have almost certainly underestimated the scope of this phenomenon by attributing it only to complex organisms like ourselves. All the indications seem to point to a simpler idea: with varying degrees of complexity, or sophistication, we should rather be looking to understand how spacetime itself—as it observes itself and feeds back storylines etched through process trajectories of modularized memory—thinks.

9

Smart spacetime

> When the stars threw down their spears
> And water'd heaven with their tears:
> Did he smile his work to see?
> Did he who made the Lamb make thee?
>
> Tyger Tyger burning bright,
> In the forests of the night:
> What immortal hand or eye,
> Dare frame thy fearful symmetry?
>
> *–William Blake*

What tumble of interactions is our world, that passes information back and forth across channels of communication, between agents of different size and unequal dimension. Communicated data are the linkage that unifies dynamical change with interaction semantics, in a dialogue that brings greater and greater complexity with increasing scale. When things grow big, they can't always sustain what they did well when they were small. When things become many, they can't easily maintain the coherence afforded as a few. But new functional possibilities become possible at each new scale. When resources are constrained, experience may be clipped and unpredictable. By adding bulk, we create a broader base to the pyramid of accomplishment for processes. These are the major lessons of scaling—and we see their effects in physics, biology, and computer science, wherever we look.

The story of those unfolding exchanges of information forms an umbrella for everything that happens around us, above us, and below us—at least, as far as any of us can possibly tell. Space and time are not so much the theatre for events as the leading players in every scene. True, there are differences of emphasis at different scales: some processes seek to divide and conquer spacetime, erecting protective silos that harness symbolic integrity. Others spread information outwards, and mix it up, bringing novelty, diversity, and recombination of existing ideas. Without the former, the latter would lead only to featureless entropy. Without the latter, the former would condemn the universe to deathly inactivity. It is within this tug of war—between interior and exterior, at each scale—that the integrity of symbols allows meaning to take shape. And it is all writ, small and large on the persistent promises that space can keep. From that starting point, memory and adaptation follow, and we may eventually judge some processes to be 'smart'.

The scope of these few ideas has an almost implausible span of applicability, from the small scale to the large and back. The model of cognitive process, described in the previous chapter, helps to put it all into perspective as a single pattern. In schematic functional terms, it holds up for small things like atoms, and for large things like people, and even for weird things like black holes. It also forms the basis of a plausible model of conceptual reasoning. It can be shored up or stripped down to reveal different features of spacetime, but it simply emphasizes how *observed* space and time work non-locally on all scales. I think it's a great tool, both to understand scaling, and even to dispel the more fantastic claims sometimes made about intelligent behaviours. If intelligence is a process, then it should also fit naturally into a scaled understanding of processes.

To help develop a sense of confidence in the cognitive model, and see through the layers of disguise, it's helpful to pick a few cognitive systems that are recognizable, and see how well they represent the picture I've teased out. When looking for a unification of ideas, one first has to compose the criteria for putting things in the same frame. Promise Theory is helpful, as it offers a general approach: one looks for the agents (locations in the space) and then for the promises made between them (representing the spatial adjacencies). Finally, we identify the 'give and take' that enable channels to bind and for information to flow. It's is a simpler agenda than either the search for a grand unified process based on token counting (energy or money), or the dismissive reduction of phenomena to patterns corralled for machine learning. Information is equivalent to inhomogeneity and broken symmetry—boundary conditions, if you will—so trying to reduce complex behaviours to smooth idealizations with a uniform geometry is unlikely to be successful for our purpose. The conventional foci of spacetime—smooth motion and dimensional symmetries—must be of secondary importance, when dealing with information rich processes. Information, after all, breaks symmetries.

In this final chapter, I want to sketch out some of the ways in which spacetime ideas play a role in worldly processes, and consider what it might tell us about our future prospects for shaping space and time in smarter ways, practically and technologically. Are there any limits to smart behaviours? Do they scale? The potential applications for ideas about space and time seem near endless, but without careful thought we may never see them—for they are never far away from the things we take for granted.

Semantic spacetime

I've said a lot about what makes space and time tick, so to speak, but what is it that makes behaviours smart? For some, being smart is about IQ tests, recognition skills, and feats of memory. Personally, I always associated intelligence with having a sense of humour! Alan Turing's test for intelligent behaviour was based on Einstein's own relativistic idea that what you experience is your effective reality—if a machine appear intelligent then it is. Popular expectations for Artificial Intelligence have been skewed by the science fiction tales of human-like robots, and mad computers. Those robots were fashioned in the shape of humans to be a more acceptable replacement for slavery, to absolve humanity of the guilt of enslaving other beings (a case of manufacturing your cake and automating the eating of it). One would hope then that we never made our robots self-aware enough to know their status and feel the same pain as their human or animal predecessors. The idea that one could conveniently remove any emotions (i.e., any bad feelings about being enslaved) also became a standard, and thus began a key misconception that emotions are pointless human failings, to be eliminated, while pure logic and process (combined with an enviable work ethic) would triumph.

There are many things wrong with this picture from scientific and technological point of view. Of course, science fiction is rarely meant to be about the future: it's an allegory of the present, anonymized to poke fingers and people who might otherwise ban the commentary. We shouldn't therefore think about engineering smartness only as the creation of a human-like intelligence. Smartness is a much simpler thing: it's what impresses us by making life easier.

The goal to make a general intelligence, akin to a human mind, is one based in curiosity. It isn't something ordinary people need or want to see. Tools, on the other hand, that expand our capabilities as humans are something we would like to develop. Smart is undoubtedly in the eye of the beholder—but, if we think about what we consider to be smart in daily life, we mean something that anticipates our needs and makes life easier, lowers cost, and is faster. That means something that adapts to context and has a model of our needs. Taking humanity out of the picture, smart means a process that can change and adapt in order to persist, by using its local neighbourhood as a stepping stone. In biology this is any successful organism that exploits an evolutionary niche. In chemistry, it could be a self-sustaining reaction, like a heat pump. In physics, it might be a wave that sustains itself by banking its energy in surrounding space, to lay its own tracks (imagine a train where the driver picks up the track behind the train and lays it in front, in order to keep going with only a limited supply of track).

Smart is not intelligent. Machines designed to perform singular tasks could never become intelligent in a general sense, simply because they would have no basis to expand their horizons and adapt themselves autonomously to goals that were not set for them, though they might become 'conscious' or self-aware on a simple level. If machines ever began to drift away from being mere proxies for human intentions, we would shut them down and consider them faulty. Autonomous adaptation is a risky business. Take an example, like self-driving cars: one can imagine experiments with autonomous learning, where vehicles learn more about the experience of driving, as they operate according to rigid algorithmic rules. This is different from allowing the vehicle itself to determine how it should learn. Without clear goals, a learning system might start avoiding potholes and swerving all over the road, or stopping under bridges to keep dry from the rain. Because science fiction has trivialized artificial intelligence for so long, most of us do the same.

As we ponder these issues, the IT industry is pushing to implicate Information Technology (IT) in all manner of processes in our human world under the banner of 'smart systems' (smart homes, smart cities, smart workplaces, etc). These marketing concepts muddle the issue of where smartness lies: it is not the device, the home, or even the city that's smart in these systems, but rather the intentions behind the behaviours for which they act as proxy that may be intelligently conceived. The goal is not to make intelligent life but to embed reasoning by proxy, by using technology to gather and exploit data in ways that go beyond the sensory capacity of a single human being—then using it to optimize the processes of human life, or provide new services for profit. We want to embed our stories and narratives into the spaces we inhabit. Technology will play an increasing role in our future, because its role is to act as a proxy for our own limited capabilities. We make technology not to replace us, but to represent us—just as we imprint technology with processes and behaviours that mimic our own.

Computer systems are already some of the smartest spacetimes ever built artificially. Computers represent a world of algorithmic or narrative processes—coupled with varying degrees of ambition. We design information technology as 'systems' that may involve collaborating parts. Their design incorporates the ability to automate and optimize their own operations, based on what they learn as they go, or what humans anticipate they might encounter, so that's no surprise. Computers manage their own timesharing, their own memories, and schedules, across a number of scales, from transistors to datacentres, to the planet-wide infrastructure of the Internet. The smart storage of data by 'paging' and 'swapping' of data, which is analogous to the logistics of valet parking, short term parking, and long term parking, is 'smart' in the sense of optimization or cost saving. Keeping important resources close at hand, while others are squirreled away for safekeeping, shows how spacetime considerations play a role in value and economics, through time, distance, and agility. The smartness of information technology is thus an ability to respond to exterior pressures of cognition in order to expand our ambition while alleviating human toil. And so we come back to the use of cognitive systems that respond and adapt to exterior change.

Smartness seems to be about finding expedient adaptations to dynamical processes that learn from their observations and recognize contexts—maps that can be used repeatedly to tag their relevance to a goal. The learning of a process can improve dramatically with its scale, as cognitive systems show us: more memory, more parallel activity, more potential for managing information. Primitive cognitive agents will likely seem erratic and limited, with 'quantum' characteristics—as experience and observations are sorted into only a few buckets—but, more sophisticated agents can benefit not only from memory but from adaptive interior algorithms—like a computer, or a biological organism—enabling motion, fine grained discrimination, and accurate recognition. Spacetime manipulation is clearly the leading thread in this story.

Smart mechanics?

Let's recap the story from the beginning. The basic mechanical notions of process, motion, and measurement, so ingeniously formulated in Newton's time are based on a few concepts. Time, space, motion, mass, momentum, energy: these have been the phenomena that characterize spacetime processes.

- *Smart Time and motion*— Time is change. As Einstein reminded us: the fundamental relativity of observation is that what is given by one process is filtered through what is accepted by another. Absolutes may exist, but if you can't experience them directly or indirectly, then they don't matter. We use time as a tool as well as a theatre. Time can be manipulated in numerous ways: by suspending a process, or bringing it into a halt or to a dynamic equilibrium, by shutting shop, or blocking interactions.

  Motion (in its three forms) is an interplay between interior and exterior process. Why would a local interior process benefit from moving? If neighbouring agents, which host the process, have beneficial properties, perhaps ordered to form a gradient, it acts as an attractive force, like a nutrient field in biology, or a force field between the plates of a capacitor. If a computer process needs more interior memory, it might relocate to a host that can offer more. If the process continued to grow, the ongoing interior process to seek out resources would keep it moving with the gradient, in a Newtonian manner. Interior processes transform agents into a vector field by their interior polarization.

- *Smart Space*— Space is about the exterior give and take between agents, established as a form of communication channel. We can forget about imaginary embedding spaces—they are only for convenience. Space can be smarter by imbuing it with more interior capabilities. The more promises a spatial region can make, the richer the semantics of its behaviours. What charges, complexes or receptors—what information can it carry? Physics is unclear about its convictions about space: the idea that there might be processes on the interior of a point could appear to be nonsense, yet elementary particles are represented precisely as a series of 'Feynman diagram' processes in a 'virtual cloud' whose size is unclear, and depends on how one renormalizes one's viewpoint. The inconsistencies are evidence that we need to rethink space as a theatre of infinite resolution.

- *Smart Mass*— Mass is about the bindings or ties to a neighbouring environment, that add costs to changing a process. The costs accrue through the processes that constitute direct adjacent connections to other agents. Mass doesn't matter when processes don't need to move, then it matters a lot for agility. Mass is related to space by channels that bind locations together and prevent the reordering of locations. We find effective masses in all kinds of processes that involve multiple agents. For example, managing connections in the dataverse is a key part of process agility and security on all levels. Internet connections or 'sessions' may be held open to optimize agility with the local environment when a process isn't moving, but if it needs to be moved, those connections have to be closed and reestablished, divesting its investment portfolio so to speak. This shows how mass connections are both an advantage and a liability with respect to different processes. As we scale up to a more human level, the relationship connections include symbioses, mutually beneficial relationships, business partnerships, herds, flocks, and shoals maintain close relationships for strategic shielding. Large agencies have a kind of gravity in that they attract stragglers for protection and stability, but shutting down connections keeps agents agile. Political consensus is the mass and the momentum that justifies democracy. Not all processes can shed their mass on demand, but smart mass would be adaptable.

The central thesis in this book has been to underline the importance of scale, and how scale acts as a distorting lens for communicated perceptions. Perception shouldn't be dismissed from scales like those involved in fundamental physics, because the cognitive model tells us that scale mismatches there must also play a role in our ability to measure exterior changes and time intervals. We shouldn't muddle subjectivity of qualitative interpretations with the intrinsic subjectivity of quantitative measurements.

It's usual to dismiss the relevance of high level ideas and 'classical' behaviours to low level 'quantum' issues, say in elementary physics, because we've mystified what we don't understand, and fenced it off as 'fundamentally different'. However, I've come to believe that we need to keep an open mind. Indeed, I've grown suspicious of the perpetuated mystique around quantum theory. There is so much about the quantum scale that cries out for explanation, yet which superficially resembles phenomena on larger scales. Physicists are obliged to respect scaling as one of the centrally important characters in phenomenology. If we were only to pay closer attention to the scaling, might these information mysteries be released from the magical realm they've been made out to occupy? This is not a mainstream scientific view, but I think it's a reasoned one. Physicists have an incentive to maintain the allure of quantum theory: it places them at the top of the scientific ladder; but they also have an ethical responsibility to forego that superiority if it is in fact artificially maintained.

Let's be clear—we don't know what elementary actually means in physics. We can mathematize theories to elevate their mystery and perceived authority, but having worked with computers for the past twenty five years, I've seen physics from another angle. Similar phenomena pop up in the dataverse, with far simpler explanations, and without the propensity for mysticism that comes with the natural world. I think we need to take that seriously and consider what it may mean.

SECRET GARDEN

As I was finalizing this book, I picked up a copy of Peter Wohlleben's book *The Hidden Life of Trees*[281], and was amazed to read about the extensive yet unseen channels of communication that pass between forest ecosystems. The communication is very hard for us to understand, and it has only been uncovered gradually because—as Wohlleben points out—trees live on an entirely different timescale to humans. Everything takes orders of magnitude longer. Trees form networks of communication through chemical excretion, fungal intermediaries, sound, and even electrical signals passed through the roots. The signals are slow and limited, but nonetheless present, and deeply affect the life of the forest organisms. One reason we have not been aware of these pathways of communication earlier, according to Wohlleben, is that modern agriculture practices isolated planting of species for maximum growth. The plants that are deliberately planted in isolation cannot communicate through normal channels, and as a result suffer from a lack of ecosystemic support. This is one reason, he argues, that modern farming methods are so reliant on pesticides. Wohlleben caught my attention immediately by using my own phrase 'give and take', and emphasizing the symbiosis of agents at work in the forest network. The Promise Theory connection was immediately called out as an important basis for bonds between organisms (ionic and covalent, involving third parties like fungi, in the sense of the previous chapter). Trees, he claimed, form superagents, held together by communication networks[282]. While different species compete, similar species cooperation and sharing the burdens of life, in an equilibration of resources that looks like a form of social welfare, or even universal income, even to the point of sustaining certain tree root systems after the death of the tree, in order to preserve the networks.

The only way we can know these things is by slow painstaking accumulation of hints from multiple observations, collected together to map out the slow timescale changes of the tree networks. The scientific process is not about individual scientists, but about a wide net of different observations accumulated over time and space, and fed into a process of learning that gradually finds patterns of behaviour across timescales unobservable by humans directly.

Wohlleben stopped short of claiming that trees might be part of an intelligent network, but it was clear from his work that trees and forests are examples of cognitive systems, described in the previous chapter. You might object to the idea—which after all is radical in its implications—by asking: if tree networks can think and communicate, why did they never talk to us? The simple answer, related to their essential timescales, will be obvious by now. Even if they could talk, they wouldn't talk to us for the same reason we never thought to talk to them. This is the same reason why a future generally intelligent 'AI' might never choose to communicate or even interact with humans. Their processes may act on entirely different scales for time and space—like the alien observer from the previous chapter, looking down on the shifting trends of the Earth. Perhaps that alien would have been able to talk to our trees[283]. Humans and trees are barely even aware of one another, as processes. This in itself should be a lesson to us.

In spite of the differences, similar themes dominate the hidden life of trees, as dominate every other process: the need for forces that sustain its processes, represented as energy budgets, the role of information in distinguishing agents and their promised functional roles.

CHOIR AND SONG, EACH TO THEIR OWN WORLD

An interesting comparison can be made between trees and particle physics, in which the respective agents are played by very different actors. There are roles for matter and radiation, i.e. for bosons and fermions—only played by trees and microorganisms like fungi and insects. Mobility of information is a proxy for communication between static locations, and thus the moving processes that involve fluids and mobile organisms are proxies for those messenger channels normally played by photons and gluons in physics. The scale and the nature of the agents are less important than their functional role in determining outcomes of interactions. The fungi play the role of the messenger bosons, in a network of process fragments not unlike the idea of Feynman diagrams. The patterns of communication form, over many layers of scaling, in almost unimaginable ways. First there is a symbiosis—an affinity for mutual benefit—that arises by coincidence between trees and fungi (recall the idea of co-activation in the previous chapter). That establishes a mutual adjacency between the two kinds of agents, like spatial adjacency. Trees are 'next to' one another, via the intermediary of channels formed by the fungi. A new layer of virtual role-playing then forms on top of this, establishing a language of concepts and meanings with respect to the larger processes. This hierarchy goes on and on, expressing both causation and proximity—defining space and time for the forest, almost like the diagrams and processes of fundamental fields. And, just as Feynman diagrams have been criticized for being too cartoonish in their representations of reality, and not necessarily representing an actual spacetime picture[284], so one might criticize a spacetime view of the interior communication of trees along different channels, as representing the actuality of the forest. But, on the other hand, we have no better idea of the causation of the process, so why not use that picture? What's clear is that forests have multiple processes, each with their own channels—airborne transmission and underground transmission; chemical transmission, electrical transmission, and even sound—all at different fundamental speeds and timescales. Compare that to quantum communication by photon and by entanglement, for instance.

THE SOCIOVERSE

The example of smart forests shows that one can scale cognitive systems by forming a larger network on a new scale, like an ecosystem, or an economy. The meta-network forms a superagent, which might expand the process capabilities of the network, or stunt them, depending on its structure. If trees can have evolving trains of transitions in their interior states—that behave as rather slow 'thoughts'—and they can feel pain (in Wohlleben's judgement) on a timescale so slow that we write them off as inanimate features of the landscape, then we surely have to expand our views about intelligence. What other tangles of interacting processes in space and time might we call smart?

As a student, I would have rolled my eyes at the idea that trees might feel pain or fear. After thirty years of working on scaling and signalling in diverse systems, I would not dismiss this so lightly. We shouldn't be too afraid of attributing sensations, or even emotions, to simple systems. Emotional responses are really just a simple precursor to more intelligent reasoning, a cheap and cheerful form of symptom recognition that cut through the need for long stories: 'If a branch falls off, fungi might infect the unprotected wound and that will prevent me from doing A, B, and C. Or change in sunlight might mean D, E, or F'. In biology, emotions are triggered by flooding organisms with chemical signals, based on patterns recognized as cognitive stimuli. They prime body processes for what might come, using pattern memory alone. Humans also have sophisticated thoughts, so the simple chemical doses from emotions trigger complex associations afterwards that we attribute meaning by more sophisticated storytelling. But for a more primitive agent, an emotional response can be the end of it. It's simply a fast alert that biases process judgements—motivating and aligning processes along certain signature lines. This is what neural net classifiers attempt to do in AI technology. Emotions evolved to signal about survival constraints and thus implicit goals—their most important function is to respond to quickly. As signals, they have a much lower effective mass than reasoning, because they don't require a network of associations.

Social networks from forests to human organisms are intricate networks. The intelligence at work in a society—if we would dare to call it that—would not be just an agent like a single tree, any more than a single neuron, or even a disembodied brain would be intelligent. Rather, it would be the whole interacting ecosystem of the forest. Optimized processes that work on fast and slow timescales are to be expected to deal with the multiple challenges they face.

This idea brings us to social interactions in general. For millennia, human activity has dominated the processes on a certain scale of our planet. In the first wave of human history, human purpose was survival and subsistence; in the second it was labour and growth for industrialized surplus; in the third wave, we are starting to employ information and automation to diversify our sense of purpose[285]. The role of humans has ever been changing, and it will continue no doubt. From solitary beginnings, the previous argument suggests that our intelligence is not only as individuals, but as groups too, where we form collaborative processes of significance to the whole. Unlike trees, our interior networks are far superior to any exterior social networks we might form—so out intelligence is dominated from within.

On the planetary scale, the idea of smart adaptation idea was captured in James Lovelock's *Gaia hypothesis*[286], about the planet's multiple processes forming a self-regulating, symbiotic feedback system. His vision certainly had some interesting and plausible aspects, but some took it too far. As an isolated cognitive system, it could only learn from itself as a society of smaller cognitive systems. Gaia is not a true cognitive entity, but more a closed society. We humans are the most obvious cognitive systems, on our own special scale, but it doesn't end there. The combined effect of all humans, interacting, sharing and manipulating space leads to new superagencies: tribes, towns, cities, countries. The new phenomena that arise, on each new scale, bring new scales into the picture. Humans experience changes on timescales from seconds to decades. Tribes survive longer, perhaps for centuries. Towns and cities may survive from centuries to millennia. We expect cognitive phenomena on those timescales.

Dynamical concepts, like mass, enter in a surprising number of places, once one is tuned in to thinking about it as a network process encumbrance. The effective mass of a city (its inertia to change, its resilience, or it's inability to free itself from all bonds and re-make those bonds elsewhere) can be defined, and is obviously much greater than the effective mass of a single agile human being, or even a tribe, or a firm, as long as we talk about the same process. What semantic extensions to spacetime show us is that there can be many kinds of mass, with respect to different interactions, and with respect to different kinds of 'motion'. Knowing that dynamical concepts like mass can be defined properly and precisely in exotic scenarios legitimizes a lot of metaphorical language we use, like weights or burdens. Notice also, the fact that a city is filled with intelligent humans doesn't make the city necessarily as smart as any of them.

Every scale has its own behaviours, and its own characteristics. It also shows that size is not everything: if the parts within a boundary are not used to the limit of possible resolution, then size will not matter. It is rather the ratio of symbols to cognitive agent which determines how sophisticated the reasoning capabilities might be[287].

As discussed, we view mass as a measure of process agility. We can alter the masses of certain processes in a society, by re-engineering the way things are done. Sometimes this happens through a new technology that enables virtualization. For instance, we can make it easier for humans to change their relationships: jobs, housing, even life partners, etc. An increasing number of jobs today is virtualized, meaning that people can maintain a bond to the same work while moving around physically or staying at home. The decoupling of work and agent, thanks to cheap communications and remote control technology, has opened a channel over which we rewire social interactions to piggyback work processes. By making a layer of pluggable connections, like a kind of switchboard for resources (or a pointer indirection table in computer parlance), we may be able to circumvent the processes for which mass applies—to use a different channel that achieves the same result, but with lower mass, or less cost. Evidence of these layers of virtual circuitry is found everywhere in science.

Layering of communications is one route to optimization, but is it smart? Adaptation that we would call smart involves spacetime too, because adaptations are processes that necessarily have to match the timescale of the pressure they are responding to. The reason for this goes back to Nyquist's sampling theorem. Any agent that can capture and respond to an exterior process, by detailed balance, needs to match the timescale of the driving force[288]. Closed loop processes can speed up or slow down time relative to some exterior process, but coupled processes have to operate on the same timescales. Mismatches of scale are a constant dilemma for the description and governance of systems. It's not uncommon that characterizations, laws, regulations, and policy—set for the average case—fail to address the needs of individual cases.

SMART PROMISES TO GIVE AND TAKE?

The replacement of *locality* as a primary concept is one of the more surprising consequences of rethinking smart spacetime. Locality is replaced by the idea of separation of promises for give (+) and take (-), across a boundary. This subtle shift better matches formulations of many phenomena, from Quantum Mechanics to biology, and it shows up elsewhere too. The separation of promises for give (+) and take (-) has played the central role in explaining how spatial networks wire processes together, and regulate information. We know that information comes from 'boundary conditions' and this is precisely the wiring of the process. The separation of give and take also releases us from the idea the causation is inevitable, that the imposition of forces and actions always mandates a response. Force is a simplifying assumption if it works, but it often doesn't—no matter the scale. Nor should we think of information as being locally sourced anymore, because it almost certainly isn't. Locality is convenient as a renormalized approximate view of spacetime interaction, but its meaning becomes increasingly unclear as scale is reduced, because it assumes implicitly the concept of a passive space and time background. As we shrink boundaries, space and time become far from passive.

From a Promise Theory perspective, every give and take process looks a bit like a quantum scattering problem. Information is promised by a sender (+) as a pretext for causal influence. It may be partially blocked or fully accepted by a receiver. If accepted, we say it propagates over the channels formed by the binding of (+) and (-), to an adjacent agent. The receiver may filter it, in order to observe different components of the information (position or momentum, for instance, like the operators of quantum observables). Filtering happens in quantum projections, impedance mismatches in electrical circuitry, and in epigenetic counter-selection—to mention just a few ad hoc cases. All the time, processes are alive within these bounded agents, causing the ticks of their clocks. We don't know why, and we know nothing about the rate at which the ticks happen, or even whether that question is meaningful to ask without a god's eye view of the universe. I think the important thing to take from this is not to think of filtering as passively projecting out symbols of information, like a camera shutter, but rather as a potentially sequenced process, like a computation with intermediate stages, in the most general case. In that way, we can scale from simple absorption to pattern recognition, language, and ultimately computation, in the exchange of discrete symbols[289].

The scaling of give and take in processes also relates to memory processes on different scales. Memory, which is a process that uses spacetime as a tool for recording information, reads and writes information in the properties of space, using agents' promises (motion of the third kind). Scalar properties of space are altered by their interactions with processes. Actionable information, in a system, comes from the edges and boundaries, and what happens between the edges is determined by the promises between the agents. Perhaps the most meaningful information is that which can change the behaviour of the agent by altering its 'programming', i.e. by rewriting the memories it relies on as dependencies of the process! As a spacetime process this boggles the mind somewhat, but actually it's just how a computer works. A single memory item might be small, like a bit (1-0) or a spin (up-down), or it might be a piece of complex imagery represented by a pattern in a neural array, or a crystal matrix.

To recognize and retrieve ideas, processes need to maintain the integrity of symbols on all scales. Memories can't always be numbered, because spacetime might not be ordered for counting. In that case we can still locate items by proximity to significant signposts—spacetime phenomena that persist as symbolic boundary conditions (the lake at the bottom of Mount Everest). Those persistent markers seed the accumulation of similarity networks, which constrain entropy, and allow an alphabet of symbols to emerge, from which language can be built.

What seems simple, unitary, and centralized at a high level, may be complex and decentralized on a smaller scale of higher magnification. It's unclear whether it is always possible to infer a single causal chain from going forwards or backwards. The famed reversibility of the laws of physics (meaning just the propagation part) may not be a feature of all scales. It depends on the extent to which we can hide the non-locality by normalizing our view in a tidy way. Newton's accomplishment in this matter is a stunning success, or perhaps he was also exceptionally lucky to be born on the scale he was.

THE CITYVERSE

Society is one of the natural world's most ingenious inventions, from the flocks and shoals of biology to towns and cities by human hand. Society is the persistence of a social cooperation process—the socioverse. The cooperation of agents to form scaled superagents brings interactions up a level, and it doesn't stop with society: towns and cities incorporate human processes with non-human processes—animals, machinery, bureaucracy, technocracy, etc. Society is a technology in its own right.

Cities are a form of spacetime too, but of course nothing like Einstein's outer space. Probably, the antithesis of that tranquil void is the assault of colour, sound, and chemicals, such as one might find in an urban hot-spot like Times Square New York, Kowloon Hong Kong, or Roppongi Tokyo. In any major city the activity level is frenetic. The number of overlapping purposes is enormous. Yet those urban processes are meager compared to what happens on a biological scale. Are towns and cities smart? I think this case can be made on a number of levels. Certainly, cities are cognitive systems, and they learn from experiences. Technology marketing would have us believe that cities can be made smart simply by equipping them with information technology—but that's a simplistic argument associated with facilities that make humans happy within the city. There is no intrinsic intelligence on the scale of the city. On the other hand, on the scale of the city, the impact of surroundings does cause changes: memories are encoded in the arrangement of the town plan, like the position of certain buildings and monuments—or even as a diagonal scar across a grid of otherwise perfect rectangles, such as the anomalous street Broadway, which marks an old Indian trail through the metropolis of New York.

The question of who benefits from the city's behaviours may influence our decision about whether we consider the city smart or not. The city can be well designed, or can have evolved into a set of good solutions that seem smart for humans within it, or it can be a smart entity that adapts for its own benefit. Smart is in the eye of the beholder—but the processes are all there. Cities modularize in space and time to protect memories, e.g. through cultural boundaries that form memorial symbols: Chinatown, Little India, The French Quarter, etc, all recall the roots of trade and cultural exchange, belonging to processes of the past, which led to migration from remote places. Those parts of town affect the processes within the city: the import of certain goods, the delivery of pizza or chow mein. The memories become encoded as living processes inside towns and cities, a bit like the way cells developed by absorbing other entities like mitochondria for symbiotic benefit.

When we walk through a city, we pass through districts and places in which space and time mean different things to different agents (people, businesses, neighbourhoods) within, whether human, machine, plant or animal. Cities operate on a virtual forest of clocks that measure a myriad timescales. The phrase urban jungle might not be a bad analogy. When we pass these spaces, we interact with them like particles in some kind of field of influence. The field does not push us crudely around; rather it impels a weak attraction (bright lights from appealing shops or irresistible smells from restaurants become messenger signals that promise a speciality within) and forces of weak repulsion like the opposite from slums and junkyards. Interactions and signals are mediated by speech, by money, and by actual visitation. Cities even join together by sharing in national grid power and water networks, and today they cooperate on goals and processes of improvement and mutual learning.

When travelling, I like to look out for those signs of adaptation. In South East Asia, for instance, I've often been struck by the attention paid to drainage and flood control. In Thailand and Malaysia, fields and towns are dominated by drainage channels to handle the Monsoon rains, in a way we just don't see in Europe of America. Taiwan is a verdant country dominated by a surprising number of tall mountains, for what is a small island. It suffers major earthquakes on a regular basis, and is no stranger to natural disasters. It has what, on first appearance, seems to be a massive overcapacity for drainage, but given the propensity for sudden flooding, Tsunami, and typhoon, it makes perfect sense[290]. Moreover, the presence of those features allows us to infer the existence of those phenomena, just as archaeologists and paleontologists make inferences about conditions in the past from layers of deposition and surviving structures, adaptations to skeletons and tools.

Suppose a city were indeed struck regularly by Tsunami waves (see figure 9.1), the responses of the city's many inhabitants, over short or long times, would encode those occurrences. On a short term, a city represents its current state of contextual 'awareness' by the evacuation of people, closing of shops, etc. In long term memory, processes unleash a chain of adaptations to automatically protect the city. The memory of the floods lies in layers of repairs, layers of depositions, reinforcements, and in replanning. The changes exhibit a primitive form of autonomous reasoning. Reasoning that happens on a larger scale would not be clearly visible to the occupants of a city, and may take place on a much slower timescale than they normally observe. Thus, a smart city might not be apparently smart to human concerns.

Spacetime adaptations affect survival in other ways too. Recall that the dimension of a spacetime process depends on the number of independent freedoms it can express. We might call a tunnel a one dimensional process, or a two or three dimensional one, depending on our scale and our vantage point. What matters more than dimensions of embedding spaces is the connectivity of the networks involved. Networks enable the transportation of information on different scales, including the transportation of goods and services in cities, or

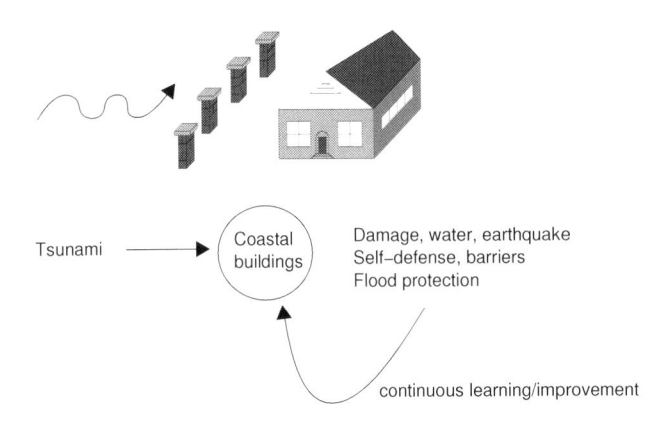

Fig. 9.1. A city learns from repeated Tsunami: human built flood defenses are the first line of memory that accumulates by interior processing, and layers of exterior deposition in the geology reflect the floods too.

blood and white cells in biology. Only recently have we come to understand how networks also underpin the efficiency of processes at scale.

SCALING OF ORGANISMS AND PROCESSES

In biology, a well-known result about the scaling of organisms is the Kleiber '75%' economy-of-scale law, named after biologist Max Kleiber (1893-1976). Since Kleiber's law was described in 1932, it has been known that the metabolic rate of a lifeform B, meaning the energy usage in most animals and plants, follows a scaling 'law' that was eventually explained by physicist Geoffrey West[291]:

$$B \sim M^{\frac{3}{4}}$$

where M is the size of the body as measured by its classical mass. This is a so-called sub-linear scaling relation. To put it simply, if you double the size of an organism, you only need 75% more energy to run it, but the downside is that you slow down, and get 75% less out of the energy (imagine a fly versus an elephant). Most other rates, such as heart rates, reproductive rates $M^{-\frac{1}{4}}$ and reproductive rates and lifetimes scale with quarter powers $M^{+\frac{1}{4}}$. The bigger you are, the slower you are, the longer your life span, and the more energy you use. Eventually, an organism would hit a size limit. Size has a diminishing return on investment, and a predictable effect on timescales for processes. The energy savings are not enough to justify the lower return. Thus there is a maximum size to organisms (bad news for King Kong and Godzilla). These patterns have their explanations in the properties of networks. West found a way to count the effect of processes in terms of variables for the scaled characteristics and showed that the scaling behaviour was a result of space-filling networks. That means, the one dimensional information channels that transport energy around organisms, were numerous enough to fold, twist, explore, and finally fill the embedding space of the three dimensional organism to an approximate degree. The embedding space model allows the circuitry of blood flow to be counted in a simple way, and compared to easily measurable dimensions in data. The results showed that energy accounting was a good representation for characterizing the timescales of biological processes.

Inspired by this simple scaling law, West went on with collaborator Luis Bettencourt to study the network influence on cities and companies too[292]. A city packs effort and communication into a small space, bringing processes together into a localized region—a superagent, in the Promise Theory sense. Their studies showed that, as a city grows, it needs fewer petrol stations per capita, and less length of roads, electrical lines, per capita. So there is an economic incentive for cities to grow. These are economies of scale (at about 85%), similar but not identical to the biological scaling (after human-machine infrastructure is qualitatively different). However, the meaningful outputs (the city's 'metabolism'), labelled semantically by wages, patents, disease, GDP, waste, police, and crime, all exhibit super-linear scaling (115%), because cities bring people into closer contact, enabling functional or semantic payoffs that could not be possible without a hierarchy of multiple, compounded efficiency gains. In short, the siloing of functions, or semantic localization around particular trades and specializations, induces a change in the scaling efficiencies of processes where specialists need to work together. In biology, semantics are locked into DNA, which is wrapped in cells, so creativity only happens through sexual reproduction. In a city, creativity is happening everywhere, and cooperation is facilitated even non-locally thanks to a broader spectrum of processes, some physical and some virtual.

Economies of scale are a result of process dependencies. A simple explanation can be given in the language of Promise Theory, which allows us to take the insights of West and Bettencourt and apply them more generally to virtual spacetimes too, without the need to refer to a physical embedding space.

Economies of scale are known in economic theory, often based on logistical supply networks, but it's rare to see an example yield to a simple accounting model in technology or biology. Understanding the counting of measures in technological and social systems is going to be of huge importance to the next generation of technologically enabled society we build. The influence of measurable data is already being felt. Some authors refer to scaling laws like these as scale-free, universal, or even fractal behaviour[293]. The term scale-free is, I think, misleading, because it deliberately neglects the relativity of observation and the granularity of processes. No processes are truly scale-free across every scale. The planet is only so big, and atoms are only so small. The fractal approximation used to count these laws is a continuum approximation, absolutely standard in physics, but like all continuum approximations it's a counting convenient. Nevertheless, counting processes is not always easy, and whatever method works is what matters. That is essentially the reason why Newton's ingenious prescriptions for synthesizing computable mechanics are so revered.

Does this pattern of scaling extend to other cases too? Today, information infrastructure is one of our main concerns. What might process scaling mean for information infrastructure and other mixtures of dynamic and semantic behaviours? Companies and computer operating systems are like very small cities, in a sense, and both show different behaviours to cities. A key difference between biological organisms, cities, and computer networks is that, unlike the former, computer networks did not evolve—they were not selected by their successes and failures, they were designed and improved by humans. This has led to a lot of convenient centralization on a geographical level. In the future this will have to change, because there is a wealth of computers all over the planet that are underused and poorly integrated. 'Edge computing' is what the IT industry is now calling all those personal computers, laptops, tablets, and phones that are located in homes and offices rather than in the large datacentres. The next spacetime challenge for the cloud is therefore to re-integrate all those edge devices with the cloud itself, enabling free movement of processes between them. Currently they are separate universes that can sometimes teleport (using FTP or some such wormhole technology). We might expect datacentres to scale according to a similar—but not identical law[294]—but, as yet, there are no data available to support this, because the large datacentres are all inaccessible proprietary organizations that are unwilling to share their interior details. They do not promise observability, so they are effectively black holes as far as science is concerned. However, it's possible to theorize. The virtual nature of processes means that their actual dynamical embedding space might be quite different from the physical container of a datacentre.

Workspaces in the dataverse

One of the big challenges for human engineering, in the decades ahead, will be to fully understand the scaling of workflow and data processing, and how this integrates into human society. The scope of that challenge stretches from humans to computers, sensing, learning, understanding, and acting together. It might seem that we've come very far in that area—with mega-companies like Google, Facebook, Amazon, and Microsoft boasting mind-boggling data processing capabilities, and extreme manufacturing capabilities using new materials, 3D-printing, and robotic factories. These accomplishments leave us awestruck, compared to what was possible even just a decade ago—but, the intelligent use of scale, with the integrated understanding of both dynamics and semantics, is really still in its infancy. Processing *pipelines*, where information is collected from sources, and fed through a kind of staged production line facility of computers to produce a result, are the new circuitry of the third wave of human civilization. They're behind the insights of data science and of data-driven businesses, as well as the darker sinister activities of viral attacks, misleading propaganda, and privacy hacks that we increasingly hear about on the news. The science to understand the technological challenges will develop alongside the wisdom to apply them ethically for the common good. Let me mention just a few points where a spacetime view is integral to the challenges.

Abstraction—or virtualization of processes, as it's known in computer science—has been standard practice in the dataverse for a long time. Of course, the beginning of virtualization precedes information technology by a long way. It probably started with the dissociation of territorial process boundaries from geographical ones, i.e. when the location in the geographical spacetime became less important than the location in the dataverse. That has been one of the drivers for globalization. Information technology has risen to the challenge, and stepped in to create a much more explicit layer of virtualization of every kind of relationship. We only need to look at Internet Domain Names to see how national and commercial boundaries are perceived in the dataverse. Initially, when the Internet was confined to the United States, there was no need for explicit national boundaries. Instead, the semantic boundaries were '.com', '.org', and '.net', for the commercial sector, the public institutional sector, and more technical network resources. In an interesting bit of politics, as the Internet spread—instead of following this pattern—the Internet Addressing Agency IANA introduced country domains: '.uk', '.fr', '.ca', '.no', etc, with commercial and academic partitions under them—reinstating a bit of nationalism to favour the United States. The model was imposed, but my gut feeling is that is was largely a failure. Few companies operate purely from one country anymore. Only schools and other public institutions really belong to single regions in the 21st century.

Spatial locations, for data and processes, are meaningful anchor-points if they allow us to describe the connectivity of circuitry. Semantically, there is a big difference—at least in the minds of users—between a network connection (like the Internet) and a data processing pipeline (like a factory) for crunching numbers and reshaping data like a production line. But, data pipes are actually just a renormalized extension of networking, which add direction—boundary conditions—that we think of as intent, like a field of force guiding the motion. The pipeline represents an 'opinionated' process, a directed intent, not merely a blank set of connections with latent possibility. If we were to embed the processing into the wires somehow, by some fancy renormalization trick, then we would see this simpler picture of the dataverse.

In the 1990s, Sun Microsystems Chief Scientist John Gage coined the slogan 'The Network is the Computer'[295], making the observation that computing is basically a networking problem (a spacetime problem), on many levels or scales. Since then, relatively little attention has been given to that observation. The pervasive nature of computing today is an imperative to renormalize the boundary between computer and network (agent and superagent cluster) to reduce the complexity. To use an expression from quantum physics, we need to look at 'dressed' services rather than microservices. So, it's not only agents that can be combined to form virtual superagents. The adjacency channels between them can also be made to tunnel from place to place, through potentially many adjacencies at lower levels. Network overlays, like virtual private networks, rewire space with tunnels, as if flying above the landscape of the Internet to teleport directly without following the usual highways. There's still enormous scope for the routing and scheduling of 'business processes' (whether in the public or private sectors) on new multi-scale platforms. Key processes want to live in two kinds of location: on the boundary, or at the edge, where information lives, and in fast dense processing cauldrons like datacentres where proximity brings efficiency. Whether proximity is the same as physical distance or latency or some other measure remains to be seen. It depends on a number of spacetime factors.

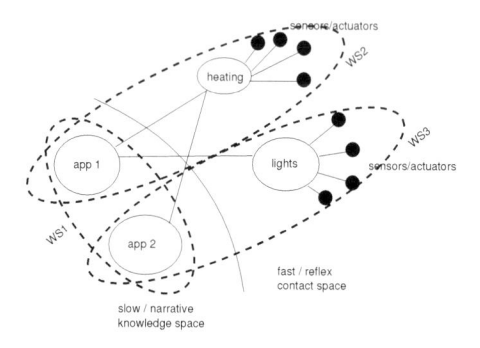

Fig. 9.2. Overlapping regions of smart spacetime adaptations, are like overlapping virtualized brains. Humans focus on working roles as their semantic boundary. The result is a patchwork of overlapping spatial regions—called Workspaces.

One way of renormalizing the plumbing of the Internet would be to develop the notion of operating systems for networks—to make the transition from pipes to plumbing. Some ideas exist for this, around Network Function Virtualization (NFV), but the idea is probably moribund—based on an old fashioned, fixed-scale notion of how a computer system should work, with one agent per function (like the service load-sharing lens spaces in chapter 5). Modern social interactions, on the other hand, create an enormous variety of tailored spaces. Companies, private and public organizations are also processes that have a spatial footprint, and they have much in common with cities. They are no longer geographically contiguous, but rather they are connected by independent messaging channels, and fed by logistical processes of transport and communications. They are part human, part machine. We are entering an age of cyborg systems[296]. In my own view, the next abstraction will not be a human one or a computational one, but a human-computer hybrid that I call Workspaces[297]. Processing needs to be embedded into our very environmental spaces, as part of our extended ecosystem.

Humans like to work together around single specialized tasks because that's how we've organized for millennia (see figure 9.2)—perhaps this is related to the efficiencies of cognitive agents. Indeed, it's interesting to see how software development recently began to shift from a generalized wizard model to a specialized tradesman model of working—developing single skills, one at a time, using 'microservices' rather than grand unified projects. It's fundamentally a social model of development, and allows a separation of concerns with 'vertical integration' so that each development team can manage the entire lifecycle of a single microservice, rather than splitting responsibility along horizontal stages of a service's lifecycle: design, maintenance, and operations. As software writing gets easier and more universal, and it is applied to increasingly mundane tasks, like heating, lighting, building services, this return to a human form of organization (rather than an industrially imposed one) is certainly telling[298].

In a sense, the goal of smart infrastructure, with processes embedded as spacetime agents at every location and coupled by smart channels, is not to just wire stuff together, but to act as coherent data circuitry—to be the 'linker' in the master software for society. Socio-economics can be viewed as the most ambitious multiscale program ever executed, spanning the world. That idea goes back to science fiction writer Isaac Asimov's concept of psychohistory, from his *Foundation Trilogy*. If we look more carefully at the processes of the world—carefully enough to see what is similar and what is different, when discriminated by other key processes—then that analytical perspective becomes a reality. It is not science fiction. This is the real story of what AI researchers and big data analysts are doing today (it has little to do with intelligence).

One interesting observation is how the layers of process in any staged process, say a company, a factory, or in cloud computer program, superficially, take form like a neural net (see figure 9.3). That's because of the implicit boundary conditions that lead to the motion of data through the system as an intended process. Staged processing, along multiple paths, is how movement of data becomes reasoning, with implied intent. Intent is the implicit interpretation of boundary conditions that have successful outcomes: a symbolic interpretation of directed behaviour.

Business semantics and Information Technology semantics are currently very different. That costs businesses a huge amount every year, because fleets of engineers are required to translate the concepts of business into the concepts of

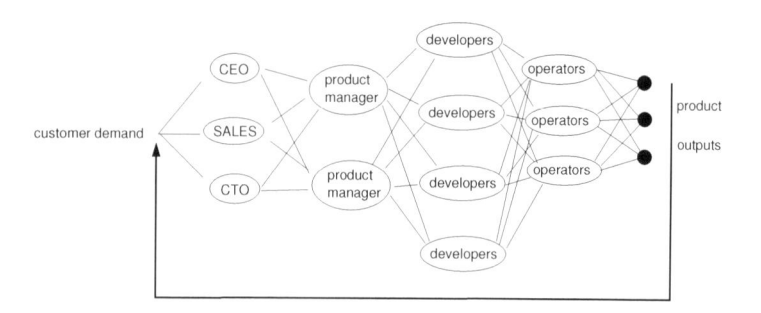

Fig. 9.3. Is a hierarchical organization like a learning semantic space? The structure is potentially similar if there are strong links between layers, without siloing. This looks quite like a neural network.

IT on an on-going basis. As long as there isn't a good alignment between the two—perhaps even a passing similarity in the process boundaries—that remains very hard. It's not merely a job for language, but for extensive meta-computation. Spacetime is polarized along a directional axis that signifies a kind of motion in a straight line. That pattern exhibits very similar scaling features to those we observe in the biology of cells. If we did not ask those basic questions—what is motion?—we might never see that there is an opportunity to learn from a larger perspective.

INTO THE NEBULA

The most marked shift happening today is surely the move from personal computing to cloud computing—or 'utility computing as a service'. Information Technology has drifted far from what it once was—when a single computer assisted a single user, who sat in front of it, by performing a sequence of steps faster than a human could calculate. It's now fast being taken over by the cloud model—a vast multilayered network of interacting scaled processes, that serve thousands of users at the same time, sometimes from half a planet away. Those channels link software agents together by multi-typed information channels, whose messages employ a highly complex and inhomogeneous hierarchy of semantics. In that sense, it looks more and more like biology every day.

The computing cloud may be logically centralized, at least for the time being—operating out of a few large datacentres, located at very specific geographic locations—but that picture will not last forever. The endpoints of cloud processes—user programs—are distributed liberally all over the planet, running on laptops and smartphones, and data have to move, to and fro, across that space, in a colossal feat of speed and coordination. Smart assistants and TV set-top boxes—home devices that we can talk to and program—give cloud technology companies a way of insinuating themselves into our homes, and one day they will likely own the technology infrastructure there too—all the way to the 'edge' of the network. The layers of virtual wrapping, with which we clothe cloud processes, in order to manage that feat, are expanding too. There is significant overhead attached to these universal conveniences.

Hardware is one thing, software isn't unaffected either. Computer programs are no longer singular listings of instructions like 'ADD', 'STORE', 'JUMP', performed one after the other, as a single clock process. In a modern computer program, we can't assume to know where a particular part of a process will end up being executed, or by whose interior clock. The paradigm for writing computer code has changed from sequences of atoms to the formation of 'groups' of concurrent mini-processes, known as coroutines, threads, or services. It's actually quite like the shift from a single cell DNA strand to a cellular model with many different strands representing different specializations that then work together on a cellular level. Those processes form new virtual agents with their own boundaries, as if spawning off a separate spacetime bubble. They are called 'threads', 'containers' or 'virtual machines', in the dataverse. Together, they form a kind of tissue that becomes a layer of technological infrastructure.

The challenges of computing in 'The Cloud' revolve around local variations: they include how to deal with semantic issues like security, 'realtime' responsiveness (or 'zero latency'), reliability as experienced by different observers, and not least the diversity or lack of standardization of environmental conditions experienced by software processes. The industry is good at marketing its offerings as 'ever better' and more competitive advances, but it's the lack of certainty about process outcomes that makes all those virtual layers of complexity problematic to the path of human purpose. This is why spacetime concerns are now paramount.

Ultimately the computing cloud's memory space spans the planet as a patchwork of cellular spacetime bubbles, joined together by meta-processes. These new layers of process form a hierarchy of renormalized effective units that perform processing, roughly analogous to a family of particles in physics, but with much more sophisticated semantics (see figure 9.4).

The shift from old-style register chip programming to a cloud programming model is a simple renormalization of the boundaries between one process and its related dependent processes, but the effect of this is far from simple. Promise Theory is helpful here to understand the effect of redefining the paradigm. In Promise Theory language, the renormalized boundaries around agents transform them into 'superagents' that make different exterior promises on behalf of a collective of agents within. Instead of treating all processes as independent, we now have to look at the hierarchy of interactions and redefine processes in terms of 'effective objects'. Clients and users no longer deal with, say, a 'mechanic', they deal with a 'department of mechanical services', which is a different kind of interaction. The processes inside that department, however, have to deal with all the old systems, the new supporting infrastructure, and the routing and load sharing of the demand. Load sharing 'lenses' are only the beginning of what is involved.

In particle physics, we might call all the layers of renormalized processes 'dressed fields', with the assumption that one never sees the undressed fields under normal circumstances. Alas, that is not yet the case in the dataverse, which is still learning how to make the dressing! Many of the wires and pipes are still exposed and have to be wrestled into submission by engineers to make it all work. Indeed, the complexity is sometimes mind-boggling. What once seemed centralized is later destined to become decentralized and eventually recentralized on a larger scale—this is the process of semantic growth. It's a discrete process. The pendulum between centralization and decentralization swings back and forth.

The subject of layered process-virtualization plumbs the depths of an effectively bottomless well of issues that I daren't venture into in detail—but, let me dip a toe into just a few of the layers for brief illustration. For software engineers—who increasingly dominate openings in the jobs market—a knowledge of spacetime concepts has become suddenly essential since cloud computing took hold[299]. Programmers would probably never use the words I'm using here—they are probably unaware of the spacetime implications—and yet, this knowledge is effectively what modern programming requires. Newton's laws don't work in the dataverse, but something a bit closer to quantum mechanics does.

The structure of layers goes something like this: spatial locations in a computer system are bounded by processes (see figure 9.4). They begin at the

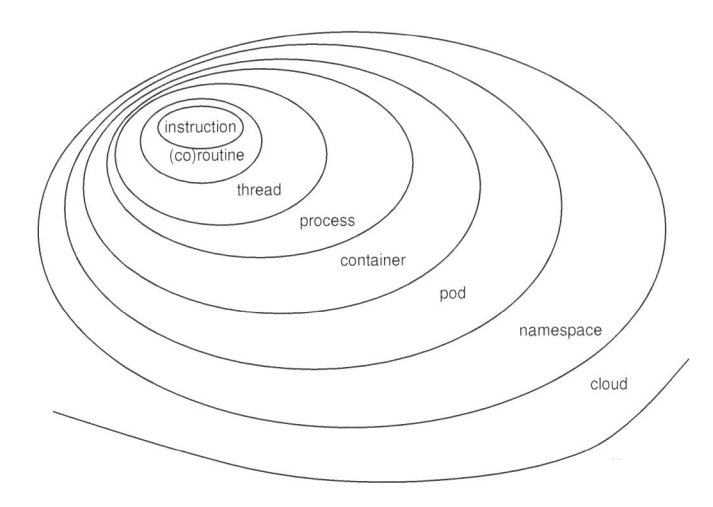

Fig. 9.4. Layers of spacetime (process) granularity in a computing cloud. Each layer has its own interior clock time. Processes are made up of atomic instructions, which are executed by CPUs. CPUs have a clock. Each instruction is a tick of a program clock (a routine or coroutine), which may involve several ticks of a CPU clock. A routine is co-located in memory space. Threads and processes are parallelized at different locations in process memory space. Containers aggregate processes, pods aggregate containers, namespaces aggregate collaborating pods, and these run in a cloud datacentre.

atomic instruction level, like 'ADD', in memory registers of the CPU chips. The next level is a layer of 'scope', which wraps a classical flowchart process with a virtual boundary—on both a dynamical level (availability and access to a spacetime location where the task is scheduled) and a semantic level (visibility of named items, and how to interface with it through APIs[300]). Process scope, container scope, namespace scope, and so on. Each layer of process has its own sense of time, at each location where it runs. A transaction received is a tick of a clock. The completion of a subroutine is a tick of a higher level virtual clock. The arrival of a message is a tick of an exterior clock, sampled according to the interior time of the process layer concerned.

Cloud-era programs have concurrent tasks on-going in partitioned 'worlds', regardless of how many physical computers they might be running on. Space and time are partitioned, shuffled, dissected, and reassembled on every level: there are interrupts at a hardware level, coroutines, timesharing, parallelism in pipelines and services, and more (see figure 9.5). To understand coordination of these multiple processes is no small feat—it involves communications using a variety of signalling channels. Physics has just four forces and a handful of representations for them, but the dataverse has many more. Mutual exclusion locks (mutex), message channels, sockets, queues, pipes, threads, processes, each of which may use the other recursively. As the arena for processes grows

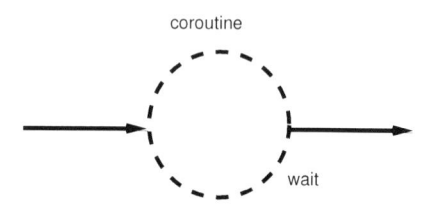

Fig. 9.5. A coroutine in programming is like a loop correction in quantum theory. It's no accident, and the reason is not really deep—it's just what processes in space and time look like.

more complex, so do the processes themselves. In earlier generations of computing, when a computer was a single localized machine, everything happened as a single serial process, or it didn't happen at all. The limitations of that approach are now unacceptable, as they cannot scale to handle the demand from online clients. There is no single way to handle a process once it becomes non-local and takes on an intrinsic process mass by defining new boundaries and interlinking the interior agents through its causal purpose. A programmer might now spawn off a small sub-task (say, 'find me the name of a server outlet from a directory listing') in order to decouple this minor task from its main storyline. It introduces another pathway in spacetime that will eventually merge again with the main thread, to keep possibilities in play. These, concurrent avenues of thought, hedge against the uncertainties of arrivals in the spacetime infrastructure—when calling out for a service that may not return, you don't want to bet the whole farm on a single rider.

One way to view the change in programming style, in the cloud, is to see it as a shift towards a pattern of employing cognitive agents in place of 'dumb' agents—these are known as 'actors' in computing. The cognitive abilities vary widely in their levels of learning ability, but they always have memory and they adapt to context. Not all of them are too smart, but they act as responsive observer processes. Moreover, the 'contexts' they accumulate from interactions, sometimes passed directly by interprocess communication, behaves approximately like the semantic contexts described in the previous chapter. Context addresses short term memory caching, which leads to 'concepts' identified from sensory inputs. Information arriving from any input channel, remote or local, past or concurrent, drives tokenization of data patterns as the outcomes—making data channels appear like inputs from smart sensors. Context can be used on the level of a single interaction or even a single transactional message to give it meaning, as well as to guarantee its short-lived applicability.

Take the recently invented programming language *Go* (or 'golang') as a symptom of the changes, and we see computational functions directly take on the role of promise theoretic agents—localized process agents that promise results from interior details. An agent can consist of numerous independent concurrent sub-agents, and the links between them (conveniently referred to as 'channels') are processes that connect them with interior queues. Programmers are encouraged to view a program as a collection of interacting agents, i.e. as a small spacetime network, rather than as an old-style sequential data flow; they can add single purpose agents to a superagent 'group' and the treat the group as a single entity, waiting for its results and ending its lifecycle as a unit[301].

Data, in a cloud, move at different speeds and with different effective masses depending on their degree of replication, relative to these agents who share a standard of interior time amongst concurrent tasks—and this leads to an overlapping mixture of concerns. To ignore the spacetime implications of this arrangement is to bury one's head in the sand. The picture feels more complex even than particle physics, but the basic understanding of has remarkable similarities, as discussed in the preceding chapters. *Go* was intended to be a more natural programming model for the realm of cloud computing than the concepts of older languages, like *C*, *C++*, or *Java*. Why do processes benefit from languages? The first reason is that language is process—symbolic process, and that's what computing is about. Another reason is that the assumptions about information, motion, synchronization, and time no longer apply in the way they did for *C* or for *Java*. How does one condense a serialized story from the outcomes of multiple distributed processes into a single causal picture (see figure 9.6)? How does a distributed cloud speak with one voice? In fact, this question is what Promise Theory originally sought to answer. Programmers

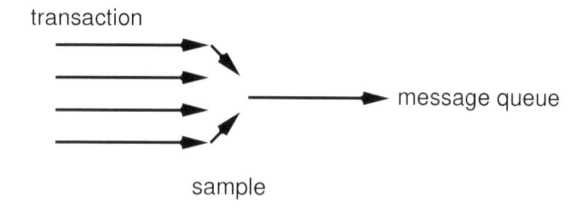

Fig. 9.6. Layers of message or interaction simultaneity (exterior time ticks).

might not realize that they are confronting fundamental spacetime issues when seeking the answers to this, but all these constraints are turning modern software engineers into spacetime relativists.

Classical desktop computers performed all processing locally, and were driven 'deterministically' by a model of programming steps that followed one another slavishly in sequence, but cloud software makes computer programs non-local, and their causality is equally complex. This already has major implications for software in the future, and it's only going to get worse. Programs are increasingly 'event driven', as if every program were a relativistic observer waiting for signals from afar. It means that systems appear stochastic on a short term level—so how can we account for intent? How do we achieve planned outcomes? Advanced boundary conditions, or 'desired state' systems are a common answer.

Chains of reasoning typically also now depend on the confluence of many signalled changes from different modular sources. Suppose you want to build a new result of a computation, or even a car on a production line: it depends on components from a dozen different providers. You know what the final outcome should look like, i.e. what the constraints are. No random outcomes are allowed, so there is a fixed boundary condition. If the manufacturing rules are fixed, then that outcome also serves to constrain the manufacturing paths going backwards to the start of the process (backwards in time, implicitly). So, is it the arriving events from sources or the final boundary conditions that matter the most in this process? The answer depends on the scale of the change. A change in the kind of car is a change in the future boundary condition. A change in the inventory to build the same car is a past boundary condition.

Machines, including computers, are only useful if users, inputs, and outputs can interact with them—but the way centralized services, like cloud computing or public services, separate those users from the machines' processes by large distances is a bit like turning technology into one of Einstein's thought experiments in relativity. The results, ironically, can turn out to be quite similar. The same is true of interactions by post, where the transport speed is slow compared to the processes themselves. The sampling rate of the receiver is distorted. The challenge that technology will face (the Elephant In The Cloud, so to speak) is going to be the unstable effect of mass and time on observability—on the effect it has on its interior time. The traceability or observability of a computer system, like any observable phenomenon, depends on the ability to get signals in and out of the region of spacetime where the relevant processes are running. Moreover, observation can't be scaled with impunity: observation is not a scale free process, because the observer breaks scale invariance.

Observations may be spread out, but they all have to be focused back onto a single location: by the process of observation by a single agent. In a cloud, there are always two kinds of interaction going on: the flood of input and output interactions, from the programs that are running, and the shadow world of diagnostic processes that allow data to be observed. While we can abstain from trying to watch the latter (to know what's happening on the interior), we can't turn off the former without sabotaging the entire point of the cloud—the running provides a service. With the programming practices described above, programs that would once have been local are now spread out non-locally, but have even more wires that hold them together. The effective mass of the program is higher. Now, as the traffic load on the cloud grows, the mass grows higher still, and the overhead of managing all the links grows at least as fast as the load[302]. As the pathways of communication get clogged up with the overhead of this scaling, two things happen: the processes run slower because the latency or Time To Wait increases for every operation, and it gets harder to get signals in and out of the interior. When a program is distributed non-locally, we need far more interactions to observe the entire process—that increases the impact of observation the process, and number of things that need to be seen. Moreover, because of the semantic separation of modules, new barriers between process components increase the overhead even more. It makes observation hard to impossible, without adding even more mass overhead.

From outside the cloud, the effective mass of processes seems to increase and the rate of time passing on its interior seems to get more and more sluggish, as the rate of traffic load reaches saturation—ironically similar to what an observer would see in a remote spacecraft moving at close to the limiting process rate there (the speed of light). On the interior of the cloud, purely local activity can't notice any difference, because everything is affected in the same way. It's a simple form of relativity, analogous to what happens in Einstein's theory and for analogous reasons. The harder we try to interact with the cloud, close to its limiting performance, the less we'll be able to see and the slower it will get. It will turn opaque, like a quantum system, and we'll know less and less about the effective locations and rates of processes going on there. It's a direct result of the scaling behaviour of processes in a finite size system.

At the other end of the spectrum, the wide area communications challenges, employed in computing clouds, have their own implications for predictability and reliability. The messages passed between agents may not be delivered, or may be delivered in unexpected order, especially as the process mass increases and fluctuates differently along different paths. Remember the cloud doesn't just have one interior time—every agent has its own interior clock. There are literally millions of different versions of time. Having multiple clocks driving a plethora of concurrent processes that may be either sender or receiver of messages only further impacts predictability—reminiscent of the Heisenberg uncertainty about positions and momenta. Again, we can learn from spacetime processes to seek a deeper understanding and a solution.

High level error correction loops, such as the Internet Transmission Control Protocol (TCP) protocol, are built into the Internet. This is the classical solution to reliable delivery of data. TCP works by trading uncertainty of delivery for uncertainty in the time of delivery (if you wait long enough, the data will eventually come)[303]—but this error correction effectively adds a non-negligible mass to communication, associated with reliability. One of the first high level promises that suffers from this uncertainty is the ability to rely on data *consistency*. Recall how institutions like banks need to maintain consistent copies worldwide of transactional information to know whether payments can be made, and who gets paid first. Protocols like TCP can perform this task, but slowly, and with overheads that could be rationalized by seeking a lower level promise of delivery[304]. A rule of thumb in the dataverse, known as the CAP conjecture, plays a role a bit like Heisenberg uncertainty but for a tradeoff between data *consistency* (common information over a scaled region) and channel *availability* (ability for data to move)[305]. Using TCP to add a layer of certainty on a longer timescale works for a dataverse that is mainly empty and has a few processes going on, but that is not always acceptable in the dataverse. Inefficiency costs.

Sparseness turns out to be the key to understanding how many spacetime processes work. As soon as activity passes a certain threshold, process scaling may become unsustainable, because of non-determinism. Put simply, uncertain processes need a wide margin to appear quasi-deterministic.

An alternative approach, that renormalizes error correction at a lower level, has been proposed by Paul Borrill and his team at Earth Computing[306]. It makes use of the classical analogue of entanglement, made famous by Quantum Mechanics, to effectively switch time on and off, as information spreads. Briefly, in order for message transactions to be coordinated in time, they need to form a single co-moving, co-dependent superagent—i.e. they need a single clock. Such a clock can be constrained by entangling the two endpoints, in a promise theoretical sense (see figure 9.7). In a transactional view of motion, when data leave one end, they should arrive at the other—at the same moment, because there is only a single clock time for them both. That isn't the case for TCP, because it is a process that relies on the voluntary cooperation of agents at a lower level than the error correction. The alternative method works from the bottom up, to prevent exterior observers from seeing anything inconsistent before it 'happens' for them. One effectively binds the two endpoints into a single spacetime point with polarized boundary conditions—amoeba style. This is rigid and would normally be considered fragile and risky, if attempted on a high level, but if applied to the lowest physical layers of communication, it can be implemented with a very low mass. The key underlying limitation on predictability is the indeterminism of network communications, compared say to the relative determinism of communication of a single computer. Causal codependence could therefore be the antidote. Whatever the agents in a distributed system promise, they must promise it together, as an indivisible causal unit in order to keep in step with one another. Entangled links might therefore be helpful in calibrating data consensus by engineering 'observability controls'.

Effectively, entanglement adds a kind of observer lock on what can be seen at each moment of exterior time. Endpoints recover synchronicity in communications, by prioritizing interior time. An entangled link is basically a renormalized error correction channel. Parties in a system can be prevented from seeing inconsistent states before clear promises can be made about their interpretation.

These random applications of spacetime promises feel like ad hoc additions to ad hoc components in a system. It isn't a scalable way to achieve certainty.

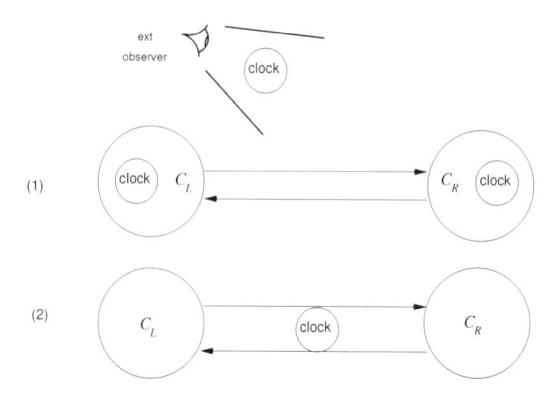

Fig. 9.7. Independent agents in (1) have independent clocks, making them asynchronous. Co-dependent or entangled agents in (2) are synchronous because they have a rigid shared time.

A better way would be to make the promises apply at every location and for all times—for all transactions—indiscriminately. Spacetime could be woven more homogeneously, like a fabric—more like something that could look like Euclidean space. Then, we would finally see a value to translational invariance, and its conservative promises. They would imply an approximate scale independence. It's not a property that can be exact, but it can be very helpful to bring predictability over a range of scales. Cloud architects and their forerunners have been designing data *fabrics* to engineer this kind of homogeneity for many years, and continue to develop the idea. I wrote about this extensively in my book *In Search of Certainty*. Every move towards a more homogeneous and regular underlying spacetime fabric makes it easier to virtualize processes with impunity.

I want to close this section by remarking that the fine granularity of processing involved in cloud computing, across multiple scales, is a huge conceptual burden on programmers, and it makes the tracing of a process storyline extremely hard too. The causality of the software behaviour becomes a multiscale interaction, involving layers of causal aggregation and many-worlds splitting. The 'scaling of true and false' effect, discussed in chapter 6 now completely disrupts programming logic on an operational level, making it vulnerable to changes of scale. The layers of wrapping have meant that the program design and packaging (development or 'dev') has become almost completely decoupled from program execution (operations or 'ops'), because they deal with different scales. One human response to that shift was to campaign for 'DevOps'—a process of dialogue meant to re-establish trust and communication between developers and operations engineers[307], with limited success. The level of awareness of detail required for a programmer to fully understand the implications of data processing in a computing cloud has been approaching a critical juncture for some time. This is witnessed by movements to try to simplify the cognitive burdens for humans, such as with the 'microservices' movement, which offers one way of breaking up human cognitive complexity into smaller, more manageable pieces. Increasingly cloud-centric programming interfaces like 'function as a service' hope to sell partial workarounds for users (after all they are businesses) but they are not too convincing yet. As an addendum, one could also note the explosion of semantic types involved in modern programming from modular systems. Whereas, in the past, programmers might have had to deal with strings, integers, and real numbers—today there is an enormous menagerie of types, and programmers are encouraged to introduce new ones for every purpose. This is helpful to compilers during error checking, but it adds a burden on programmers to know and manage these types. Moreover, there is no natural coordinate system for a computer program: computational space is not Euclidean, Cartesian or polar: it is a scaled promise graph of boundaries within boundaries, connected along channels like a mesh[308]. There is no simple ordered numerical vector that describes the spacetime coordinates of a process—programmers generally have to erect signposts around in their code, and react to significant index events. This is an additional mass on the process agility of developing and maintaining code.

The added complexity from all the considerations just mentioned is a major headache. It's remarkable how spacetime scaling is involved at all levels of the problem. Building homogeneous, isotropic fabrics (as they say, in cosmology) can help us to simplify the issues to some extent, by renormalizing away details and providing simpler more deterministic interfaces—just as effective determinism emerges from scale in physics. Cloud providers are trying to simplify matters by offering simplified services[309]. Personally, I am betting on the 'Workspace' abstraction, mentioned above, along with new layers of human specialization that simplify information infrastructure for ordinary folk. That would be the strategy biology has taken—and it seems to work quite well, though you never know quite on what scale understanding and success will emerge.

Either way, the current trajectory for technology is unsustainable, in terms of the human mental feats required. There may be room for Artificial Intelligence to alleviate some burdens, but we are unlikely to give up the sense of control in designing new systems. Architects have embraced new methods with smart materials, without having to know about complex chemistry. Similar advances in reorganization and renormalization will no doubt occur for our information services. After all, they too will become a part of the architecture[310].

For agent and country

Let's change gears now, and move up a scale to our old and almost forgotten friend—geography. Historically, sovereignty is associated with geographical territories, marked out in two or three dimensions, within physical spacetime[311]. Separating territorial regions as countries, airspace, fishing zones, and over-arching empires, is a legacy that remains with us, for better and for worse. Political boundaries are closely associated with trust, and trust is the basis of all cooperation[312]. The boundary of a region is dominated by the semantics of self and non-self[313]. We trust self and mistrust non-self. This simple rule explains far too much of the conflict in human life, e.g. in the disputes amongst the world's main continents and their fragile attempts at union. Tribes, on the inside, will always look to lay blame outside of their borders. It's a natural consequence of a cognitive system: the observation of exterior from interior will always be projected onto the values of the interior as its sole measuring scale—'collapsing the wavefunction of diversity' across the eigenvectors on the interior.

On top of this, political differences are cultural differences codified into laws and decrees (impositions, in Promise Theory speak). If we imagine that there is a kind of average state of political opinion, calibrated into a single region, then we could call the space of all such regions the 'politiverse'. On the other hand, I already dispelled the notion that that calibrated statistical opinion is a straightforward stable measurement in the voting example of chapter 6, so as the politiverse tries to reduce complex information into a few simple 'eigenstates' it will flap around like a quantum system, because it has a similar mismatch of information between input and output.

The agents of the politiverse—countries or regions, like the United States of America and European Union—have varying degrees of equilibration, or mixing, in terms of norms, opinions, laws, culture, and so on. The effects of scale are therefore engaged in a deep conflict of interest: dynamically there is momentum in numbers, but consensus is unstable and difficult to maintain. As we pretend to distill unique conclusions from the superposed possibilities, there is inevitable discord in the perceptions of individual locations on the interiors of the boundaries. Countries vote to elect a trusted leader, and sometimes on they vote on key decisions, but democracy is not scalable or pervasive, for the rational reasons already mentioned. The mismatch between alternatives and voters leads to disjoint storylines that flip between disconnected outcomes like quantum states: indeed, both are the effects of an information mismatch just at different scales.

We misrepresent the scaling of democracy: it is not a property of an agent (a country) but of a process. Even the self-proclaimed 'free and fair-minded' nations don't apply the same processes to all scales of activity. Most of us go to work every day to be told what to do by a boss, which is far from a democratic process. In Western Europe, I've heard many complaints from companies and organizations, in countries where a culture of trying to be democratic at all levels paralyzes all change: the endless consultations with workers and unions introduce an effectively high process mass by insisting on democratic involvement. When you need to ask everyone's opinion and get a consensus or even just be heard, the overheads come at the expense of agility—and you may miss your opportunities. It seems we only like democracy when it seems worth the investment in cost.

Governments, in the modern world, are conceived of as centralized services whose role is to plan and afford infrastructure for the common good. Part of that task involves the redistribution of wealth. If all the resources end up in the hands of a few, society stops working. Today, it's easy to see how the rapid acceleration of pace caused by information technology has left governments flailing. Society works on a faster timescale than government can cope with, and issues (like money and banking) that were previously delegated to private institutions (like banks) could now be re-absorbed for a more just and efficient approach to promise-keeping. The optimum approach always depends on an arms race of scales: who can act faster and shoulder more burden? If any nation could re-engineer government today, in the interests of its people, I hope they would do it quite differently. Technology affords new possibilities for managing the 'detailed balance' of fluctuating information.

At the time of writing, there is much political ado about globalization and one-to-one trade deals between independent countries, or sometimes blocs, seeking to maintain trade barriers and protections against cheap imports. Economists have commented that the best thing for the world at large would be for everyone to follow a common set of rules (say the World Trade Organization (WTO) rules). The weight of bilateral relations needed for these adds considerable mass to trading, not in tariffs but in diplomacy: maintaining many separate relationships instead of having a single common standard (e.g. World Trade Organization) is quite expensive.

The politiverse of countries and territories, even organizations, is a conglomeration of thinking processes, initially separated over 'many worlds', but increasingly coming together by globalization. It's tempered at the boundaries by interactions and trade relations. But where actually is the edge of a country? Once upon a time, it used to be the geographical border, or even a 'bloc' boundary like the Iron Curtain. Today, how can we tell? The world is porous, criss-crossed by numerous tunnels of information transfer. Individual humans may act as representatives of the bounded region, wearing their allegiance like a charge. A country's language may be its most important barrier. As they move from geographical region to geographical region, countries become porous and distended through the movements of their globalized diaspora. Interior communications channels involve tunnelling from region to region by aircraft and through telecommunications.

The cost of agreement in a political process, famously analyzed by Robert Axelrod in his studies of game theoretic cooperation[314]: a form of political mass, on the scale of human and business participants. From spacetime considerations, it is easiest to achieve consensus through a single point of calibration, or authority. However, we also know—from cloud computing—that this is not a scalable answer if you want to maintain speed and agility. World domination is no doubt the most efficient system of reaching agreement, but it takes a long time, and doesn't address other symbolic concerns, like fairness. The alternative is to reach a kind of consensus, by mutual interaction. This is much more expensive. Consider the World Trade Organization (WTO) and the United Nations (UN) where over a hundred countries trying to reach agreement. This has to take a long time. Large companies operate almost like as nation states, with their own systems and flexible economies—but semantic earmarking can also sabotage that capability. So labelling (such as in double entry accounting) puts up barriers to such redefinitions. This is why companies get embroiled in accounting scandals. Taxation rules may erect barriers that get in the way of an essential redistribution of wealth, so simplifying taxes can benefit free flow of trade. On the other hand, losing the ability to make labelled distinctions makes it harder to trace processes—and this leads to white washing or money laundry. This also shows how interacting with a complex agent can exert a powerful constraint. It's not hard to understand why parties would want to wriggle out of regulations and rules—even when they are designed to address stability and wellbeing on a larger scale. Information breaks homogeneity.

The economics of money is superficially a diffusion process, but also shares something with Quantum Mechanics, because of the need to match give and take at every link in the economic network. Money has to keep moving to prevent stagnation of an economy, because although money is approximately conserved, the underlying driver of transactions is a dissipative process (food is not conserved, goods are not conserved). If all the money became concentrated at certain locations, like a static quantum wavefunction, the economic world would stop turning. Diffusion and trade are based on necessary inequalities and the flows that tend to even them out. Promise Theory reveals what makes the economy somewhat quantum-like, by revealing the hidden barriers of give (+) and take (-). An economic wavefunction is more like the field of promises satisfying all the complex boundary conditions—the actual transitions of money are not described directly in that picture. It's a coarse grained, large scale approximation to the information. There are more turtles underneath to be discovered by observation.

Artificial cognitive systems may allow us to win back some insight and agility in these processes, by extending simple cognitive systems to perceive trends and detect patterns on a greater scale. But it's how we engineer the pathways and stories from inputs to outputs that will determine how smart that makes us. We know by now that finding all possible outcomes is not the hard problem. Understanding what and how to select and block certain outcomes may be the more important technology to make human spaces smarter at scale.

Agility is a key concern in business today. Companies talk about 'the lean manifesto' and 'agile' methods, while other management systems focus on procedural correctness or reliability of outcome[315]. From a spacetime perspective, the mass concept is clear to see in the network of promises that weigh organizations down. Established companies are quite slow to adapt. Learning from experience and training their networks of regimented processes takes a long time. Responding to changes, as a cognitive system, always lags significantly behind that learning—partly because most organizations underestimate the need for learning. The exception is startups, which is not regimented and have few ties. The obligations they service are low in overhead cost. Their agility may make them undisciplined and haphazard. For innovation, that's a perfect environment—one that's impossible to achieve for an established well-oiled machine. For execution of a long term plan, with the momentum to carry it to completion, it's a disadvantage. Thus, large established companies tend to absorb startups in order to attempt to innovate.

A sizeable democratic vote may provide the political mass to ensure the momentum of a political process's ideas, but surely its intransigence too. If an opinion carries sufficient weight, its impact makes it hard to ignore, but also hard to change. One should also not forget the cost of building such as critical mass, forging every connection, one at a time. Consensus need not be as rigid as disciplines like Computer Science seeks to ensure, but it needs to yield an approximately consistent agent that acts as a single entity. This is the challenge of a society. Modules of differing opinion form in order to protect special interests against the ravages of entropy from outside, but they cannot isolate completely, else they lose their source of information and their relevance. Finding the balance between separation, mixing, and conquering of opinions is the crux of information mechanics, for ecosystems and for society. If all the modules coalesced into a single collective superagent, that agent would only avoid the self-annihilation of equilibration by developing its own cognition on a larger scale, to re-differentiate its interior. If we apply that to the Gaia hypothesis: what does a planet talk to, or learn from to avoid that fate? There might be a need for nationalism after all, at least on some levels. What politicians often fail to realize, on the other hand, is that there are many semantic labels by which to define modular regions: nation states may not be the most important. We shall hopefully continue to live multi-dimensional lives, where the boundaries between concepts and allegiances don't completely separate us all.

Even size matters for global cooperation. A superagent's rate of time is its rate of interior activity. The challenge for a society with a small population, for instance, is that nothing much happens, and there is less demand for diversity, because there is a smaller audience. There is no point oversampling events that never happen, so the ticks of institutional time are slower in a small community than in a large one, and change is slow, compared to the wider world. This mismatch of impedances can be frustrating for processes that cross the boundary. If you are a businessman looking for a market, a small town is a bad bet. This is one reason why countries advance by forming alliances and larger trade zones. Globalization helps to stabilize a common standard of time for industry, and a regular pulse to drive progress. When large and busy countries (like the United States or China) need to work with smaller and more sleepy countries (like, say, Norway or Cuba), there is an impedance mismatch in the clock rates of processes. The faster country has its own clock that goes on ticking impatiently, while the slower society ignores most of that change.

As the speed of information transfer increases, so society speeds up. People who live in the big financial cities feel it the most. They are under constant pressure to work over-time and deliver results that are limited by the timescales of serialized human processes. This has already reached crisis point in cities like Tokyo, Hong Kong, London, and San Francisco. It seems that human rights, as we understand them, are likely to be one of the first casualties of civil society, as the pace of the economy increases. The mass exerted by consensus voting is too high an overhead by conventional means, so there will be a temptation to short-circuit the wires of consensual society to eliminate resistance. Autocracy is the norm in many corners of the world, but it needn't be inevitable if we understand how to scale decision making. To do that, we need to understand the mechanics of economic processes too. The scale of globalized markets will be a forcing function impelling us to operate on an unnatural timescale, unless we can shift to a much faster lighter form of decision making and also decouple individual responsibilities from global processes. We need to separate timescales to couple humans only weakly to larger process flows. A goal must be to use one technology to counter the mismatch produced by another—choosing technologies whose timescales are matched.

It might seem strange to express these matters as spacetime concepts, but the grand manifesto is to understand the universal mechanisms of processes, no matter what the agents. Perhaps the most surprising spacetime connection of all is found in the channels of information that pass between us as individuals in a society, and the most important of those channels is something all of us both crave and fear: money.

Is time money, or is money spacetime?

Money might just as well be the fifth force of nature. It's practically a cliché that money makes the world go around, but there is a real sense in which monetary interactions are analogous to a kind of physics for society[316]. Money is not really the main force that drives human intent—at least not for most of us—but it promises a couple of roles that make it a virtual proxy: that of a measuring stick for interactions, and that of a network messenger, binding together an economic spacetime in a role somewhat analogous to messenger particles in physics.

As a technology, the nature of money has changed slightly over recorded history: from cash to 'I owe you' promissory notes, to credit, and so on[317]. Money acquired more functions as it developed. One of the impediments to looking at money objectively is that discussions of money frequently get muddled together with discussions about perceived value, wealth, and other relative concepts that are derivative of the primary networking role it has[318]. Strangely, economists have a history of ignoring the role of money, and treating their physics of society as an abstraction of bulk quantities, where distribution in space and time are absent and balance is an intrinsic property of their models[319]. A growing band of newer economists, especially those following John Maynard Keynes (1883-1946) and Hyman Minsky (1919-1996)[320] has begun to criticize these glaring omissions. However, one role of money has been constant: it serves to bind people and organizations (scaled people) together into a network of economic cooperation that shifts and changes, like a cognitive system. Money is, in a short-term sense, transactional and one-to-one, but—through its memory traces—it defines on-going relationships, like a lasting business-adjacency between agents. Money is a proxy for trust, and trust is built on the expectation of promises kept between agents.

Smart means the capacity to adapt relationships and build contextual models by networking the concepts learned from interacting with an exterior set of processes. In a spacetime sense, the role of money is to exchange information— a bit like having the locations of space exchange messages, as financial gravitons or photons. They walk the repeated path of long term adjacency relationships between agents, wiring spacetime locally together and allowing other processes to ride virtually on top of them through their long term patterns. Money has layers of virtualization too. In the businessverse, being next to a neighbour means having a business relationship with them. In business, anyone who neglects the long term social network aspect of money is unlikely to do well, as it is the relationship that is the source of economic value, not the money itself[321].

It's sometimes hard to step back and think about money as a technology, let alone as a spacetime process. We need to look at the structure of the spacetime it connects together. The model we have of economics is that goods and services are exchanged between human parties, with money as an intermediary. A 'good' (in the sense of goods and services) is a spacetime agent that can be owned—and it is, in fact, the legal ownership of goods rather than the goods themselves that changes hands in an economic transaction[322]. The physical movement of the goods happens in a different set of processes, through different 'dimensions' of interaction. Money and ownership exist purely as information, documented in bookkeeping ledgers. Transportation of the physical good agents happens in physical spacetime. Exchange of goods is handled as motion of the second kind. Goods take up space, so they are spatial agents too. Goods may have mass in the physical universe, so they are not effective messengers for a relationship. Money, ownership deeds, and payment receipts are more or less massless, and don't take up space in the same dimensions as the goods (they are the voices of the choir), unless there is bureaucracy or bank processes involved in transference—so money's figurative role is to be the massless messenger particle of the economic binding force.

Services are processes that might involve goods. Those interactions are like scattering processes in physics: goods and money go in, goods and money come out and something happens in between. Services may change the state of agents involved in the interaction. It's superficially like particle physics, and that explains why economists were drawn to the similarities to physics in the past. But there are key differences between economics and physics. The goods and services can be considered almost entirely a side-effect of monetary exchanges. Money becomes a proxy for almost everything, economically speaking. Double entry bookkeeping registers money transactions as semantic *types* of payment, or accounts, lending privacy and modularity to interactions, and thus adding a much broader layer of semantics to interaction types than one has in physics.

As money's fast multiplying faces change in character, we need a way to understand it independently of these superficial appearances[323]. The tool I've used to compare spacetime concepts throughout this book is Promise theory, and we can apply it to money too. Promise theory asks us to consider: who and

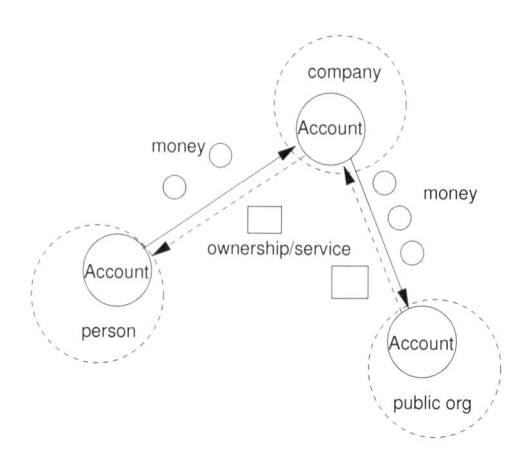

Fig. 9.8. Money is one channel of an economic network of business relationships between economic agents such as bank accounts. Money is promised by economic agents, from people to nation states. Money's promises have numerous representations from cash to bank transfers. Money is offered and accepted independently of a promise of service of ownership being offered and accepted in return. The give and take of economic relationships may involve many transactions of different types.

what are the key agents in a network of interactions, and what promises do they make to one another? From a spacetime perspective: what are the locations and what are the messages between them?

The locations in the econoverse are economic agents, which may also have interior structure (see figure 9.8). The messenger bosons are the money channels. Today, bank accounts are the actual agents that transfer money, while tax registered organizations are the agents that transfer ownership, but an account is only a virtual representation of people and businesses whose intentions are behind the economic transactions. Clustered associations between the different channels form superagents which are the virtual agents of the economy. Physics is relatively simple by comparison to the scale of social science. Remarkably, we can't even capture all the kinds of interaction ongoing in a social-financial network properly to describe a model of what goes on. The kind of simplified idealizations we can work with in physics are harder to come by, and more subtle in their interpretations.

Agents at a microscopic level may be individual persons, households or firms, interacting amongst themselves. The active endpoints of a payment channel are actually bank accounts—as money rarely leaves the confines of banks in our increasingly cashless world. However, like energy, money comes in several forms. There is cash: in each nation state, coins and notes, though the amount of cash is small compared to the total amount of money in circulation, and it's being replaced by electronic forms of money. Then there are cheques/checks which are variably denominated cash. Finally, one has direct ledger entries to transfer funds at the bank. The foregoing methods apply to a single currency at a single bank. in several banks are involved, then they have a 'clearing' process, in which each holds an account at each others' banks. At certain times, physical cash or gold reserves may have to be transported physically from bank to bank for legal reasons.

How different currencies are exchanged, from interior to interior, is even more interesting. Agents at the macro level may include nation states, governments, and central banks. The promises macroscopic agents make at the macro level cannot directly or immediately influence microscopic scales (just as weather does not directly change people's behaviours to any large extent under normal conditions), except as an effective (non-linear) boundary condition; but, with a powerful information technology, multi-scale agents could self-govern by detailed balance, countering every debt with a payment (like Maxwell's Daemon).

THE FIFTH FORCE: MONEY'S PROXY CARRIERS

Agents, working together in clusters, may use intermediaries like delivery agents and banks to delegate the keeping of promises—forming covalent interactions[324]. Even if they don't, a single payment for a cup of coffee fans out into a myriad tentacles to supporting relationships: your coffee is 4% wages, 10% shop rent, 15% price of beans, 3% delivery, etc. The agents involved in monetary interaction promises are often only information references in transactions made between bank accounts. It's now rare for them to be involved in transacting physical money. The endpoints of payments may now just be conditionals or dependencies parameterizing the payments. The channels for money and goods are in different universes altogether! Apart from cash, money never comes into contact with the physical presence of actual people or institutions. Banks or other currency hubs mediate in making payments, because money is now almost entirely pure information.

Money flows through intermediaries, which are the transactors of currency. These agents can be called *accounts*. In the capitalist economy accounts usually belong to banks. Banks form hubs. Recently decentralized currencies, based on 'blockchain' technology, have been experimented with. In this case, the blockchain acts as a kind of bank, often with a limited array of functions compared to private and national banks—its non-local network of peer members forms a huge democratic mass that keeps it from deviating ad hoc, but which also makes it painfully slow and expensive to run.

As a form of communication, we like to think of money as being conserved—like energy, it's helpful to assume that it doesn't just appear and disappear, because this documents the continuity of process. Obviously this is a convenient fiction. We can easily destroy money, by burning it or burying our treasure. At its best, it passes from agent to agent, location to location, and it needs to be observed at the right times to drive processes between those agents. Like forces, money's effect is only felt when it is exchanged between agents. Money that is trapped or burned at one agent location might as well not exist. This is one reason why accumulation of wealth is such a problem for the economy: the accumulation of wealth takes money out of circulation, where its usefulness is limited.

Economic activity is just another spacetime process. Its accounting degrees of freedom (money) are separate from its physical realizations, just as energy is a separate accounting measure that unifies the counting of radiation, and matter, etc. We identify proxy agents (see figure 9.9), like coins and notes, that represent money help to account for it by being more or less robustly conserved—authorized notes and coins are deliberately difficult to forge, and there is an incentive to not destroy them.

The concept of a mass for money turns out to be quite important. In economics it is related to what is referred to, inversely, as 'liquidity'. A high mass means a low liquidity, and vice versa. In fact, the transactional nature of money makes it more like Quantum Mechanics, where mass is not really important to transactions; instead one has tunnelling out of captive states. For longer term recurrent payment relationships, a continuum approximation makes sense though. The free exchange of money is facilitated by its having a low 'mass' in the processes of the econoverse. The velocity of money matters to processes in the same way it matters in physical mechanics. Arrival times may matter—and strict determinism is far too rigid to offer a stable platform for society to rely on.

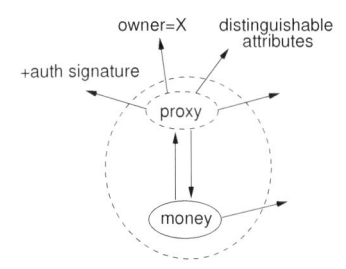

Fig. 9.9. A money proxy cluster consisting of the money and the vessel, from reference [5]. The number of promises involved in proxying for money is non-trivial, even for the simplest coin. There is money represented by the coin, and the acceptance and authorization of the representation, as well as the coordination of the measure with the proxy's public appearance.

Thus, the acceptance of debt, borrowing, and delayed payment all turn out to be necessary inventions—both in economics and physics.

Money needs to be able to be a lingua franca for any type of goods or service, virtually, i.e. be interconvertible to and from it—else we would need different kinds of money to buy different kinds of things. At the same time, it has to maintain relationships in a cooperative network of parts that forms an ecosystem or society. Superficially, it looks like a one-to-one exchange, but behind the scenes payments form a wide support network for society. So money can't be too connected to any particular kind of good. Eventually, computers will likely renormalize our present currencies away, and we won't even notice what kind of money can buy different things. Types or semantics reduce the liquidity of money, so they need to be hidden in side-channels. The more semantics, or types we attach to things, the less likely we are to be able to use it. The best money, like featureless 'energy' has virtually no semantics and can be used as a measure of all things.

MAKING MONEY

If money didn't exist, we would surely have to invent it! In fact, forms of money have existed almost as long as humans: it seems utterly inseparable from the workings of fair society[325]. Indeed, we take money so much for granted that its true roles are practically invisible to us. In spite of that, I venture to say that few people, outside the finance industry, probably have a clear idea about what money is, how it works, or where it comes from. The phrase 'to make money' is so common that we have come to accept a narrative about money that is entirely false. Unless you are counterfeiter or a bank (an authorized 'printer' of money), you don't make money—you only try to attract money from someone who already has it—we are all beggars, with different cups[326]. And I'm not just talking about fiat paper currency authorized by a central bank. It doesn't really help to substitute gold or silver for money either. Few of us would be willing to accept gold in exchange for goods or services today. Money is an authorized, labelled standard of interchange.

The idea that labour makes money goes back to the Marxist narrative of 'economic value' being the result of labour—a natural view for someone who lived during the industrial second wave[327]. However, there's a difference between making money, making something of value to someone else, and working hard—making this connection, at best, oblique and, at worst, misinformed. The act of working does not lead to there being more money or even more wealth. Even if you mine gold and diamonds, or manufacture computers for a living, you create something but you don't create the money to buy anything–nor do you 'create wealth'. All you do is transform a thing for sale. If new markets get interested in this thing, then there may be a semantic basis on which to sell it, and that is what creates the opportunity for wealth. To create wealth for someone, you then have to sell it, by attracting money from someone who already has it. However, with a lot of new things to buy in the world, those original hundred notes the bank printed are not going to be enough, so we would need to expand the money supply. Each time there are more things in the world to buy, someone has to pour extra money into the economy in order to allow someone else to pay for those activities. That's what banks do—by lending.

In principle, anyone *could* arbitrarily create a form of money of their own, just as anyone can create their own language—but what use would it be with a network of one? The trick would be to get your money accepted, or your language spoken, in a substantial network that binds an economy as a coherent massive body. If everyone did create money, its role as a measurement system would be compromised. Measurement is based on the conservation of units, i.e. of having the same number of ticks on your ruler or clock each time you use it—on the persistence of units over long times. Banks have a government license to create money. Money is created by borrowing from the bank, creating equal amounts of positive liquid funds and negative debt. Ordinary people are not authorized to create legal tender or currency money, but rather we create outcomes and resources that can be sold, and for which we might need to expand the money supply. Outcomes can be traded without money. Many refer to this as the creation of wealth, and even "making money" but this is an inaccurate colloquialism.

It cannot be free to borrow money to acquire something either, just as particles of fermionic matter cannot be massless, else the world would suddenly be flooded with every imaginable outcome. Barriers to spending—like potential wells in physics—are an important regulator of meaning. Without them, their entropy would increase to a maximum level very quickly.

Money promises both a measurement system for activity, and a network of non-local influence—a social force. Unlike the physical world, there is no 'natural' measure of socio-economic activity (except, of course, the energy of physical processes, but socio-economics is a turtle that rides well above the physical universe). Some economists believe that energy should be used to account for processes in the economy too, as this is a key cost both for now and future impact on the environment. Unlike energy, the presumption that money is conserved is curiously inconsistent, as we'll see below.

No one has any idea what motivates the physical universe—but the promise of increased total wealth motivates the econoverse. It provides a kind of forcing field for exchanges, polarizing a space for motion to take place. Without that motivation, societies stagnate and have no reason to cooperate. The social cohesion and scaled sense of common purpose is largely related to profit, i.e. the promise of a better life. Money is a proxy for thinking about present and future goals. It forms a cognitive system that drives global markets. I wouldn't call it intelligent, but it seems to have good days and tantrums.

UNIVERSAL EXCHANGES AND THE RACE AGAINST THE ECONOMY

Physical interactions are spacetime transactions that can be counted in energy units. Economic interactions are exchanges that can be counted in money. Money can be carried as a number of proxy types (cash, ledger, credit, debt, etc), just as there are *types* of energy (kinetic, potential, matter, radiation, etc). In both cases, the interchange values are not supposed to be created and destroyed *ad hoc*—but they may be moved around and interchanged. Energy plays the role of money in that sense (see figure 9.10).

Both money and energy play a network role. We have no idea if the structure of physical spacetime has something like routers and switches of the Internet—i.e. any privileged hubs at a sub-quantum level, but we do know that there are such hubs for money, in the banking system. Banks are critical hubs in monetary communication, because they are in a real sense the source of money itself. They are not only interaction hosts, they can be amplifiers. Banks can create liquid money and debt, side by side, so that the sum is zero in principle, but then they *sell* that as a service. That makes an interesting, if not paradoxical, accounting problem. The outstanding detail, which makes economics apparently unlike physics, is *interest*—the cost of borrowing money. Interest on savings and loans is a curious invention—one that is both socially despicable—as it drives inequality—and one that has been absolutely necessary to make the economy work—because it drives motivation.

In physics, when energy is appropriated as a loan, our theories say that it's paid back precisely. There is no interest on borrowing, as far as we know. In capitalist macroeconomics, however, banks are private businesses that take fees for different services, including loaning money. That leads to some curious paradoxes. How can anyone pay for banking services if they have to borrow money to pay for them? They could, of course, get a job and go to work. But how does work pay them? It has to borrow from a bank. Okay, work could sell some goods or services, but how could any customers buy them? They would have to go to the bank. It's turtles all the way down. Clearly that's something of a paradox. It can be sustained only by continual growth—which is one reason why economists are so obsessed with growth.

All money begins with borrowing from the bank. But how can that be? The accounting could never balance if banks added interest charges to what was paid back and the calibrated measures of money remained the same over time. In physics, Noether's theorem tells us that the invariance of the detailed balance of energy conservation over short times implies energy conservation on the same timescale. There is no such law for economics, because the relativity of prices and monetary amounts is being renormalized irregularly around the economic network, in different countries, regions, and banks, all the time. Money is not conserved. It's a small wonder that the global financial system is stable at all. In fact, we have ample evidence that it isn't t^{328}.

It means that there is always a deficit of money in the system, unless someone keeps borrowing. The incentives to borrow are primitive but effective on a slow timescale: interest rates set by banks are the only knob economists can tune to lure people to borrow or return money to the banks. The economy is a kind of pyramid game in which someone is constantly borrowing money to settle debt. The economy operates on trust in advanced boundary conditions: a gamble that, if we borrow today and have to pay for it later, it will all be worth it in the end. If the economy ever stopped growing or inflating this would all come crashing down, and the system would fail[329]. .

Luckily the exchange of money is only a proxy game for the proper exchange of resources, represented as property, goods, and services. If money couldn't be repaid, it would not be the end of the world. The sovereign authority controlling money could choose to reset the system, write off debts, and start again, just as long as all the buyers and sellers survived.

The economy is both a spacetime process and a set of cognitive processes. As a cognitive system, the role of monetary transactions is to communicate the impulses from an exterior world of goods, and, services, and human concerns, to an interior memory system that documents business relationships through prices, discounts, contracts, and other instruments. Prices on goods and services are the memory cells that learn from these signals and imprint the memories of events. Stocks and shares further document ownership relationships that explain intentions, in a stigmergic trail of storyline—held together by the lingua franca of money[330]. The key features that turned out to be missing from our descriptions of both physics and economics, in the 20th century, were *scale* and *indeterminism*, i.e. the role of large number approximations and loss of distinct information about a system (entropy)[331]. What holds a stable picture together is the ability to buffer the storylines of process by playing with spacetime scales.

Inflation is a renormalization concept in economic counting. If agreed market prices increase, then fewer goods can be bought for the same calibrated amount of money. If supply is scarce, then prices may be increased to limit the demand. Prices act as a potential barrier, in physics parlance. The relation between price and purchasing power may eventually erode debts too[332], so borrowing against the future can be renormalized by increasing economic growth. Conversely, growth in demand may 'cause' relative supply bottlenecks—a lensing effect in the econoverse's logistic networks. Putting up prices acts as a rate limiting feature, effectively increasing the mass of processes, by rescaling the burden to each transaction. One role of inflation is to continually renormalize the value of money, so that debts become smaller relative to new borrowing. This has saved people and countries from ruin throughout history. Economics is a bizarre Red Queen race to nowhere—a distorting lens for global allowance. What's truly important is to manage the outcomes it inflicts in the physical world—which the monetary part of the economy ought merely to be a shadow of.

As network hubs, banks also have the role of central calibrator, or lingua franca for standardization that defines the measure of money. It's true that there is not only one form of money—there are many currencies and money comes in various forms, as cash, transfers, electronic money, and so on. Other dimensions of money are also 'typed': currencies are one type, but then we have loyalty cards, flight miles, petrol stamps, coffee cards, and so on. Loyalty points are money that can only be exchanged with certain agents. But these are not important differences. What makes money ultimately somewhat unstable is that giving or taking money can cost money. Money is non-linear, superficially like gravity—and partly for the same reason. Establishing a connection may not be free. If a message can be piggybacked on top of an existing connection, then there is no cost of setup, but if changes have to be made to connections, then this costs. An exchange hub for the calibration of units and measures (figure

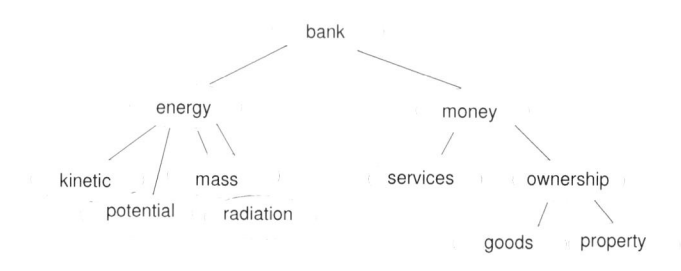

```
                          bank
                 _____/    _____
              energy                money
             / / \ \              /      \
       kinetic / | \ mass     services   ownership
          potential  radiation           /    \
                                      goods   property
```

Fig. 9.10. The transmutation hierarchy of measurement: energy and currency have the role of lingua franca for measuring and accounting for phenomena, using a universal interchange model. We take the currency to be approximately conserved, by design, to maintain the integrity of the accounting—but occasionally the scales may all be renormalized for convenience.

9.10) is not the same as an exchange hub for routing of communication, and yet banks currently take on both roles in the econoverse.

Banks used to be a major centre for communication and commerce. Today, alternative channels for communication have made banks less relevant as communication hubs, and less powerful as lenders. Money can now be 'crowdsourced' from willing donors, thanks to the Internet, by casting a net to a wide region of spacetime, not just from banks. But crowds still can't make their own money yet, to increase the money supply (figure 9.11).

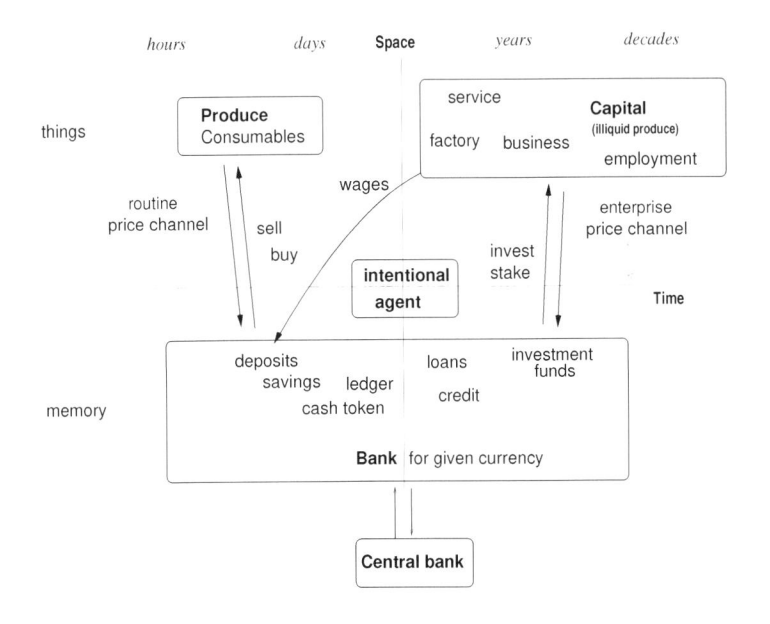

Fig. 9.11. The agents of money. The economy does not pass directly through humans, except for cash transactions and direct swaps. Humans are only interior authorizers of transactions that happen between different accounting ledgers at the locations in the economy.

In many ways, the accounting aspect of money has become secondary to the networking aspect, especially given access to financial services. This is similar to the way the rigid accounting of energy in classical physics is of secondary importance to the transactional properties that allow spacetime locations to be connected in the quantum world. Money plays a role roughly analogous 'messenger bosons', in interactions between goods and services. In order to facilitate the universal interaction of processes, it's necessary for the messenger between points in a space to be universally accepted—what physicists would call a global symmetry. Similarly, in economics, money has to be a medium that's universally accepted.

Cognitive aspects of money

We've established that money has some mechanical properties. It can move around, and it has different effective masses in different forms, just like energy. But can we say if the economy smart? Like all the spacetimes discussed so far, the econoverse has a centric observer-agent arrangement. Agents define a boundary between interior and exterior on a variety of scales, so does the econoverse also have cognitive functions, on some timescale? It has no obvious translational invariances, except over trivial promises about common practices, so we don't expect to learn anything from a kind of classical mechanics. The econoverse has memory, in companies, stocks, shares, products, and more. It has concepts like brands, types, commodities, derivatives, and so on. It acts on itself with investments. Like Gaia, it is unlikely to make sense as a single system, because the world economy can only be introspective. What anchor points would it learn from? Rather, it makes more likely sense as a community of separate systems—which may or may not be the economies of different nation states, different currencies, or different commodities, in dialogue. I am not an expert in economics, but I suspect that divisions lie somewhere in these concepts, and possibly fluctuate.

Economies certainly have some simplistic programmed responses. We might turn up our noses when news readers tell us that the markets are unsettled or nervous—a silly anthropomorphism. But as Turing taught us: if it looks, walks, and quacks like a duck, then you might as well call it a duck, because you've got nothing else to go on. So the econoverse might be something like a small cluster of symbiotic cellular organisms.

Money relies on being accepted without obstacle, by agents who trust its promises. That makes it effectively an interior form of communication. There is a locality to money. It works within a certain boundary and not outside. True, we exchange currencies of different types, but money doesn't really cross borders because it has no semantic value there—rather, banks borrow from one another in different currencies on trust. This is paradoxical, and it's one reason why promise to give and take have to be established. Money only makes sense if there is an explicit promise to honour its implications between autonomous locations. This applies to every transfer of money. In physics we have the tacit assumption that energy is accepted by every location in physical spacetime processes with an impartial exchange rate. That makes physics much easier than economics, except that no one knows why can we attribute to such acceptance?

As a measurement system, interior monies would be better separated if there were separate currencies for interior and exterior transactions. In a sense this is what foreign exchanges enable, but there is a problem which has become quite clear with the advent of cryptocurrencies like Bitcoin. Currencies are constantly being renormalized by one another. Bitcoin has no fixed value relative to anything, because it is an international, exterior currency relative to nation states. That means it operates more like a commodity than as a promised, sovereign currency. It is a leaf in the wind of bidding races to buy it using other currencies. Exterior money becomes a commodity. Interior money is fiat. If we extend this to think about other exterior concepts, what does this say about other semantics between political boundaries?

Is money smart? It might be, in some sense, but a fickle and unstable soul it is. When we speak of the markets reacting, of being unsettled, this is a superagent collective reaction. We interpret it as emotional behaviour, at a local stock exchange. That picture of reality could be the distortion of a single observer, who projects observations onto its own small set of concepts—collapsing the exterior 'wavefunction' onto a set of interior concepts. The short term appears simplistic and emotional, but a long term storyline can be more rational. Money and credit allow us to play with time and space in adaptive ways. The econoverse is slippery and fluid, not like the more rigid predictability of the dataverse. Borrowing smooths over the continuity.

Spending in the economy is analogous to 'thinking' in regional economies in the sense that continued spending on a certain good or service establishes and deepens that signals that link products to purchasers in the memory structures of the economy. Customers numbers are the conceptual mass of the brands in economic memory[333].

SPENDING FAST AND SLOW

The wealth of semantic distinctions in socio-economics allow us to slice the economic cake along many different criteria to make many different kinds of map. Semantics dominate these fields of study, which has almost certainly harmed the development of the field. Economists look for distinctions and patterns based on semantics, rather than on dynamical scales and variables, which inevitably mixes it with political intent. That complication, though certainly unavoidable, makes economics feel unnatural from a physics point of view. All cultural and technical jargons feel arbitrary, and forces us to remember the concepts rather than intuit their meanings from simpler generalities like time, space, and movement, which are more universal and more familiar. As fields mature, semantic separations of concerns tend to be replaced by a unified picture, in which distinctions are described in relative terms (by measure in duration or size). Economists will therefore speak of spending versus investment, using semantic terminology to distinguish what a physicist might call fast and slow spending, or short term and long term money interactions. Economists speak of commodity goods and capital, where a physicist might only note the timescale over which these goods are held and are used up. Capital items (long term goods) tend to be more expensive purchases, and they tend to be usable for processes that can lead to rent collection or manufacturing, whereas consumer goods tend to be eaten or used quickly (and although they fuel all our activity, we choose not to see them as wealth creating).

So, the main distinction between *capital* from *commodity goods* is a timescale for the keeping of its promises, but an economist would be unlikely to pay attention to that—yet it has immediate process implications. Capital is the sum of all persistent promises over long times, while commodities tend to be transitory consumables. That makes capital an illiquid asset (gives it a high mass), whose resale price decays slowly, perhaps after appreciating in a specialist market. A commodity's life cycle is short and liquid, with therefore little opportunity for resale. These are different processes on different timescales. But do they ever mix? Surely the world is more complicated than this cartoon two-category model. Luxury goods lie somewhere in between these cases: if not immediately consumed, they might be resold (other than wholesale/retail distribution). Figure 9.12 attempts to show how the separation of concerns in economic semantics broadly follows implicit spacetime scales, where space refers to aggregation of instances, and time refers to accumulation of duration.

In computer science, the concept of financing is only just getting started with cloud computing. In the past, if you hit a limit, it was a hard limit. That made IT systems quite fragile. It was online retailer Amazon's engineers who

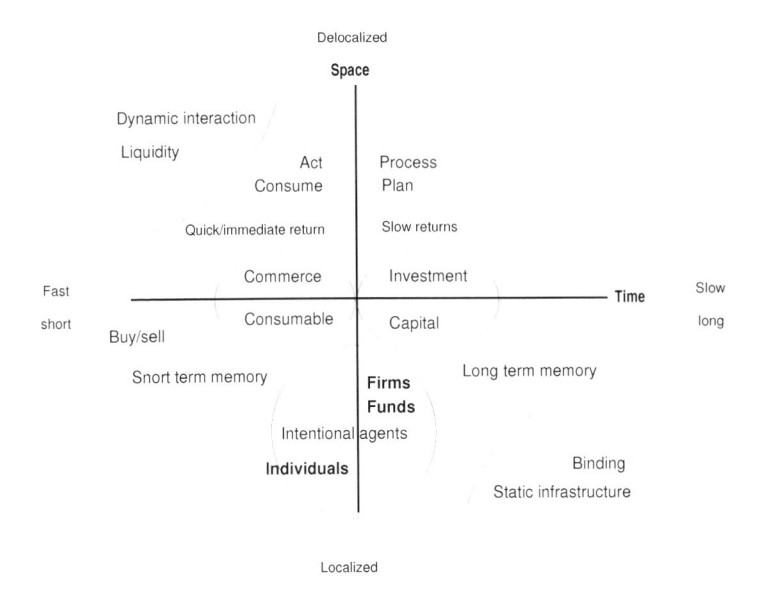

```
                        Delocalized
                          Space
                            |
 Dynamic interaction        |
   Liquidity                |
                     Act    |  Process
                     Consume|  Plan
                            |
           Quick/immediate return | Slow returns
                            |
  Fast       Commerce       |  Investment                 Slow
 ─────────────────────────── ───────────────── Time ─────
  short                     |                             long
            Consumable      |  Capital
         Buy/sell           |
                            |
    Snort term memory       |      Long term memory
                            Firms
                            Funds
               Intentional agents
                       Individuals
                            |            Binding
                            |     Static infrastructure
                            |
                         Localized
```

Fig. 9.12. The interaction scales for economic processes can be organized by spacetime scales. The traditional separation into capital investment and commodity consumption is not significantly different in semantics, but different in timescale.

invented the computing cloud as a pool of fluid resources where As long as cloud resources inflate continuously, like an economy, there will always be enough coverage on average, subject to some spare capacity that can be rented some of the time. A stable margin of error will always involve some sparseness of usage. This is almost a law of stability: each independent silo can be unaffected by others as long as there is sufficient separation between them—like interstellar space. Cloud resources can be scaled up on demand, like borrowing 'elastic resources', but they have to be paid for in by detailed balance of payments, on the billing schedule of the cloud provider. That gives the borrower time to earn back the cost through its own business. There is an extra indirect step involved, but that is the nature of networks. In matter-antimatter particle creation in physics, borrowed processes can be reified if they are paid for. This is how CERN creates particles from huge amounts of energy.

The loaning of money enables agents to overcome obstacles that would be 'classically' impossible to overcome. A credit card buffers monetary flow, allowing payments to flow smoothly to the payee, without delay, even no money has yet arrived in the payer's account. The schedule for repayment is an arbitrary limit, usually thirty days or 'end of the month'. So time granularity is related to the uncertainty of energy and money, or causal network communication[334].

As a tool, conservation—the strict accounting of amounts—is a useful abacus for tallying, and it can play a wider role on the scales of environment and society, as well as in virtual realms. But accounting is simplistic, because counting is a process that is not instantaneous. How shall we define conservation? On what timescale? By inference, how long can we accept debt? The repayment of debt by relative processes is a direct function of their clocks and the relativity between them. Suddenly observer relativity plays a role in matters we have taken to be straightforward and Newtonian.

Interacting agents are cognitive agents, with interior and exterior time. In principle, no agent would be able to predict *a priori* how many ticks of its own clock it would take for an exterior party to reply by making a payment, whether it had the money or not. So when we speak of laws of conservation of energy or money, we've been spoiled by the stabilizing effects of scale when formulating laws of conservation. As agents get smaller, it becomes impossible to measure or predict the spacetime volume over which a region of spacetime will balance its books, in terms of a set of commodities we are able to account for. This is not usually a problem for physicists or accountant, both of whom are capable and ready to invent new items to explain away the losses until they resolve. Is that a trick, or is it just a natural part of scaling? Whatever fundamental phenomena energy might be the measure of, in terms of physical reality, as far as science is concerned, it is little more than a universal bookkeeping value to balance the spacetime accounts across neighbouring local regions. We say that energy takes on many forms, but we really mean that each phenomenon can be accounted for by some amount of something else, in the currency of energy. At the level of human concerns, we make the same separation between *types* of phenomena and a simple proxy for accounting that's stripped of semantic distinctions.

How long should a repayment be allowed to go unpaid before conservation is violated? This depends on the accounting interval—a discrete time interval over which aggregate accounting is settled. The observer always has the privilege to determine that interval—so it's the observer's clock that matters. Naturally, the counting is a form of coarse graining, or renormalization. The semantic types or posts introduced by double entry bookkeeping all have to be settled independently. Sometimes this involves introducing new interaction types to make the books balance. Third parties, like tax authorities, complicate the accounting, making the econoverse much more complex than the physical universe.

In quantum tunnelling, there can be a finite probability of finding a particle far outside the places allowed by naive energy accounting. In a sense, the energy is borrowed to escape the trap of the well as long as the particle can pay its smugglers on the other side, without interest. Think of a prisoner who knows the location of a treasure chest through his prison bars, just out of reach. If he could borrow enough to pay his way out of prison, he could escape and still be rich even after paying back the money. This is how loans work. In the econoverse, financing is an important tool to overcome barriers that enslave people in 'poverty traps'—the financial equivalent of a potential well or bound state in physics. The invention of financing, allowing customers to do the kind of borrowing of energy that virtual particles represent in quantum field theory: temporary, short-term loans. This is essential to the continuity of businesses, and it might be central to the continuity of material processes too.

If we were to believe the news media, we could practically take it for granted that simple-minded economic concerns motivate commerce and shape the world through raw Pavlovian incentive. Yet socio-economic well-being seems greatly complicated by both observer relativity and semantics and symbolism, on all levels. Symbols are not orthogonal to economics, they are a part of its conceptual memory, through brands, companies, people, habits, crashes, and so on. The chief architect of economic gravity was probably the city—not only for the economic advantages of being within a tight localized cooperative ecosystem, but because people are attracted by gods, leaders, and other symbols that stimulate emotions in a positive way. Unique or rare cultural symbols of status can also attract the attention and investments of swarming agents, leading to their own kind of 'gravity'. No one except the bank makes money, they try to attract money. That brings me to the final piece of the puzzle at the scale of human interest: hopes and fears.

Artifex principalis

Throughout history, humans have viewed activity and purpose through the lens of divine meaning. It's reflected in the things we've built: from stone circles, representing the shifting positions of heavenly bodies, to churches and towns built in the shape of Christian crosses, to the cities—whose rectilinear forms signified sacred geometries to mimic Ezekiel's linear city, or the centre of Jerusalem—to the concentric orbital patterns, whose onion layers signified a hierarchy of closeness to a God at the centre[335]. The shapes and angles of the pyramids of Egypt, as well as Mayan structures, and other civic edifices, have been attributed to astronomical patterns pertaining to various gods.

Space and time have held their own place in the pantheon of supernatural forces. Ouranos (Roman Uranus) is the Greek (night) sky god was father to Chronos, the god of time. Then, of course, there was Aether, the god of the sky after which the luminiferous aether is named. This scarcely scratches the surface. Many of the symbolic structures built by humans served as representations of space and time. The arrangement of parts within the human body was once believed to have a divine significance too. These ideas harken back to Plato's work *Timaeus*, in which he attributed geometry and form to divine intent. As his work was carried forward and translated by Roman philosophers Cicero and Calcidius, and seized upon by Ptolemy for his Geocentric view of the universe. The tradition seeped into culture after culture, and was brought into Christianity in the 12th and 13th centuries, through the translations of those earlier Greek and Latin texts. In Italy, the philosophers in the time of Thomas Aquinas (1225-1274) in Florence, Milan, Pisa, used geometry as a literal and metaphorical interpretation of scripture[336]. This signified the hold religious doctrine had on allowed thought during the dark ages, well into the enlightenment. Aquinas came to the belief that the pathways of reasoning were themselves an expression of the divine—lending early acknowledgment to the connection between stories and geometry.

As mathematical sophistication entered the consciousness of town planners, particularly during the enlightenment, architects began to shape towns using spacetime concepts too. Major towns invariably had a clock tower, sometimes attached to a church, as if to act as the pacemaker for the community's interior time. Timekeeping longer than a day influenced cultures through the calendars they used—counting the dominant processes ruling their environments. In equatorial regions, where the solar driven seasons don't change much, lunar calendars have been common. The moon affects the tides more visibly, and is a visible marker in the clear night sky, whereas the length of a day doesn't change much. In polar regions, on the other hand, the solar year certainly dominates, with large changes in the hours of daylight between summer and winter, with accompanying temperature variations. For the arrangement of space into semantic districts, a radial rather than a rectilinear form became common. Castles and forts act as both protective barriers between interior and exterior, as well as observational vantage points. At the centres of towns, market places often formed central exchange hubs for town communications—where the town crier would read the news—and, of course, the people would perform their economic transactions. At the same time, the structures surrounding a centre were often organized geometrically.

Nicolaus Copernicus (1473-1543), the polish renaissance astronomer, who was lucky enough to escape persecution for his heliocentric view of the cosmos, completely changed the perception of human space, from a purely observer-centric Ptolemaic arrangement to a subordinate cyclical view. Perhaps the shift he initiated was an uncomfortable one for us, because it stripped the world of a personality—a central character in the story of the world. Or perhaps, his more 'cognitive view', with humans looking out from a protected interior where we reside together as a single unified people, was preferable to a view that demoted humans before something greater than themselves. Galileo (1564-1642), who *was* indeed imprisoned for providing the evidence championing Copernicus's view, was the master architect of a rethink of space—not in terms of observer-centric shells, but in terms of motion in straight lines. Even Newton, who perhaps did the most to depose the observer-centric viewpoint, was fully motivated by a narrative of finding the beauty in God's work. The rise of geometry as a pursuit with its own divine meaning led philosophers away from an Ptolemaic or Einsteinian view of the world as looking out through the senses of one's own cognitive world. Compare, for instance, the criss-cross coordinate system of New York to the orbital rays of Paris.

Culture is a strong force in human consciousness. I imagine that many physicists reading the sentences above let out a scream at the comparison between Ptolemy and Einstein, even though both underlined the importance of observer centricity, albeit for different reasons. In modern times, the cultural underpinnings of our ideas about the meanings of space and time have persisted in the ways buildings have taken shape. Housing and town planning were often built using rounded forms, until the Christian cross influenced architecture and rectilinear forms came to dominate through Christianity. Today, squares and rectangles have become a habit—think again of the grid-locked New York. Only recently, in our more secular era (and as building techniques have advanced to allow daring artistic possibilities) have architects begun to diverge from centuries of conceptual gridlock.

Cities grow in rings, like trees, from the clock of accumulated layers on their exteriors—unless they are constrained by area, like New York or Hong Kong, say, in which case they expand upwards. The tall buildings in these cities were built from necessity, unlike the tall buildings built for symbolic supremacy in many other cities. Just as the structure of the Earth, from its mantle and crust, to the layering its tiny gaseous atmosphere and larger magnetosphere, creates a safe space in which complex chemistry can form and remain stable, so the accretion of people into cities led to stratifications of functional significance. Power (the divine ruler) ruled at the centre, then scholars and aristocracy closer to centre (if not in possession of their own castle). These symbolic totems provided the 'gravity' to attract workers, who would come to subsist, as symbiotic working classes, in the essential trades and factories around the outskirts. This left the outer defensive walls and barracks at the strategic edge to consolidate the gravitational pull of the city through defense. Cities were fashioned on functional and cultural enclaves, separated by class, occupation, and race. This changed somewhat when ships and railways displaced rulers as the primary focus of power. It's fascinating how the quantitative economic processes win out over symbolism when there is competition for priorities. The hubs for trade and commerce, like market places, may have begun close to civic centres, but have eventually expanded and dislodged old town centres to new centres of gravity. Economic communication between cities dominated through the development of commerce, and as humans entered an industrial age, so the pragmatic dynamics of processes (involving money and technology) began to retake their dominance over semantics and doctrines of power.

The connection between meaning and spacetime structure seems almost a tautology, at least on some level, but that might just be coincidence. Using modern ideas, like Information Theory and Promise Theory, we can understand the mechanics of meaning more critically. In particular, we should pay attention to the binding of (+) and (-) promises, at any scale or level of complexity—what Darwin might have called them the components of a process of natural selection. This pattern persists into the artificial worlds of the dataverse and beyond. Through the discovery of these crucial bindings that propagate information by transaction, nearly everything that goes on in spacetime can be interpreted as an extension of semantic space and time, in some guise. Every structure is the result of a process that tells a story, pieced together from parts that play a particular role.

One outstanding city that marked almost a 'coming of the ages' in the history of smart spaces is Venice—a city not built for God, but for commerce. Venice is a waterborne city, perched atop a raft of wooden piles, driven into the soft muddy sand of numerous tiny islands off the coast of Italy, in the Adriatic sea. It's an astonishing feat of civil engineering that seemingly defied gravity for many centuries, instead of sinking into the quicksand beneath the waves. The city is built, brick and mortar on top of this platform, with a rich variety of architecture to rival any major city. Politically, Venice was a smart space, modularized by functionality as a trading port, governed by a symbolic head of state, the Doge, and had a political system built on council and distributed institutions, deeply afraid of giving too much to any single individual. Venice is divided into Boroughs, which originally segregated major functions and even cultural groups, with direct access by boat through the canal network that penetrates deep into its heart. Water is the simplest approach to mass transportation of heavy loads. It's accompanied by land-based adjacency channels, formed by a network of bridges that connect the different islands. Venice has both direct adjacencies and back-channel 'tunnels' by waterway.

Venice had the first banks, the first industrial production lines for its military arsenal, and was a major world power until its conquest by Napoleon Bonaparte. Its transportation by canal, and the semantically rich use of its highly limited space are remarkable innovations, absolutely functional and rich with hidden meanings[337].

THE BEGINNING AND THE END OF REASON

In the examples above, we could see that all space is not equal. Some is actively involved in playing a role in collaborative process, while other space simply provides the clean separation to keep symbolic phenomena in sharp distinction. We shape and manipulate spaces for abstraction, to accentuate or de-emphasize meaning (play with entropy), both inside and outside of our various agencies. These are the tools for describing the world.

The ability to construct a coherent narrative from a collection of parts is what we call reasoning. Memories give meaning to the persistent processes of the past, but the lessons of the story model for cognitive processes suggests that we need to know how to stop irrelevant stories from dominating those memories. The smart part of searching for answers in a database or memory archive lies in setting up the boundary conditions for the search process, i.e. in *posing the right question*: what pattern are we looking for, and at what scale? Scale plays a special role in reasoning. Time is not merely the accumulation of large scale entropy, as some authors have claimed. It is a process of ordering. It's true that epochs throughout the state of decay of the world have an absolute order, according to an average view of the macrocosm, but events can be ordered by many other means too. Time is a story—a history—as well as an inexplicable broken symmetry. Memory plays its role in this, because without it we can't distinguish states at all. Order matters to the outcome of evolving processes. If you wear out your shoes on one side because you favour one side, you will not fix the shoes by deliberately favouring the other. The symmetry can't be restored because the memory of the broken symmetry is already stored in the shoes. Similarly, we can't really turn back time; the laws of physics may be reversible by some trivial definition of process, but contextualized processes aren't.

There is a lesson to be learned here, from the analogue of entropy in the dataverse. In recent years, it has also been popular to try to answer questions simply by trawling 'big data' for unspecific artifacts of information, collected from a wide range of sensory sources, in hope of discovering interest concepts and relationships by introspection of clustered events[338]. Applying brute force searches to data can speed up those selections as intelligent assistants, but without the seed of a boundary condition to select certain stories, there can be no way to connect the dots into a coherent story—one that we can attribute symbolic meaning. Statistical tools have become our allies in the modern world: they identify relationships that we miss with our limited human faculties, but human expertise will still offer helpful input, because humans are the arbiters of significance, by virtue of cultural context and societal memory.

As we move into the 21st century, the use of information technology to enhance spaces, using special sensors, machine learning, and training algorithms, is clearly a popular idea—but what principle shall we use to guide these incursions? How shall we choose the knowledge that is relevant in each context, and make the best use of it? Indeed, what will the dataverse look like once every cubic centimetre of physical space is equipped with sensors that connect interior processes to the exterior physical world? The complexity of the resulting cognitive system will be vast.

Locality (including personal privacy) will start to mean more to us. We shall have to embed the laws of the dataverse into every region independently, not merely try to enforce them from giant datacentres in the clouds. This is why Promise Theory has started to play an increasingly prominent role in the understanding of systems—it is about how to understand those scalings from the bottom up (where everything is local and designed for the few or the individual) to the top down (where larger non-local concerns are represented for the many).

Will we recognize it, will it recognize us?

Could a smart molecule, a smart material, a software system, or a smart city—a smart process—actually perform functions of an intelligent entity? Obviously, the smaller a cognitive agent gets, the less capacity it has for enacting complex processes. Yet, for some set of agent characteristics on some scale, we know that this must be true—humans, after all, are amongst those agents. Animals and perhaps plants too show capabilities of communication, adaptation, and reasoning. High profile demonstrations of software have demonstrated the ability of software machinery to beat human masters at game play, using memory techniques, and rule sets. They show an approach to solving problems that is probably nothing like what a human brain does. Most likely, 'these are probably not the intelligences you're looking for'. Yet, at some scales, and somewhere in the gulf between pristine logical distinction and the messy interference of parallel storylines, there lies some collection of approaches that yield intelligent behaviours. Like the Turing test, it may well be that the answer—to what intelligence is—is whatever gives the answer we most like the sound of.

My favourite example of an intelligent dumb system is the immune system one finds in the higher vertebrates. Our disease fighting processes are remarkably sophisticated reasoning systems, yet look nothing at all like a brain. In spite of their consisting of a plethora of seemingly uncoordinated cells, none of which seem too remarkable, an immune system is a complex reasoning engine based on an ecosystem of almost completely distributed, fluid-phase cells. It has no wires, no CPU, and yet it can decide what is friend and what is foe, albeit with a few bugs from time to time (no pun intended); it can remember past foes, it can work with the brain to manage the body's larger ambient respiratory complex; it can mass-produce silver bullets that single out specific molecular signatures (even ones that have never existed before in the history of the world). And, most importantly, like all superior intelligences, it knows when to shut up.

This book has not been primarily about Artificial Intelligence. It's about understanding the entire spectrum of behaviours that can emerge from spacetime processes, culminating in intelligence and perhaps beyond. Nevertheless, intelligence is one of the most remarkable outcomes of those spacetime processes—and, we expect it to require a demanding explanation. Like the other mysterious aspects of the world, including Quantum Mechanics and the dark forces of Cosmology, intelligence is a phenomenon that may tax our descriptive power, but I'm adamant that we should not resort to explanation by the mystical properties of neural networks, emergence, and 'big data' unnecessarily. The impetus for this book came from my experiences with space and time while developing smart processes in the dataverse. Those experiences led me to believe that there might be a different kind of unification of ideas available to us than the ones we've been searching for in the particle accelerators of high energy physics. That unification revolves around a centric observer viewpoint—a viewpoint that we've striven to exorcize for the past five hundred years.

There is no description that can be applied to processes at all levels, or for all intents and purposes. In these chapters, I've applied the Promise Theory because it abstracts exactly those minimal characteristics—and it has served me well in the very practical cases of technology development. In that view, processes become smart when they promise outcomes that anticipate the needs of a particular context. That is usually a human judgement, so we need to be careful not to get lost in our own prejudices. For us, a process may seem smart because it is tailored to specialized context at just the right moment—one that gives great service, providing safety and preparation before we know that we need it. We may not find intelligence in smartness, but one always hopes for the reverse.

As humans have developed our modern concept of society, we find that we need artificial systems to expand the repertoire of scales over which we hold dominion—so that police can watch security video footage at superhuman speed, and without fatigue to quickly locate criminals, so that humans can pilot jet fighters safely, or so we can detect illnesses from a single bad cell amongst a billion billion. These are simple tasks that we will soon take for granted. They are brute force processes, not leaps of imagination—and that is where ordinary 'dumb' processes excel. As we develop more cyborg extensions to expand our human cognition, these will help us diagnose and anticipate competing processes that we consider harmful—including faults, errors, pathogens in biology, ecology, instability in economics, and technology. Countering these errant forces is one of the hallowed problems to solve in AI, because it has immediate application to the understanding of systems, and of collective distributed intent. It is a difficult problem because intent is not universal, and the definition of a 'fault' or a condition is both subjective and relativistic to what was promised and by whom.

The systems we describe are usually artificial (intentional) designs: we can try to apply diagnosis to more directly 'natural' phenomena too, but this is often more speculative, because nature has no obvious intent (i.e. 'advanced' outcome), without knowing precise context. Nonetheless, the situation is not hopeless. There are ways of addressing these challenges. We can decompose nature, and then project it into a pattern of our own intentions (e.g. we can simply claim the viewpoint that illness and disease are 'faults' in the biology of humans, to be repaired or avoided, even while biology looks on, from its own perspective, disappointed that we didn't like its new cool experiment). This is all consistent and fine; however, it brings up an interesting point, which, although many have written about over the years, is still widely ignored in the search for simple deterministic answers.

Numerous authors have written about the hopes and fear about automation and machine intelligence in recent years—from robots taking our jobs to the idea that the emergence of a super-intelligence through AI could spiral out of control and ride roughshod over humanity continues to fascinate. I'm not too concerned by these hopes or fears, because I've been outsmarted myself by the unpredictability of human choices, in the past, when trying to improve the human condition through automation. We often reject an obvious path in favour of a workaround that better suits human needs. Rational optimization is not an inevitability. When we can't appreciate the possibilities of scale properly, we tend to focus on simple processes built on skills on our own scale, and we may miss a completely new way of thinking that came from out of the deep. Events rarely unfold the way we expect when it comes to human involvement, because there is never only one possible future.

Aside from the obvious research challenge, and exploratory curiosity, we need to find a good ethical answer for pursuing the AI goals. Of course, there are ways in which artificial systems can be harmful. I hope we will not surrender our human role in decision-making, simply out of laziness, or that we do not engineer an economic model so inhuman that it can only be managed on our behalf by brute force machinery—that could do real harm on a human level. Heaven forbid that weaponization of AI (for high speed economic warfare, or for cyber warfare) could be the principal motivation. Could, for instance, a smart city reach a point at which it started to adapt to goals that did not favour its occupants? Unless the scales were similar or coupled, there is no reason to suppose that we would even interact strongly with an artificial intelligence, whether it be a city or a nano-scale adaptive material. If the scales did interact, then the pinning of that scale by sensory interaction would inevitably limit the growth of that intelligent entity. Once again, scale could be the main limiting factor between dumb and smart.

The question of whether we will recognize intelligence seems related to whether or not we choose to mystify it or even deify it, like another quantum theory, relegating its study to an elite, or whether we can cut through some of the mystery by rethinking our biases. It is not always clear what it means to recognize intelligence in humans, never mind simpler agents. Talk of AI and robots displacing humans is just lazy thinking, in my view, but we are at a dangerous juncture in which governments and institutions are clinging on to old second wave purposes, adapted for the industrial age and poorly adapted to the current one. As Machiavelli wrote[339],

> "There is nothing more difficult to take in hand, more perilous to conduct or more uncertain of its success than to take the lead in the introduction of a new order of things, because the innovator has for enemies all those who have done well under the old scheme and lukewarm defenders in those who may do well under the new..."

Change is always difficult, but my intuition tells me that a broader understanding of how we can use space and time effectively may help us to see where some of the hurdles lie, and perhaps even to overcome a few of them.

There is an alternative future available to us, free from the most bitter struggles for survival, and from the distinctions of class superiority within the turbulent socioverse. Humans and machinery will both play a role in that transformation, but neither one can continue without the other. The real issue is: what will that state look like? Smart spaces will undoubtedly be engineered for humans, but they are unlikely to treat all humans equally, because context is not equal. Will there be a meta process to handle that? The laws that claim to check us today are too slow for a machine-enhanced world. They were written for a different kind of society, and are falling behind. Processes far simpler than human thoughts and dreams may rule the simple facilities of that world, but they will need to mix knowledge of physics, chemistry, biology, economics, alongside concepts represented as knowledge and art. Without a broader view of process—of space and time—how shall we not be caught unawares?

Finally, what shall we do with our time? What stimuli will drive our clocks? Intellectuals, artists, and athletes may find their own motives to occupy them, even without a full time job to give them a path, but those pursuits will not be the ones that lead to a cohesive society. Above all, the scaling of society has its strengths for stability and innovation in diversity and number. A major challenge will surely be to scale education for the coming age, so that our handover of key roles to artificial systems does not simply make us stupid again. That could be the biggest risk of delegating to processes that don't adapt in line with expectations[340]. Today, we are trying to automate those processes too—through the delivery of information to every possible location where it may have an audience.

We humans envy machines their simplicity, their grace and power, their perfect equanimity and flawless execution under pressure. We mimic them and suppress our human instincts for self-preservation, and for the larger goal of producing an industrialized surplus. We are drawn to a greater simplicity than ourselves, because we need to keep our grasp of the world within the reach of stories we can fit inside our heads. No one person understands the theory of relativity, or economics, or cookery, but together our collective society understands it better. This is a hard idea to swallow—yet, in the future, there will be ideas that don't fit into society without the help of machines. More of our relationships in the future could be made through the proxies and intermediaries of machinery, as we rewire spacetime to redefine distance. Is that good or bad? If we rely too much on assistants, we will become passive and surely perish. This is why humanity will not easily retreat into isolation, being fed by the welfare of free food and trinkets from machines. We are smart enough to adapt and recapture a sense of human purpose. Perhaps the future of AI will be the re-engineering of human working life. What network will be doing the learning then? Our cultural, generational memory is already the seat of human identity. If we're lucky, we won't try to transplant that existing memory with a simple device forced upon us by the economic pressures of mismatched scales.

10
Epilogue

1. When a distinguished but elderly scientist states that something is possible, he is almost certainly right. When he states that something is impossible, he is very probably wrong.

2. The only way of discovering the limits of the possible is to venture a little way past them into the impossible.

3. Any sufficiently advanced technology is indistinguishable from magic.

–Arthur C. Clarke's three laws

The purpose of this book has been to better understand some wide ranging concepts that involve space and time. Even though the book is quite long, we've barely scratched the surface of the issues spacetime encompasses. The simple ideas of connection, direction, and expression of information are universally important, and span all scales—but also have vastly different characters as they scale. The scaling of what I call cognitive systems may be the great neglected issue that unifies many different phenomena we observe, from the measurement anomalies of physics to the emergence of organisms and intelligence: relativity is not just about mismatched speeds, but also about mismatched scales. Scaled cognition allows us to demystify the role of intent and 'agency' in phenomena—one of the great philosophical dilemmas. Promise Theory and Shannon's model suggest that even primitive processes may be scaled models of cognitive agents, with interior memory or 'states'. This is speculative, of course, but follows

10 Epilogue

from the consistency of how we view observation. Far from being unique to the higher animals, cognition may be the inevitable outcome of semantic scaling in interacting. It's about the information. The simple counting principles of physics, for example, are effective on scales where labels and distinctions are few, and we are mainly tracking amounts. This leads to simple rules of thumb, or 'laws of nature' at that scale. Biology works on the very same principles, but its behaviours are dominated by a multitude of labels and distinguishable types, with diverse semantics—making such laws far harder to describe. Biology is dominated by information complexity. Some prefer to separate these regimes, but they can also be unified—in what may be the proper approach to Grand Unification of nature.

If we think of space more like a network than a continuum, many apparent differences and peculiarities about spacetime at different scales become not only explicable but quite reasonable. If we pay attention to discrete scales, and limitations on information transmission—especially the complementary roles of give and take—then the strange distortions, raised in the modern physics of quantum theory and relativity, become natural and more common than we might think. Spacetime is no longer the sole domain of physicists and astronomers. We have much to learn from the phenomena of networks.

As we build artificial processes in the human world—especially in the metaphorical cloud of computers we call the Internet, and on a greater scale than ever before—technology has begun to stumble across phenomena like those already witnessed in physics, chemistry, and biology. This is surely no coincidence. We could try to separate those phenomena, and pretend they are merely analogous to the natural world, or we can use physics itself to see them as rescalings of the same fundamental kind of phenomenon. Some physicists will embrace that agenda, while others will feel their position at the top of the scientific ladder threatened by the incursion of the mundane. I believe the methods of physics are strong enough to withstand some reinterpretation, not least in its shifting philosophical underpinnings, and that the tendency towards deliberate mystification of quantum theory and relativity, in particular, needs to give way to a reexamination of more fundamental ideas like scale. It's not enough for theories to make quantitative sense—we have to strive to find qualitative meaning too.

In a more general view of the world, spacetime needn't be the preserve of historical origins and standardized methods in physics textbooks. That which becomes frozen in tradition, becomes vulnerable to reinvention—as Euclid, Galileo, Newton, and even Einstein discovered. To bend the precision terminology of mathematics for a moment, in the interests of simplicity, a 'space' (really a network) can be viewed as any kind of canvas, formed from process agents, rather than the more usual collection of static points. Space and time are really inseparable at all scales. The snapshot states of those agents form observable patterns that characterize all observable phenomena. Meanwhile time advances for each observer by the interior process of sampling, which counts changes experienced by those agents. Finally, classical ideas of motion, mass, momentum, and so on, are derivative phenomena that result from the scaling of communication between the agents—ultimately related to the sending and receiving of information. The basis for all the phenomena we see is information and its transactional communication.

The running thread throughout this book, for all its detail, diversions, and complexity, has been to show that there are limitations to any observer's sampling of the world, and that those limitations offer plausible explanations for the unusual behaviours we see around us on all scales—including physics at the quantum and relativistic levels, and cognitive recognition and reasoning. Those distortions are amplified by sparse network characteristics, and rate limits. This is not so much a new idea, as it is a fuller commitment to that simple principle of scaling and measurement—magic free! Information, which labels and is promised by locations in space, equips space with memory, and—in turn—those memorized patterns feedback onto dynamical processes to influence their behaviours as they move through them. What observers, inside and outside a space, can see of those processes depends on how information is able to travel between them and the observer. Sometimes what they see will appear disjoint and distorted: time may appear to stop, space may appear to shrink because the information is transformed along the way, or distorted by the receiver. This needn't mean that the underlying processes themselves are weird or non-deterministic, only that their interaction with observation leads that distorted perception, because the autonomous and independent agents that harbour the locations of the world become busy or fall asleep now and then.

Basic processes themselves can be understood in terms of changing information too, without a requirement to adorn them with distinctions like matter and energy—though sometimes, semantic distinctions are helpful, even if they are empty, so there is no need to reject them completely. Space has properties at each scale, but the exact nature of those properties may change with scale. The unifying thread then, which weaves all of the above together, is how interacting processes are altered by *scale*. The simplicity of this idea is immensely powerful, and surely worth some attention, no matter how attached we might feel to the many beautiful and ingenious constructions of classical science and philosophy. No one knows, and we may never know, whether the physical universe is truly bizarre and weird, or whether it is only our limited perception of it, but—as the medical profession likes to say—when you hear the sound of hooves, think horses not zebras. The simplest explanations may be the best.

In the last chapter I proposed that the culmination of all the ideas about processes is a particularly useful process that I defined as a 'cognitive system', from its simplest sense as a local receiver on the end of an information channel, to the complexities of a human being, and beyond—in other words, a system that observes and learns cumulatively. As we develop information technology and equip our private and public spaces with it, there are many lessons to be learned about scaling. Our explorations of basic science must be applied to these; there is no sense in reinventing what is already known. Perhaps the lessons we learn from the dataverse can also inspire new ideas about the quantum and sub-quantum world too. This cannot be ruled out as far as I know.

Throughout these difficult chapters, I've explored the nitty gritty of space and time, using the idea of processes, and how they are viewed through vantage point of boundaries that maintain and protect the distinguishability of information. Spacetime is process, and process draws new virtual process layers on top of old. The very concept of a location, and of 'being next to' somewhere, is process dependent—not a universal constant. The impact of relativity on those observations is both dynamic and semantic. Spacetime forms memory banks, patterns, symbols, self-altering processes, and therefore—ultimately—thinking minds. Scale plays the crucial role in understanding the symbolism that emerges to operate at the heart of it. The symbols we seek start with events, matter and energy, atoms, molecular structures, DNA, cells, phonemes of speech, characters of writing, countries, planets, and so on. The range of scales may boggle the mind, but the economy of ideas is compelling.

What I hope comes out of these chapters is the simple message that 'we' (i.e. humanity) have spent a lot of time thinking incorrectly about space and time—for very natural reasons, but nonetheless. When we think of a place, it's not really the coordinate location that we are thinking of—it's the process that's happening there. Think of the town where you grew up. It changes slowly or quickly over time, and it is constantly revolving around the Earth and moving through outer space. It is never truly in the same place at all. What if you transplanted your entire town to another country (as certain monuments and buildings have been moved and reconstructed in museums etc)—is it the same place? Is it the same process? What really matters is not location, but the pathways of interaction between processes. How do we access a certain process? What steps do we have to undertake to reach it? Those channels involve not only quantitative measures (which are the results of other processes of measurement) but also semantic distinctions and characteristics that address different process channels.

Semantics take the shape (literally) of observable properties that we can identify repeatedly as symbols. Symbolism runs far deeper than the one-way process we often portray in science. A symbol is a necessarily discrete pattern, but if you don't treat it as a unit, its meaning is lost. That's why scale is crucial to meaning. Half an atom is not an atom. Part of a cell is not a cell. The horses and monkeys carved into hillsides by ancient peoples around the world were only peculiar trails until humans were able to view them from the air and perceive them in their fullness. We want to think of symbols and numbers as atomic, but unless the pattern is scanned (-) atomically, leaving the pattern lying around does not yield transmission of information. So even smart spacetime characteristics (+) have to be recognized and processed (-) in a smart way to be smart.

Part of this book has served to answer the question: what if the universe actually looked more like a computer network than the continuum of Euclidean space? The answer is quite subtle, and far more scale dependent than physicists tend to prefer, but ultimately not too different at large enough scale. This is not to be taken as evidence that spacetime necessarily has a particular network structure at the quantum level, but it could be enough to consider it as a plausible alternative to present formulations that have deep rooted problems associated with their continuity. Perhaps more interesting to me is the idea that physics, with its rich battery of methods for analysis can also be applied to computer science at the level of information technological infrastructure. The biggest challenge with distributed systems is that they quickly expose issues of relativity, analogous to those pointed out by Einstein in our astronomical universe. We engineer systems with certain assumptions about synchrony and determinism, but we that is what fails as the speed of change approaches the speed of our ability to observe. The problem of distributed consensus, or reaching a mutually agreed understanding of the state of something, like a bank account, is a non-trivial problem in any space. It is even harder in a computer network, which has only sparse paths and is full of holes.

Traditionally, physics has distinguished matter from space, treating the two as qualitatively separate phenomena. Matter is assumed to exist in the form of space-filling bodies or entities that sit *within* spacetime, as if it were a container. At the same time, Newton showed that, as far as mechanics and motion go, we can usually reduce a three dimensional body down to a single point for the purpose of describing its ideal motion: centre of gravity, centre of mass, axes of inertia, etc. The problem with this separation of something and nothing is that, if space is truly 'nothing at all', then how can there be more of it between some things than others? How do we explain distance? Moreover, if the quantum view of massive particles' locations, as the results of measurement, is the fundamentally correct view (which seems unlikely), then how can where exactly is the source of a gravitational or electric field? In practice, measurable quantities end up being defined through the effective constraints of source (+) and observer (-) symmetries, relative to inertial frames of reference. These points of view can only be reconciled if our current understanding of fields is an average approximation to spacetime[341]. In a computer, we have entities or 'agents' that do things, and we think of them as the locations in a space, because they are the 'centres of intent', something analogous to the centres of mass or charge in networks. But they have positions in an absolute space, which can nonetheless change relative to one another. Phenomena we associate with relativity in physics do still happen, but on an entirely different (much higher) level.

Space and time are at least as old and wide as the universe, so you might think that there would be nothing new to say on the subject today. What has changed now to prompt this reopening of the case? There are surely those who think it unnecessary to rethink orthodox ideas about spacetime, especially as a multiscale, coherent concept, or without a veil of high powered mathematics to legitimize it; but, whether one likes it or not, the reinvention of spacetime is something that is happening around us, thanks to information technology. As we bury the modern world under increasing layers of information, the need to maintain important distinctions only increases. Significance and meaning are undergoing renormalizations. The eminent philosopher Alfred North Whitehead wrote[342] that the goal of science and philosophy is:

> "to see the general in the particular
> and the eternal in the transitory"

Today we build complex artificial systems some of which rival the complexity of the natural world. They allow us to see the scaling of processes from an entirely different perspective, and they have reminded us about some of the basic lessons of dimension and scale. New perspectives lead us back to questions we've encountered before, in subjects like biology and quantum mechanics.

At the very least, the dramatic changes in scale that are taking place in computing infrastructure, with components getting smaller and faster, and deployments getting larger and denser, mean that we need to modify the way we describe space and time in the technological realm. The distortions that occur from looking through a lens that is barely able to keep up with the changes it funnels to us are frustrating our attempts to hold onto traditions and use the world of information as a source of universal truth. Concepts like true and false, so prominent in scientific culture are finally revealed for what they are: scale dependent[343].

Einstein argued that there was nothing to indicate the presence of an absolute space. That is also true of any virtual process, but underneath the computing cloud there is such an absolute. It strikes me that we shouldn't be so quick to form unflinching judgements about the universe each time we learn something new. Everything in the physics we observe may seem to be relative. On the other hand, no experiment has really disproven the existence of an absolute space in the sense of the dataverse[344]. We make much ado about the weirdnesses of quantum mechanics and relativity, but we make those judgements based on a view of space, time, and motion that's simply wrong—a convenient fiction that works for a medium sized, slow world with very fast signals. Once we pay more attention to scaling our regular understandings of phenomena, the weirdnesses of quantum theory and relativity seem quite natural. The problem is one of access to information, and there may be nothing weird about the quantum world at all.

As far as intelligent behaviours are concerned, we need the perspectives that scale brings more than ever. Reasoning goes far beyond logic. It seems to be closer to the exploration of multiple trajectories through a memory network

The interference between those paths seems to be important to selection and elimination of junk—as we learn from semantic models, and see apparently in evolutionary epigenetics. As F. Scott Fitzgerald wrote in *The Crack-Up* (1936):

> The test of a first-rate intelligence is the ability to hold two opposing ideas in mind at the same time and still retain the ability to function.

Finding a single correct 'logical' outcome is a matter of mathematical interest, but it is not an efficient way of automated modelling, because it is far too fragile. A smart, adaptive process that can keep all possibly explanations open in order to refine its understanding of a situation. It can be of two minds at once, like any other process with multiple evolutionary pathways. Present day artificial intelligence research has most recently focused on the advances in artificial neural networks, or so-called 'Deep Learning' using fixed connections and pre-trained semantics. While fascinating, to be sure, this approach may not be the route that respects scaling in the long run. I'm concerned that our views on intelligent behaviours are too narrow, and too obsessed with human skills. Just as we may have become too rigid about fundamental physics, so we may be stuck our ideas about other processes too. If our interest in smart and intelligent behaviour only begins at the level of human intelligence, then we are fundamentally missing an opportunity to expand our own minds. The potential to enrich and develop human life may just depend on how we choose to deal with that issue.

If you are still thinking about spacetime just as the theatre we live in, that physics already knows the literal truth of space and time, and all I've done is to cough up a few analogies to the bastion of truth that is modern physics, then I've failed to convey the lessons that I believe are staring us in the face. If we can't explain the basic semantics of processes, even those immutable ones like the conservation of energy, without resorting to tautologies (or turtle-ologies) like Noether's theorem, then no new story about unification will add much to the present one. I have come to believe that returning to basics could offer a better chance at understanding the remaining mysteries of our world—and that the depth of similarity between phenomena, across all the scales we observe, should be taken seriously.

Keep your eyes open, and you will see the ideas I've described in action—they are everywhere around us. Every process traces out a path, a story that connects information to an audience of cognitive receivers. We will probably never know why all this is possible, but we can try to better understand how.

Chapter Notes

Notes

[1] The same story applies to the quantum theory and its unsatisfactory assumptions too, and with astounding mathematical successes.

[2] Thales and Pythagoras are widely thought to be the first real mathematicians. And mathematics was developed gradually in Mesopotamia, Egypt, China, and Ancient Greece. See the nice review and story of geometry in [1, 6].

[3] The term Cartesian coordinates refers to Descartes, though the allusion to the French word for 'maps' is fortuitous.

[4] Euclid took on the task of stating clear definitions and built new theorems on previously established facts. In this way, he developed the beginnings of formal logic, or that special kind of disciplined storytelling that simultaneously acts as an accounting ledger for what has already been established.

[5] Einstein himself struggled with this notion of empty space. In a fit of reasoning that both dismisses Newton and adopts his own methodology, Einstein wrote [3]:

> Descartes argued somewhat on these lines: space is identical with extension , but extension is connected to bodies; thus there is no space without bodies and hence no empty space. The weakness of this argument lies primarily in what follows. It is certainly true that the concept of extension owes its origin to our experiences of laying out or bring into contact solid bodies. But from this it cannot be concluded that the concept of extension may not be justified in cases which have not themselves given rise to the formation of this concept. Such an enlargement of concepts can be justified indirectly by its value for the comprehension of empirical results ... the storage possibilities that make up the box-space are inde-

pendent of the thickness of the walls of the box. Cannot this thickness be reduced to zero without the space being lost as a result?

[6]There are good reasons to believe that our capacity to imagine and comprehend other abstract ideas stems from our capacity to perceive and navigate habitats in that wider world[4, 7]. After all, space and time are the most basic things any animal has to contend with, as it evolves its repertoire of sensory equipment. If there were ever a plausible seed for evolving conceptual thinking it would surely lie in the ability to form maps of terrain, recall journeys and paths for navigation (stories). What else could explain our ability to think in abstract, other than the capacity to recall and learn from past adventures or imagine hitherto unvisited realms and places in our dreams?

[7]This was the question Descartes asked at length.

[8]There are many versions of this story, and I have no idea which is the original. See the notes on Wikipedia.

[9]Major religions have been accused of abusing this insistence of faith as the ultimate answer (or non-answer) to promote ignorance and control beliefs for the purpose of political power.

[10]When teaching at the university, I can attest to the fact that a single step justification is often enough for many students. Only a small number will persist in their inquiry.

[11]The term bootstrap is used quite a lot, especially in computing, to refer to a process of getting started without prior help. It comes from the idea of lifting oneself by the bootstraps—obviously something that violates the laws of physics. But the expression is also used for '(re)booting' computers, where getting the computer started requires the computer to start reading some instructions, i.e. already requires the computer to be started! The term bootstrapping is this about getting a concept to stand on its own legs without tautology, by not using the very concept it is trying to explain in order to explain it.

[12]For a discussion of space filling networks see the excellent review by Geoffrey West [8, 9].

[13]Indeed, I wrote a book about this in 2003[10], fully on board with this idea. Only by working in computing have I spent that past 15 years unpicking that belief system.

[14]The essence of relativity (whether Galileo's, Newton's or Einstein's) is that different observers experience different versions of what happens in space and time. That seems to suggest that space and time stand apart from what happens within them, but that turns out to be too simplistic.

Relativists talk about *frames of reference*. A frame of reference is a version of spacetime experienced by an individual observer, i.e. a point of view. Why do we have to include space and time in something as straightforward as a viewpoint? The reason is that our notions of distance and time are measurable things, and different observers can also measure one another by observation of each others' rulers, clocks, and speedometers. What is unexpected is that what these instruments will seem to disagree with one another.

This is not about a free-to-choose difference of opinion, or a simple illusion (at least no more than we choose to believe that the whole world is an illusion). No observer could escape having their views of the world altered by a change of circumstances. Two observers could start at the same place and time, calibrate their measuring devices and then set off along different paths, watching one another all the way and still disagree. The main reason this feels wrong is that the peculiarities only happen on a scale far too big or small for us to experience directly.

It turns out that, in the world of computer science, the effects of relativity are much more 'in your face' than they are in physics. The average person roves around the entire world of the Internet, without a rocket, and therefore may experience effects that seem quite exotic, compared to what an Earthbound traveller experiences of our universe.

Technically, a frame of reference is imagined as a kind of roving system of coordinates that observers are imagined to drag around with them, along with a clock that tells their version of time. This idea is a very mathematized view of reality, but it is the view that is handed down to us from antiquity. So, it is true that the idea of a reference frame has become cartoonish after a century of popularizing the ideas of Einstein's carefully tested explanation of relativity, with reference to spaceships travelling very fast, or passing close to black holes. This can easily give the impression of effects of relativity as something fanciful and far removed. This means that one of the things we need to rethink more carefully in a generalized view of space and time is: what would constitute a frame of reference?

[15] Space and time are the prototypical concepts that underpin almost everything we talk about. Linguistics is full of examples of metonyms and metaphors for space and time. When we say that someone was a huge influence, we don't mean that the person was physically large. Our sense of measurement is based on the world we know, and our understanding of space and time dominates.

[16] The precise dates of Euclid's time are unknown.

[17] See the excellent story told in [1].

[18] See the discussion in [11].

[19] By the age of 15, Gauss became the first mathematician in history to realize that one could construct a logically consistent geometry in which Euclid's assumption that parallel lines never meet is false. But that was only the beginning of the revolution in ideas about space and time. From there, it took the technology revolution, in the 19th and 20th centuries, starting with thermodynamics of steam engines, electricity, and later computers to see the world through a different lens.

[20] I recommend the fine and entertaining guide [1].

[21] I'm grateful to Dr. Michael E. Smith (@MichaelESmith) and Martijn Storms (@MapStorms) for helping me with this question. See the excellent History Of Cartography [12].

[22] See the article in [13]. I'm grateful to Alexa Rose for this reference (@doubleAroseshow).

[23] See [3], Appendix V.

[24] Locations on a map need names to call them by. Before the invention of numbers, people could compare things in relative terms (i.e. relativistically): your rabbit is less than a cat but more than a mouse; my town is beyond the forest, before the mountain, and so on. By creating a set of names or symbols with a definite ranking or ordering, numbers made it easy to quantify and name things both relatively and absolutely at the same time. Simple maps can be made in this way, without attending to accurate distances. Absolute locations can be placed relative to a fixed origin, like a significant landmark.

Numbers might not be absolutely necessary, but they add a finesse to these stories, and eventually allow us to go from topology (relationships between places) to geometry (distances and directions). Distances are assisted by semantic markers like mountains and rivers, or if you are a technological species you can invent signposts, like ants do, laying down trails. Distances

[25]Tensor networks [14] are a form of circuit diagram for causal processes that have certain properties useful in computation and quantum mechanics. They only exist in a few dimensions. A tensor network diagram represents a matrix operator transformation. This is quite like the data transformations used in big data pipelines and machine learning applications. The importance of circuits as a model of causal evolution has grown over the past decade, especially in response to scaled computational methods. My friend Petar Maymounkov developed a model of this called Euclid, which evolved into Ko, a derivative written in Go, inside Google.

[26]A single mathematical point has no internal degrees of freedom, and could therefore not discriminate states arriving. It could participate in the signals by altering its own state, but it couldn't compare now and then, so it cannot measure or see change.

[27]The irreversibility of non-deterministic systems was shown in [15] and related to the non-invertibility and the problem of division by zero.

[28]IP stands for Internet Protocol address, the system of labelling computers normally used in global communications today.

[29]In practice, experiments in space are performed by assuming that the speed of light is constant, and that we can measure distances and thus lay out an imaginary coordinate system based on fixed objects, like the very distance (called 'fixed') stars that effectively don't move on any human timescale.

[30]Coordinates are a simple form of information: a pattern we can use to calibrate another pattern. Again, the theme is that space only works if we can contrast states. If empty space is the quality of being uniform and featureless, and information is the quality of being distinct and measurable, then they are surely opposites.

[31]Calculus is based on the idea of defining smooth properties by allowing the differences between points to become 'vanishingly small'. This is an idealization with many useful properties in the imaginary world of mathematics. It is highly unlikely the real spacetime is such a smooth manifold with infinite smoothness, nevertheless the utility of the construction wins over all objections in common usage.

[32]See [16].

[33]For those of us who have studied geometry in school, the concept of dimension is more or less clear, except when it's not. There are three (or six) dimensions: up and down, forwards and backwards, left and right. These apparent facts are so universal that they are almost beyond question. We are so conditioned by this description that we scarcely seem bothered by the abstract idea that left and right are part of the same direction (i.e. left is the inverse of right), rather than being an independent direction. Nonetheless, this is not to be sniffed at. The number of dimensions in space is not as clear as we have come to believe.

[34]In the overlap between scaling of functional promises and the counting of spacetime states, the concept of fractional dimensionality arises—fractals. When results are presented as spacetime scaling ratios, with approximate form of a scaling law in a one-dimensional spacetime $D = 1$, with Hausdorff dimension $D > 1 + \delta$ and $\delta < 1$. This indicates that a serial process, with some parallelism, is essentially a one dimensional problem, with some

fractal complexity in its trajectory due to parallelism. Alternatively, if we think about the problem graph theoretically, we can also say that it behaves like a $D = N$ dimensional space, and a trajectory with Hausdorff dimension $H = \pi/\sigma$. In a graph, the node degree $k = N$ is the effective dimension of spacetime at the point[17]. Interestingly, as the parallelism increases, the duration of the fractal dimensionality shrinks to nothing. Thus the large N limit for serial processing tends to squeeze the degrees of freedom in the system.

[35] It has even been suggested that we should look at this statistically. See the work by Norwegian physicist Jan Myrheim. [18].

[36] Membership in a basis set is a semantic convention used by observers. It cannot be imposed. A point in a space need not promise its role in a coordinate basis, because that information is only meaningful to an observer, and could simply be ignored by the observer. An agent can promise to be adjacent to another agent, but to propose its own classification as a member of some basis would be to impose information onto others from a different viewpoint. That violates locality. By autonomy, each agent is free to classify another agent as a member of an independent set within a matroid that spans the world it can observe.

The dimensionality of spacetime, perceived by any observer belongs to the rank of the matroid it chooses to apply to the agents it can observe. The consequence of this is that spacetime can have any dimension that is compatible with the adjacencies of the observer. Indeed, the notion of dimensionality experienced by the elements of a promised space is different for every agent, at every point. The observer with n outgoing adjacencies may regard each independent adjacency as a potential basis vector or direction.

[37] This is actually a major philosophical point that is greatly underplayed. We have come to believe in a reality of spatial dimensionality that may actually be misleading. I'll come back to this point.

[38] This counting of dimensions can be seen very clearly in the approximate scaling expressions derived for economic output of cites: see the work on the scaling of cities in [19].

[39] Time is directional at the small scale level of individual agents, and also at the statistical level through the second law of thermodynamics. Physicists tend to focus on the latter, and deny the former, invoking 'reversibility', but this is a misunderstanding of scale and a misrepresentation of the laws of physics.

[40] The air is also ubiquitous on our planet, but we can evacuate small spaces of air, increase or reduce its pressure, and so on.

[41] This is related to Mach's principle that we'll come back to into chapter 7.

[42] The justification for this idea is that we know that influence travels at a finite speed. This seems to not be the case for quantum entanglement but it might still be true that there is a finite speed hidden channel linking quantum states, one that is simply infinite with respect to our ordinary spacetime dimensions, and observations that we can make at the speed of light.

[43] This includes the spatial part of Minkowski-spacetime in Lorentzian relativity, for the purpose of this book.

[44] See the theory of the effective action in [20], for instance.

[45] See reference [21].

[46] It's easy to see why. in IT systems, there is no natural definition of motion. Most dynamical activity is of the form of finite state transitions, more like quantum mechanics than Newtonian

mechanics. Emission and absorption from an atom is more like data arriving and leaving from a bounded region. What happens to the emitted data, or where they came from is not usually the focus of attention. In other words, the idea of data being 'in motion', i.e. of information travelling in a straight line at a certain velocity is not something anyone thought too much about, until quite recently when data 'big data' processing through pipelines was moved into the 'computing cloud'. In fact, I'll return to this topic in chapter 5. The focus on translational invariance in the physical view of spacetime is based on the preeminence of Newton's laws of motion, which for centuries were the principal phenomenon for philosophers to investigate, thanks to endless cannonball warfare and slightly more peaceful astronomy.

[47] In mathematics the term for a network is 'graph', and graph theory is the study of graphs[22, 23, 24]. The term graph is also commonly used for diagrams with axes that depict trends and relationships in data, but that is an unrelated usage.

[48] Mathematicians have different conventions for defining dimension in a graph. Often it is based on the dimension of an embedding space required to hold the graph without any lines having to cross.

[49] Graphs can also be used to describe towns and cities. The roads and utility networks for water and sewage form networks with this kind of space filling form. Space filling networks are found in the capillary systems of our bodies, and turn out to explain the scaling of many of the functional properties of organisms. For an excellent summary of this work, by Geoffrey West and collaborators, see [8], and [9, 25].

[50] The Border Gateway Protocol black holes are a troublesome phenomenon in Internet routing, which sometimes causes blackouts of the Internet, when mistakes are made in the engineering of network interconnection rules. Luckily software black-holes can be opened, even without Hawking Radiation.

[51] In Euclidean space, we have to supplement the degrees of freedom in spacetime by equations of motion (constraints) which are then solved for the final trajectories.

[52] For example, expander graphs ([26]) are a special type of graph that is both well-connected (mesh), meaning you cannot disconnect points without cutting many edges. But at the same time, most points are connected to very few other points. As a result of this last trait, most points end up being far away from each other (because the low-connectivity means you have to take a long, circuitous route between most points). Some links in a graph might not be allowed in all circumstances, especially in information technology, where access controls are used for security. There are different criteria for distance. Why would distance matter? Ideally it shouldn't unless there is a defect or a dislocation like an entry point to a datacentre. What does length mean? It is easier to push off this problem to a definition of time, since it is process that makes the universe work. Time seems more fundamental than space in that respect, although one still needs a spectrum of states in order for time to exist, and this is what ultimately generates space.

[53] Category theory has acquired a huge following since it was invented by Samuel Eilenberg and Saunders Mac Lane in the 1940s as a way of expressing certain constructions in algebraic topology. There are many introductions to it, for instance [27, 28, 29].

[54] This turns out to be related to reversibility once we account for time, but a priori I believe it is more fundamental.

[55] Physicists might enjoy musing over whether anthropomorphism is a kind of semantic

renormalization group transformation. Or they might simply shudder and put the book down, depending on how many degrees of freedom they have in their brains. Leonard Mlodinow's excellent book[1] has the following anecdote:

> To paraphrase the great Göttingen mathematician David Hilbert, "One must be able to say at all times—instead of points, circles, and lines—men, women, and beer mug. Then, ..., Euclid's first three postulates:
>
> 1. Given any two men, a woman can be drawn with those men as its endpoints.
>
> 2. Any woman can be extended indefinitely in either direction.
>
> 3. Given any man, a beer mug with any radius can be drawn with that man at its center.

This is precisely what we mean, in promise theory, by semantic scaling. It is a way of keeping one's abstractions under some semblance of control.

[56] If you have read any popular accounts of physics you will know that not only does physics have plenty to say on the subject of spacetime, it can't stop coming up with ways of bending current theories into explanations for these topics. I want to try to avoid most of this, in this book, for a couple of reasons. The first is that I don't really believe half of what theoretical physics says about spacetime, especially in connection with String Theory. But a more important reason is that there is more than enough to learn about spacetime that doesn't need any kind of speculation at all: we can build it an measure it through the wonders of information technology. This is, for me, one of the most exciting things about studying computers.

[57] See the review and discussion in my book In Search of Certainty[11]

[58] Our prevailing idea about physics is that it is inevitable. We speak of physical law, as if it were still a theological imperative. There is another way of looking at interactions that is complementary to this view. If the inevitable deontic view of physics is non-local law, then a more local interaction viewpoint that matches the aims of relativity better is the idea of 'autonomous cooperation'. This latter view is what is called Promise Theory. In philosophy there are various theories of promises, but don't be misled into thinking that promise theory is about moral philosophy, whatever the name might sound like. In fact, the promise theory I'm referring to, is very like a relativistic physical theory, talking about clear interactions and localizable semantics. In fact, it can supplant a lot of the moral versions of promises too. Promise theory is an idea that I began to develop around 2004, and later developed in collaboration with my friend Jan Bergstra [30]. It has proven to be remarkably versatile and resilient, with few assumptions. Although it has limited predictive power, it has a way of organizing one's thoughts into necessary and sufficient criteria, for a wide range of issues. We say the charges are attracted to one another. One of them wants to move in a certain direction. This is highly anthropomorphic language, which students of physics are chastized for using in so-called scientific discourse. We are supposed to say that there is an attraction as a matter of law. But how is this any less anthropomorphic. Physics is supposed to be inevitable, a matter of physical law. This is because at some time in the past, we emphasized the regularities

of behaviour of elementary things at a particular scale. Later, this got undone by Quantum Mechanics, but at some level there were regularities, so the idea of a law persisted. Particles and things in physics must do as they are told! Told by whom? This sense of must and law is built into our culture, since the religious edicts told to Moses on the mountain. It is an affectation that comes from dogma. If we've learned anything in the 20th century it must surely be that nature is not deterministic except in quite exceptional circumstances, over quite particular scales. If we start again, there are ways to explain these matters. Instead, physicists have chosen to mystify them, thrilled by the sense of the unknown. In my view this is something physicists (theoreticians) need to get over. More and more authors of my generation, who grew up on these mystical mythologies, are beginning to question these received truths about the mystical qualities of modern physics. See *The Trouble with Physics*, Lee Smolin[31], and *Lost in Math*, Sabine Hossemfelder[32]. Like other religious dogmas, physicists are very particular about which mystical ideas can be the right ones, not that they all agree, but each faction has its own belief systems. It is all very anthropological.

The question I'm keen to ask in this book is the following: is our paradigm about space and time simply ill suited to describing the kinds of phenomena we are starting to be interested in? The old theatrical view of spacetime worked on a certain set of scales, for a certain set of problems, but we move on. Is there a different formulation, which is better suited? What might be the benefits to such an altered view?

[59] Promises made also have to be kept, and their outcomes assessed. Assessments are continuous, because they are imaginary, rather than measurable. There is a kind of continuum of assessments about semantic interpretations. Promise Theory has proven useful because it tries to keep the assessments pragmatic, and functional in context.

[60] Hackers and crackers with bad intentions might be able to disrupt and even destroy parts of the Internet, but this is not the same as controlling it like a machine. The behaviour of the network is no longer a simple deterministic machine. See the discussion in my book *In Search of Certainty* [11].

[61] In fact these events are not technically 'transactions' in the sense of computer science, only in a popular sense.

[62] We don't necessarily know enough about what spacetime looks like below the level of subatomic particles. Who knows what underlying complexity might be buried inside the interior of space?

[63] We have to call it something, and this pokes just a little fun at ideas like the multiverse, and the metaverse, and other subversive ideas.

[64] When spin pairs are co-dependent, and have zero initial spin in total, both spins are in an undetermined state. However, when one is measured, the other's must be 'instantaneously' determined, according to some internal clock of the entanglement. In spite of the endless references to Bell's inequalities as proof of no hidden variables, there is a profound sense of denial around entanglement, and how influence travels. I am still hoping for a pragmatic explanation, and one that makes sense rather than appealing to quantum mysticism.

[65] The network is solid state in two senses. Anyone who grew up in the 1970s will remember that 'solid state' was a marketing term that was meant to signify that an appliance was 'transistorized' as opposed to using vacuum tubes. Transistors were new at the time. Vacuum tubes are

filled with a rarefied gas state, while transistors are entirely made of solid crystalline germanium and silicon. Solid state is used to mean crystalline in physics, and the solid state of the network also means that it has a fixed crystalline structure. The atoms or computers have fixed locations relative to one another; they don't float around like a gas.

[66] In the 1972 episode of Dr Who, The Time Monster, an experiment to transfer matter from one spacetime location to another called TOMTIT, or Transmission Of Matter Through Interstitial Time, where interstitial time was described as 'the space between now and now'. In an interview with a science advisor on the re-release of the DVD, the advisor rejected this idea as silly. This follows the orthodox idea that what we understand as spacetime today is continuous and real, but based on the discussion here of embedding space, that feels like a naive view based on current doctrine. In fact, interstitial time would simply be the places in the embedding space for our spacetime, which rather prophetically was considered to be discrete in the episode. All in all, I was impressed by the thinking of the non-scientist writers of the episode.

[67] It turns out that cabling is the most expensive part of datacentres. Cabling is extremely complex, and very expensive. When cables break, or need upgrading, locating the culprit amongst literally thousands of similar looking wires is next to impossible. Eliminating physical cabling and replacing them with non-interfering wireless light could lead to enormous advantages.

[68] See the discussion of cellular automata in [11], or the works [33, 34] and [35, 36].

[69] See Paul's company Earth Computing, and his many patents and writings.

[70] Technically this called the first normal form in database theory. The normal forms for databases straddle dynamical and semantic concerns. See the discussion in [37]. The *first normal form* says that we should try to make the structure of every spatial location in a database similar. This is homogeneity, or translational invariance of the dimensionality of space, but not necessarily its content. Somehow this is like saying that spacetime is a fundamental uniform substrate, and any material excitations within it are just 'data', or local differences in the scalar properties expressed at certain points. The *second normal form* says that, if an agent of space has patterns that are common to it, then they should be factored out and identified as an independent pattern, or an independent kind of space, i.e. an independent dimension, possibly with different scalar properties. The *third normal form* says that locations should not be connected redundantly. There should be only one planet Earth, not endless copies, even if it would be useful to have a local branch in different parts of the universe. Third order normalization is like a matroid basis, every property is defined as a unique node, and references to the property.

[71] There are many ways to differentiate cells in biology, so this likeness is on an abstract level. It could be a Major Histocompatibility Complex (MHC) tag, or a genetic expression, etc. In Promise Theory terms, it's about what exterior promises a region of space (a cell) makes to its environment.

[72] This is the problem of database consensus, championed by Paxos, Zookeeper, and Raft-based Consul, etcd, and others [38, 39, 40].

[73] A square array can be reformatted into a single sequence of numbers, by repacking the dimensions, and can be transformed into alternative views by a well understood mathematics of objects called matrices and a special subset called tensors, as used by Einstein in his description of spacetime.

[74] To count over the full geometrical space, in a computer program representation, one uses 'loops', or processes that count like a regular clock, modulo some fixed dimension. For example, consider a cubic data array:

```
array[x][y][y]
```

This array represents a three dimensional space of locations, each with unit volume. In Euclidean space, the size or volume of a spatial region in three dimensions is a volume equal to the product of the distance intervals in each dimension. In a discrete space, like the dataverse, the size of the container at each location is independent of the perceived distance between the points. They are independent things—which makes spaces, like Euclidean space, rather special and constrained to a particular set of promises.

An array, in the dataverse, does not have to be limited in this way. Each location can be filled with numbers, and the locations represent a three dimensional shape. The 'for' or 'for each' loops found in nearly all programming languages can step through all of these locations, scanning like a digital clock,

```
.for x = 0 to dim(x)
.  (
. for y = 0 to dim(y)
.    (
.    for x = 1 to dim(z)
.      (
.       print array[x][y][y]
.      )
.    )
.  )
```

in order to output the values. The 'for' loops act like a digital clock. If we set all the dimensions to 12, and write

```
.for hrs = 0 to 11
.  (
. for mins = 0 to 11
.    (
.    for secs = 0 to 11
.      (
.       print clock[hrs][mins][secs]
.      )
.    )
```

The seconds count from 0 to 11 and then on wrapping around to zero again, the minutes counter advances by one. When the seconds have passed 12 time 12, then the minutes have counted from 0 to 11 and back to zero again, and now the hour counter increases by one. The loop above therefore covers half a day (12 hours) second by second.

Arrays like these have many uses and purposes in mathematics of space and time—going far beyond what I intend to discuss in this book. The *covariance*, or ability to transform matrices and tensor tables is used in 'machine learning' as a central part of pattern recognition used in Artificial Intelligence. If we can understand spacetime patterns regardless of how they might be distorted and rotated, then we have the ability to recognize shapes automatically. This is the essence of facial recognition, predictive analysis, and many other data applications: curve fitting, path recognition, etc. They embody the essence of how computer games manipulate three dimensional imagery, for instance, as well as how spreadsheets and accounting ledgers are computed.

The for loops in the computer programs have an analogue in the calculus of Newton and Leibniz, used in physics, from classical to quantum mechanics. Only the notation is different. The analogue of the for-loop above in calculus would be:

$$\int_0^{11} dh \int_0^{11} dm \int_0^{11} ds \ \text{clock}(h, m, s)$$

The notation is irrelevant. The important point to note is that the same spacetime processes are at work across physics, computers, hardware and software, meaning that it is spacetime properties that we see in action across all these phenomena.

[75] Schwinger was quick to point out that the spacetime labels in quantum field theory are really dummy variables in a generating functional[41, 42]. Compare this to the graph generating functionals in [43]. The result, however, is an expansion that has the form of a number of graphical processes[20]. Feynman took these graphs to be models of spacetime processes[44], but in fact they are only 'virtual processes', analogous to indices in a for-loop. They have no significance other than as a parameterization of the field data.

[76] See, for example, the classic reference to regularization of infinities in quantum field theory [45].

[77] The existence of scale-free phenomena has been a big deal in modern 'complexity science'. The case has been made for scale free graphs, and scale free fractal processes. In practice, these scale invariant features can only persist over a few levels of hierarchy, depending on how semantically leaden they are with respect to their environments.

[78] Whether there is ever a smallest elementary scale no one really knows, but we assume there is, else it's only turtles underneath.

[79] This reminds of the old joke about the man who asks a local for the way to the Post Office, and receives the reply: 'Well, if I were going to the Post Office, I certainly wouldn't start from here!'

[80] Surely everyone has heard about black holes in outer space, but not many know that there is also a phenomenon called a 'black hole' in the dataverse. A dataverse black hole is a region of the network that becomes either unreachable or inescapable, or both. If data go into a region but there are no links out. The dataverse gets a kind of semi-permeable membrane. This is hopefully a temporary effect, usually caused by a mistake in the configuration of software called the Border Gateway Protocol (BGP). This is the software that connects the space of the dataverse into semantic regions known as 'ASes', originally short for Autonomous Systems. BGP black holes lead to certain places losing contact with the Internet for a certain time, until the horizon becomes permeable again. From a network perspective, we don't have to postulate the existence of a force out of control to arrange the geometry of spacetime to prevent signals from getting out.

In General Relativity, Einstein pointed out that the geometry of spacetime is sufficient to explain these effects in the universe too, and that the geometry of spacetime is an alternative explanation to the existence of a force. So the thing that corresponds to a force in the dataverse is the rulesets that preferentially route messages around. Gravity is a kind of preferential routing of matter, or the migration of information states associated with matter, according to the geometry. The shape of spacetime in the universe is related to the localization of mass. Mass is therefore also related to the shape of or non-local state of a spacetime region, and we'll return to discuss it more in chapter 7.

[81] RCS was an early system, based on CVS the Concurrent Version System. Today 'git' is a popular version control system used widely.

[82] See the discussion in [11]

[83] This quote is from 'How fear came', the Second Jungle Book, by Kipling, reprinted as a poem in numerous collections.

[84] See the discussion in [11].

[85] The term agency is used in politically charged ways, to imply a kind of free will, rather than just an autonomous freedom from exterior constraints. What Promise Theory shows is that these two things are largely scaled versions of one another. If one is able to let go of the metaphysical traps of free will, and view behaviours as patterns of give and take, in complex circumstances, then these philosophical mysteries begin to melt away. In the western scientific tradition, no human concerns should enter into scientific explanation, especially those that have to do with opinion or purpose. This started with the good intent (ironically) to avoid partial concerns. Science is supposed to be an objective reality. However, we now have a more sophisticated understanding of the difference between impartial and objective. Science cannot be objective, but it can be disconnected from ambition and personal gain. As it turns out, the concept of intent is far more important to impartial science than many classical scientists appreciate. Intent captures the idea that regions of space can embody patterns of behaviour that have a function, within the scope of a particular context. Intent is then about compatibility and specificity of context rather than free will.

[86] For a discussion of dynamical similarity, see *In Search of Certainty* [11].

[87] In mathematical terms, we say that the computational complexity of a hierarchy is an $O(N)$ process, whereas a mesh network is $O(N^2)$.

[88] See the brand new biography [46] for a fascinating historical account. The original work

by Shannon is found in [47].

[89] Continuum mathematics has always been a highly convenient way to count things in large numbers—so much so that many scientists have come to assume that it is the fundamental way to describe the universe. However, Shannon's work points to the unlikelihood of that being the case.

[90] See Weaver's contribution to [47].

[91] On a philosophical level, this questions the basic assumption in modern physics that symmetry should be the natural state of a system. Symmetry might be the simplest state, i.e. the state of least information or maximal information in some cases, but that's not necessarily a reason to believe it to be fundamental.

[92] It is perfectly possible to interpret quantum theory in this way, as well as computer science, and other sciences too. However odd it might seem, it actually says very little that is controversial, because it almost says nothing at all.

[93] The overlap construction is basically like the 'open ball' construction in topology. The overlap region is called mutual information, but this belies that fact that the information goes in one and only one direction, by assumption.

[94] This is what happens in Ethernet or Wireless networking.

[95] In my book *In Search of Certainty*[11], I related the story of how information theory is linked intimately to quantum physics.

[96] The assumption here is that each sample is a discrete event, but this is not important, because the same argument can be made in a continuum description using Fourier modes, as Nyquist himself showed.

[97] It is not even clear what it would mean for one point to sample faster than another, because that assumes the existence of a neutral third party clock by which to measure both. The fact that no such clock can be known from within the system means that agents see what they see, and the concept of time is just whatever it appears to be on whatever scale we are measuring.

[98] This is reminiscent of Harlan Ellison's famous story: *I have no mouth, and I must scream*, only with touch instead of voice.

[99] See the next note.

[100] The switch is probably the simplest and most revolutionary technology in history. It allows us humans to control electricity, or the flow of rivers for transport and irrigation; it allows biological processes to classify and separate resources and develop specializations by determining when something can and can't pass.

A switch is a valve that connects resources (+) to potential users (-). Data processing is literally all about the management of switching at the appropriate time and place. In computers, switching begins at the level of transistors, in the nano-wafers of silicon chips. Particular arrangements of circuitry create 'logic circuits' which are the basis for memory and computation. These are combined to create 'pipelines' of bit transformations that result in addition and subtraction, multiplication and division, and so on.

At a high level, data processing that makes the world of commerce work by connecting different sources of data together with process components, passing data from location to location along adjacency channels, from input to output. Data tagged (+) (with a meaningful identity credentials) are recognized and selected (-) in order to filter out and distill information of particular interest. The need to match compatible semantic criteria in data pipelines goes all the way to the highest level, where *types* of information, e.g. numbers or text, have to be matched correctly when offered (+) and received (-) in order to make sure they lead to something meaningful. If a process were to add a number to a name, the result might be meaningless (indeed, many IT disasters have been the result of such mismatches between intent and actuality).

[101] China is the one country that tries to maintain a strict one-to-one link between its geographical border and its Internet border, treating both as political regions.

[102] Passports and identity cards virtualize boundaries. Members of organizations are still members when they leave the club. Even prisoners are now being tagged electronically instead of being kept behind a physical barrier. Managing distinctions is also about managing spacetime, because it is always spacetime processes that are the discriminators of information.

[103] A similar argument was made in the early 20th century about the luminiferous aether: a medium in which light could travel. Famous experiments like the Michaelson Morley experiment, which were used to confirm Einstein's view of matter and energy showed that there was no such aether in the sense imagined. However, no experiment can rule the idea of absolute space if it uses only the virtual processes that run on top of it. Absolute space is just that embedding space on the turtle's back that is so impossible to eradicate.

[104] Physics savvy readers may already be suspicious. This seems to violate Newton's third law, and conservation of momentum, energy, and so on. I'll return to this in later chapters. For now, we can see this as a mystery. Such one-way spaces can and do exist in nature, so we need to understand what this tells us about laws like Newton's Eroica.

[105] In physics, it became the accepted doctrine to pay one's deepest respects to invariances and symmetries of phenomena within regions of space, but to only mention the dirty business of boundary conditions (which break those lovely symmetries) if we actually have to solve an actual special case. The way this is presented is a bit dishonest. The obsession with beauty and symmetry may have blinded us to the simple truths.

[106] General relativity is derivable in essence from such an ordering assumption [18].

[107] Before promise theory there seemed to be no obvious distinction between dynamics and semantics in accident or fault investigations. This led to a muddled picture about causation, based on politics and causality denial.

[108] The second law of thermodynamics is somewhat reminiscent of Mach's principle, which claims that acceleration involves the positions of the fixed stars, i.e. a global state of the universe. The second law suggests that the entropy of the universe influences the direction of time on a local scale. This violates locality—but then there are plenty of concepts in physics that violate locality.

[109] See the excellent book by Judea Pearl on this topic [48].

[110] What lies beneath it has to be something else altogether—something that ejects the classical concepts of particle and momentum altogether for something related to pure information. The key concepts in canonical Quantum Theory are position and momentum, and are derived concepts. They rely on the continuum spacetime model to scale indefinitely, but this is what quantum theory itself rejects, so the assumption is inconsistent. It is therefore bizarre how discussions of the uncertainty principle are often used to define the necessary behaviour of

quantum theories of spacetime.

[111] Going back to the notion that energy might in fact be the substance of the message channel corresponding to adjacency, the structure is at least compatible with that view.

[112] There are two kinds of observer relativity in quantum theory: operator measurements as a model of 'wavefunction collapse' and Einsteinian relativity describing the experiences of different observers. These two operate on different scales. The incorporation or Einstein's relativity into quantum theory led the quantum field theory is a large scale, low energy effective approximation (divergent at high energies) to spacetime as a continuum. The operator measurement theory is a smaller scale, individual entity approximation for closed systems, e.g. the Hydrogen atom, which describes how uncertainties in space and time affect energy accounting.

In both cases, there are two sets of coordinates: one set that refers to measurements (operators and scattering boundaries) and one set that refers to a kind of configuration for energy exchange processes, In the field theory, interior energy process details, called 'loops', use a spacetime-like representation that imagines processes in a real space, but these processes are 'interior details', not exposed to the exterior of an interaction region, so their status as a model of spacetime is unclear. Feynman diagrams turn these interior processes into compelling visualizations that mimic a classical reality, but they are not usually representative of actual particles.

[113] Restricting positions and momenta followed from the need to make electrons occupy only allowed orbits, consistent with observations, but it led to the view of the interior state space for electrons as agents being distributed with respect to the coordinates for location of the continuum limit. The form of the boundary between interior and exterior took the form of *matter waves*: a distribution of energy states with respect to spatial locations that looked a lot like light. Matter waves in Quantum Mechanics are also called probability waves, because that interpretation seems to work to express the non-local range of possibilities for space and time in terms of the wavefunction.

[114] Asimov's positronic brains could possibly have referred to holes in conductors, but antimatter brains would be quite explosive.

[115] This situation is not really complete without some way of accounting for the particle-antiparticle pairs. It was by linking the Dirac theory to electromagnetic fields that it was all made consistent in the theory of Quantum Electrodynamics for which Feynman and others won the Nobel prize. For more details see [11].

[116] Computers can multitask, or share their time between tasks, but only one at a time per processor.

[117] There are fancy encodings of memory, such as RAID storage, that seem to overlap memory, but this doesn't increase the density of storage.

[118] Bosons are named after physicist Satyendra Nath Bose (1894-1974), who developed their theory together with Einstein. Fermions are named after physicist Enrico Fermi (1901-1954) who did important work on atomic energy and radioactivity, and named the 'exclusion principle' for particles now known as Fermions. Bosons and fermions are traditionally though misleadingly called 'particles'. Actually, the possible details of the kinds of particle depend on the number of spatial dimensions in which the particles are localized. In two dimensions there are more kinds of particle, called 'anyons' which can have any variety of topological type, but their behaviour

is otherwise similar to fermions in that they cannot occupy the same space at the same time.

[119] The theory of this scaling is the key to reconciling dynamical and semantic notions of scaling in Promise Theory. See reference [49] for its outline.

[120] See the description and references in [50].

[121] This interpretation was pointed out to me by my friend Paul Borrill, on reading my work on Promise Theory. As he was right to point out, there is an uncanny compatibility between the promise theoretic structure and the components of the transactional interpretation. I trace this back to the role of information channels. See the book [51] for a summary, exposition, and references.

[122] See a summary and references at [51]

[123] The probability cloud for a static system is just an equilibrium energy distribution, without exterior time. Interior time is irrelevant because it is effectively stopped by the scale of the equilibrium. Thus Quantum Mechanics has a clock that spans the whole of space, but which is stopped unless there are separated modular systems, in which one acts as receiver/observer.

[124] The so-called wavefunction of the universe, and the Wheeler-DeWitt equation are described, for example, in [52].

[125] There are several different notions of time in Quantum Mechanics. The time that appears in the time dependent Schrödinger equation is a different process than the time at which the wavefunction is sampled by measurement.

[126] This would be essentially the Hilbert Space vector.

[127] The wavefunction is normally considered to be a probability amplitude, whose square is a probability for localizing a measurement, but another way to look at it is for ψ to be the generalized probability for a source $P^{(+)}$, and ψ^{dagger} to represent the probability of the receiver $P^{(-)}$.

[128] Actually, we really mean that it is a product state. In the case of a symmetry between sender and receiver, the strict adjoint symmetry makes it a square modulus. If that symmetry is broken, as it may be, then it becomes a transition overlap function.

[129] See the discussion in [11].

[130] See, for instance, discussions in [37] or [53].

[131] One way to think of the Schrödinger equation is as a dimensional reduction processor that takes inputs as spacetime field values over parameters we call positions and times, and maps them to a discrete set of eigenvalues that represent invariant states of the symmetries residual in spacetime, once the wavefunction has broken the symmetry: $(x, t) \to E_n$. This makes it roughly analogous to a neural network.

[132] The wavefunction in Quantum Mechanics is based on distributions that are closed and conserve energy. A datacentre is a fundamentally non-conservative (dissipative) process, so there is no analogue of Schrödinger's equation, even though the Schrödinger equation has a character similar to a diffusion equation. That's not to say that we will not one day find a way to express processes at the virtual level in similar terms.

[133] This is presumed to be the Planck length of around 10^{-35} metres—some twenty five orders of magnitude. It turns out that one can combine three dimensional scales, believed to be universal constants into a value with the dimensions of length: the speed of light in a vacuum c, the gravitational coupling strength between any two bodies G, and Planck's modified constant

$\hbar$:

$$L = \sqrt{\frac{\hbar G}{c^3}} = 1.6 \times 10^{-35} \text{metres}.$$

[134] Calculating the uncertainty of tasks in computer science is known as estimating the 'makespan' of the task, see e.g. [54, 55]. Almost without exception there is no attempt to estimate uncertainty on a theoretical level. Some authors try to measure the uncertainty and even dismiss it as unimportant [56, 57]. Variations of up to 15 percent [58] have been measured—which is not that small.

[135] The Many Worlds interpretation has received a lot of popular attention. See the original paper [59], or the discussions in any number of opinioniations.

[136] The most intelligent and interesting rendition of the problem is championed by Rich Hickey in his talks [60, 61].

[137] The link between the transactional interpretation and our promise computational analogy is that everywhere we look there are channels of information short circuiting simple minded propagation of waves in the spacetime sense. There are multiple embeddings, so we need to be open minded about what processes are occurring rather than trying to imprint inappropriate rules invented for different cases.

[138] H. Andersson of BBC Panorama made an interesting documentary about how Facebook and other companies use this method to effectively sell advertising attention, by driving addiction to uncertainty, feeding dopamine hits to us through automated approval ratings.

[139] A natural question is: how does this work in physical spacetime? Why isn't there an explosion of information? The answer is that nature has a way of limiting the number of signals, even though they seem potentially to spread to all possible locations. This is one of the mysteries of quantum mechanics and the wavefunction.

[140] There is a longstanding question in astronomy about so-called dark matter and dark energy that theory predicts must exist to explain the observed movements of galaxies. To date no one knows why such material does not respond to queries.

[141] Magnifying lenses bend spread the light paths more evenly across the finite sized retinal cells in our eyes. Without a higher density of information, with no more paths to spread, then we would simply see a kind of pixellated image, as we do when trying to expand computer images.

[142] In fact, those who criticize the reliance on those pictures end up embedding alternative formulations in spacetime too. By the end of this book, readers might have changed their minds about the possible realism of Feynman diagrams. They may indeed be more fundamental than spacetime, in an Einsteinian sense.

[143] I published this discussion originally in [17].

[144] Virtual motion includes systems like Labelled Transition Systems and Petri Nets of computer science.

[145] Rods and cones, place cells, and others have adapted specially to service the detection of motion. Sensors are often designed, shaped, or honed, over their evolution, specifically to register particular spacetime phenomena. This seems to be what Ashby called homomorphic machines[62] in his work on cybernetics. The smart sensor 'designs' in our eyes favour the

recognition of patterns by virtue of their structural properties, and minimize the need for data analysis and post-processing of interpretation. For example, the ability to judge distance might be enhanced by a sensor array with constantly spaced pattern between locations, acting as a graduated measuring stick. Through this specific adaptation, sensors are made efficient as pattern discriminators, by learning that is cached by evolutionary selection, over longer timescales. There is evidence that this is true in the human senses: eyes and ear do not merely pass data to the brain but perform staged preprocessing and limited 'type' recognition, through the existence of specific cells for the measurement of motion. Research has further identified fabrics of place cells, grid cells, head direction cells, border cells for seeing navigation of spacetime, and even facial recognition regions, and so on[63, 64, 65, 66, 67].

[146] In physics, we tend to think that the brain is not special. It is just a process like any other, so the fact that it can produce illusions by approximation and limited information is not a surprise. In cognitive sciences, on the other hand, we become aware that our very thoughts and perceptions are entirely coloured by our brains' abilities (or inabilities). So we have a conundrum: we desperately want science to be objective and universally 'true' in some sense, but it is not clear that such a sense exists in any meaningful way. It is a challenge to all of science to try to construct experiments that refer as little as possible to human perceptions.

[147] This advanced view caused plenty of controversy over the twenty or so years[68, 69].

[148] That of course, is negative feedback. Positive feedback is also possible, where there is superposition of current and new, leading to a chain reaction that explodes throughout the whole system).

[149] A stack is formally known as a Last In First Out structure. It's like a pile of reminders, where you put new ones on top each time you are interrupted and pull from the top to finish each task in order.

[150] See Richard Dawkin's book, for instance [70].

[151] Actually, there are many definitions of entropy—all related, but different in application. See [71, 72].

[152] Physicists usually define entropy as the opposite of information, whereas Shannon defined it as I do in the text. This is just a question of framing. See the discussion at [72].

[153] You might also notice, in passing, that this law is another case of using non-local phenomena as a relative baseline to determine local properties, like Mach's principle.

[154] See any book on irreversible systems and non-linearity, e.g, [73, 74].

[155] It's a curious irony, because entropy, as originally described, is an equilibrium property, so it automatically determines spacelike hypersurfaces. In a simplistic view of mechanics, the only degree of freedom left to associate with entropy is thus time.

[156] Actually, this is a simplification that relies on the assumption of weak coupling. The bulk properties of a medium may decouple from the size of the medium provided the coupling is weak enough to neglect boundary effects and non-linearities, so one hopes for a linear medium at all costs. This linearity also comes back under the heading of superposition of signals.

[157] Actually, there are two kinds of velocity for waves: group and phase velocity, which represent different aspects of the process, and go a bit beyond the scope of this book.

[158] This inbuilt human observability has not been questioned, because it is key to how experiments are performed, and therefore the predictions that are relevant for us, but it will be

interesting to see whether this is a 'real' view of spacetime, or whether something more fundamental also exists, i.e. are there phenomena that do not depend on light for their measurement by us.

[159] Time critical means the ratio of the data processing time to the Time To Failure has to be close to zero.

[160] Yellow Pages is a trademark, so the analogy has not been without legal challenges. The idea is accurate though—you look up a function and get a possible provider.

[161] The Norwegian company Numascale tries to extend the reach of the chips and their local caches to extend locality, analogous to consensus technology.

[162] One might call this the giant lie of cloud computing. It may be neither cheaper than owning your own computer (because price depends on supply and demand, and it may also not be faster because there are many sources of latency beyond your control.

[163] The balancing of multiple criteria is a form of Nash equilibrium, in a game theoretical sense. See the discussion in [11].

[164] See microsharding at Facebook in [75].

[165] If you meet a wave without knowing its origin, then the uncertainty in the relative motion of the sources translates to an uncertainty in the frequency of the waves, not an uncertainty about the speed of the wave. You can suck up information faster than the source emits it, eventually running into the source, but you can't sample the information faster than the wave fluctuates, because it's virtual. So your queue length might be shorter or longer at speed. When the speed of the source approaches the speed of the wave, there is a shock wave, in which all the outstanding information gets compressed into a very short interval. For sound, this called a sonic boom. When motion approaches the speed of light in a medium, there is an analogue known as Cherenkov radiation.

[166] The theory of this scaling was worked out by Geoffrey West and colleagues, and was extended to other networks in cities to explain spacetime constraints on throughput of different processes. See [9, 8, 76, 25, 19].

[167] See the explanation in [11], or [44].

[168] See [77].

[169] Kleinrock did much of his PhD work under Claude Shannon, as mentor, and later wrote the two volume treatise on Queueing Theory [78].

[170] Paxos was the first discussion about information in spacetime, but it was not the last. The problem of scaling certain information in the dataverse continues to be a basic research question that seems to be far from solved, mainly because computer scientists will not take no for an answer.

[171] See the unpublished CERN preprint about Statistical Geometry [18].

[172] See [18].

[173] Some people in the age of social media appear to be entirely dormant until they receive a notification their social media account. They are powered by a kind of photosynthesis, when photos appear on their screens.

[174] In Special Relativity, there is the twins paradox. Identical twins are divided by a journey. One travels close to the speed of light, and returns. One seems to be older than the other due to the difference in rates of time. But naively the situation seems symmetrical. How can this be?

No one goes forwards or backwards in time, but the sampling rate is slowed, and the energy cost is the acceleration of gravitational distortion required. This, alone, should be a clear warning that the ideas about spacetime, acceleration, and gravity are inextricably linked, and that gravity is not simply a field of force that rides on top of spacetime, like the other forces. We expect its influence to bend the way the other forces behave. All transitions or clock ticks in matter (from subatomic interactions to biology) rely on these other forces. So at the scale of general relativity, the predictions made by Einstein about different rates of aging amount to literal distortions of all physical processes that measure time due to the underlying distortion of the theatre in which they occur. Is it the gravitational field that slows time, or is it the slowing of time that leads to the relativistic illusion of a field? Is gravity caused by mass, or is the mass an effective signature of the geometry of spacetime distribution? The Einstein equations have the form of a source response equation – that one is the source of the other, but which way around does it go? There is no delayed response unlike a response equation, so this suggests that the two are merely correlated aspects of the same phenomenon.

[175] This point seems important, indeed the very essence of Einsteinian relativity, yet is not true of all theories. String theory is formulated in an embedding space, for example, and the discrete nature is superposed through a kind of uncertainty relation for position. Reformulating string theory to observe the rules of interior relativity may not be possible. According to theorist and physics popularizer Lee Smolin[79]. However, this is exactly what the programme on which Promise Theory was based, and without any reference to Quantum Mechanics. The reliance on embedding spaces may make matters easier to picture but that is not an acceptable argument. Loop Quantum Gravity on the other hand builds a version of spacetime built on interior principles, apparently compatible with promise theory arguments, and is therefore capable of representing locality in a fundamental way.

[176] In the Weeping Angels episode of Dr Who, the episode turns this idea of time upside down by imaging statues that can only move when they are not observed. This makes a very frightening experience because it is so counter-intuitive.

[177] Nothing in physics contradicts this idea for the spacetime of our universe either, but we have currently no way of knowing how to probe that question. The kind relativity corresponding to Einstein's theory is thus what can be observed between the virtual locations that act as agents for the promises.

[178] This was discussed in the second of the papers on semantic spacetime [49], and was used extensively in the promise theory of money [5].

[179] In Einstein's theory of gravitation, the speed of light is assumed fixed, but gravity acts as a lens for motion, refocusing space around it into a rescaled region around a massive body. The presence of the lens effect imposes a scale distortion on the spacetime region. This means that events tend to queue up at that scale, so time seems to run slower than it would in free space, in order to process all the signals that are forced through its bottleneck at a fixed speed. How can time possibly run at a different rate? Well, it doesn't—not at the location itself—we only see it run at a different rate from a distance, because the spacetime processes that reveal what happens there run basically like a movie, preserving the order of queued events but slowed down.

[180] It was the German Hermann Minkowski who observed that space and time could be combined into a geometrical point of view, but the popularity of the message was no doubt a

result of Einstein's own popular appeal.

[181] The snapshots, in turn, could only be taken if time were globally a discrete series of records, like still frame pictures in a movie, with a single unifying clock—like the 'wavefunction of the universe'.

[182] This would be a kind of fake time travel, and yet there are theorists who spend an incredible amount of time of this issue, inventing folded spacetime topologies that defy the symmetries of a pointlike big bang. Cosmologists and relativists in physics speculate on the idea of time travel using wormholes, but there is some fudging going on—Einsteinian bridges between different times are also between different places, so they have no common time in any meaningful sense. You could take two twins and send them on different journeys, slowing down all processes for one along a shorter path, and meeting up with a normal speed process taking a longer route. The twins would have different ages because they experienced different processes. This is analogous to the branching and merging of timelines in version control systems of the dataverse, discussed in chapter 3.

[183] When scientists and technologists talk about entropy and information, it can be confusing, because physicists and computer scientists define information differently. Information in physics is about uniqueness of an experiment within an ensemble of experiments, so it measures the uniformity of statistical patterns across many episodes (external time). Information in Shannon's Theory of Communication is about uniformity of states within a single episode, thus it measures the uniformity of patterns within a single experiment (over internal time). Thus, information scientists would say that information is entropy, and why physicists say that is it the opposite of entropy. According to Schmitz, there are at least five interpretations of entropy [71] in science.

[184] The much over-mentioned and misrepresented reversibility of the laws of physics in no way makes this a trivial matter—even by trying to account for locality through bookkeeping tricks like summing 'new kinds of entropy' at event horizons. Moreover, there is deterministic chaos to contend with. Undoing would require perfect accuracy, which in turn requires perfect memory.

[185] The correlation between national 'modules' and timezone calibration is quite ad hoc, illustrating how absolute time is just a convention. China keeps over three normal timezones on the same clock across its huge area. Europe has three timezones, and the US has five (including Hawaii and Alaska).

[186] Modules may be atoms, cells, organs, houses, businesses, countries, etc.

[187] In the heart of Silicon Valley, Paul Borrill has been advocating the importance of a more realistic model of time for over a decade. Computer science is potentially overflowing with different ways to define and measure time, but these are mainly swept away by very simple ideas we currently use to put a wall clock onto our computers. When we want to remember when something happened, we record timestamps by looking at this wall clock. But these clocks rarely take into account relativity, in any form. My friend Paul Borrill has championed this idea [50].

[188] In [15] I I showed that this is effectively like dividing by zero.

[189] For a full description of this story, see [11].

[190] In topology this kind of multivalued time would be characterized by time modulo one week, and a winding number.

[191] Paul Steinhardt of Princeton in New Jersey, Neil Turok of the Perimeter Institute in Ontario, and Roger Penrose of the University of Oxford, have all envisioned ways in which the Universe might be renewed in fresh cycles of time. See [80] and [81] for a review.

[192] See [82, 83].

[193] Computer scientists might call this a global variable, or a mutable state, which is considered to be risky in computer programming.

[194] I recall being asked by a colleague to convert a spreadsheet made with Windows Millennium Edition (a one time data format that couldn't be read by later versions of the Office software). When you upgrade software, you might have to reformat the data encoding. This could be a continuous process. A creeping entropy of relative semantic changes.

[195] There are so-called versioning databases that can keep every version and every change, like git and Datomic, but these are not the usual state of affairs. With those databases, your closed world really can go back in time.

[196] The term 'bugs' originates from genuine bugs that could get inside the machinery of early computers leading to non-deterministic mixing of causal influence!

[197] There is a large literature, even to this day, on what it means to ask for a collection of agents to be able promise a consistent outcome when sampled. The issue simply boils down to what it means for samples to overlap at different locations and 'times', but there is seems to be an almost stubborn refusal to acknowledge the role of space and time in these matters, and so the story continues. See, for example, references [84, 38, 85].

[198] Programming languages try to account for this by freezing data coming in as a snapshot of time each time it's called up, but this is not enough for a consistent outcome, because the next time the function is used the result could be different. The only reliable information in a distributed setting is to define data processes such that they converge to a 'fixed point' by design. A fixed point is an attractor in the dynamical sense: a state that all states tend to aggregate to, with advanced boundary conditions. Then, order no longer matters, because you are going to eliminate it anyway at the fixed point.

[199] An example experiment showing the uncertainty in causal ordering was reported in [77].

[200] If a computation relies on information send over channels, perhaps from multiple store branches, or traffic information from different vehicles, the data arrive at different rates depending on the balance of give and take over the information channel. A channel that has reached its maximum allowed filling will not fill up further. Queued items may have to wait, if they contend with others and end up in the next longitudinal batch. In the mean time, the current data may be sampled and instigate new processes. How does a causally dependent process, a pipeline, know when its inputs are ready to be used? When is it safe to compute a function $f(x)$ of x, without being afraid that x will change 'in the meantime'? Modules and cells are used to help isolate processes from influences of this type. In a timeline, a single variable would ideally evolve in a predictable way, not appear as a random variable—but if you don't know the data are immune to change, then it is effectively a random variable, and you can't rely on its constancy.

[201] For data dissemination, where information becomes widely available, as with ambient light, the structure of spacetime adjacencies is key. The telephone network, international flight networks, and other point to point graph structures, the standard method for scaling 'reach' is to create a hierarchy as in figure 2.18(b), where we first send information to a regional hub and

then onwards from there. Regions are assumed to contain most traffic.

[202] See the story in [11].

[203] I'm grateful to Daniel Mezick for telling me about this term.

[204] There are, by now, numerous attempts to look for something fundamental about the laws of direction and flow. The second law of thermodynamics is often the head of this church, with its simplicity. But we attribute enormous semantic baggage to this simple principle. The Constructal Law, by Adrian Bejan (1948-) a form of thermodynamic argument which he calls Constructal Law...[86], claims that systems tend to evolve in such a way that their constraints eventually become trivially satisfied, or impotent over time. He uses the phrase 'access to flow', which implies a relationship between give and take. It's a kind of restatement of the second law of thermodynamics. It's a statement about the emergent configurations expressed by a dynamical system. It attributes to them a form of semantics 'unmistakable appearances' and relates these to the concept of design, in a universal meaning. It's a bold attempt to reconcile semantics entirely as a side effect of dynamics. In my opinion, it does not feel convincing. It attempts to give a mesoscopic interpretation to the arrow of time. It refers to concepts like fitness, also used by Kaufmann, Prigogine and others[87, 88].

[205] See the discussion in [11].

[206] Actually, more specifically it involves the inertial mass. Mass also plays a secondary role as a source of gravitation. The masses are found to be numerically equal, but have very different semantics.

[207] In computing, it is not the case that information continues to travel. It has to be emitted and absorbed, more like light from an atom. Routers absorb and relay. So spacetime points must also absorb and relay in some sense. That makes their promised behaviour very different from matter, which acts like a singularity. When one mass strikes another, momentum is exchanged. In Promise Theory, this tells us that one body promises a property which is recognized by the other, much like charge.

$$S \xrightarrow{+p} R \qquad (10.1)$$

$$R \xrightarrow{-p} S \qquad (10.2)$$

We do not write this with a plus and a minus sign in physics, but in PT electromagnetism and mass do look more alike. When we write that force is a rate of change of momentum:

$$F = \frac{dp}{dt} = \frac{d(mv)}{dt}, \qquad (10.3)$$

the momentum p is really a field, and thus the mass is also a field. This interaction process involves one mass striking another mass, so it is the combined non-local region that acts irreducibly for a moment. This is very like the concept of entanglement in quantum theory. If readers know this equation, then they also know that the implication is that mass is not necessarily constant. Rockets work by pushing mass away from them, and moving forward as a result of the backreaction.

[208] Theoretical physics has effectively reinvented such substrates in the scalar fields of the Higgs mechanism, for instance. These are exactly what we need to model mass in the standard model.

[209] The DNS root name servers translate Internet names into IP addresses. These root servers are a network of hundreds of servers in countries around the world. However, together they are identified as 13 named superagent servers, labelled A-M, in the DNS root zone.

[210] This is not entirely true, as there is the possibility of gravitational waves, which are incredibly weak.

[211] The MHC or Major Histocompatibility Complex is a protein signature that marks cells, for instance, defining their recognizable type.

[212] The implication is that there would not be independent 'gravitons' that carry the gravitational field as a force, different from spacetime itself, but that spacetime adjacency is the binding that already adjusts self-consistently with non-linear feedback, because it's a many body equilibrium problem. Call that the graviton if you like! Attempts to 'quantize gravity' using the concepts of canonical quantization, with generalized positions and momenta feel like a misplaced application of procedure without due consideration. Gravity seems qualitatively different, its granularity likely on an entirely different scale. It has no local charge. Or, turning it around, should we really view the other charges as being more like mass, and be looking for the underlying clustering agencies that lead to distortions of spacetime?

[213] The lack of trust in banks after the financial crisis of 2008 led to the invention of cryptocurrencies based on a distributed consensus process called a 'blockchain'. Those involved in blockchain technologies, like Bitcoin and Ethereum and others, like to claim that by involving many voting members in every decision, there is no need to trust anyone anymore—as long as there is an acceptable quorum. This is quite misleading though—now you have to trust everyone to a lesser degree for the decisions, but you also have to trust the software that makes it work. You also decide what kind of quorum is acceptable. The same possibilities for distortion exist. Indeed, the major distortion is that the formation of a single serial channel of change—which also implies a single notion of time—over a wide area means that the equilibration time for the network is much longer and payments are very very slow.

[214] While still controversial, there seems to be a growing acceptance of the communication channels between plant organisms [89].

[215] The mathematical basis for this biological phenomenon was established by Alan Turing in his 1953 paper 'The chemical basis of morphogenesis' [90]

[216] See the discussions in [11, 33, 34, 35, 36].

[217] John Horton Conway's Game of Life, in 1970 was published in Scientific American 1970 as a popular article [91]. It can simulate complex emergent pattern generation, and display a rich set of behaviours including fixed points and limit cycles. Automata are used as models for cellular life, robotics and even studying material properties: for example, there is a simple sense in which the propagation of a crack through a solid could be viewed as a straightforward computation in a cellular automaton. Hårek Haugerud, Are Strandlie, and I used them to show vortex phase transitions in magnetic crystals. See the article [92].

[218] The literature is effectively unbounded, see for instance [93, 94, 95, 96, 97].

[219] For an excellent historical account, see [98].

[220] As a student, I preferred the abstract beauty over the mechanical view of Feynman diagrams, but as the years pass I find much virtue in Feynman's initial intuitions.

[221] The history of this name goes back to Fermat's discussion of light and the path it takes.

His idea that light always takes the shortest path, i.e. that it travels in straight lines is an early expression of the philosophical conundrum we are still exploring. Today we assume the existence of straight line motion, as part of spacetime.

[222] The action is an integral over a function called the Lagrangian, which is formally the difference of the kinetic and potential energy—at least in classical mechanics. In field theory, its form is based more on invariant algebraic forms, constructed from the variables of the problem at hand, and the argument about energy interpretations is secondary to the assumption of the method. When we use the action principle, we basically promise the primacy of continuity and symmetry as generators of expected behaviour[10].

[223] We write down a differential invariant form, and attempt to vary it with respect to different parameters. By asking when the variation has no effect, we determine the conditions for invariance or conservation of key quantities based on the field[10]

[224] Some early considerations on this topic were expounded by S. Harnad[99] in 1999, but he built on the classic philosophical arguments by John Searle, which were ill-equipped to understand concepts like scaling.

[225] See the discussion in my book In Search of Certainty [11], and [82, 83] for overviews.

[226] See the discussion in my book In Search of Certainty [11], for a longer explanation.

[227] I feel that the role of logic is quite widely misunderstood in reasoning: something like confusing chemistry with the periodic table.

[228] See [7].

[229] In 2013, as I was writing *In Search of Certainty*, an idea which had been lingering in my inner own inner network for a while, finally crystallized into sufficient clarity to try to explain it in terms of Promise Theory. This led me to draft a series of papers called Semantic Spacetime I, II, and III. See references [17, 49, 4]. It took several hundred pages of painstaking definitions and descriptions just to sketch out the outlines of a hypothesis, but I think it was worth the effort. Along the way, I rediscovered plausible explanations for a few aspects of artificial reasoning that had been taken on faith without theoretical support. It led to some testable ideas, though testing proved harder than I had hoped.

[230] See the extensive discussion in [11].

[231] See reference [100].

[232] See the discussion in [4].

[233] If we take two persons, there is no symmetry. But if we take an entire country, there may be. It's normal that when locals perceive people from foreign places as a group (e.g. as a race or a country, etc) they perceive them as all looking alike, because they are focused on the similarities rather than the differences. The observation is sometimes used as a racial slur, but it is based in simple scaling law.

[234] The result is something analogous to the optical experiments used by astronomers, passing light through multiple lenses and diffraction gratings.

[235] Perceptual manifolds arise when a neural population responds to an ensemble of sensory signals, on the same object. They are forms of invariant classification[101].

[236] The effect of effective weighted averaging result might be something analogous to a Principal Component Analysis, leading to a kind of statistical sorting.

[237] See the discussion in [102]. The status of these studies is currently under scrutiny, at the

time of writing, due to the discovery of software bugs going back several years, that may be showing false positive results.

[238] See the discussion in [4, 7].

[239] Once a cognitive system, approximating a thinking mind, has integrated enough concepts, and blended them into enough contexts, it is possible for an agent to 'just see' ideas in a flash of complex associations. This is not like a linear computer process, even if it could be simulated by one. Moreover, it's not the same as being able to explain those flashes, or be able to tell a story about them. The latter is a very different process, like the one we try to emulate and sanitize through formal systems like logic.

[240] This is the basis for taxonomy by the idea of 'prototypes'[103]. In the somewhat sidelined field of cognitive linguistics, concepts are treated as 'semantic units', (i.e. signal agents in PT parlance) that exist independently of language[103]. Although the concepts of cognitive linguistics are not well developed in a formal sense (even by their own criteria), they seem to be broadly compatible with a promise theory view in which concepts act as spanning sets for words in contexts. This is the view I take here, based on experiences with systems of agents in CFEngine[104], etc. Langacker is vague on what constitutes the spanning sets, but indicates that the following general containers are involved in classifying concepts: space, size, shape, orientation, function (container, force, motion, role), material nature or composition, and also economic issues like cost. These regions are called 'spaces', and have a close resemblance to the notion of namespaces[49] as containers of concepts, analogous to directories of files on a computer. In Langacker's view, metaphors like in the associative overlap between conceptual classifications. Langacker riles against the notion that concepts must be discrete structures, presumably in protest against the discrete mutually exclusive treelike structures of generative linguistics[105], suggesting instead that concepts might overlap; however, there is no contradiction between discreteness of superagency and the possibility of overlap. There is a superficial similarity in structure between Cognitive Grammar[103] and the spacetime view taken in this book. It is plausible that this spacetime view could provide a more rigorous notion of conceptualization that is compatible with it.

[241] There is apparently some evidence for this from research done at Carnegie Mellon University and the University of Pittsburgh showing how generally short term context connects similar long term concepts. So when a tennis player plays squash, the similarity of context fires up memories and concepts relating to tennis first, biasing the outcome. It's hard to lose old habits because context addressing is imperfectly segregated.[106].

[242] See the discussion in [7] for one version of this. Neuroscientist and computer scientist Jeff Hawkins has a representation of a semantic network in terms of low level bits, on a substrate that looks like an Artificial Neural Network, based on his work at Numenta[107]. Apart from the encoding method, his model of Sparse Distributed Representations and Hierarchical Temporal Memory agree well with the model I describe in this chapter.

[243] Recommendations by one member of a group get passed on to others until a group also represents an interest in that topic, by epidemic transmission[108, 109]. Members who are not interested in the mentioned topic will fall away or start new groups. Thus overlapping memory locations cluster around topics to which they feel an affinity, forming concepts. Their affinity depends on a context in the wider world that brings about the topics in question. Parts of the

concept cluster may come at the topic from different angles, triggered by different atoms of context during a concurrent activation.

[244] I first used this story in a magazine series for USENIX ;login: magazine in July 2001, Volume 26, Number 4.

[245] Other representations of knowledge, like the RDF semantic web, may try to represent this differently, but will lead to an explosion of new types and rules for combining them.

[246] For a beautiful introduction to Jeff Hawkins' model, see [107]. Although it leaves many questions unanswered, especially with regard to scaling, this provides a much deeper dive and concrete mechanism than I can provide here.

[247] There might be several channels to language, like body language, inflections, etc, but the dynamics of transmission from sender to receiver are essentially one dimensional, since one could easily interleave the streams without any advanced structural encoding, such as a metagrammar. That is not to say that language is purely one dimensional. Its fractal dimension is doubtless higher than one due to the presence of parenthetic remarks and inflections.

[248] It is practically a cliché from historical literature that devout believers would lift their sights to the heavens and cry out to God: 'Give me a sign!' The sign would not be a banner in the sky, or a voice from the clouds, but something more mysterious. Whether a swaying tree, a strange encounter, or a bolt of lightning, the ability to attach contextual significance to an event is seen as symbol, like an obelisk in the desert. Any spacetime event that stands out against its background in an identifiable way can be used to convey meaning or nonsense, depending on its context.

[249] See [110].

[250] The technical term for this comprehension is parsing, and the computational processes were different levels of transition system, from finite state machines all the way up to Turing complete systems.

[251] Chomsky is also known for the more controversial idea that the structure of language is an innate capability of the human brain, but that's not of interest here. Although there are structures in our brains that facilitate language, it's almost certainly a multiscale learning process, whose memory is encoded in genes, society, and individuals.

[252] Chomsky showed that grammars are associated with transition systems operating on discrete linear patterns[110]; from this we have a theory of signalling and data encoding[111] that goes beyond Shannon's basic information theory[47].

[253] See the lucid exposition by Pinker in [112], as well as critiques in [103], and an endless variety of other popular works.

[254] The evidence for the evolution of these basic spacetime concepts has advanced in recent years and even won a Nobel prize in 2014, see [64, 65, 66].

[255] This process is described in the excellent book [113].

[256] See the fine essay in [114].

[257] See the tale in [46].

[258] For the most convincing account of this, see [115].

[259] The field of cognitive linguistics deserves a mention here, as it invokes many of these ideas, albeit in a completely different language and mode of reasoning! See [103, 116].

[260] This, in fact, is what I've tried intermittently to do since 2006. It turns out to be a non-trivial

challenge, but not an impossible one, and the entire sequence from end to end requires a lot of pieces to fall into place. Today, I would say that the status of the model is 'very plausible' but far from validated.

[261] Deep learning is a separate kind of learning that corresponds to evolutionary learning for the adaptation of hardware. It shouldn't be confused with the cognitive learning that leads to reasoning.

[262] Some AI researchers now dismiss expert systems (knowledge bases with helpful recognition features) as not true AI, but I think this is wrong. Expert systems are as much a part of learning and reasoning as neural nets, for instance. They operate on a different scale, with different optimizations, and more akin to semantic networks that I'll discuss later. I believe they play a central role, on a completely different scale than neural networks.

[263] Semantic maps are not new, of course—others have developed technologies for engineering 'Topic Maps' and 'RDF semantic webs', for descriptive ontologies, but neither of those approaches can be made to work without manual curation by humans, because they rely on already meaningful relationships being known.

[264] Perhaps the connection between these process interpretations may eventually be understood from studying the relationship between path integrals and semantic graphs [117].

[265] The number of euphemisms for toilet related concepts is quite fascinating, and embody all manner of historical references: spend a penny, pull the chain, etc.

[266] One approach is to try to abandon linguistic meaning and use some form of unique hash. This approach is used in natural language processing[118] with some claimed success. However, hashing is a one-way transformation, which is unsuitable for our associative map. To create an interactive system of knowledge, we need to be able to read back concepts in a form parsable by humans.

[267] Readers who studied maths in school probably know of Venn diagrams.

[268] There is such a wealth of literature on this topic, but I like these references [113, 119, 112, 120].

[269] In biology, trauma is seen by cell necrosis, apoptosis, floods of signal cells—including hormones, antigen presenting cells as part of immune responses, and so on. The signals are picked up by a variety of other processes like lymph nodes that may produce antibodies, or macrophage garbage collecting cells. In computer science, trauma is based on measurements of system behaviour—how hard the CPU is working, how much memory is busy or available, how many messages are being received from neighbouring processes. In physics, nothing is automated, but we can describe phenomena quite well. There is the concept of a catastrophe, which is a sudden discontinuous change (one that violates Noether's theorem for instance). This involves an underlying concept of spacetime, so we can identify it with a boundary or the insertion of information.

[270] Every since I was a child, I've had a habit of going dizzy to the point of fainting when I stand up too quickly. When that happens, I sometimes experience varying degrees of loss of consciousness, from almost complete to light-headed. I've learned a lot from this somewhat inconvenient habit though. First of all, it's not high level reasoning that fails, reducing me to lesser animal instincts. The first thing to go is balance—a profound dizziness—and then motor functions and vision. Muscles begin to flap intermittently, and I'm afraid I'll drop what I'm

holding and fall over. Then as my senses fail one by one, I am still observing what's going on and thinking about it. In a bad case, there are also flutters of consciousness. I know this because time seems to jump in small discontinuous gaps. So my storytelling organ survives seems better protected from these episodes than my motor or sensory functions.

[271] It's as if our brains are filled with thoughts accreted from floating detritus that exterior spacetime washed up onto our senses. Perhaps that's taking it too far. It makes sense for there to be an element of unknown in cognition, but there has to be some order to the process of memorization too. We exploit that when we want to know the reason for events. For some years, I've believed—based on work initially done with colleague Alva Couch or Tufts University—that explanations are simply causally oriented stories we tell until we dare to stop—that there is no *a priori* rational or logical basis to explanation, unless we impose one as a discipline.

[272] Author Nessa Carey wrote an excellent overview of a change that is sweeping through biology in her book *The Epigenetics Revolution* [121].

[273] Social insects have developed methylation coding in different ways. Male and female DNA chromosomes haploid, are necessary for a successful reproduction. The degree of methylation seems to indicate semantics for gender. Female DNA is heavily methylated, while male DNA is stripped of its own during conception. Inherited attributes are thus passed on along gender lines. See [121] page 126 and p289 honeybee memory.

[274] See the account of this work in [46].

[275] Shannon did not make any such observation, but nevertheless came up with an idea which is familiar in physics of materials too as part of his study of connectivity in genetic processes.

[276] See [122, 123, 48, 124].

[277] Before stumbling onto this work in epigenetics, I had already arrived at a similar conclusion about negative switching in a quite different context: knowledge representation for artificial reasoning.

[278] See [65, 66, 64, 63].

[279] Interesting work in the study of chanting or texts in Sanskrit and tribal stories and its relation to neuroscience help to make this plausible[125].

[280] The idea that consciousness is a result of a super-bus communication channel has been expounded in [126].

[281] See the citation [89].

[282] See the mathematical models recently proposed in [127].

[283] This fulfilling Steve Hillage's prediction, from his wonderful if flower-powered album *Green*.

[284] Because the coordinates are interior virtual coordinates within a path integral process, summed over all possible values, not exterior spacetime coordinates.

[285] The three waves were discussed in detail by Alvin Toffler [128, 129, 130].

[286] See [131, 132].

[287] See [11] for a discussion of dimensionless ratios.

[288] See [11].

[289] Each filter is a process with its own interior time, because measurements are not atomic. The inputs are 'crowdsourced' on some kind of scale that relates to the way the information channel is defined.

[290] We can compare this to the negative mutations, such as the paving over of gardens and grassy patches that drain rainfall into the water table, in urban areas of the West. To save the work of gardening, or to park a surplus of cars in driveways, Westerners have paved over large areas of urban land in cities. During heavy rains, this has led to flash floods that have caused millions worth of damage.

[291] See the excellent story of this work in [8, 9].

[292] Again, see the excellent story of this work in [8, 9].

[293] Networks that fill their embedding spaces by having a significant density of points led to the idea that the number of dimensions could be fractional. For example, a straight piece of string is one dimensional, but if you fold a string back and forth on a sheet of paper, in a zig-zag pattern, then it covers a two dimensional planar region. So isn't it then two dimensional? And if it only winds a little bit, isn't it then somewhere between one and two dimensional? Curves and networks that seem to fill the gaps in their embedding spaces, according to measurements of the embedding, are called *fractals*. Fractal structures seem ubiquitous in nature, leading some scientists to believe that they are deep. Mathematically, the notion of fractional dimensions is another example of a concept that is based on a continuum approximation to a discrete problem. Instead of dealing with networks at a fundamental level, it is sometimes easier to think of an embedding space with an effective dimensional parameter if you are trying to count how much you can fit into the space. Space-filling fractals express the way in which a network can be viewed as filling a container, without having to admit to the fundamental lower dimensional restriction to the network's interior. Fractals are a veil of abstraction that mathematics throws over the complex details of reality in order to unify and classify a variety of cases on a simple scale. The fractional dimension is known as the *Hausdorff dimension*. In a sense the usefulness of the method is a testimony to the idea that it is the effective counting of *dynamical activity* that governs the true dimension of processes, not the dimension of a theatrical frame in which the activity plays out.

[294] See the remarks and arguments in [19].

[295] See the Wired article [133].

[296] I wrote about this in a blog on my website markburgess.org, called the Cyborg Compulsion, in 2015.

[297] This argument goes back to the 1990s and pervasive computing, and my work with CFEngine [104], but I originally proposed this in a formal document in 2015 with some colleagues, see [134, 135].

[298] In the past, the rich or landed gentry were named according to the lands they owned, the family names of working people often reflected their occupations: Smiths, Carpenters, Coopers, even Burgesses—pick a surname. Who knows, in the future, perhaps someone will be named Rachel Dnslookup, or Simon Microservice.

[299] This has been a hobby horse for me ever since I began to teach computing 20 years ago. Universities have done little to expand the scope of relevant teaching

[300] Application Programming Interfaces (API) are parameter sets that mediate communication channel semantics between processes.

[301] Groups and ErrorGroups in Go offer different interfaces to do this, but the essence is the same: wrap a number of processes in a superagent boundary, combining their promised outputs

(+), and define the criterion for waiting until accepting a completed outcome (-).

[302] For a group of programs of size N, the effective mass grows as much as N^2 because of the connectivity of the network. As the load gets bigger, more processes are needed to cope with it.

[303] Conventional computer communication is 'unreliable', meaning that they make no assurances of delivery[136, 137]. Higher level constructs, such as TCP reconstruct reliability with acknowledgment circuits, but the application streaming abstraction limits one's ability to reason about failures, and pushes recovery to the application layer: if a packet does not arrive within a certain time, one is not sure why or what to expect. In telephony, lower layer protocols like the ITU's Public Network Signalling SS7 protocol, MTP layers 2 and 3 offer lower level reliability and recovery quite similar to the scheme described here[138]. Of the many others, we mention ATM[139], Virtual Synchrony[140], Distributed State Machine Replication[141], etc. However, these protocols work atop opportunistic signalling mechanisms, and lack the ability to say precisely when a packet was delivered, from one end to another, at each leg of a journey.

[304] The well known Fischer, Lynch and Patterson (FLP) result[142], which details how consensus is impossible in an asynchronous system if agents only *might* be unreliable, is one of the most discussed issues in system reliability. Consensus protocols[38, 39, 143] make a variety of promises concerning transactional ordering and versioning races.

[305] The CAP conjecture—sometimes called a theorem, but unproven. See the discussion in [84].

[306] See the definition of entanglement, according to Paul Borrill in [50].

[307] See the summary in these blogs[144, 145, 146].

[308] The only work on the scaling of graphs that I know is in [49].

[309] These are called things like 'function as a service' or 'serverless', such as Amazon's Lambda service.

[310] Perhaps there is a lesson there for physics too: witness all the processes that have to come together to create a state of virtual uniformity. It could give some clue as to the number of layers of process between apparent Euclidean space and something discrete.

[311] Some have seen changes afoot [147, 128, 129, 130].

[312] See the discussions in [11, 148].

[313] The self, non-self distinction was also used as an explanation of immunology too until the Danger Theory challenged it[149].

[314] See [150, 151] and my discussion in [11].

[315] The literature on this topic is unlimited. See, for instance, the discussions in [11, 152, 153].

[316] This part is based on the notes I originally wrote and published on my website. More details can be found in [154, 155, 156, 157].

[317] The history of money is long and involved[158, 154, 159, 160, 161, 162, 163]. It is a study in technologies for exchange. From the 19th century, economics was developed alongside (and often in the image of) the physics [164]. It is surely no accident that Newton worked at the Royal Mint. Economics borrows freely of the concepts and language of continuous causal transfer, like the Newtonian mechanics, which predates the modern non-deterministic understanding of systems. The language of differential calculus is thus ubiquitous, yet this already conceals significant assumptions about scale, continuity, and the smoothness of change. The differentiable picture, with infinite accuracy, contrasts with the day to say transactional

nature of the economy, where certain cashflow payments may be blocked for want of sufficient funds: money is everywhere discrete at the scale of human concerns, and where numbers are rounded to the smallest convenient monetary unit. We also note, in passing, that non-linearities in any economic network may potentially lead to the uncontrolled amplification of these approximations.

[318] As far as I can tell, these confusions were identified many times in the history of money. See, for instance, the terse if not turgid philosophy by Simmel [165].

[319] This is a large topic, not for this book, but I recommend the discussions in [166, 167, 168].

[320] Minsky was one of the first to develop the ideas behind a dynamical and properly scaled model of macroeconomic processes, see [169, 170, 171, 166, 167].

[321] See the discussions in [5, 11] about how the Axelrod and Dunbar views of relationship play into the definition of economic value.

[322] See the discussion in [5, 155, 156]

[323] A simple Promise Theory of money clear up these problems of nomenclature, in particular to separate the agents from the promises and observations: money is fundamentally about information channels—see [5, 155]. Is it information, a phantom of value, or something else entirely? Standard lore suggests that money, goods, capital, and the other denizens of economic theory have more than just one role to play. For some excellent historical reference, see [163, 161].

[324] See the delivery problem in [30]).

[325] See the great history in [163, 158].

[326] The view most of us have been handed down about personal wealth is that it is a product of work. This belief persists in the developed world, in spite of plenty of evidence to the contrary, and seems to be a hangover from the largely bygone era of subsistence living. Although working for money brings transfers some regular amounts of money from employer to employee, or from customer to seller, modern studies have shown that, in the developed world, at least, significant wealth cannot easily be earned as the product of individual work, rather it is accumulated as 'capital' over generations, often fortuitously, passed on through family dynasties by inheritance, and amplified by rent collection[172]. Thus the ability to grow one's wealth significantly depends on what 'rentable' assets one has. Such rentable assets are what are known as 'capital'.

[327] See the works of Karl Marx if you dare, or take a simpler introduction, perhaps [168].

[328] See for instance [166, 167, 173, 168, 174].

[329] Microeconomics is the 'kinetic theory' of the economy, ballistic and transactional, with payments for goods and services firing back and forth like billiards or particle physics. The analogy to physics is a matter of trivial accounting, simple and measurable. But what about the macroeconomy? The flows of money are sometimes treated as simple physical analogies in macroeconomics. As the economic studies developed, during the industrial revolution, its philosophers went to some lengths to model the capitalist macroeconomy on the successes of the macroscopic physics of the day, namely the equilibrium thermodynamics of heat engines. This was an intentional design, based on 'physics envy'. For an excellent historical review, see [164].

[330] We could still give away goods for free to survive. It's really military power—the threat of violence—that keeps the monetary system working fairly. Whoever wields that power can

force anyone to absolve others of debt, and start again, or remove the problem by killing off the debtors. If that sounds shocking, in a modern economy, it shouldn't. Killing and money have gone hand in hand throughout history.

[331] It was during the twentieth century that the role of *indeterminism* or *uncertainty* was given proper credence, in physics, but this did not percolate into economics or a long time. Keynes was apparently the first to point to the need to take uncertainty and indeterminism seriously[169, 170], but his work was only partially heeded and later misrepresented for decades[171].

[332] See the extensive arguments and data in [172].

[333] The role of money as a memory system was explained long ago by Hart in his excellent book [158].

[334] We know this from the uncertainty relation in Quantum Mechanics too. Instead of thinking about this as some kind of quantum weirdness, we can also see it as an inability to sample states accurately when there is a non-local spacetime distribution.

[335] The story of this religious inspiration is remarkable in its detail. See the fascinating book [175].

[336] Again, see [175].

[337] Our modern term 'quarantine' refers to the period of 40 days that foreigners would have to spend isolated to prevent the transmission of disease, are modular and functional. Venice may have been the very first smart city.

[338] Generalized clustering techniques, like Principal Component Analysis[176, 177] can find conspicuous patterns, but cannot attach significance to them.

[339] This quote adapted from the English translation of The Prince.

[340] I wrote about this in my novel Slogans, about how technology makes us stupid through complacency[178].

[341] You might have a certain preference for calling these aspects of our world material or immaterial. In fact the whole description of space as a measure of distance, defined through edges, containers, boundaries, and so on, seems to have been displaced, in modern science, by the ritual worship of symmetry and invariance of the empty stuff. The alternative view of matter and space, could easily have been that matter is simply an excited state of a spacetime fabric, like a coat of paint. But, in our minds, we have already de-unified them.

[342] I have known this quote for years, but am unable to find the source of its original usage. Needless to say, Whitehead expressed similar sentiments throughout many of his writings.

[343] In *In Search of Certainty*[11] I explained how probabilities and other statistical results are properly interpreted as being analogous to other dimensionless ratios in the theory of scaling.

[344] What the Michaelson-Morley experiment looked for was a form of fluidic space in which matter floated, like boats on water. However, nothing in those experiments excludes the possibility of a world in which matter is an excited state of an absolute space. In some ways, this is a picture that we regularly assume in modern physics.

References

[1] L. Mlodinow. *Euclid's Window: The story of geometry from parallel lines to hyperspace.* Penguin, 2001.

[2] P.C.W. Davies. *Space and Time in the Modern Universe.* Cambridge, 1977.

[3] A. Einstein. *Relativity, The Special and General Theory.* Methuen & Co., 1920.

[4] M. Burgess. Spacetimes with semantics (iii): The structure of functional knowledge representation and artificial reasoning. *http://arxiv.org/abs/1608.02193*, 2016.

[5] J.A. Bergstra and M. Burgess. Money and ownership as an application of promise theory. Published online at markburgess.org.

[6] N. Calder. *Einstein's Universe.* BBC books, 1979.

[7] M. Burgess. A spacetime approach to generalized cognitive reasoning in multi-scale learning. *https://arxiv.org/abs/1702.04638*, 2017.

[8] G. West. *Scale: The Universal Laws of Life and Death in Organisms, Cities and Companies.* W&N, 2017.

[9] G.B. West. The origin of universal scaling laws in biology. *Physica A*, 263:104–113, 1999.

[10] M. Burgess. *Classical Covariant Fields.* Cambridge University Press, Cambridge, 2002.

[11] M. Burgess. *In Search of Certainty: the science of our information infrastructure.* Xtaxis Press, 2013.

[12] J. B. Harley and D. Woodward, editors. *THe History of Cartography, Vols 1,2,3*. University of Chicago Press, 2011.

[13] G. Land. Ancient maps: How did the romans see the world. *HistoryHIT*, July 2018.

[14] J. Biamonte and V. Bergholm. Quantum tensor networks in a nutshell. *arXiv:1708.00006 [quant-ph]*, 2017.

[15] M. Burgess and A. Couch. On system rollback and totalized fields: An algebraic approach to system change. *J. Log. Algebr. Program.*, 80(8):427–443, 2011.

[16] K. Pardo, M. Fishbach, D.E. Holz, and D.N. Spergel. Limits on the number of spacetime dimensions from gw170817. *Journal of Cosmology and Astroparticle Physics*, 2018(07):048, 2018.

[17] M. Burgess. Spacetimes with semantics (i): Notes on theory and formalism.
http://arxiv.org/abs/1411.5563, 2014.

[18] J. Myrheim. Statistical geometry. CERN preprint TH.2538, August 1978.

[19] M. Burgess. On the scaling of functional spaces, from smart cities to cloud computing. *arXiv:1602.06091 [cs.CY]*, 2016.

[20] L.F. Abbott. *Acta Physica Polonica*, B13:33, 1992.

[21] R. Milner. *The space and motion of communicating agents*. Cambridge, 2009.

[22] D.B. West. *Introduction to Graph Theory (2nd Edition)*. (Prencctice Hall, Upper Saddle River), 2001.

[23] C. Berge. *The Theory of Graphs*. (Dover, New York), 2001.

[24] D.J. Watts. *Small Worlds*. (Princeton University Press, Princeton), 1999.

[25] L.M.A. Bettencourt. The origins of scaling in cities (with supplements). *Science*, 340:1438–1441, 2013.

[26] K. Hartnett. Universal method to sort complex information found. *QuantaMagazine*, 13 August 2018.

[27] B.C. Pierce. *Basic Category Theory for Computer Scientists*. MIT Press, 1991.

[28] F.W. Lawvere and S.H. Schanuel. *Conceptual Mathematics: A First Introduction to Categories*. Cambridge, 1991,1997.

[29] B. Fong anf D.I. Spivak. *Seven Sketches in Compositionality: An Invitation to Applied Category Theory*. arXiv:1803.05316v3 [math.CT], 2018.

[30] J.A. Bergstra and M. Burgess. *Promise Theory: Principles and Applications*. $\chi tAxis$ Press, 2014.

[31] L. Smolin. *The Trouble with Physics*. Penguin, 2006.

[32] S. Hossenfelder. *Lost in Math: How Beauty Leads Physics Astray*. Basic Books, 2018.

[33] J. von Neumann. The general and logical theory of automata. *Reprinted in vol 5 of his Collected Works (Oxford, Pergamon)*, 1948.

[34] J. von Neumann. Probabiltistic logics and the synthesis of reliable organisms from unreliable components. *Reprinted in vol 5 of his Collected Works*, 1952.

[35] S. Wolfram. *A New Kind of Science*. Wolfram Media, 2002.

[36] S. Wolfram. Statistical mechanics of cellular automata. *Rev. Mod. Physics*, 55:601, 1983.

[37] M. Burgess. *A Treatise On Systems Volume I: Analytical Description Of Human-Information Networks*. $\chi tAxis$ Press, 2017-.

[38] Leslie Lamport. Paxos Made Simple. *SIGACT News*, 32(4):51–58, December 2001.

[39] Diego Ongaro and John Ousterhout. In search of an understandable consensus algorithm. In *Proceedings of the 2014 USENIX Conference on USENIX Annual Technical Conference*, USENIX ATC'14, pages 305–320, Berkeley, CA, USA, 2014. USENIX Association.

[40] F.P. Junqueira, B.C. Reed, and M. Serafini. Zab: High-performance broadcast for primary backup systems. *IEEE/IFIP 41st International Conference on Dependable Systems and Networks*, pages 245–256, 2011.

[41] J. Schwinger. Theory of quantized fields i. *Physical Review*, 82:914, 1951.

[42] J. Schwinger. Theory of quantized fields ii. *Physical Review*, 91:713, 1953.

[43] M.E.J. Newman. The structure and function of complex networks. *SIAM Review*, 45:167–256, 2003.

[44] R.P. Feynamn. Space-time approach to quantum electrodynamics. *Physical Review*, 76:769, 1949.

[45] W. Pauli and F. Villars. On the invariant regularization in relativistic quantum theory. *Reviews of Modern Physics*, 21:434, 1949.

[46] J. Soni and R. Goodman. *A Mind at Play*. Simon & Schuster, 2017.

[47] C.E. Shannon and W. Weaver. *The Mathematical Theory Of Communication*. University of Illinois Press, Urbana, 1949.

[48] J. Pearl. *The Book of Why*. Allan Lane, 2018.

[49] M. Burgess. Spacetimes with semantics (ii): Scaling of agency, semantics, and tenancy. *http://arxiv.org/abs/1505.01716*, 2015.

[50] P. Borrill, M. Burgess, A. Karp, and A. Kasuya. Spacetime-entangled networks (i) relativity and observability of stepwise consensus. *arXiv:1807.08549 [cs.DC]*, 2018.

[51] R.E. Kastner. *The Transactional Interpretation of Quantum Mechanics*. Cambridge, 2013.

[52] J. Barbour. *The End of Time*. Phoenix books, 1999.

[53] J. Bjelland, M. Burgess, G. Canright, and K. Eng-Monsen. Eigenvectors of directed graphs and importance scores: dominance, t-rank, and sink remedies. *Data Mining and Knowledge Discovery*, 20(1):98–151, 2010.

[54] M. Burgess and F.E. Sandnes. Predictable configuration management in a randomized scheduling framework. *Proceedings of the 12th internation workshop on Distributed System Operation and Management (IFIP/IEEE).*, INRIA Press:293, 2001.

[55] F.E. Sandnes. Scheduling partially ordered events in a randomized framework - empirical results and implications for automatic configuration management. *Proceedings of the Fifteenth Systems Administration Conference (LISA XV) (USENIX Association: Berkeley, CA)*, page 47, 2001.

[56] Q. Liu, W. Cai, D. Jin, J. Shen, Z. Fu, X. Liu, and N. Linge. Estimation accuracy on execution time of run-time tasks in a heterogeneous distributed environment. *Sensor*, 16(9):1386, 2016.

[57] Maciej Drozdowski. Estimating execution time of distributed applications. *LNCS*, 2328:137–144, 2002.

[58] Ilia Pietri, Gideon Juve, Ewa Deelman, and Rizos Sakellariou. A performance model to estimate execution time of scientific workflows on the cloud. In *Proceedings of the 9th Workshop on Workflows in Support of Large-Scale Science*, WORKS '14, pages 11–19, Piscataway, NJ, USA, 2014. IEEE Press.

[59] H. Everett. *Theory of the Universal Wavefunction*. PhD thesis, Princeton, 1956.

[60] R. Hickey. Are we there yet? Talk at QConn, 2009. http://www.infoq.com/presentations/Are-We-There-Yet-Rich-Hickey.

[61] R. Hickey. Simple made easy. Talk at QConn, 2011. http://www.infoq.com/presentations/Simple-Made-Easy.

[62] W.R. Ashby. *An introduction to cybernetics*. J. Wiley & Sons, 1956.

[63] R. Desimone. Face-selective cells in the temporal cortex of monkeys. *Journal of Cognitive Neuroscience*, 3(1):1–8, 1991.

[64] T.L. Bjerknes, E.I. Moser, and M.B. Moser. Representation of geometric borders in the developing rat. *Neuron*, 2014.

[65] M.B. Moser, D.C. Rowland, and E.I. Moser. Place cells, grid cells, and memory. *Perspectives in Biology*, 2016.

[66] C.F. Doeller, C. Barry, and N. Burgess. Evidence for grid cells in a human memory network. *Nature*, 463:657–661, 2010.

[67] M.D. Fox, A.Z. Snyder, J.L. Vincent, M. Corbetta, D.C. Van Essen, and M.E. Raichle. The human brain is intrinsically organized into dynamic, anticorrelated functional networks. *Proceedings of the National Academy of Sciences of the United States of America*, 102(27):9673–9678, 2005.

[68] S. Traugott. Why order matters: Turing equivalence in automated systems administration. *Proceedings of the Sixteenth Systems Administration Conference (LISA XVI) (USENIX Association: Berkeley, CA)*, page 99, 2002.

[69] L. Kanies. Isconf: Theory, practice, and beyond. *Proceedings of the Seventeenth Systems Administration Conference (LISA XVII) (USENIX Association: Berkeley, CA)*, page 115, 2003.

[70] R. Dawkins. *The Blind Watchmaker*. Penguin, 1986.

[71] G.J. Schmitz. Entropy and geometric objects. *Entropy*, 20(453), 2018.

[72] M. Burgess. In search of science behind 'complexity'—deconstructing infamous and popular notions of complexity theory. markburgess.org, 2015.

[73] J.H. Holland. *Emergence: from chaos to order*. Oxford University Press, 1998.

[74] S. Johnson. *Emergence*. Penguin Press, 2001.

[75] M. Annamalai, K. Ravichandran, H. Srinivas, I. Zinkovsky, L. Pan, T. Savor, D. Nagle, and M. Stumm. Sharding the shards: Managing datastore locality at scale with akkio. In *13th USENIX Symposium on Operating Systems Design and Implementation (OSDI 18)*, pages 445–460, Carlsbad, CA, 2018. USENIX Association.

[76] L.M.A. Bettencourt, J. Lobo, D. Helbing, C. Hühnert, and G.B. West. Growth, innovation, scaling and the pace of life in cities. *Proceedings of the National Academy of Sciences*, 104(107):7301–7306, 2007.

[77] K. Goswami, C. Giarmatzi, M. Kewming, F. Costa, C. Branciard, J. Romero, and A. G. White. Indefinite causal order in a quantum switch. *Phys. Rev. Lett.*, 121:090503, Aug 2018.

[78] Leonard Kleinrock. *Queueing Systems: Computer Applications*, volume 2. John Wiley & Sons, Inc., 1976.

[79] L. Smolin. *Three Roads to Quantum Gravity*. Basic books, 2001.

[80] R. Penrose. *Cycles of Time*. The Bodley Head, 2010.

[81] P Halpern. Time after time. *Aeon*, 2018.

[82] E. Bonabeau, M. Dorigo, and G. Theraulaz. *Swarm Intelligence: From Natural to Artificial Systems*. Oxford University Press, Oxford, 1999.

[83] J. Kennedy and R.C. Eberhart. *Swarm Intelligence*. Morgan Kaufmann (Academic Press), 2001.

[84] M. Burgess. Deconstructing the 'cap theorem' for cm and devops. *Blog at markburgess.org*, 2012.

[85] H. Howard and R. Mortier. A generalized solution to distributed consensus. *arXiv:1902.06776 [cs.DC]*, 2019.

[86] A. Bejan and S. Lorente. Constructal law of design and evolution: Physics, biology, technology, and society. *J. Appl. Phys.*, 113:151301, 2013.

[87] S. Kauffman. *The Origins of Order*. Oxford University Press, Oxford, 1993.

[88] I. Prigogine. *The End of Certainty*. Simon & Schuster, New York, 1996.

[89] P. Wohlleben. *The Hidden Life of Trees*. Greystone Books, 2016.

[90] A. Turing. The chemical basis of morphogensis. *Philosophical Transactions of the Royal Society of London*, B237(641):37–72, 1952.

[91] M. Gardner. Mathematical games - the fantastic combinations of john conway's new solitaire game life. *Scientific American*, 223:120–123, 1970.

[92] M. Burgess, H. Haugerud, and A. Strandlie. Object orientation and visualization of physics in two dimensions. *Computers in Physics*, 12:274, 1998.

[93] X. Li and A.G. Yeh. Neural-network-based cellular automata for simulating multiple land use changes using gis. *Int. J. Geographical Information Science*, 16(4):323–343, 2002.

[94] R. White and G. Engelen. Cellular automata and fractal urban form: a cellular modelling approach to the evolution of urban land use patterns. *Environment and Planning A: Planning and Design*, 25:11751199, 1993.

[95] R. White, G. Engelen, and I. Uijee. The use of constrained cellular automata for high-resolution modelling of urban land use dynamics. *Environment and Planning B: Planning and Design*, 24:323343, 1997.

[96] F. Wu. An experiment on the generic polycentricity of urban growth in a cellular automatic city. *Environment and Planning B: Planning and Design*, 25:103–126, 1998.

[97] M. Batty and Y. Xie. From cells to cities. *Environment and Planning B: Planning and Design*, 21:531548, 1994.

[98] M.S. Longair. *Theoretical concepts in physics*. Cambridge University Press, Cambridge, 1984.

[99] S. Harnad. The symbol grounding problem. *arXiv:cs/9906002 [cs.AI]*, 1999.

[100] W. McCulloch and W. Pitts. A logical calculus of ideas immanent in nervous activity. *Bulletin of Mathematical Biophysics*, 5(4):115–133, 1943.

[101] SY Chung, D.D. Lee, and H. Sompolinksy. Classification and geometry of general perceptual manifolds. *Phys. Rev. X*, 8(031003), 2018.

[102] A.G. Huth, W.A. de Heer, T.L. Griffiths, F.E. Theunissen, and J.L. Gallant. Natural speech reveals the semantic maps that tile human cerebral cortex. *Nature*, 532:453458, 2016.

[103] R.W. Langacker. *Cognitive Grammar, A Basic Introduction*. Oxford, Oxford, 2008.

[104] M. Burgess. A site configuration engine. *Computing systems (MIT Press: Cambridge MA)*, 8:309, 1995.

[105] I. Heim and A. Kratzer. *Semantics in Generative Grammar*. Blackwell, 1998.

[106] J. Rennie. Brains cling to old habits when learning new tricks. *Quantamagazine*, 2018.

[107] J. Hawkins and S. Blakeslee. *On Intelligence*. St. Martin's Griffin, 2004.

[108] G. Canright and K. Engø-Monsen. A natural definition of clusters and roles in undirected graphs. *Science of Computer Programming*, 53:195, 2004.

[109] R. Pastor-Satorras and A. Vespignani. Epidemic spreading in scale-free networks. *Phys. Rev. Lett.*, 86:3200–3203, 2001.

[110] N. Chomsky. On certain formal properties of grammars. *Information and Control*, 2(2):137–167, 1959.

[111] H. Lewis and C. Papadimitriou. *Elements of the Theory of Computation, Second edition*. Prentice Hall, New York, 1997.

[112] S. Pinker. *The Language Instinct*. Penguin Press, London, 1994.

[113] G. Deutscher. *The Unfolding of Language*. Academic Press, 2005.

[114] L. Chittka and C. Wilson. Bee brained. *Aeon*, 2018.

[115] L.L. Cavalli-Sforza. *Genes, Peoples and Languages*. Allen Lane, 2000.

[116] V. Evans. *Language and Time: A Cognitive Linguistics Approach*. Cambridge, 2013.

[117] R.J. Rivers. *Path Integral Methods in Quantum Field Theory*. Cambridge, 1987.

[118] Qifan Wang, Dan Zhang, and Luo Si. Semantic hashing using tags and topic modeling. In *Proceedings of the 36th International ACM SIGIR Conference on Research and Development in Information Retrieval*, SIGIR '13, pages 213–222, New York, NY, USA, 2013. ACM.

[119] G. Deutscher. *Through the Language Glass: Why the World Looks Different in Different Languages*. Academic Press, 2011.

[120] S. Pinker. *How the Mind Works*. Penguin Press, London, 1999.

[121] N. Carey. *The Epigenetics Revolution*. Icon Books, 2011.

[122] J. Pearl. *Probabilistic Reasoning in Intelligent Systems: Networks of Plausible Inference*. Morgen Kaufmann, San Francisco, 1988

[123] J. Pearl. *Causality*. Cambridge University Press, Cambridge, 2000.

[124] J. Pearl. Theoretical impediments to machine learning with seven sparks from the causal revolution. *arXiv:1801.04016 [cs.LG]*, 2018.

[125] C. Sheppard. Ancient indigenous memory systems. *Uplift*, 2016.

[126] S. Reardon. A giant neuron found wrapped around entire mouse brain: 3d reconstructions show a 'crown of thorns' shape stemming from a region linked to consciousness. *Nature (News)*, 2017.

[127] F. Tedone, E. Del Dottore, B. Mazzolai, and P. Marcati. Plant behaviour: A mathematical approach for understanding intra-plant communication. *bioRxiv*, 2018.

[128] A. Toffler. *Future Shock*. Random House, 1970.

[129] A. Toffler. *The Third Wave*. Bantam, 1980.

[130] A. Toffler. *Power Shift*. Bantam, 1990.

[131] J.E. Lovelock. Gaia as seen through the atmopshere. *Atmospheric Environment*, 6(8):579–580, 1972.

[132] J.E. Lovelock and L. Margulis. Atmospheric homeostasis by and for the biosphere: the gaia hypothesis. *Tellus. Series A. Stockholm: International Meteorological Institute.*, 26, 1974.

[133] S. Reiss. Power to the people. *Wired*, 1996.

[134] P. Borrill, M. Burgess, M. Dvorkin, and H. Wildfeuer. Workspaces. Technical report, 2015.

[135] M. Burgess and H. Wildfeuer. Federated multi-tenant service architecture for an internet of things. `draft-burgess-promise-iot-arch-00`, October 2015.

[136] A.S. Tannenbaum. *Computer Networks*. Prentice Hall, 2003.

[137] P. Borrill, M. Burgess, T. Craw, and M. Dvorkin. A promise theory perspective on data networks. *CoRR*, abs/1405.2627, 2014.

[138] L. Dryburgh and J. Hewett. *Signaling System No. 7 (SS7/C7): Protocol, Architecture, and Applications*. Cisco Press, 2003.

[139] R.J. Vetter. Atm concepts, architectures, and protocols. *Commun. ACM*, 38(2):30–ff., February 1995.

[140] K. Guo, W. Vogels, and R. van Renesse. Structured virtual synchrony: Exploring the bounds of virtual synchronous group

communication. In *Proceedings of the 7th Workshop on ACM SIGOPS European Workshop: Systems Support for Worldwide Applications*, EW 7, pages 213–217, New York, NY, USA, 1996. ACM.

[141] M. Poke and T. Hoefler. Dare: High-performance state machine replication on rdma networks. In *Proceedings of the 24th International Symposium on High-Performance Parallel and Distributed Computing*, HPDC '15, pages 107–118, New York, NY, USA, 2015. ACM.

[142] M.J. Fischer, N.A. Lynch, and M.S. Paterson. Impossibility of distributed consensus with one faulty process. *J. ACM*, 32(2):374–382, April 1985.

[143] S. Nakamoto. Bitcoin: a peer to peer electronic cash system. http://nakamotoinstitute.org/bitcoin/, October 31 2008.

[144] M. Burgess. The promises of devops. markburgess.org, 2012.

[145] J. Sussna. Promise theory, devops, and design. Web document `http://blog.ingineering.it`, 2013.

[146] J. Sussna. Promise theory, devops, and design. Web document `http://blog.ingineering.it`, 2013.

[147] J.D. Davidson and W. Rees-Mogg. *The Sovereign Individual*. Touchstone, 1997.

[148] J.A. Bergstra and M. Burgess. Local and global trust based on the concept of promises. Technical report, arXiv.org/abs/0912.4637 [cs.MA], 2006.

[149] P. Matzinger. Tolerance, danger and the extended family. *Annu. Rev. Immun.*, 12:991, 1994.

[150] R. Axelrod. *The Complexity of Cooperation: Agent-based Models of Competition and Collaboration*. Princeton Studies in Complexity, Princeton, 1997.

[151] R. Axelrod. *The Evolution of Co-operation*. Penguin Books, 1990 (1984).

[152] J. Sussna. Scaling agile projects using promises. Web document `http://blog.ingineering.it`, 2013.

[153] W.E. Deming. *Out of Crisis*. Massachusetts Institute of Technology, 1982.

[154] N. Ferguson. *The Ascent of Money*. Penguin, 2008.

[155] M. Burgess. Notes on macroeconomic modelling from a promise theory viewpoint. Published online at markburgess.org.

[156] M. Burgess. Notes on wealth and banking. Published online at markburgess.org.

[157] J.A. Bergstra and M. Burgess. Blockchain technology and its applications—a promise theory view. Technical report, 2018.

[158] K. Hart. *The Memory Bank: money in an unequal world*. Profile Books, 2000.

[159] Y.N. Harari. *Sapiens*. Vintage, 2011.

[160] E.D. Beinhocker. *The Origin of Wealth*. Random House, 2005.

[161] M.E. Smith. The archaeology of ancient state economies. *Annual Review of Anthropology*, 33:73–102, 2004.

[162] J. Suzman. When a 200,000-year-old culture encountered the modern economy. *The Atlantic*, July 24 2017.

[163] D. Graeber. *Debt: the first 5000 years*. Melville House, 2011.

[164] P. Mirowski. *More Heat than Light*. Cambridge, 1989.

[165] G. Simmel. *The Philosophy of Money*. Routledge and Kegan Paul, 2011.

[166] S. Keen. *Debunking Economics (2nd ed)*. Zed, 2011.

[167] S. Keen. *Can we avoid another financial crisis?* Polity, 2017.

[168] Y. Varoufakis. *The Global Minotaur*. Zed, 2011-2015.

[169] J.M. Keynes. *A Treatise on Money*. Harcourt Brace, 1930.

[170] J.M. Keynes. *The General Theory of Employment, Interest, and Money*. Harcourt Brace, 1964.

[171] N. Minksy and P-P. Pal. Law-governed interaction: A coordination and control mechanism for heterogeneous distributed systems. *ACM Transactions on Software Engineering and Methodology*, 9:273–305, 2000.

[172] T. Piketty. *Capital in the twenty-first century*. Belknap, Harvard University Press, 2014.

[173] S. Keen. Finance and economic breakdown: modelling minsky's financial instability hypothesis. *Journal of Post Keynesian Economics*, 17(4):607–635, 1995.

[174] R. Foroohar. *Makers and Takers*. Crown Business, 2016.

[175] K.D. Lilley. *City and Cosmos: The Medieval World in Urban Form*. Reaktion books, 2009.

[176] R.O. Duda, P.E. Hart, and D.G. Stork. *Pattern classification*. Wiley Interscience, New York, 2001.

[177] K. Begnum and M. Burgess. Principle components and importance ranking of distributed anomalies. *Machine Learning Journal*, 58:217–230, 2005.

[178] M. Burgess. *Slogans: The End of Sympathy.* χt-axis Press, 2005.

Index

"AI", 333

A quiet belief in angles, 278
A universal prototype space, 70
Absolute Position, 186
Absorption and emission,
    120, 330
Absorption of light, 508
Abstraction, 73
Acceleration, 22, 320
    Newtonian, 306
Acceleration by slowing
    your clock, 320
Access
    Motion, 306
Access Control, 233
    Not in Euclidean space,
    233
Action principle, 221, 323, 526
Actor model, 452
Address, 18, 41, 88, 315, 357
    By context, 359, 360
    Computer, 18, 43, 45, 315
    Indexing, 392, 405
    IP, 18, 41, 43, 45,
    315, 506
        Semantic, 357, 392
        Signpost events, 405

Adjacency, 33, 41, 48,
    79, 102, 445
    And Information, 37
    As interaction, 112
    Dataverse, 80
    In a graph, 59
    In forest, 433
    Relation to boundary,
    115
    Types of, 48, 362
    Virtual, 102, 219
Advanced-retarded boundary
    conditions, 133, 198, 275, 404
Agency, 110, 514
Agent
    As a model for location,
    67
    Cluster, 111
    In a graph, 59
    In Promise Theory, 71, 514
    In promise theory, 67
    Superagent, 85, 111,
    225, 277, 338
Agents and promises of
    language mechanics, 377
AI, 222, 333, 530
Airports, 221
Algebra, 346, 367

Algorithms, 35, 208
Alien shopping story, 298
All Our Yesterdays, 247
Alphabet, 368
Amazon, 215
Ambient light, 177, 354
An abstract belief in motion, 195
An information process view, 289
Anderson, Jon, 329
Andrews, Julie, 293
Anisotropy, 65
    Neural net, 346
Anthropomorphism, 509
Antigen, 208, 530
Ants, 261
Apple cost story, 294
Aristotle, 37
    On time, 38
Array, 96
Artifex principalis, 483
Artificial Intelligence, 222, 333, 416, 490
Artificial Neural Network (ANN), 341
Artificial reason, 394
ASCII code, 378
Asimov, Isaac, 447
Assessment, promise theory, 510
Astronomy, 77
Atlas, 31
Attractor, 203
Autonomous agents, 5, 427
Autonomy, 66
    Blockchain, 282
Autopilot, 215
Awareness, 408
Azure, Microsoft, 218

B-cell, 93
Babbage, Charles, 84
Back to the future: recursive evaluation, 198
Backpropagation, 347
Balance, 321, 436, 461, 469, 474
Ball
    Open, 16, 515
Bandwidth, 112
Bank
    As network hub, 469
Bank consensus, 235
Bank payment, 256
Barrier
    Language, 462
BASIC, 200
Bejan, Adrian, 525
Bend, 22
Bergstra, Jan, 509
Bettencourt, Luis, 442
Big Bang, 14
Bigraph, 61, 86
Billiards, 288, 305
Binary alphabet, 37
Bing, 391
Biology, 35, 76, 77, 192, 198, 225, 296, 308, 494, 519
    B-cell, 93
    DNA, 277, 338, 368
    Emergence, 330
    Give and take, 202, 208, 283, 296
    Scaling, 100, 192, 519
    Scaling of organisms, 441
    Sexual mixing, 277, 331
    Sperm and egg, 331
Bioverse, 308
Birds of a feather, 166
Bits, 112
Black box
    Flight recorder, 265, 269
Black holes
    BGP, 508, 514
    Dataverse, 514
    Outer space, 514

Index

Blake, William, 424
Blockchain, 280, 371, 526
Blocking
    Neurons, 264, 416
Blue shift, 224
Boiling water story, 169
Boole, George, 266
Boot, 15, 504
Bootstrap, 15, 504
Border Gateway Protocol
    (BGP), 508, 514
Borrill, Paul, 90, 511, 523, 533
Bosons, 147, 227, 293, 433
    Messenger, 304, 309, 327, 433
Bottleneck, 216, 223, 521
Boundaries, information, 128
Boundary
    And Euclidean space, 370
    And symbolism, 404
    Between interior and
    exterior, 44, 60, 115, 203, 244,
    Between net and subnet, 46
    Cognition, 333
    Concentric, 60
    Condition, 133, 198
    Condition advanced, 133, 198, 200
    Condition retarded, 133, 198
    For locality, 40, 115, 163, 403
    Process, 273
Boundary conditions
    As information, 327
    Ignoring, 269
    Statistics, 266
Breaking time symmetry
    in the thinking process, 403
Bush, Vannevar, 84, 415

Camera
    As cognitive system, 340
Canonical variables, 287
Cantonese, 387
CAP conjecture, 456
Carey, Nessa, 531
Carroll, Lewis, 28, 285
Cartesian square arrays, 96
Catalogue, 97
Causality, 51, 114, 130, 198
    Arrow of time, 199
    Dataverse, 322
    Downstream, 135, 203, 225
    Reverse, 134, 199
Cause
    Single point of, 132
CDN, 216
Cells, 16, 338
    Motion, 308, 309
    Place, 192
    Place and grid, 519, 529
    Polarization, 309, 371, 457
Cellular automaton, 312, 322, 385
Central system, 109, 110, 115, 234
Centralization, 109, 110, 115, 234
Centre of mass, 192
CFEngine, 264, 282
Chain, 371, 422
Change, 1
Channel capacity, 217
    Superposition, 228
Channel for information
    Half process, 114
Channel for information, 112, 332
    Causality, 114
    Dimensionality, 326
    Superposition, 228, 348
Charge
    Electric, 319
Charge to the rescue?, 319

Chemical bond, 302, 320
Chemistry, 340
  Genetics, 413
Cher, 247, 256
Cherenkov radiation, 521
Chess, 334
Chinese medicine, 419
Choir analogy, 227
Choir and song, each to their own world, 433
Chomsky
  Innate controversy, 380
Chomsky, Noam, 362, 379, 396
Cinderella, 259
Circuit, 33, 58, 79, 108, 214, 263, 415, 447
  Electrical, 36
Cities, 72, 91, 339, 438
  Clocks, 439
  Learning, 440
  Modularity, 339, 438
  Renormalization, 31
  Smart, 440, 487
  Space, 486
City
  Smart, 427, 440, 487, 491
Cityverse, 438
Clarke
  Arthur C., 494
Classical Covariant Fields, 3, 504
Classical physics
  Concepts, 162, 327
  Motion, 195
Cleopatra's needles, 365, 404
Clock
  Synchronization, 105
Clock tick, 22, 274
Clocks, 12, 21, 39, 102, 174
  Atomic, 241
  City, 439
  Computer, 80
  In cognition, 168, 337
  Logical, 237
  Matter and energy, 55
  Scaling limits, 120, 241
  Synchronization, 17
Clockwork universe, 21, 276
Cloud computing, 17, 159, 162, 218, 231, 291, 448, 508
Cluedo story, 371
Cluster of agent locations, 111
Coarse graining, 333
Cocteau, Jean, 403
Cognition
  Defined, 332
Cognition and causal reasoning, 416
Cognitive aspects of money, 478
Cognitive system, 76, 331, 336, 452, 497
  Artificial, 463
  Evolution, 381
Cognitive systems, 336
Collective behaviour, 109, 212
Colour space, 47
Communication make us closer, 124
Complex numbers
  In Quantum Mechanics, 138
Complexity
  Computational, 514
  Fractal, 507, 532
  Properties, 8
  Scaling, 110, 168
  Science, 257, 513
  Semantic, 77
Compound concepts, 395
Computational chemistry, 86
Computer
  Actor model, 452
  Analogue, 84
  Cloud, 17, 159, 161, 162, 218, 231, 291, 448
  CPU, 18, 41, 90
  Functional programming,

202
- Kernel, 254
- Memory, 18, 41, 94
- Network, 17
- Programming, 200
  - Stack, 520
- Reset, 260

Computer science
- Relativity, 504

Computer system, 16, 72
- Spacetime properties, 76

Concept, 352, 365
- Drift, 404
- Invariant, 360, 365
- Representation, 366, 378, 529
- Scaling, 360, 365

Consciousness, 193, 408

Consensus
- Distributed, 234, 268, 282, 296, 464, 524, 526

Conservation of momentum, 305

Conserved measure, 325, 474

Constructal Law, 525

Containment
- In computing, 97

Content Delivery Network (CDN), 216

Context, 352, 356, 365, 390
- Short term memory, 405

Context addressing, 359

Continuity
- Of process, 289, 323

Continuous motion, 179
- Insufficient sampling, 192, 297, 326

Continuum
- Approximation, 41, 42

Conway, James Horton, 526

Cookbook story, 200

Coordinates, 22, 31, 43, 171
- x, y, z, 22, 24, 31, 43, 48, 86, 259, 404
- And names, 97, 357, 402
- Axes, 42
- Cartesian, 22, 31, 43, 48, 57, 96, 97
- City, 45
- Cognitive, 404
- Computer memory, 41
- Dataverse, 79, 80
- Hotel room, 258
- Linguaverse, 382
- Patches, 42, 98, 269
- Semantic, 97, 357
- Time, 259

Coroutines, 449

Cosmogony, 14

Cosmology, 490

Couch, Alva, 531

Counterfactuals, 416

Country, 91, 460

Covalent bond, 302, 308, 320

CPU, 18, 80, 90, 94, 225, 254

CTRL-Z, 40

Curvature, 19, 22

Dark Matter, 519

Dark matter, 295

Darwinian evolution, 107, 208, 414, 486

Data
- Compression, 409
- ordering, 271

Data in motion, 107

Data pipeline, 195, 352, 444

Data type, 87
- Double entry bookkeeping, 467

Database, 66
- As spacetime, 92
- Dimensionality, 94
- Distributed, 280
- Normalization, 92
- Scaling, 100
- Schema, 93

Stigmergy, 262, 272
Databases, 91
Datacentre, 87, 231
Dataverse, 17, 77, 216
    Directions, 90
    Embedding, 80
    Language, 401
    Locations, 86
    Position, time, velocity, 108
    Sampling, 194
    Scaling, 84
    Time, 83
Dawkins, Richard, 520
Decision, 276, 526
Deep learning, 341, 390, 501, 530
Defining motion, 171
Degrees of freedom, 19, 48
Democracy
    Scaling, 267
Denver, John, 253
Dependency, 131, 524
    And time ticks, 274
Descartes, René, 8, 29, 48, 88, 95–97, 402
Desired end states, 134, 199
Detailed balance, 321, 436, 461, 469, 474
Determinism, 199
Differential calculus, 33
Dimension
    Hausdorff, 506, 532
Dimensionality
    Neural net, 341
Dimensions, 19, 45, 48, 94, 362
    Euclidean versus network, 49
    Matroid, 507
    Statistical property, 507
Directed Acyclic Graph (DAG), 59, 64, 134, 262, 274
Direction, 48

Dataverse, 90
Directory, 97
Disambiguation, 373
Disaster recovery, 256
Discretionary Access Control (DAC), 233
Discrimination, 349
    Unstable, 360
Dissipative process, 463, 518
Distance
    Dataverse, 211
    In a graph, 65
Distinguishability, 42, 221, 261, 265, 269, 315, 349,
Distribution of energy, 162
Distribution of money, 463
Division by Zero, 523
DNA, 277, 338, 368
DNS, 293, 526
Doll
    Russian, 15, 45, 98
Domain Name Service (DNS), 293, 526
Doppler effect, 224
Dopplergang—playing with time distortions, 223
Double entry bookkeeping, 467
Downstream
    Causality, 135, 225
Dr Who, 511, 522
Dressed fields, 450
Drift of concepts, 404
Dynamics, 25, 35

EBCDIC code, 378
ECG, 93
Economy, 428
    Global, 216, 229
    Of scale, 442
Econoverse
    Locations, 468
Ecosystem, 76, 283, 336
Ecotone, 283

Effective Mass
  City, 435
  Cloud computing, 455
Egypt, 503
Eigenstates, 157, 360, 460
Einstein, Albert, 1, 20, 25, 34, 98, 212, 247, 327, 503,
Einstein-Rosen Bridge, 523
Election story, 268
Electrocardiogram (ECG), 93
Electromagnetic field, 210
Electromagnetism, 21
Electron, 71
Elementary phenomena, 103, 513
Elements, 8
  Earth Air Fire Water Gold, 418
Embedaverse, 15
Embedding, 15, 80, 243
Emergent behaviour, 330
Emission of light, 508
Emotion
  Role in reasoning, 353, 422
Empty, 11, 26, 34, 87, 300
Encapsulation, 97
End justifies means, 418
Energy, 26, 186, 245
Entanglement, 40, 162, 317, 457
Entropy, 40, 204, 260
  Dataverse, 488
  Defined, 523
  defined, 204, 520
  Distinguishability, 221
  Division by Zero, 523
  In orchestra, 264
  Reversal, 277
  Semantic specificity, 272
  Shannon, 365
Epidemic process, 234
Epigenetics, 411
  Selection, 437
Episode, 24, 265, 405

Equations of Motion, 112, 206, 208, 242, 251, 317
Equilibrium, 280, 520
  Nash, 521
  Semantic consensus, 236, 464
  Thermodynamic, 236
Error correction, 200, 405, 456
Ethernet, 218
Euclid, 8, 29, 95, 402
  Elements, 8, 29
Events, 370
  From sensors, 405
  Significant, 405
Everett, Hugh, 81, 106
Everything and nothing, 37
Evolution
  And language, 388
  Darwin, 107, 208, 414
Exclusion Principle, 227
Expanse, 87
Expert systems, 390
Exterior
  And dimensions, 95, 325
  Time, 39, 56, 92, 225

Fabric of spacetime, 16
Fabric of spacetime, 458
Facebook, 341
FCFS, 220
Feedback, 200, 360, 520
Fermat's principle, 221, 323, 526
Fermions, 147, 227, 293, 433
Feynman diagram, 183, 318, 379, 433
Feynman, Richard, 48, 59, 97, 183, 230, 318, 324
Field
  Electromagnetic, 52, 210, 212, 308
  Quantum, 52, 308
Filesystem

History, 106
Financial crisis, 280, 406
First Come First Served (FCFS), 220
Fitzgerald, F. Scott, 501
Fixed point, 524
Fixed stars, 291
Flawed reasoning or poor taste?, 418
Flight recorder, 265, 269
Flight simulator, 172
Fly story, 193
Folder, 97
For agent and country, 460
For whom the clock ticks, 167
Force
    Application of, 110
    Concept, 189
    Newtonian, 21, 286, 306
    Of nature, 7, 362
    Strong nuclear, 308
    Weak nuclear, 308
Force and acceleration, 306
Four promises
    Of spacetime, 362, 370
Fractal, 506
Fractals, 532
Frame of reference, 504
Freedom
    Degrees of, 19, 48
From mass m to momentum's process p, 300
From space to spacetime, 20
Function, 18, 110, 213
Functional role, 97, 110
Fungi, 433

Gage, John, 445
Gaia hypothesis, 435, 478
Game of Life, 312, 526
Game of life, 526
Garbage
    Separation, 277

Garbage in garbage out, 135
Gauge symmetry, 50
Gauss Law, 129
Gauss, Friedrich, 32
GDP, 198, 442
Genetics, 412
Geoglyphs, 261, 498
Geography, 448, 460
Geology, 77
Geometry, 29, 193
git software, 514
Give and take
    Mass, 294
Give and take, 109, 167, 316
    Biology, 208
    Blockchain, 281
    Causality, 130, 131, 281, 283
    Dependency, 131
    Genetics, 412
    Mismatch effect, 188, 193, 294
    Mixing, 274
    Neural net, 347
    Scaling of, 437
    Stigmergy, 261
    Symbolism, 368
    Time, 168, 275
    Wavefunction, 153, 463
Give and take in ANN–
    -from discrimination to reason
Give and take in the unfolding of events, 283
Global Positioning System, 32, 192, 200, 282
Global trade, 229, 461
Global variable, 524
Globalization, 461
GMT, 105, 258
Go
    Game, 334
    Programming language, 453
Goal

Football, 387
God, 14
Gods of spacetime, 484
Goldilocks zone, 77
Goods and services, 467
Google, 216, 218, 341, 391, 506
Google Docs, 40
GPS, 32, 192, 200, 282
Grammar
    Chomsky, 379
Grand Unification, 495
Graph
    Hierarchy, 64
    In mathematics, 58, 64
    Process, 64
    Star topology, 64
Graphs
    Expander, 508
Graviton, 304
Gravity
    As routing of adjacency, 301, 304, 323, 347, 514
    Einstein's General Relativity, 292, 301
    Privileged role in physics, 323
    Quantum, 304, 526
    Spacetime lensing, 308, 347
Greece, 503
    Ancient, 29, 31
Greek theatre, 29
Grid, 45
Grid cells, 519, 529
Gross Domestic Product (GDP), 198, 442

Hall of mirrors, 189
Harry Potter, 96
Haugerud, Hårek, 526
Hausdorff dimension, 532
Hawking radiation, 508
Hawking, Stephen, 1, 508

Hawkins, Jeff, 373, 528, 529
Hearsay, 271
Heartbeat, 21
Heat reservoirs, 235
Hebb, Donald, 341
Heisenberg, Werner, 5, 243, 289
Hickey, Rich, 519
Hierarchical Temporal Memory, 528
Hierarchy, 14, 90
    Of meaning, 374
Hierarchy of meaning, 374
Higgs model
    Dataverse, 295
Higgs model
    Physics, 295, 302, 320
Hilbert, David, 509
History, 24, 208, 248
History of spacetime, 7
Holes in the dataverse, 104
Homeostasis, 282
Homogeneity, 69, 92, 313, 346
    As data normalization, 92
Honeycomb, 312
Hotel room coordinates, 258
How many dimensions are there to information?, 94
How many dimensions?, 19
Hypersurface
    Spacelike, 25, 94, 520
Hysteresis, 392

If a tree falls....., 174
IKEA, 36
Immunology, 197, 208, 282, 316, 406
Immutability, 232
Imperative
    program, 200
Imposition, 460
Incredible shrinking time, 54
Inertia, 291
Inference

Pattern recognition, 416
Information, 93, 94
    And motion, 170
    And spacetime, 37
    And symbols, 366
    Bandwidth, 112
    Bits, 112
    Channel, 112, 312
    Dimensionality, 94
    Immutability, 232
    Incomplete, 213
    Receiver, 112
    Sender, 112
    Too much, 404
Information Theory, 112, 116
Infoverse, 75
Inhomogeneity, 65
Insect story, 193, 531
Insects
    Social, 531
Instability
    Reasoning, 268, 360
Integrity
    Modules, 279
Intelligence, 27, 330
    And language, 402
    Artificial, 333
    Compare bio and data, 376
    Swarm, 334
Intent, 110, 281, 520
    Quasi-intentional
language, 283
Intent with probable cause, 130
Interaction
    As adjacency, 112
Interest, 474
Interior
    And dimensions, 95, 325
    Time, 21, 39, 56,
92, 105, 164, 225, 407
Interior and exterior time
    —the end of privilege, 240
Interleaved processes, 272

Intermediate agent, 304,
307, 309, 327
    Bank, 469
Internet, 74, 78, 189
    Addresses, 315
Interpolation
    Continuum approximation,
192, 297
Into the nebula, 448
Invariant
    Concept, 360, 365
Ionic and covalent mass:
why gravity
is not like the o
302
Ionic bond, 302, 308
IP Address, 18, 41, 43,
45, 315, 506
Is language intelligence?, 402
Is time money, or is money
spacetime?, 466
Islamic Empire, 35
Isotropy, 69, 313, 346

Jacquard Loom, 84
Jungle Book, 109
Just in time, 133, 198,
229, 316

Kaluza, Theodore, 45, 362
Kaufmann, Stuart, 525
Kipling, Rudyard, 109
Kleiber's law, 441
Kleiber, Max, 441
Kleinrock, Leonard, 521
Knowledge
    Representation, 27
Kripke, Saul, 81

Labelled Transition System, 519
Lamport, Leslie, 235, 239, 282
Language, 386
    And evolution, 388
    As a barrier, 462

Bootstrapped by spacetime, 384, 529
   Danish, 377
   Semitic, 377
Last In First Out (LIFO), 134, 520
Latency
   And cloud, 455
   And interior time, 322
   Defined, 214
Latent speed limits—thunder and lightning, 213
Learning, 337
   Cities, 440
   Deep, 341
   Economics, 478
Left
   Politics, 387
Leibniz, Gottfried Wilhelm, 139, 364, 513
Lens, 32, 179, 215, 519
Lenses, prisms, and load balancers, 215
Lensing, 219, 290
   Gravitational, 224
LIFO, 134, 520
Lifting cognition by its bootstraps, 332
Light
   Ambient, 177, 354
Light cone, 107
Line
   Defined by Euclid, 29
   Straight, 192, 288
Linguaverse, 98, 387
Linguaverse phenomenology, 387
Linguistics, 71, 202, 373, 377, 505, 529
Load balancer, 215, 216, 219, 229, 348
Load sharing, 215
Local distortions of spacetime, 228

Locality, 23, 40, 56, 68, 163, 489
   Management in cloud, 232
   Replace with give and take, 436
Logic, 501
   And reasoning, 355, 409, 417
   Boolean, 267
   First order, 334
Logical clocks: the Treaty of Common Time over space, 232
Logistics, 198, 316
London, 42
Long Range Order, 110
Looking forward or looking back, 133
Loop Quantum Gravity, 522
Loops
   Computer programming, 512, 513
Loops and multidimensional time, 257
Losing your mind to logic, 266
Lovelace, Ada, 84
Lovelock, James, 435
LTM, 352, 360

M Theory, 81
Mach's principle, 56, 291, 516, 520
Mach, Ernst, 291
Machiavelli, 492
Machine learning, 222
Machine learning, 333
   Superior to algorithm, 380
Magic, 282, 348, 431, 494, 496
Magnetic Resonance Imaging (MRI), 350
Magnification, 179, 215, 519
Major Histocompatibility Complex (MHC), 419, 511, 526
Makespan, 519

Making money, 471
Mandatory Access Control (MAC), 233
Manifold
　Perceptual, 345, 527
　Spatial, 506
Many worlds theory, 81, 106, 163, 346, 519
　Interleaved processes, 272, 451
　Version branching, 523
Map-Reduce, 159
Maps, 34, 505
　Location names, 35, 505
　Semantics, 34
Mass
　As baggage, 300
　as charge, 320
　Economic, 467
　Effective, 435, 455
　Gravitational, 292, 525
　Higgs, 293
　Inertial, 287, 291, 525
　Newtonian, 287
　Polaron, 293
Mass and motion to alien observers, 297
Mass and the dataverse, 295
Mathematics, 29
Matrix, 96
Matter, 26, 37, 69, 186, 245
　and energy, 27
　Dark, 519
Maxwell's daemon, 251
Maxwell, James Clerk, 212, 317
Maymounkov, Petar, 506
McCulloch, Warren, 341
Meaning, 365, 366, 391
　Scaling, 365
Measurement
　And trust in instruments, 171, 243, 306
　Process, 25, 177, 210, 330, 504
Membership
　Topology, 507
Memory, 18, 41, 254
　And space, 91, 376
　Arrow of time, 209
　Cities, 339, 440
　Genetic, 412
　Long term, 352, 360
　Paged, 94
　Renormalization, 475
　Shared, 280
　Short term, 352, 354
　Sideways, 94
　Stigmergic, 261, 339, 386, 475
Metal, 87
Metaphor, 398, 418, 505
Metaverse, 510
Metonym, 398, 505
Mezick, Daniel, 525
MHC
　Biology, 511
Michaelson-Morley experiment, 535
Microservices, 165, 283, 295, 445, 459
Middle man, 174, 276, 304, 306, 307, 309, 327
　Bank, 469
Milner, Robin, 59, 61, 74, 86
Minimum requirements
　Dependency, 275
Minkowski, Hermann, 3, 20, 22, 247, 323
Mobile devices, 78
Mobility, 171
Modular time—seeing the wood and the trees, 265
Modularity, 10, 60, 70, 166, 265
　Cellular structure, 277
　Cities, 339, 438

    Cognitive system, 330
    Semantics, 329
    Software, 278
Modules, 10, 60, 70, 198, 254, 523
    As database, 272
    Integrity, 279
    Locality, 326
Molecule, 340
Momentum, 243, 286
    Classical, 287
    Conservation of, 288, 291, 305
    In Quantum Mechanics, 287
    Non-local quantity, 288, 291
    Vector field, 525
Money
    And time, 466
    As a network, 469
    As network, 469
    Capital and commodities, 480
    Conservation, 474
    Creation, 474
    Give and take, 478
    Inflation, 475
    Interest, 474
    Making, 471
    Measurement, 479
    Payment, 256, 466
    Semantics, 471
    Thinking, 479
    Types, 473
Moore's Law, 224
Motion, 21, 51, 170, 297
    Amoeboid, 309, 313, 457
    Brownian, 171
    Continuous illusion, 179, 190, 191
    Defined, 171, 195
    Discrete, 179
    First kind, 184
    Give and take, 242
    In dataverse, 172
    Information, 170
    Rights to, 306
    Scaling of, 213, 242
    Second kind, 184
    Third kind, 184
Motion and velocity on a limited network, 179
MRI scans, 350
Muddling observation with continuous inference, 191
Multiverse, 510
Mutation, 354
Myrheim, Jan, 239, 507, 521

Names
    As coordinates, 97, 357
    Numbering, 401
Namespace, 98, 357
Namespaces, scope, and rings around the target, 97
Napoleon Bonaparte, 487
Nash equilibrium, 521
Navigating language with dimension and coordinates, 382
Nazca, Peru, 261
Netflix, 165, 215, 216
Network, 17, 49, 85
    Centrality, 391
    Clos, 88, 89
    Dimensionality, 49
    Fractal, 442, 532
    Hierarchical, 90
    Hub, 356, 469
    Money, 469
    Petri, 519
    Roads, 104
    Routing and switching, 78, 110, 413
    Scale free, 442, 532
    Star, 90, 234, 356
    TCP, 230, 456

Tunnel, 234, 483
Network is the computer, 445
Networks, 36
    Tensor, 506
    Trees, 431
Neural net, 341
    Backpropagation, 347
    Dimensional reduction, 352, 390
    Dimensionality, 341, 346
    Give and take, 347
    Tokenization, 352, 390
    Training, 347
Neuron
    Inhibitory, 416
Neuron dead time, 264
New York, 45
Newton's achievement, 287
Newton, Isaac, 5, 139, 327, 364, 513
Newtonian
    Centre of mass, 192
    Clockwork universe, 21
    Differential calculus, 33
    Reversibility, 206, 438
Newtonian physics, 21, 192, 206, 276, 287
    Laws of motion, 75, 79, 112, 171, 206, 242, 285, 49
Next to, 33, 48, 79
    Promises, 115, 373
    Ways to define, 115
No agent is an island, 330
No U turns from equilibrium, 280
Noether's theorem, 323, 501
Noether, Emmy, 323, 501
Non-local, 56, 520
    Distribution of energy, 162
    Paradoxes, 284
Normalization
    Of database, 92

Numascale, 521
Numbers
    As names, 401
    Real, 192
Numenta, 373, 528, 529
Nurture, 413
Nyquist and Shannon, 112
Nyquist sampling theorem, 118, 168, 169, 186, 208, 229, 28 320, 436
Nyquist, Harry, 112, 208, 289, 320, 436

Obelisk, 366
Objectivity in science, 112
Obligation, 132, 133
Observability, 163, 213, 520
    Dataverse, 173, 196
    Density of paths, 179
Observer, 24, 43, 54
Observer-centric view, 68
One clock to rule them all?, 244
Ontology, 357, 530
Oolong, 387
Orchestra, 168
Orchestra story, 264
Order, 109, 110
Organism, 330, 336
Overlapping locations, 115, 269
    Memory, 528
Overlapping states, 168, 269
    Relevance, 391
Overlapping times, when pasts collide, 269

Paging
    Memory, 94, 428
Parallel dimensions and superposition, 346
Parentheses, 379
Paris, 42
Parsifal, 170
Partial ordering, 239

Particle physics, 77
Partitioned worlds in cloud, 451
Passport control, 348
Past
    Causal, 276
    Observational, 276
Path, 24
    Blocking in reasoning, 392
    Shortest, 220, 324
Path integral, 324, 420
Paths, 51
Pathways of explanation, 395
Pattern
    Information, 93
    Recognition, 341, 416
    Scaled agent, 265, 350
    Scaling, 350
Pattern recognition by spatial imprint (ANN), 341
Patterns, 27
Pauli Exclusion Principle, 227
Paxos, 237, 282, 511, 521
PCI bus, 218
Pearl, Judea, 417, 516
Peer to peer system, 234
Penrose, Roger, 319, 422
Petri Net, 519
Phantoms of continuity, 187
Phase transition, 109, 110
Philosophy, 28
Phonemes, 377
Physics
    Material, 87
    Particle, 7, 9, 77, 197, 225, 450, 510
Pictures at a database exhibition, 249
Pipeline
    Data, 195, 278, 352, 444
    Sales, 262
Pitts, Walter, 341
Place cells, 192, 519, 529
Planck scale, 519
Playing with time consistency, 238
Plumbing, 33, 36
Point
    In space, 41
Polarons, 293, 302
Politics, 283
Politiverse, 460
Post Office story, 513
Predicting speed or velocity, 209
Prey-Predator model, 311
Prigogine, Ilya, 525
Principle of least action, 221
Principle of least action, 323, 420, 526
Prism, 214, 215
Probability interpretation, 196
    Scaling, 227
Process
    Makespan, 519
Processes, 6, 13, 495
    And information channels, 114
    Continuity, 289, 323
    Cyclic, 257, 262
    Epidemic, 234
    Events, 370
    Functional, 18
    Measurement, 25, 210, 330, 504
    Virtual, 444, 513
Promise Theory, 66, 339, 425, 514, 522
    Acceptance (-), 67
    And influence, 167
    And Quantum Theory, 166
    Assessment, 510
    Money, 463, 467, 534
    Observer-centric view, 68
    Offer (+), 67

Principles, 166, 490
Promising cooperation, 116
Promising the details, 317
Proper time, 107
Protecting time with cells
   and modules, 273
Protein, 338
   And memory, 359
Ptolemy, 115, 163, 324, 485
Ptolemy returns: many worlds
   and the dataverse, 163
Publish-Subscribe model, 316
PubSub, 316
Pythagoras, 7, 29, 95, 503

QED, 9
Qualitative behaviour, 110
Quantization procedure, 526
Quantum
   Tunnel, 483
Quantum computers, 362
Quantum Electrodynamics, 9
Quantum Field Theory, 308
Quantum fluctuations, 69
Quantum give and take, 141
Quantum Gravity, 304
Quantum measurement, 136
Quantum Mechanics, 18,
   76, 81, 162, 195, 230, 318, 4
513, 522
   Complex numbers, 138
   Exclusion Principle, 227
   Many worlds, 81, 106,
   163, 519
      Sender-Receiver, 155
      Spin, 318
      Times (different), 518
      Transitions, 163
      Virtual process, 159, 513
      Wave-particle duality, 322
      Wavefunction, 138,
   155, 162
Quantum Probability, 227

Cloud, 159, 161, 295, 518
   Potential well, 295
Queue, 217, 220, 224, 226,
   246, 520
      Dependency, 524
Queueing versus superposition?,
   227
Queuing Theory, 220, 226
Quorum, 268, 296, 526

Racing change with change, 172
Radar, 177
Raft, 511
RDF, 529, 530
Reasoning
   And relativity, 335
   And story paths, 392, 409
   Backwards (advanced), 409
   Counterfactual, 416
   In concepts, 374
   Instability, 268, 360
   Neural networks, 347
Receiver of information,
   112, 155
Recipe
   As a story, 35
Recoil and conservation
   of momentum, 288
Recommendation engines, 66
Recursion, 134, 199
Recursive, 134, 199
Recycling, 277
Redshift, 224
Regularization of sums, 97, 513
Relativity, 25, 504
   Absolute Motion, 186
   Dataverse, 105, 232
   Einstein's theory
   of, 26, 38, 185, 227
   Einstein's Twins paradox,
   521
   General, 107, 514
   In computing, 38

Relativity shaken and stirred, 25
Relevance, 391
Renormalization, 9, 23, 509
    And infinity, 97, 513
Reordering of data, 271
Repeatability, 269
Replication, 282
Representation
    Concept, 366, 378, 529
Reproducibility, 269
Responsibility, 132, 133
Reversibility, 114, 199, 206, 207, 275, 520, 523
    Equilibrium, 520
    Undo CTRL-Z, 251, 252
Revert
    And time reversal, 40
Revision control system (RCS), 106
Riemann, Georg Bernhard, 32, 98, 323
Riemannian geometry, 98
Right
    Politics, 387
RNA, 338
Roads, 104
Robots, 416
Rods and cones, 192, 519
Roles
    Functional, 97
Roles and maps, 31
Rollback, 253, 256
Rollback, playback, and kernel passport control, 252
Rotation, 12
Routing
    Information, 246
    Network, 78, 110, 183, 413, 525
Russian doll, 15, 45, 98

Same time, 254, 282

Sampling
    Population, 191
Sampling rate, 92, 118, 178, 186, 408
Scalar, 43, 46
Scale
    And observability, 163, 188
    And spacetime, 188
    Defined, 100
    Dynamic, 100
    Illusions of, 297, 363, 422
    None in continuum spacetime, 239
    Renormalization, 9, 37
    Semantic, 100
Scale dependence, 49, 101
    Of dimensions, 52
Scales and renormalization, 8
Scaling, 37, 102
    Of processes, 422
    Of true and false, 266, 458
Scaling all the way down, 102
Scaling in the dataverse, 84
Scaling of organisms and processes, 441
Scaling patterns into concepts, 350
Scattering process, 195
Schema
    Database, 93
Schrödinger equation, 153, 161
    For datacentre, 518
    Wavefunction, 138
Schrödinger, Erwin, 327
Scientific method, 265
Scope
    In computing, 97, 164
Search
    Brute force, 411

Search me, 389
Searle, John, 527
Secret garden, 431
Secret Life of Trees, 431
Security camera story, 119
Seeing may not be believing, 121
Seeing through a network, darkly, 89
Selection, 354
  Darwin, 414
Self and non-self, 460, 533
Semantic names and addresses, 357
Semantic spacetime, 75, 111, 426
Semantics, 6, 35, 71
  Bounded regions, 329, 403
  Error of, 192
  Ignoring, 265, 269
  Labels, 315
  Maps, 34
  Promise theory, 71
Sender of information, 112, 155
Sensor, 32, 381, 390
  Observations, 405
Sentences, 377
Separation of concerns, 74, 351
Separation of timescales, 351
Service directory, 216
Service Level Objectives, 213
Shannon
  Error correction, 200, 405
Shannon, Claude, 84, 112, 200, 312, 332, 365, 369, 387, 51 521, 523
  Genetics, 415
Sharding, 223, 521
Shared memory, 280
Shop delivery story, 274
Shopping brand alien story, 298
Shopping mall, 18, 91
Shortest path, 220

Significance, 366, 391
Simultaneous, 39, 254, 282
Single point of causation, 132
Single point of entry, 216
Slogans novel, 535
Smart
  Not intelligent, 427
Smart cars, 335
Smart city, 427, 440, 487, 491
Smart infrastructure, 447, 489
Smart mechanics?, 429
Smart promises to give and take?, 436
Smartphones, 78
Smith, Michael. E., 505
Smolin, Lee, 522
Social insects, 531
Social media, 521
Society, 438
Socioverse, 434
Software, 18
Software update, 278
Solid
  Crystal, 87
Sonar, 177
Sovereignty, 460
Space
  Anisotropy, 65
  As emptiness, 11
  As gaps, 28
  Colour, 47
  Continuum, 57, 104, 227
  Curvature, 19
  Defined, 28
  Dimensions, 19, 45, 48, 94, 362
  Embedded, 15, 80, 243
  Euclidean, 16, 49, 57, 86, 104, 178
  Fabric, 16
  Four dimensional?, 20
  Hidden dimensions, 45, 46, 362

Inhomogeneity, 65
Point, 41
Tuple, 97
Space and time different, 54
Spacetime, 21
    And scale, 188
    As a computer, 24
    Defined, 22, 28
    Discrete, 64, 66, 313
    Fabric, 16, 458
    Four promises, 362, 370
    Limits on density, 228
    Minkowski, 22
    Passive, 71
    Semantic, 111
    Spacetime, 75, 426
    Summary, 429
    Theatre, 5, 178, 326
Sparse information, 331
Speed, 17, 209
    Constant, 211
    Non scalable concept, 213
    Of light, 17
Spending fast and slow, 479
Spin, 318
Spooky action at a distance, 317
Stack
    Computer, 520
Standard Model, 293, 301, 327
Star Trek, 230, 247, 273
    Captain Kirk, 288
Staring into spacetime, 11
State machine, 199, 367, 379
Stateful program, 284
Statistical recombination
    of DNA, 414
Statistics
    Aggregation, 267, 360
    Boundary conditions, 266
    Meddle with time, 266
    Patterns, 288
    Principal component
    analysis, 535
Stigmergy, 261, 302, 316, 386
Stimulus-response, 335
STM, 352, 354
Stories, 35, 382
    As thoughts, 374, 382
    Generating, 390, 411
    Pruning too many, 411
    Recipe, 35
    Wondrous, 329, 421
Stories, histories, and
    paths are semantic spacetime, 383
Story, 24
    As explanation, 395
Storyline, 208, 248, 422
Storytelling, 353
    Organ, 408, 422, 531
Straight line, 192
    Motion, 288, 299
Strandlie, Are, 526
String Theory, 13, 16, 76, 362, 522
Sun Microsystems, 445
Sundial, 21
Super-cells, 277
Superagent
    Promise Theory, 85, 111, 225, 277, 338
Superhighway
    Information, 217
Supermarket checkout story, 220, 225
Superposition, 227
    And parallelism, 228, 326
    Neural nets, 348
Swarm intelligence, 334
Switching
    Network, 78, 110, 183, 413
Symbols, 366
    Alphabet, 368
    And virtualization, 368
    Give and take, 368

Symbolverse, 367, 368, 385
Symmetry, 12
    Breaking, 206, 208, 310, 403
    Continuum spacetime, 57, 325
    Rotational, 57, 310
    Translational, 57, 97
Syslog, 273
System
    Cognitive, 76, 336
    Computer, 16, 72
    Functional, 18, 110, 213

Tax, 295
Taxonomy, 357
TCP/IP, 230, 456
Temperature, 235
Tensor, 48, 415, 506
Thales, 29
The 'give and take' of epigenetic reasoning, 411
The action principle for boundary continuity, 323
The arrow of time versus the arrow of direction, 203
The beginning and the end of reason, 487
The cityverse, 438
The dataverse, 78
The depth of a point, 143
The Emperor's New Mind, 422
The fifth force: money's proxy carriers, 469
The four promises of spacetime processes, 362
The infoverse, 75
The large scale structure of the dataverse, 101
The last turtle?, 185
The long night, 420
The non-locality of mass, inertia, and exterior time, 291
The scale dependence of dimension, 48
The scaling of adaptation and memory, 338
The scaling of meaning, 365
The socioverse, 434
The sound that falls on deaf ears, 118
The stigma of stigmergy, 261
The Three Laws, 286
The Time Tunnel, 247
Theology, 14
Thermodynamics, 235
    Entropy, 204, 366
    Second law, 204, 516
Thinking, 353
    And language, 395
    as stories, 374, 382, 395
Three processes that create motion, 183
Time, 11, 38, 54
    Arrow of, 203, 209, 240
    As a dimension, 248
    As a flow, 54, 525
    As a tunnel, 248
    As version sequence, 165
    As version sequence, 105
    Cyclic, 260, 524
    Defined, 28
    Exterior, 39, 56, 92, 225
    Finite or endless?, 55
    Give and take, 275
    Greenwich Mean (GMT), 105, 258
    Halting and locking, 282
    Human, 493
    Human perception, 208
    In technology, 57
    Interior, 21, 39, 56, 92, 105, 164, 225, 407
    Involuntary, 54
    Measuring, 211
    Modular, 265
    Multidimensional, 259, 279

Ordering, 135, 240
Past, 276
Proper, 107
Relative, 237, 238
Reversal, 40
Runs slower, 188, 224, 320
Sampling rate, 181,
186, 187, 326, 408
Scaling limits, 120, 241
Sequence, 43
Source of, 55
Synchronization, 105
Thermodynamic, 240
Travel, 523
Universal Standard
(UST), 105, 258
Vector, 279
Wall clock, 105, 258
Zone, 523
Zulu, 258
Time as a series of versions,
105
Time backwards
Impossibility, 251
Time for change, 23
Time for stories—processes,
24
Time reversal, 251
Time to live, 392
Timeline, 24, 107, 273
Times like places, 249
Timescales
Separation of, 168,
337, 351, 360, 385
Timestamp, 258, 271
Toffler, Alvin, 531
Topic map, 530
Topological, 50
Topology, 16, 33, 515
Traceroute, 176
Trajectory, 24, 175, 382
As memory, 261
Transaction, 75, 253, 288,
458, 510
Transactional interpretation,
153
Transactions, 281
Transitions
Scaling of, 305
Translation, 12, 97
Trees, 431
True and false
Scale concept, 266, 458
Tsunami, 406
Tunnel, 234
Quantum, 483
Tuple Space, 97
Turing machine, 334, 369,
376, 385
Turing, Alan, 84, 310,
369, 376, 426, 526
Turn back time
Impossibility, 251
Turtle story, 14, 102,
185, 242, 513
Turtles all the way down, 14
Twistor, 319
Type Theory, 87

Uncertainty, 161
Position, 102, 161, 181
Principle, 243, 289
Uncertainty is a principle, 161
Undo, 40
Undo undo undo, 251
UNICODE code, 378
United Nations (UN), 462
Universal exchanges and
the race against the economy,
473
Universal Resource Indicator
(URI), 98, 127, 357
Update
Software, 278
URI, 98, 127, 357
UST, 105, 258

Vacuum fluctuations, 26
Variables
    Canonical, 287
Vector, 43, 46, 52, 182
Vector field, 195
Vector space, 44
Velocity, 209
    Group and phase, 520
    Non-local phenomenon, 290, 291
Venice, 487
Venn diagram, 115, 402, 530
Version control system, 40
Version control system (VCS), 106, 523
Version time branches
    As Many Worlds, 165, 523
Version tracking, 105, 106
Virtual, 17
Virtual information physics?, 159
Virtual machines, 162, 444
Virtual Private Network (VPN), 102, 234
Virtual processes, 295, 312, 444, 513
Virtualization
    Symbolic scaling, 368
Virtuverse, 82
Virus
    Biology, 316
    Computer, 316
Visualization or processes, 183, 519
VLSI, 224
Void, 6, 11, 26, 43, 69, 87, 300
Von Neumann, John, 90, 112, 385
Voting
    And reasoning, 355, 391
    And scaling, 267, 391
Voting story, 268

Wagon wheel story, 237, 238

Warran, Dianne, 247
Watched pot story, 169
Wave-particle momentum
    from the bottom up, 322
Wavefunction, 138, 155, 162
    As analogue computer, 153
    Collapse, 151, 152, 163, 517
    Complex, 138
    Give and take, 153
    In economics, 463
    Many worlds, 163
Waves, 32, 115, 322
    Collective excitations, 212
    Particle duality, 322
    Physics of, 228
    Propagation, 212, 224, 521
    Sound, 224
    Superposition, 227
Waving the messenger through, 211
Weak coupling, 520
Wells, H.G., 247, 253, 297
West, Geoffrey, 441, 504, 521
Wheeler, John Archibald, 48
Whitehead, Alfred North, 500
Why make it harder?, 69
Wiener, Norbert, 112
WiFi, 214
Will we recognize it, will it recognize us?, 489
Wohlleben, Peter, 431
Wolf story, 109
Wolfram, Stephen, 90, 312, 385, 511
Words, 377
    Compound scaling, 395
Working backwards from the answer, 409
Workspace concept, 444, 459
Workspaces in the dataverse, 444

World line, 24
World Trade Organization (WTO), 461
World Wide Web, 282
Wormhole, 273, 443
Wormholes, 249
WTO, 461

Yellow pages, 216

Zeno's paradox, 29, 30, 45, 54, 420
Zodiac, 35
Zookeeper, 511

About the author

Mark Burgess is a British theoretical physicist, turned computer scientist, who now lives in Oslo, Norway. He holds degrees in theoretical physics from the University of Newcastle Upon Tyne, and was Professor of Networks and Systems at Oslo University (MET), until abandoning the constraints of academia to get closer to the cutting edge of technology.

He originated a number of widely used technologies for computing infrastructure, including the CFEngine software, which played a central role in bringing about the expansion of modern datacentre infrastructure, and was co-founder of CFEngine Inc. and Aljabr Inc, and ChiTek-i where he acts as an advisor to major Information Technolgy companies around the world.

He worked on Kaluza-Klein theory, the Quantum Field Theory of Spontaneous Symmetry Breaking, Topological Field Theory, Fractional Statistics, Analytic Networks and Systems, Machine Learning, Graph Theory, Promise Theory, and Semantic Spacetime.

He is the author of many books and scientific publications, and is a frequent speaker at international events. Mark Burgess may be found at *www.markburgess.org*, and on Twitter under the name *@markburgess_osl*.

Printed in Germany
by Amazon Distribution
GmbH, Leipzig